TEACHER EDUCATION AND TEACHER PROFESSIONAL DEVELOPMENT IN THE COVID-19 TURN

These proceedings present a selection of papers from the ICCTE 2021 conference. While face-to-face classroom instruction is brought back, there are a lot of lessons learned from the COVID-19 pandemic that schools, teacher training and education institutions, and government have to take into account. There is a need to reconsider what additional knowledge and skills pre-service teachers and in-service teachers need to be prepared for to anticipate such a similar unexpected situation in the future. Additionally, there is also a need to listen to in-service teacher' experiences during the emergency remote teaching and integrate the positive lessons that they have gained, such as the use of technology, into the current post pandemic face-to-face classroom instruction. This proceeding is designed for teacher educators, researchers, in-service teachers, and pre-service teachers the field of language education, math and science education and social science education, who are interested in these topics.

PROCEEDINGS OF THE INTERNATIONAL CONFERENCE ON TEACHER TRAINING AND EDUCATION (ICTTE 2021), SURAKARTA, INDONESIA, AUGUST 25–26, 2021

Teacher Education and Teacher Professional Development in the COVID-19 Turn

Edited by

Nur Arifah Drajati
Universitas Sebelas Maret, Indonesia

Kristian Adi Putra
Prince Sattam Bin Abdulaziz University, Saudi Arabia

Routledge
Taylor & Francis Group

First published 2023
by Routledge
4 Park Square, Milton Park / Abingdon, Oxon OX14 4RN
e-mail: enquiries@taylorandfrancis.com
www.routledge.com – www.taylorandfrancis.com

Routledge is an imprint of the Taylor & Francis Group, an informa business

ISBN: 978-1-032-38970-7 (hbk)
ISBN: 978-1-032-38973-8 (pbk)
ISBN: 978-1-003-34779-8 (ebk)

DOI: 10.1201/9781003347798

Typeset in Times New Roman
by MPS Limited, Chennai, India

Teacher Education and Teacher Professional Development in the COVID-19 Turn – Drajati & Putra (Eds)
© 2023 The Editor(s), ISBN: 978-1-032-38970-7

Table of contents

Math and Science Education

Social Science Education

In-Service Teacher Professional Development

Teacher Education and Teacher Professional Development in the COVID-19 Turn – Drajati & Putra (Eds)
© 2023 The Editor(s), ISBN: 978-1-032-38970-7

Preface

All of us, teacher educators, in-service and pre-service teachers, school administrators, students, and parents, witnessed and experienced the many sudden changes that happened in the classrooms before and during the pandemic. It was undoubtedly unexpected and unprepared. In Saudi Arabia, for instance, the conventional face-to-face classroom was switched to an emergency remote teaching overnight. On Monday morning and afternoon, March 2, 2020, students were still in the classroom with their teachers. After the announcement from the government on Monday night, March 2, 2020, the students started studying fully online on the following day. Ready or not, it needed to be done. Saudi Arabia government took this step as a precautionary measure against the COVID-19 outbreak that was detected in the country late February 2020.

All around the world, there were reports about challenges that education institutions, teachers, students, and parents faced at the beginning of the emergency remote teaching, especially on the availability of facilities and the readiness of teachers and students. However, the education institutions, teachers, and students eventually got used to the rhythm and learned a lot of new things during the pandemic. Many teachers, for instance, had an opportunity to teach online, which they had never done before. For almost two years, they have learned how to prepare and deliver their teaching materials in online settings, communicate with students and engage with them in the teaching and learning process in different way, and asses the academic progress of their students remotely.

As the governments have felt more confident to lift all the restrictions after the massive vaccination program, the conventional face-to-face classrooms have also been brought back. The question that raises is, therefore, whether or not the teachers will go back again to their old practice of teaching as before the pandemic or integrate the new experiences that they got during the pandemic into their post pandemic pedagogical practice. In the current proceedings, we include the articles from the three keynote speakers, Dr. Ju Seong Lee (The Education University of Hong Kong), Dr. Hyo Jeong So (Ewha Womans University), and Dr. Lynde Tan (Western Sydney University), to discuss this topic. The three keynote speakers highlight the importance of teacher agency, identity development, and resilience during and after the pandemic.

The current proceedings also include 29 other articles written by colleagues from 12 countries. We categorize the subtopics into five, which are 1) English language education, 2) math and science education, 3) social science education, 4) in-service teacher professional development, and 5) pre-service teacher training and education. We hope that such a diverse background of authors and sub-topics will provide us an overview of the common issues in the current post-pandemic pedagogy and what we need to do to better support pre-service teachers and in-service teachers.

Surakarta, June 7, 2022
Nur Arifah Drajati
Kristian Adi Putra
(Editors)

 ISBN: 978-1-032-38970-7

Acknowledgements

On behalf of the committee of the International Conference on Teacher Training and Education (ICTTE) 2021, we would like to extend our gratitude to the Rector of Universitas Sebelas Maret (UNS), Prof. Dr. Jamal Wiwoho, S.H., M. Hum. and the Dean of the Faculty of Teacher Training and Education of UNS, Dr. Mardiyana, M. Si. for their moral and financial supports for this conference. Their advice, encouragement, patience, and enthusiasm made the success of this conference possible.

We are also indebted to all the invited keynote speakers, Prof. Dr. Andayani, M.Pd. (Universitas Sebelas Maret, Indonesia), Nadiem Anwar Makariem, B.A., M.B.A. (Ministry of Education of Indonesia), Dr. Lynde Tan (Western Sydney University, Australia), Prof. Hyo-Jeong So (Ewha Womans University, South Korea), Dr. Ju Seong Lee (Education University of Hong Kong), and Dr. Punya Mishra (Arizona State University, USA) for their willingness to share their knowledge and experiences with all the participants of this conference.

We also would like to thank all the committees of this conference. Without them, this conference would not be possible. We are extremely thankful with our colleagues, especially Sri Haryati, S.Pd., M.Pd., Dr. Djoko Sulaksono, S.Pd., M.Pd., Chafit Ulya, S.Pd., M.Pd., Anis Rahmawati, S.T., M.T., Mila Purani Sistiyan, S.E., Hefi Sugiyarti, Figur Rahman Fuad, S.Pd., M.A., and Murni Ramli, S.P., M.Si., Ed.D., as well as our students and former students for their tireless commitment and dedication for this conference to be successful from the beginning till the end.

The Scientific Committee of the ICTTE 2021 International Conference

 ISBN: 978-1-032-38970-7

The Committee of the ICTTE 2021 International Conference

Patron

Rector of Universitas Sebelas Maret (UNS)	:	Prof. Dr. Jamal Wiwoho, S.H., M.Hum.

Steering Committee

Dean of the College of Education, UNS	:	Dr. Mardiyana, M.Si.

Organizing Committee

Chair	:	Dr. Nur Arifah Drajati, M.Pd.
Vice Chair	:	Dr. Djoko Sulaksono, S.Pd., M.Pd.
Secretary	:	Chafit Ulya, S.Pd., M.Pd.
Vice Secretary	:	Anis Rahmawati, S.T., M.T.
Treasurer	:	Sri Haryati, S.Pd., M.Pd.
Coordinator of Event	:	Murni Ramli, S.P., M.Si., Ed.D.
Coordinator of Public Relations	:	Figur Rahman Fuad, S.Pd., M.A.
Coordinator of IT Support	:	Mila Purani Sistiyan, S.E.
Coordinator of Documentation	:	Kristian Adi Putra, S.Pd., M.A., Ph.D.
Coordinator of Publication	:	Hefi Sugiyarti

A Note from the Editors

Teacher Education and Teacher Professional Development in the COVID-19 Turn – Drajati & Putra (Eds)
 ISBN: 978-1-032-38970-7

Moving ahead: Challenges and opportunities for teachers in post-pandemic pedagogy

Kristian Adi Putra
Prince Sattam Bin Abdulaziz University, Saudi Arabia

Nur Arifah Drajati
Universitas Sebelas Maret, Indonesia

ABSTRACT: In this chapter, we will introduce the three keynote speeches and the five topics discussed in the articles in the proceedings related to teacher education, professional development, and agencies in post-pandemic pedagogy. They are English language education, science education, social science education, in-service teachers' professional development, and pre-service teachers' training and education. In this article, we will first introduce the issues and changes in the field of education soon after the outbreak of the COVID-19 pandemic and the return of the face to face classroom instructions. Finally, we will continue with the discussion of how all these issues are addressed in the articles in the proceedings, starting from the keynote speeches to the selected participants' articles.

Keywords: Post-pandemic pedagogy, teacher agency, teacher education, teacher professional development

1 INTRODUCTION

Every cloud has a silver lining. (Anonymous)

The quote above has been the spirit of the International Conference on Teacher Training and Education (ICTTE) 2021 and the publication of the selected articles in this proceedings. When the outbreak of the COVID-19 pandemic began in mid-March 2020, many teachers and students, especially in third-world countries, were still unfamiliar with the online instructions. However, ready or not, they had to do it anyway. In the end, everyone eventually felt comfortable with it and learned a lot of new other things during the period of the emergency remote teaching. Today, as many governments worldwide have begun to loosen the COVID-19 pandemic restrictions and allow students to study again in class, teachers need to adapt to the new routines and think of how to integrate what they have learned during the pandemic into their current face-face classrooms.

At the beginning of the COVID-19 pandemic, the situation was almost the same as mentioned by Egbert (2010) in her edited book a decade ago, "CALL in Limited Technology Contexts." While it has been a while since it was published, it still seems relevant. Egbert identified three areas of challenges to CALL (Computer Assisted Language Learning) implementations that were common in the contexts where access to technology was limited, namely infrastructure, execution, and ideology. In many contexts, it was reported that during the COVID-19 emergency remote teaching, many education institutions, teachers, and students did not have the required infrastructures, such as hardware, software, internet access, etc. The execution was also complicated, as teachers had never been trained to do it and experienced it themselves. The situation was unpredicted and unprepared. It took a while for academic institutions to ensure emergency remote teaching could run, especially in preparing all the supporting infrastructures and the readiness of the teachers and students to do it. However, there were no issues related to the ideological challenges, such as parental objections,

DOI 10.1201/9781003347798-1

socio-political restrictions, and misaligned curricula. Everyone agreed that it was something that must be done. Everyone's safety was a priority, so the government, parents, school principals, and teachers agreed that emergency remote teaching was the only solution.

In many countries, it was not only the academic institutions and governments that moved fast. In Indonesia, for instance, many professional organizations and academic institutions organized online webinars to help university instructors and school teachers to adapt to the new challenging situation. The topics varied, starting from preparing online teaching and assessment materials, selecting relevant technologies for online teaching, engaging students in online classes, conducting online classrooms, and conducting assessments online, among many others. It was also the time when teachers started to realize that there was a "blessing in disguise," in which, because of technology, they had opportunities to learn many new things from their colleagues and different experts from other countries regardless of their time and geographical differences. Professional organizations in Indonesia, such as TEFLIN (The Association for the Teaching of English as a Foreign Language in Indonesia), ITELL (Indonesian Technology Enhanced Language Learning), and IERA (Indonesian Extensive Reading Association), for example, organized many series of webinars attended by English language teachers from all over Indonesia and other countries. They invited speakers from different regions in Indonesia and countries and collaborated with various government agencies, such as British Council and non-government organizations. It helped English language teachers in Indonesia and beyond to get a clue of what they needed to do with their students during the emergency remote teaching.

In 2020, we started to think about what to do next soon after the COVID-19 pandemic ended or at least after the governments are confident to bring the students back to the class. It inspired us to organize an international virtual conference, namely the Language Teacher Training and Education (LTTE) 2020 conference, in which we invited keynote speakers and participants from Indonesia and other countries to share their ideas about post-pandemic second language pedagogy. After the conference, we also published the selected articles in the proceedings of the conference. Later, the positive outcomes and feedback from all the keynote speakers and participants in the LTTE 2020 conference encouraged us to have a wider audience, not limited to English language teaching but education in general. In 2021, we finally conducted the International Conference on Teacher Training and Education (ICTTE) 2021, which included the field of English language education, science education, social science education, in-service teachers' professional development, and pre-service teachers' training and education. We brought the theme, "Teacher Education and Professional Development in the COVID-19 Turn" to focus specifically on the importance of training and education of pre-service teachers and professional development of in-service teachers in the post-pandemic pedagogy. We are able to select 32 articles, including three keynote speeches, to be included in the current proceedings. In what follows, we will describe all the included articles to provide a map for the readers about the contents of the proceedings.

2 KEYNOTE SPEECHES

Three keynote speakers of the ICTTE 2020 conference, Ju Seong Lee from The Education University of Hong Kong, Hyo-Jeong So from Ewha Womans University, and Lynde Tan from Western Sydney University, wrote the three articles in the keynote speeches section. The three articles similarly focus on teacher education, professional development, and agencies in post-pandemic pedagogy and bring up three related and important points. Ju Seong Lee, in line with his expertise in informal digital learning of English (IDLE), highlights the importance of integrating students' informal online engagement into formal education and the need for teachers to be proactive agents of change in their classrooms. Additionally, during the pandemic, students' online engagement increased significantly not only because they wanted to but because all their classes, which included communication with their teachers and classmates, classroom meetings, and assignments, were online. In his previous studies on IDLE (see Lee, 2019, Lee & Drajati, 2019, Lee & Hsieh, 2019 as examples), Lee argues that informal digital learning provides many opportunities for students to learn from arguably limitless learning resources available online. Therefore, teachers need to be able to find ways to integrate it into their classrooms.

The second article, written by Hyo-Jeong So, stresses the need for teachers to have the ability to adapt to any kinds of changes and demands and recover quickly from the difficulties that they and their students are facing, such as in the case of emergency remote teaching during the COVID-19 pandemic. The drastic switch from the traditional face-to-face classroom to emergency remote teaching at the beginning of the COVID-19 pandemic has given teachers an experience of such a need. While they did not expect and had not been prepared for such a scenario, they had to do it anyway and learned fast how to do it. As teaching and learning activities have started in the classroom, teachers are also demanded to continuously be adaptive and resilient. Not going back to their old ways of teaching, So argues that teachers should integrated the positive things that they have learned during the pandemic into their face-to-face classrooms, such as the integration of technology and the management of students' engagement. Additionally, So also highlights the need for teachers to be committed to their responsibilities as professional teachers, in this case, those related to the commitment to continue developing their professional, social, pedagogical, and personal competences. To do that, they need to continuously learn and look for opportunities to take part in teacher professional development programs either those facilitated by their institutions or external professional organizations.

The third article, written by Lynde Tan, connects the theories of teacher identity, teacher agency, and teacher competencies. Tan starts with the argument that during the COVID-19 pandemic, teachers need to learn something fast in order to have the required knowledge and skills to teach online. Such a proactive teacher agency is an indicator of their professionalism and identity as a professional teacher. However, Tan also highlights that teacher identity and agency can only be negotiated and constructed through positive relationships, social interactions and supportive contexts. Therefore, teacher educators, mentor teachers, and school leadership teams play an important role in shaping both pre-service and in-service teacher identity and agency. Once teachers are able to position themselves as a professional teacher and to critically self asses and reflect upon what they need to do and learn more to manage the changes that are taking place in their classrooms, they will be able to facilitate successful educational experiences and supportive learning environments for their students.

3 ENGLISH LANGUAGE EDUCATION

There are six articles in the English language education section. The articles discuss various topics, which include technology, globalization, and collaborations in ELT classrooms. Ma'rifatul Ilmi et al., for example, raise an interesting case of the mushrooming business of informal online courses, including English, in Indonesia even before the COVID-19 pandemic. The pandemic made this business grow more significant. It raises the questions of how the teachers teach the courses and what the best online teaching models are available for them to follow. Using an autobiographical narrative inquiry and adopting the Technological Pedagogical Content Knowledge (TPACK) framework by Koehler & Mishra (2009), the study focuses on what needs to be considered and done during the planning and implementation stages of an informal online TOEFL preparation course. The study highlights the need for teachers to be adequately trained, assisted, and given constructive feedback. They must also be continuously encouraged not only to take part in internal teaching professional development (TPD) sessions but also to look for external TPD opportunities to develop their teaching competencies further.

Concerning the issue of globalization in ELT classrooms, Sri Imelwaty et al., for instance, wrote about the lack of materials that depicted the use of English by the speakers of English from the outer and expanding circle of English in three English language textbooks in secondary schools in Indonesia. Written by local authors and officially endorsed by the Indonesian government, the three textbooks only had examples of communication between Indonesians, between Indonesian and native speakers of English, and between native speakers of English. Imelwaty et al. argued for the need to include the materials from the speakers of English from the outer and expanding circle of English, such as the Middle Eastern, East Asia, and Southeast Asian countries, in the textbook referring to Kachru's model of Concentric Circles (Kachru, 1985; 1992). It aligns with the idea that English is a global language (Baker, 2011). The owners of English are also not only in English-speaking countries, notably the United Kingdom, the United States, and Australia but also in other countries. While English might not be their first language, they are also users of English who might encounter English language learners in real-life conversations.

4 SCIENCE EDUCATION

There are three articles in the science education section. The first article by Anwari Adi Nugroho et al. focuses on the development of interactive multimedia based teaching materials in a biology subject for senior high school students in Indonesia. More specifically, the study looks at the characteristics, feasibility, and practicality of such a teaching medium. It was found that besides being feasible and practical, the interactive multimedia also increased students' motivation and interest in learning biology. Additionally, it also improved the quality of the teaching and learning process. The second article by Endang Susantini et al. looks at how to develop a valid instruments that can be used to assess Higher Order Thinking Skills (HOTS) based mobile learning applications used in biology subjects. Such an instrument is needed as it can provide the users and developers information about the characteristics of good mobile learning applications. Finally, the last article in the section by Lina Listiana et al. analyzes the correlations between students' metacognitive skills, learning motivation, and communication skills in a science class. The findings show that both students' metacognitive skills and learning motivation contributed positively to their communication skills. Separately, students' communication skills correlated with their learning motivation more than with their metacognitive skills. This study implies that teachers need to positively encourage students' learning motivation and train their metacognitive skills to make them have better communication skills, which are essential for their active classroom participations.

5 SOCIAL SCIENCE EDUCATION

In the social science education section, we included eight articles covering diverse topics, such as Geographic Information System (GIS) and project based learning, learning style and motivation in a geography class, guidance and counseling, puberty and sexuality in autistic adolescents, innstructional designs for hard-hearing students, e-comic based local history lessons, and entrepreneurship. We will start with the study by Siti Dwi Lestari et al. that explores students' motivation in learning geography using Android based mobile learning applications. The study found out that students who used the applications were more motivated to learn geography than those who did not. It was also observed that students' intrinsic motivation as well as active parental involvement played an important role in the students' success during online learning. Another study by Ema Agustina and Theresia Manalu describes the story of a history teacher who created e-comic for his history classes. The e-comic was based on the local histories included in the school curriculum. While highlighting the challenges, particularly length of time spent to learn how to do it and create it, the use of e-comic in the class was able to stimulate students' curiosity as well as improve students' interest and motivation in learning history. This study implies on the need for teachers to be creative and innovative, as well as continuously find ways to create a fun yet effective learning environment for students.

6 IN SERVICE TEACHERS' PROFESSIONAL DEVELOPMENT

There are, in total, eight articles on the in-service teachers' professional development section. As the main actor in the classrooms, teachers, together with students, have gone through a lot since the very beginning of the emergency remote teaching until the return of the classroom instructions. The eight articles here explains various important issues that teachers dealt with during the pandemic and what needs to be anticipated in the future, such as the need for teachers to have continuous TPD, TPACK competencies, digital literacies, assessment literacies, and self-reflection skills. Written mostly by teacher educators, the studies shed lights on what the college of education need to formulate in their teacher training and education curriculum as well as what the academic institutions need to consider when designing internal TPD sessions. Tri Murwaningsih et al., for instance, pointed out the current implementations, problems, and needs of TPD for vocational high school teachers in Indonesia. While the TPDs had been well implemented, there were still many problems during the implementation, particularly those facilitated internally by their affiliated institutions and done in turn by every teacher at the schools. The problems included the lack of TPD materials, supporting facilities, qualified trainers, teachers' interest in scientific works, and available days for every teacher to take part in the sessions. They also reported the many administrative burdens that they needed to finish that made them have no time to think about taking part or looking for TPD opportunities inside and outside their institutions. All these problems imply the things that schools need to consider and solve when organizing TPD and encouraging their teachers to take part in external TPD sessions.

7 PRE-SERVICE TEACHERS' TRAINING AND EDUCATION

We also include four selected articles on pre-service teachers' training and education as we believe that teacher educators need to reconsider their existing curriculum. One of the challenges that in-service teachers faced during the emergency remote teaching was that because they were not prepared to do it since they were in college. Nobody expected it to happen, so that it reflected in the curriculum being used. The four articles included here focus on listening to the voice of the pre-service teachers on such a possible change of directions, e.g. the need to TPACK and pre-service teachers' emotional experiences, agencies, and identity construction during emergency remote teaching. A study by Banun Havifah Cahyo Khosiyono and Nico Irawan, for example, showed how the COVID-19 pandemic has affected pre-service teachers' learning process, psychological mental state, and communications with their lecturers and classmates. Another study by Intan Pradita et al. similarly pointed out the issues happening with pre-service teachers during their mandatory teaching practicum activity. The minimum physical contacts with their mentor teacher and students affected their confidence and resulted in confusion. Intan Pradita et al., thus, highlight the need to have a more flexible and informal, instead of rigid curriculum that allows the pre-service teachers to have abilities to do self-assessment and critical reflective practices, informal dialogue with teacher educators on campus and mentor teacher and students at schools, and informal peer discussions with their colleagues.

8 CONCLUSION

Regardless of the diverse fields and topics covered in the current proceedings, the articles similarly describe the experiences of pre-service teachers, in-service teachers, and teacher educators dealing with the COVID-19 emergency remote teaching and the lessons learned for the post-pandemic pedagogy, specifically focusing on pre-service teacher training and education and in-service teacher professional development. As mentioned by the three keynote speakers, there is a need for teachers to be adaptive, resilient, and committed as a professional teacher. Teachers need to be supported and continuously learn to develop their teaching competencies, as they are required to be ready to teach their students in any unprecedented situations and to be an agent of positive changes in their classrooms.

REFERENCES

Baker, W. (2011). From cultural awareness to intercultural awareness: Culture in ELT. *ELT Journal,* 66(1), 62–70.

Egbert, J. (Ed.). (2010). *CALL in limited technology contexts.* Computer Assisted Language Instruction Consortium.

Kachru, B. B. (1985). "Standards, Codification and Sociolinguistic Realism: The English Language in the Outer Circle." *In English in the World: Teaching and Learning the Language and Literatures*, edited by R. Quirk and H. Widdowson, 11–30. Cambridge University Press.

Kachru, B. B. (1992). World Englishes: Approaches, issues, and resources. *Language Teaching*, 25 (1), 1–14.

Koehler, M., & Mishra, P. (2009). What is technological pedagogical content knowledge (TPACK)?. *Contemporary Issues in Technology and Teacher Education*, *9*(1), 60–70.

Lee, J. S. (2019). Informal digital learning of English and second language vocabulary outcomes: Can quantity conquer quality?. *British Journal of Educational Technology*, *50*(2), 767–778.

Lee, J. S., & Drajati, N. A. (2020). Willingness to communicate in digital and non-digital EFL contexts: Scale development and psychometric testing. *Computer Assisted Language Learning*, *33*(7), 688–707.

Lee, J. S., & Hsieh, J. C. (2019). Affective variables and willingness to communicate of EFL learners in in-class, out-of-class, and digital contexts. *System*, *82*, 63–73.

Keynote Speeches

Teacher Education and Teacher Professional Development in the COVID-19 Turn – Drajati & Putra (Eds)

Teacher as change agent: Integrating informal learning into formal education

Ju Seong Lee
The Education University of Hong Kong, Hong Kong, China

ABSTRACT: With technological advancement and its growing affordances, coupled with the ongoing suspension of face-to-face classes, English learning beyond the classroom with technology has received increasing attention. The presenter calls it Informal Digital Learning of English (IDLE) or "self-directed English activities in informal digital settings, motivated by personal interests and undertaken independently without being assessed by a teacher" (Lee & Lee 2021, p. 359). This talk highlights what new roles language teachers should play during the ongoing and post-pandemic teaching environment and how they, as change agents, can leverage the power of IDLE to complement the limitations of formal language education.

Keywords: Informal Digital Learning of English (IDLE), Teacher change, Language education

1 INTRODUCTION

In recent years, English learning environments have been drastically changed due to the new roles that the English language, affective turn, and technological advances play. In line with these moving trends, Informal Digital Learning of English (IDLE) has emerged as a new phenomenon. Extensive studies have shown that those who engage regularly in IDLE activities tend to enjoy higher levels of affective, cognitive, and linguistic aspects of English learning outcomes. Despite several pedagogical advantages of IDLE, frontline teachers often find it difficult to incorporate IDLE into the classroom. I call this phenomenon a "pedagogical dilemma." From the perspective of Bronfenbrenner's (1979) ecological systems theory, several factors stemming from micro-, meso-, exo-, and macro-systems seem to affect the current pedagogical dilemma. Nevertheless, I argue that teachers (as change agents) can play a vital role in integrating informal learning into formal education.

1.1 *Changing English learning environments*

In today's globalized digital world, we are increasingly, virtually interacting with diverse English users to perform various personal and professional activities. Friedman (2005) calls this phenomenon a "Flat World." In the fields of teaching English to speakers of other languages (TESOL) and applied linguistics, critically oriented researchers on World Englishes, English as a Lingua Franca, Translanguaging, and English as an International Language (EIL) have revealed that conventional English language teaching has not yet fully reflected the changing face of English and its users (Galloway 2017). In the same vein, English learners have typically learned English in order to integrate into an English-speaking society or community (Gardner 1985). However, since English is currently used as a lingua franca or international language, English learners learn that particular language to become their ideal selves or to achieve their ideal goals (Dörnyei 2005). For instance, one Korean EFL university student said that she practiced more English to become a translator between BTS (a K-pop group) and its American audience (Lee & Lee 2020).

DOI 10.1201/9781003347798-2

Although cognitive dimensions have dominated the fields of TESOL and applied linguistics during the past five decades, research on L2 emotions has consistently been conducted (Barcelos 2015). The concept of 'affective turn' (Pavlenko 2013) has prompted scholars to acknowledge the vital role that emotions play in learning a foreign language. Notably, in line with the movement of positive psychology, researchers have begun to take the holistic perspective of focusing on both positive (e.g., enjoyment) and negative emotions (e.g., anxiety). This is an important issue because Asian L2 learners, most of whom were considered EFL learners, had the lowest levels of L2 enjoyment and the highest levels of L2 speaking anxiety among foreign language learners from around the globe (Dewaele & MacIntyre 2014). This may be because English is part of high stakes tests such as in-school tests and the national college entrance examination (Lee 2020). Kim's (2019) research supports this view, namely, that Korean students' English learning motivation plummeted when they took the first formal English test during their 8th grade year. They showed gradual declines in English learning motivation, marking the lowest motivation during their high school years.

Levy (1997) defines Computer Assisted Language Learning (CALL) as the "study of applications of the computer in language teaching and learning" (p. 1). In tandem with the development of technologies and their pedagogical benefits, the field of CALL has also evolved (Chun 2019). During the 1970s and 1980s, mainframes (e.g., language labs) were used for teaching grammar and translation skills. During the 1980s and 1990s, audio-visual tools (e.g., PCs) were used for improving learners' comprehension and communication. During the 2000s, more innovative types of technology (e.g., the Internet) were used for facilitating active and authentic L2 input and output. During the 2010s, ubiquitous technology (e.g., mobile phones) has been integrated, which allows students to use L2 beyond the classroom. Against this backdrop, researchers have begun to shift their attention from classroom-based language learning to out-of-class language learning (Richards 2015; Reinders & Benson 2017).

1.2 *Informal digital learning of English (IDLE)*

'Language learning and teaching Beyond the Classroom' (LBC) has received an increasing amount of attention from CALL and TESOL communities during the past decade (Benson & Reinders 2011; Reinders & Benson 2017; Sundqvist & Sylvén 2016). LBC can be classified into two subcategories (Lee & Sylvén 2021): 1) LBC offline and online (an informal L2 activity that takes place in face-to-face and digital settings); and 2) LBC online (an informal L2 activity that occurs exclusively in digital environments). IDLE is a growing phenomenon in EFL contexts (e.g., South Korea and Indonesia), in which language learners rarely use English in everyday life. To illustrate, Zhang and her colleagues (2021) found that 88% (or 29 out of 33) of papers done on the topic of extramural language learning investigated the English language. Soyoof and his team (2021) reported that nearly 80% of papers on the subject of IDLE had been published in Asian and European EFL contexts.

Defined as "self-directed English activities in informal digital settings, motivated by personal interests and undertaken independently without being assessed by a teacher," IDLE can be practiced through consuming English content (e.g., watching YouTube videos) or producing English content (e.g., using English through gaming) in digital settings (Lee & Lee 2021, p. 359). Overall, there are 12 key characteristics of IDLE, all of which can complement the limitations of formal non-digital education in English:

1. Autonomy: Students feel a strong sense of control about their IDLE activity because they can make decisions about their own materials and strategies.
2. Authenticity: IDLE learners can be exposed to authentic English while interacting with other English users in real-life situations.
3. Identity: IDLE learners often assume a new identity as English users rather than English learners.
4. Motivation: IDLE activities are generally triggered by intrinsic motivation (e.g., interest) rather than extrinsic motivation (e.g., test).

5. Investment: IDLE learners, who perceive themselves as English users, are likely to invest in practicing English in extramural digital contexts.
6. Flow: IDLE learners (e.g., watching Netflix in English) tend to lose track of time and experience a flow state.
7. Grit: IDLE learners exert continuous efforts in learning and using English while maintaining their personal interests.
8. Community of practice: IDLE learners feel connected to others from online communities (e.g., fan fiction or gaming communities) that share the same interest.
9. Affective filter: Since there are no teachers and classmates, IDLE learners often practice English in a low affective filter situation.
10. Multimodality: IDLE activities often involve various modalities, such as text, image, audio, video, and music.
11. Comprehensible input: IDLE helps learners grasp the meaning of words and sentences.
12. Accessibility: Since there are no time or space limitations on the Internet, English is continuously available to IDLE learners.

Empirically, IDLE has been found to be positively associated with affective, cognitive, and linguistic aspects of L2 learning outcomes (Lee 2022). For instance, in a review of 33 papers on extramural language learning, more than 70% of the studies (24 of 33) reported that IDLE-related activities played a positive role in developing learners' linguistic and affective aspects of acquiring the language (Zhang et al. 2021). Lee (2020) provided additional evidence by presenting a contrasting case of two Korean IDLE learners, Ju-no and Jin-young. At the time of the study, both EFL learners had been learning English through the formal school system, without any overseas experience. However, Jin-young regularly practiced 18 different types of IDLE activities, while Ju-no did only three activities. Jin-young performed better than Ju-no in willingness to communicate in the classroom, productive vocabulary knowledge, and speaking proficiency. Lee and Dressman (2018) showed a longitudinal effect of IDLE on the development of English skills in one Korean EFL learner. In this particular study, Su-ja had been regularly engaging in a variety of IDLE activities (e.g., playing online games in English) since early childhood, and she acknowledged that all of the English inputs from IDLE have become a strong foundation for her improvement in English.

1.3 *Pedagogical dilemma*

As a teacher educator, I have been sharing this emerging phenomenon with front-line English teachers. In general, English teachers are well aware of the pedagogical advantages of IDLE. And yet, ironically, they seem hesitant to incorporate IDLE into their classroom, for various reasons. Lee (2022) calls this phenomenon 'pedagogical dilemma' or, a gap between how teachers view IDLE and what they actually do in the classroom. To understand this issue systematically and comprehensively, it seems useful to adopt Bronfenbrenner's (1979) ecological systems theory, which is composed of four systems or levels. Microsystem (the first level) refers to an immediate environment that is close to an individual. Mesosystem (the second level) refers to relations between an individual's microsystems. Exosystem (the third level) represents more distant influences on an individual but which nonetheless impact him or her. Lastly, macrosystem (the fourth level) indicates the influence of even more distant elements. Reflecting on his Russian background, Bronfenbrenner illustrates the model by using these 'nested' systems, which are enclosed by other systems in a manner similar to a set of Russian matryoshka dolls.

Bronfenbrenner's nested ecological systems theory provides a useful visual guide by explaining the various factors that affect teachers' pedagogical dilemma. At the individual level, past IDLE experience, attitudes toward IDLE, constructivist beliefs, and pedagogical or technological knowledge appear to influence teachers' pedagogical dilemma. At the microsystem level, teacher autonomy, limited resources, and students' learning needs and styles seem to be attributed to pedagogical dilemma. At the mesosystem level, colleagues' resistance to radical ideas (e.g., IDLE) and pressure from parents seem to be relevant factors. At the exosystem level, principals' perceptions, curriculum design, and obsolete teacher professional development models seem to affect the pedagogical challenge. At the macrosystem level, more distant factors, such as societal values (e.g., collectivism) and educational policy (e.g., high-stakes national tests) seem to influence the pedagogical dilemma. Overall, this ecological approach to understanding the pedagogical dilemma suggests that a teacher is embedded in and affected by various levels of environment.

1.4 *Teacher as change agent*

The Cambridge Dictionary defines agent as "a person or thing that produces a particular effect or change." Based on this definition, I argue that we should view teachers as change agents that are actively participating and negotiating within their own teaching context. Ultimately, these interactions will lead to changing their teaching practice as well as their teaching environment.

Here, I would like to offer some implications for teachers about how to bring an innovative idea (e.g., IDLE) into formal education. In Verplanken et al.'s (2008) research, habit discontinuity is described as follows: "When a context change disrupts individuals' habits, a window opens in which behavior is more likely to be deliberately considered" (p. 121). We often act out of habit or routine without thinking too much about either. It may range from the way we brush our teeth to the way we teach in the classroom. But when your context changes, even to a small degree, we must think about what that entails. We no longer act out of habit. That is why this is a window of opportunity for change. Now you can rethink behavior (e.g., teaching practice) that you have heretofore taken for granted. This suggests that changing our context can drastically alter our behavior, routines, and habits (e.g., in our teaching practice).

Heatherton and Nichola (1994) provided evidence to support this view. The researchers asked 119 students from the Harvard Extension School to describe changes in their health, careers, education, and relationships. They found that those who had successfully changed their behavior tapped into the power of habit discontinuity. They changed contexts by going away for the summer, moving, leaving their jobs, or finding a new group of friends. This suggests that they stepped out of their normal environment (or comfort zone) and embraced change. In contrast, those who had failed changing their behavior tended to have mixed feelings about change and continued in the same role. This suggests that they remained in a familiar environment and hence resisted change. The nature of habit discontinuity is closely synonymous with characteristics of teachers as change agents, which have been identified by Van der Heijden et al. (2015). For instance, these teachers are eager to learn new things, are reflective about their current teaching practices, and are innovative in experimenting with new ideas.

Wong (2011) has provided 50 teacher professional development tips to provide ways to nurture these attributes of teachers as change agents. These methods may help teachers experience and leverage the opportunity of habit discontinuity. Some of Wong's ideas include:

1. Observe other teachers
2. Find a mentor
3. Attend a conference
4. Take an online or distance course
5. Enroll in an MA or PhD program
6. Launch a teachers' reading group
7. Engage in a case study
8. Present at a conference
9. Become active in organizations
10. Volunteer to serve
11. Keep a journal and reflect on teaching
12. Start a special interest group
13. Engage in action research

I would like to illustrate how teachers can integrate an innovative idea (e.g., IDLE) into the formal education system through action research. Action research is understood as a spiral of movements between teachers' action and research (Kemmis et al. 2014). To be precise, action refers to interventions into existing practices in the classroom. Research indicates a systematic observation and analysis of change as a result of taking actions. Therefore, teachers play an active role in planning, acting (to implement a plan), observing the effects of the action, and reflecting on these effects for further planning.

Early this year, two Hong Kong secondary school teachers and I conducted an action research project titled "30-day Extramural English (or IDLE) challenge" (Lee et al. 2021). To this end, we invited students to choose their own favorite English content and practice it for 30 consecutive days. We began small: students choose their own learning goals (e.g., speaking), materials (a dubbing app), and strategies (e.g., imitating other persons' pronunciation). For the ensuing two weeks, they were asked to showcase their performance each day. We also collected the students' feedback. Surprisingly, most of the students continued with the challenge for more than the required 30 consecutive days. For instance, one student stayed committed to the challenge for 82 days. Based on the observations and reflections, we are currently revising the second stage of the IDLE challenge program. Consistent with Van der Heijden et al. (2015), these two in-service teachers have characteristics of teachers as change agents: they are open to new ideas, are reflective about their current teaching, and have taken initiative in designing and executing action research.

2 CONCLUSION

Godwin-Jones (2019) predicted three possible scenarios for the future of language learning. The first scenario is that due to the development in technology (e.g., AI-generated language assistants), we may not need to learn another language. The second scenario is that we may learn a foreign language from the Internet (e.g., Netflix and YouTube) rather than from a formal classroom. The third scenario is that we may learn another language from formal L2 teachers who are able to integrate SMART technology into formal education. The author envisions that the third scenario is most likely to happen because stakeholders are increasingly expecting it. In that regard, Godwin-Jones argues that language teachers will need to adapt to new language environments.

I first started learning English beginning in the 7th grade. Like many typical Korean EFL students, I mainly learned English through drilling and memorizing for the test. I was reticent and silent in the English class because I had no reason to speak English. Consequently, I found learning English to be boring and stressful during secondary school.

However, there was one high school teacher, Mr. Tae-hwan Choi, who totally changed my view of the English language. His teaching practice was different and rather unusual: sometimes, he played English pop songs in the classroom. During one particular English lesson, I was deeply touched by the beautiful lyrics of the song, "My Heart Will Go On," performed by Celine Dion. Mr. Choi also arranged a penpal exchange with American students as part of an after-school program. Personally, it was a critical moment, because for the first time in my life I made a decision to do something for my learning on my own: I went to a bookstore and bought a Korean-English dictionary.

Thanks to that extraordinary teacher, English became fun and meaningful in my life. I also became a more autonomous learner. Now I am teaching and researching autonomous English learners (or IDLE learners). I could not have been who I am today without Mr. Choi, who integrated informal learning (e.g., pop songs and penpals) into formal teaching. To me, he is a change agent.

The same is true for today's language teachers. Despite the pedagogical dilemma that they face, I strongly believe that teachers (as change agents) can play a vital role in integrating innovative ideas (e.g., IDLE) into formal education, thereby influencing students' mindsets, and bringing about changes in their behavior. If you leave this paper with one word or one idea, I hope it is 'agent'. Teachers as change agents should not underestimate their ability to effect change.

FUNDING

This work was supported by the Seed Funding Grant Project [grant number RG 54/2019-2020R] from the Education University of Hong Kong.

REFERENCES

Barcelos, A. M. F. (2015). Unveiling the relationship between language learning beliefs, emotions, and identities. *Studies in Second Language Learning and Teaching, 5*(2), 301–325.

Benson, P., & Reinders, H. (Eds.). (2011). *Beyond the language classroom*. New York, NY: Palgrave Macmillan.

Bronfenbrenner, U. (1979). *The ecology of human development: Experiments by nature and design*. Cambridge, Massachusetts: Harvard University Press.

Chun, D. (2019). Current and future directions in TELL. *Educational Technology & Society, 22*(2), 14–25.

Dewaele, J.-M., & MacIntyre, P. D. (2014). The two faces of Janus? Anxiety and enjoyment in the foreign language classroom. *Studies in Second Language Learning and Teaching, 4*(2), 237–274.

Dörnyei, Z. (2005). *The psychology of the language learner: Individual differences in second language acquisition*. Mahwah, NJ: Lawrence Erlbaum.

Friedman, T. L. (2005). *The world is flat: A brief history of the twenty-first century*. New York: Farrar, Straus and Giroux.

Galloway, N. (2017). *Global Englishes and change in English language teaching: Attitudes and impact*. New York, NY: Routledge.

Gardner, R. (1985). *Social psychology and second language learning: The role of attitudes and motivation*. London, United Kingdom: Edward Arnold.

Godwin-Jones, R. (2019). In a world of SMART technology, why learn another language? *Educational Technology & Society, 22*(2), 4–13.

Heatherton, T. F., & Nichols, P. A. (1994). Personal accounts of successful versus failed attempts at life change. *Personality and Social Psychology Bulletin, 20*(6), 664–675.

Kemmis, S., McTaggart, R., & Nixon, R. (2014). *The Action research planner: Doing critical participatory action research*. Singapore: Springer.

Kim, K. J. (2019). Changes in English learning motivation of high school students: Motivation, demotivation, and remotivation. *English Teaching, 74*(4), 249–274.

Lee, J. S. (2020). An emerging path to English in Korea: Informal digital learning of English. In M. Dressman & R. Sadler (Eds.), *The Handbook of Informal Language Learning* (pp. 289–302). Hoboken, New Jersey: Wiley-Blackwell.

Lee, J. S. (2022). *Informal Digital Learning of English: Research to Practice*. New York, NY: Routledge

Lee, J. S., & Dressman, M. (2018). When IDLE hands make an English workshop: Informal digital learning of English and language proficiency. *TESOL Quarterly, 52*(2), 435–445. doi:10.1002/tesq.422

Lee, J. S., & Lee, K. (2020). Role of L2 motivational self system on willingness to communicate of Korean EFL university and secondary students. *Journal of Psycholinguistic Research, 49*(1), 147–161.

Lee, J. S., & Lee, K. (2021). The role of informal digital learning of English and L2 motivational self system in foreign language enjoyment. *British Journal of Educational Technology, 52*(1), 358–373.

Lee, J. S., Osburn, M., & Yeung, E. (2021). *Integrating Extramural English in improving English learners' speaking*. Paper presented at the Public Lecture Series 2021, The Education University of Hong Kong.

Lee, J. S., & Sylvén, L. K. (2021). The role of Informal Digital Learning of English in Korean and Swedish EFL learners' communication behavior. *British Journal of Educational Technology, 52*(3), 1279–1296.

Levy, M. (1997). *CALL: Context and conceptualization*. Oxford: Oxford University Press.

Pavlenko, A. (2013). The affective turn in SLA: from 'affective factors to 'language desir' and 'commodification of affect'. In D. Gabryś-Barker & J. Bielska (Eds.), *The affective dimension in second language acquisition* (pp. 3–28). Bristol: Multilingual Matters.

Reinders, H., & Benson, P. (2017). Research agenda: Language learning beyond the classroom. *Language Teaching, 50*(4), 561–578.

Richards, J. C. (2015). The changing face of language learning: Learning beyond the classroom. *RELC Journal, 46*(1), 5–22.

Soyoof, A., Reynolds, B. L., Vazquez-Calvo, B., & McLay, K. (2021). Informal digital learning of English (IDLE): A scoping review of what has been done and a look towards what is to come. *Computer Assisted Language Learning*, 1–33.

Sundqvist, P., & Sylvén, L. K. (2016). *Extramural English in teaching and learning: From theory and research to practice*. Basingstoke, London: Palgrave Macmillan.
Van der Heijden, Geldens, J. J. M., Beijaard, D., & Popeijus, H. L. (2015). Characteristics of teachers as change agents. *Teachers and Teaching: theory and practice, 21*(6), 681–699.
Verplanken, B., Walker, I., Davis, A., & Jurasek, M. (2008). Context change and travel mode choice: Combining the habit discontinuity and self-activation hypotheses. *Journal of Environmental Psychology, 28*, 121–127.
Wong, M. S. (2011). Fifty ways to develop professionally: What language educators need to succeed. *Language Education in Asia, 2*(1), 142–155.
Zhang, R., Zou, D., Cheng, G., Xie, H., Wang, F. L., & Au, O. T. S. (2021). Target languages, types of activities, engagement, and effectiveness of extramural language learning. *PLoS ONE, 16*(6), e0253431.

Teacher Education and Teacher Professional Development in the COVID-19 Turn – Drajati & Putra (Eds)
 ISBN: 978-1-032-38970-7

Teachers' adaptive expertise and autonomy for resilient teaching and learning in the face of Covid-19 pandemic

Hyo-Jeong So
Ewha Womans University, Seoul, Korea

ABSTRACT: With the full or partial closure of schools during the COVID-19 pandemic, the main learning environment has shifted from physical schools to online learning environments. We have witnessed how teachers reacted to the unexpected changes such as teachers who adapted quickly to the new normal and also those who were faced with difficulties to make such a shift to online teaching. Then, what would explain the difference underlying teachers' adaptive behaviors? In this presentation, I will discuss how the perspective of adaptive expertise and teacher autonomy can be employed to explain the development of teachers' adaptive behaviors, especially resilient teaching and learning with technology. First, the development of adaptive expertise involves two dimensions: efficiency and innovation. The two dimensions are complementary but can be antagonistic. When teachers attempt new approaches, they may experience their practices being less efficient than previous approaches. For teachers to be adaptive experts, however, they need to develop dispositions to perceive such initial experiences, not as a failure but as a valuable and productive process of learning. Second, teacher autonomy is a way to allow and engage teachers in situational experimentation with new ideas, free from the bound of institutional practices. During the recent pandemic, teachers given with a certain degree of autonomy showed adaptive behaviors to create resilient and flexible learning environments and also influenced learner autonomy. With that, I will present strategies concerning how teacher education and professional development programs can be reframed to foster teacher autonomy and to help teachers grow as adaptive experts. The strategies are a) to focus on doing with understanding where teachers are continuously exposed to better understand the complexity of teaching and learning with technology, and b) to develop creative thinking skills for innovative ideas to grow. These strategies will be illustrated with the examples found in the existing cases to elucidate the importance of teacher autonomy and adaptive expertise as sustainable elements to reframe teacher education in the 'new normal' of the ongoing pandemic.

Keywords: teacher adaptive expertise, teacher autonomy

1 INTRODUCTION

With the full or partial closure of schools during the COVID-19 pandemic, the main learning environment has shifted from physical schools to online learning environments. We have witnessed how teachers reacted to unexpected changes such as teachers who adapted quickly to the new normal and also those who were faced with difficulties to make such a shift to online teaching. Then, what would explain the difference underlying teachers' adaptive behaviors? In this paper, I discuss how the perspective of adaptive expertise and teacher autonomy can be employed to explain the development of teachers' adaptive behaviors, especially resilient teaching and learning with technology. And then, the paper presents the principle-based approach for good online teaching to help teachers better adjust to the resilient teaching and learning environment.

 DOI 10.1201/9781003347798-3

1.1 *Teachers as adaptive experts*

Since the seminal article on '*Two Courses of Expertise*' by Hatano and Inagaki (1986) and the conception of Preparation for Future Learning (PFL) by Bransford and Schwartz (1999), adaptive expertise has been used to explain the differences in novice and expert learners and further the importance of developing transferable knowledge and skills. While the conception of adaptive expertise is grounded in cognitive research on student learning, a growing body of teacher education research suggests the need to reconceptualize teacher education from the perspective to help teachers become adaptive experts. Hammerness, Darling-Hammond and Bransford (2006) suggest that the development of adaptive expertise involves two dimensions of expertise: *efficiency* and *innovation*. First, the efficiency dimension means "greater abilities to perform particular tasks without having to devote too many attentional resources to achieve them" (p.361). Second, the innovation dimension involves "moving beyond existing routines and often requires people to rethink key ideas, practices, and even values in order to change what they are doing" (p.361). The two dimensions are complementary but can be antagonistic when they create conflicts. For instance, in the initial stage of learning innovative strategies and knowledge, teachers may experience their practices being less efficient than previous approaches. For teachers to be adaptive experts, they need to develop dispositions to perceive such initial experiences, not as a failure but as a valuable and productive process of learning.

The discussion on adaptive expertise naturally leads to the question of why some teachers are more adaptive than others. Fairbanks and colleagues (2011) suggest that there is a strong need to move beyond knowledge levels to better define teacher expertise. They define teachers' adaptive expertise from self-knowledge and a sense of autonomy perspectives, which are necessary capacities for teachers to negotiate personal and professional goals in complex contexts. Similarly, Lin, Schwartz, and Hatano (2005) define teachers' adaptive expertise from adaptive metacognition perspectives where teachers can adapt themselves metacognitively in response to social and pedagogical variability.

During the recent pandemic, teachers given with a certain degree of autonomy showed adaptive behaviors to create resilient and flexible learning environments. While some teachers adapted themselves quickly to such unpredictable changes to create resilient environments, there were also a great number of teachers who struggled to meet the new demands of teaching, especially the transition to online teaching. The experiences gained from the COVID-19 pandemic highlight the importance of developing teachers' expertise to become more adaptive to unpredicted situational changes. Teacher education programs and professional development programs need to help teachers better follow the '*Optimal Adaptability Corridor* (OAC)' to become an adaptive expert, rather than remaining as a routine expert or a frustrated novice (see Figure 1). Guiding teachers through the

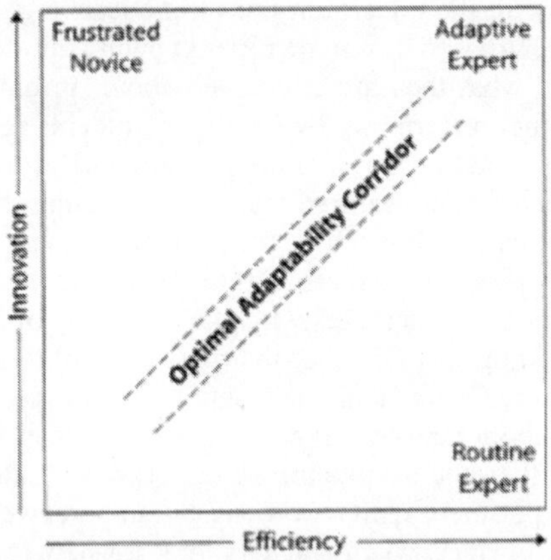

Figure 1. Trajectory toward adaptive expertise (Bransford et al. 2005).

OAC means helping teachers develop "new interpretations instead of assimilating new experiences to old ways of thinking" (Schwartz et al., 2005). That is, teachers should give new interpretations about their online teaching experiences during this pandemic, rather than assimilating them to traditional teaching in schools that they are more familiar with and become routine practices.

1.2 *Principle-based approach for good online teaching*

While the COVID-19 pandemic came to us without any clear signs and most teachers were unprepared to deal with such challenges, the fact that teachers have experienced new ways of teaching and learning can be a great asset for innovating education practices and culture. In this section, I discuss the principle-based approach for good online teaching to help teachers become more adaptive experts in this new environment. As shown in Figure 2, the proposed approach includes three main principles: (1) building personal beliefs and attitudes about good teaching, (2) empathy-centered design, and (3) experimenting with ideas.

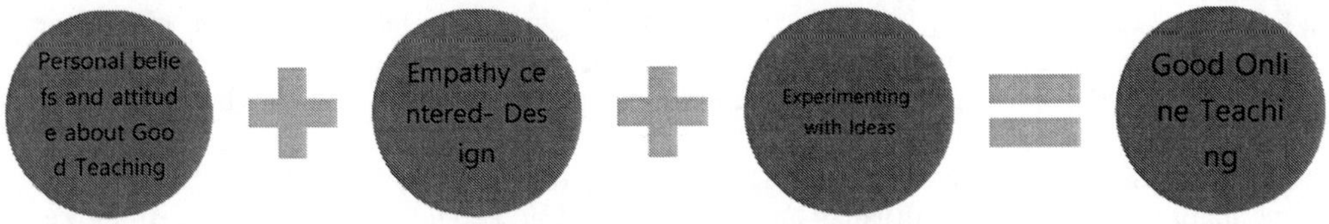

Figure 2. Principle-based approach for good online teaching.

First, teachers need to develop and articulate their personal beliefs and attitude about good teaching. This principle implies that teachers tend to enact practices based on their personal beliefs and attitude gained from past experiences, such as what they believe in good teaching and learning and their experiences as students. Teachers who believe that teaching is about transferring knowledge to students will design and enact their online lessons to be delivery-centered, giving less autonomy to students When teachers believe that meaningful learning happens through students' active participation, their design of online lessons also reflects various activities that engage students to share ideas. This principle also emphasizes the importance of good teaching that transcends online and offline teaching situations. Often, good teaching practices in classroom situations include (a) promoting interaction (student-content, student-student, student-instructor interaction), (b) designing knowledge-centered lessons for deeper understanding, (c) providing clear guidelines and criteria for evaluation, and (d) giving prompt and productive feedback. These practices are also applied to the design of online classes.

Second, good online teaching is based on empathy-based design. During the COVID-19 pandemic, students had to deal with the uncertainty of online learning, which caused psychological and motivation issues. Empathy-based design involves knowing the students such as how they feel, what they want to know, and what they are concerned about. In online learning, it is important to reduce a degree of emotional uncertainty by involving students in decision-making processes. Some strategies for empathy-based design include providing clear evaluation criteria and checklists to reduce emotional stress about required tasks, and opening discussion forums and online collaboration platforms (e.g., Padlet, Jamboard) for students to share their feelings and ideas.

Third, teachers need to experiment with ideas. These days, since there are abundant technological tools available, teachers may feel frustrated with the selection of relevant tools for their own teaching. It would be misleading to think that there are perfect tools that meet the diverse and unique needs of students and teachers. Thus, teachers need to explore and experiment with new tools. As mentioned earlier about adaptive expertise, teachers may feel their practices with new technology are less efficient than the usual approaches. However, such initial experiences should not discourage teachers to experiment with new ideas, but to perceive them as an essential process of learning new approaches. Further, when teachers look for new technologies, they should start from a pedagogical question, rather than technology itself.

2 CONCLUSION

In this paper, I discuss the importance of teachers becoming adaptive experts, given that teaching practices are no longer static and teachers need to deal with unexpected changes. The COVID-19 pandemic has been definitely challenging experiences for many teachers, but also provides opportunities for us to reflect on past teaching practices and to be better equipped with the demands in the 'new normal' of the ongoing pandemic. Further, teacher education and professional development programs need to be reframed to foster teacher autonomy and to help teachers grow as adaptive experts. In 1966, architect Cedric Price posed the question "Technology is the answer, but what was the question?" and argued for the importance of having a clear purpose of using technology. The same question goes to the current situation during the pandemic. As teachers are embracing online learning and other technological tools, they need to have an objective judgment of what is possible and impossible through technology and a clear sense of purpose in using technology. Such efforts will help teachers move beyond vague expectations and illusions of using technology for resilient teaching and learning.

REFERENCES

Bransford, J., & Schwartz, D. L. (1999). Rethinking transfer: A simple proposal with multiple implications. *Review of Research in Education, 24*, 61–101.

Bransford, J., Derry, S., Berliner, D., & Hammerness, K. (2005). Theories of learning and their roles in teaching. In L. Darling-Hammond & J. Bransford (Eds.), *Preparing teachers for a changing world: What teachers should learn and be able to do* (pp. 40–87). San Francisco, CA: Jossey-Bass.

Darling-Hammond, L., & Bransford, J. (with LePage, P, Hammerness, K., & Duffy, H.). (2005). *Preparing teachers for a changing world: What teachers should learn and be able to do*. San Francisco: Jossey-Bass.

Hatano, G., & Inagaki, K. (1986). Two courses of expertise. In H. Stevenson, H. Azuma, & K. Hakuta (Eds.), *Child development and education in Japan*. New York: Freeman.

Lin, X., Schwartz, D. L., & Hatano, G. (2005). Toward teacher's adaptive metacognition. *Educational Psychologist, 40*, 245–255.

Schwartz, D. L., Bransford, J., & Sears, D. (2005). Efficiency and innovation in transfer. In J. Mestre (Ed.), *Transfer of learning* (pp. 1–51). Greenwich, CT: Information Age Publishing.

Teacher Education and Teacher Professional Development in the COVID-19 Turn – Drajati & Putra (Eds)
© 2023 The Author(s), ISBN: 978-1-032-38970-7

Teacher identity, teacher agency and teaching competence in Covid-19 turn

Lynde Tan
Western Sydney University, Penrith, Australia

ABSTRACT: The COVID-19 pandemic has highlighted the pressing need for teachers to develop competence in harnessing digital technologies for teaching and learning. This urgency was exacerbated at the start of the pandemic when remote teaching was non-negotiable. At an unprecedented scale, online learning is accelerated and creates changes to teachers' roles and their identities. To better understand how teachers enact professionalism in their practice, examining teacher agency is crucial. Its role in assisting teachers to transit through uncertain times to minimise educational disruptions not only demands respect from the public, but also necessitates further inquiry into teachers' professional learning. This paper raises issues related to evolving teacher identities and teacher agency in the COVID-19 turn. It particularly points out the importance of making a distinction between content provision and education as teachers create their online presence in remote teaching. Given that teacher identity and agency are negotiated and constructed through relationships, social interactions and contextual factors, this paper uses the formative and design experiments approach as an example to discuss these two concepts. It offers this approach as a way of strengthening teacher identity and agency whilst developing teacher competence in integrating technology into teaching. The paper concludes with a discussion on future research agenda on teacher identity, teacher agency and teacher competence with the goal of creating positive social futures in the COVID-19 turn.

Keywords: teacher identity, teacher agency, teaching competence

1 INTRODUCTION

March 2020 marks the beginning of the COVID-19 turn when school gates all over the world are shut. Principals are plunged into abject ambiguity of running a school outside a school and teachers are required to flip their classes immediately using new modalities (Jones & Kessler, 2020). The COVID-19 pandemic has highlighted the pressing need for teachers to develop competence in harnessing digital technologies for teaching and learning. This urgency was exacerbated at the start of the pandemic when remote teaching was non-negotiable. At an unprecedented scale, online learning is accelerated and creates changes to teachers' roles and their identities.

1.1 *Teacher agency and identity*

Teachers are agents of change as they expand their roles to take actions which are necessary for the continuity of student learning in the global pandemic (Allen et al. 2020). When seeking to better understand how they enact professionalism in their transitions to remote teaching on digital platforms, the notion of teacher agency needs to be attended to. Teacher agency can be understood as "the capacity of [teachers] to critically shape their own responsiveness to problematic situations" (Emirbayer & Mische 1998, p. 971) and this agency involves forms of action that are more oriented toward the past (*iterational* dimension), the future (the *projective* dimension), and the present (the *practical-evaluative* dimension).

 DOI 10.1201/9781003347798-4

There are dissenting views about the relationships between agency and identity. Teacher agency plays an essential role in constructing and negotiating teacher identity. Teacher identity is "a key means through which [teachers] care about and care for what is going on around them" (Holland 1998, p.5). In times of changes when teachers have to make choices and changes to their practices, they develop conceptions of what it means to be a teacher (identity) and their identity is a prerequisite for them to develop agency to develop appropriate actions (Benson 2001; Holland 1998).

Agency and identity are inseparable aspects of teachers' professional development and educational reforms (Teng 2018; Voogt et al. 2015). Current body of literature that reports on teachers' challenges in the global pandemic suggests that teacher agency is strongly influenced by the new identities expected of them, particularly being an online content creator and online learning instructional designer (Adedoyin & Soykan 2020; Beach 2018; Breslin 2021). Yet, not many teachers around the world are well-equipped to deliver remote teaching on digital platforms effectively (Dabrowski 2020). There is an urgent need to expand teachers' competencies to deepen students' online learning and develop ways of building teachers' professional knowledge towards this direction in a sustained manner.

1.2 *Develop teaching competence using formative and design experiments approach*

Teachers no longer depend on participation in professional development workshops as the sole approach to further their teaching competence. Effective professional development encompasses the seven characteristics: it is content focused; it incorporates active learning utilising adult learning theory; it supports collaboration, typically in job-embedded contexts; it uses models and modelling of effective practice; it provides coaching and expert support; it offers opportunities for feedback and reflection and finally, it is of sustained duration (Darling-Hammond et al. 2017, pp. v- vi). For these reasons, there are teachers who develop their teaching competence by participating in research projects as co-researchers and co-designers of pedagogical interventions to address specific educational challenges. The formative and design experiments approach is an example that provides teachers with such opportunities to advance this professional goal.

Reinking and Bradley's (2008) formative and design experiment is a widely cited research methodology which encompasses the following key characteristics that are applicable to education:

1. Formative and design experiments are grounded in developing understanding by seeking to accomplish practical and useful educational goals.
2. They are focused on less controlled, authentic environments instead of tightly controlled laboratory-like settings.
3. They use and develop theory in the context of trying to engineer successful instructional interventions.
4. They entail innovative and speculative experimentation.
5. They are interdisciplinary, employing multiple theoretical and methodological perspectives and orientations.
6. They seek understandings that accommodate many complex, interacting variables in diverse contexts.
7. They seek generalisations from multiple examples rather than from random samples and controlled experimentation. (pp. 10 – 11)

Given that teacher identity and agency are negotiated and constructed through relationships, social interactions and contextual factors, the formative and design experiments approach is a recommended approach for teachers to develop their teaching competence when they engage in designing curricular materials to improve student learning with researchers. As teachers involve in collaborative reflective inquiry with other teachers and researchers, they share knowledge and capitalize different levels of expertise to bring about positive changes that are pragmatic and contextual (McKenney & Reeves 2012; Reinking & Bradley 2008; Voogt et al. 2015). In this approach, teachers' beliefs are challenged and are likely to change when they co-design pedagogical interventions whilst shaping their agency and identity in the process of making changes (Chai at al. 2017; Koh et al. 2015). Such an approach has proven to be effective to "help teachers learn to better understand the innovation and provide them with competencies in instructional planning and enactment" (Voogt et al. 2015, p. 277).

1.3 *Concluding thoughts on future research*

The recommended formative and design experiments approach has traces of Dewey's (1933) notion of reflective inquiry which he believed should be the anchoring point of designing educational experience. Dewey put forward five key aspects of reflective thought, namely intellectualization (problem identification), guiding idea, reasoning, testing, and ending with a satisfactory resolution. Individuals' ability to question through daily activities experienced in the social environment that they locate themselves in makes inquiry essentially a social activity (Glassman 2001; Swan et al. 2009). This notion of educational experience advocated by Dewey has given rise to a social constructivist approach of learning and teaching, particularly the *community of inquiry* (CoI) framework that warrants further exploration in remote teaching (Swan et al. 2009).

As teachers take agentic actions and adopt new identities to create their online presence as an educator, the importance of making a distinction between content provision and education should be made (Kellermann 2021). The COI framework is promising for making this distinction by promoting three crucial elements that facilitate successful educational experiences in online learning environments, namely the *cognitive presence* (students engaging in meaning-making, reflection and productive interactions), *social presence* (the teacher and students engaging in a learning community) and *teaching presence* (the teacher's design, facilitation and setting of purposeful directions for students to achieve learning outcomes) (Garrison et al. 1999). Further work is needed to enhance teachers' competence in integrating technology to develop a CoI in the online environment. Teachers would benefit from having access to shared knowledge of best practices of how the CoI framework is applied across contexts. Although the global pandemic may have presented teachers with challenges, it also offers opportunities to awaken the soul of education as teachers continue to increase their agency, change their identity and enhance their teaching competence to create positive social futures in the COVID-19 turn.

REFERENCES

Adedoyin, O. B., & Soykan, E. (2020). Covid-19 pandemic and online learning: the challenges and opportunities. *Interactive Learning Environments*, 1–13.

Allen, J., Rowan, L., & Singh, P. (2020). Teaching and teacher education in the time of COVID-19. *Asia-Pacific Journal of Teacher Education*, 48(3), 233–236.

Beach, P. (2018). Examining Elementary Teachers' Use of Online Learning Environments: An Exploratory Study. *Journal of Research on Technology in Education*, *50*(1), 34–47.

Benson, P. (2001). *Teaching and researching autonomy in language learning*. Pearson Education.

Breslin, T. (2021). *Lessons from lockdown: the educational legacy of COVID-19. Routledge.*

Chai, C., Tan, L., Deng, F. and Koh, J. (2017). Examining pre-service teachers' design capacities for web-based 21st century new culture of learning. *Australasian Journal of Educational Technology*, *33*(2), 129–142.

Dabrowski, Anna. (October 2020). Teaching over technology: Educational priorities during COVID-19. *Discover*: https://www.acer.org/au/discover/article/teaching-over-technology-educational-priorities-during-covid-19

Darling-Hammond, L., Hyler, M. E., Gardner, M. (2017). *Effective Teacher Professional Development.* Learning Policy Institute.

Dewey, J. (1933). *How we think: A restatement of the relation of reflective thinking to the educative process.* D. C. Heath and company.

Dewey, J. (1938). *Experience and education*. The Macmillan Company.

Emirbayer, M., & Mische, A. (1998). What Is Agency? *Journal of Sociology*, *103* (4), 962–1023.

Garrison, D. R., Anderson, T., & Archer, W. (1999). Critical inquiry in a text-based environment: Computer conferencing in higher education. *The Internet and Higher Education, 2(2)*, 87–105.

Garrison, D. R., & Arbaugh, J. B. (2007). Researching the community of inquiry framework: Review, issues, and future directions. *Internet and Higher Education, 10*(3), 157–172.
Glassman, M. (2001). Dewey and Vygotsky: Society, Experience, and Inquiry in Educational Practice. *Educational Researcher, 30*(4), 3–14.
Holland, D. C. (1998). *Identity and agency in cultural worlds. Harvard* University Press.
Jones, A.L., & Kessler, M.A. (2020). Teachers' Emotion and Identity Work During a Pandemic. *Frontiers in Education, 5*(5583775).
Kellermann, D. (2021, March 8). *Academics aren't content creators, and it's regressive to make them so. Discover:* https://www.timeshighereducation.com/opinion/academics-arent-content-creators-and-its-regressive-make-them-so
Koh, J., Chai, C., Hong, H., & Tsai, C. (2014). A survey to examine teachers' perceptions of design dispositions, lesson design practices, and their relationships with technological pedagogical content knowledge (TPACK). *Asia-Pacific Journal of Teacher Education, 43*(5), 1–14.
McKenney, S., & Reeves, T. (2012). *Conducting educational design research. Routledge.*
Reinking, D., & Bradley, B. (2008). *Formative and design experiments: Approaches to language and literacy research.* Teachers College Press.
Swan, K., Garrison, D. R. & Richardson, J. C. (2009). A constructivist approach to online learning: The Community of Inquiry framework. In Payne, C. R. (Ed.) *Information Technology and Constructivism in Higher Education: Progressive Learning Frameworks* (pp. 43–57). IGI Global.
Teng, M. F. (2018). Autonomy, agency, and identity in teaching and learning English as a foreign language. Springer.
Voogt, J., Laferrie're, T., Breuleux, A., Itow, R. C., Hickey, D. T., & McKenney, S. (2015). Collaborative design as a form of professional development. *Instructional Science, 43*, 259–282.

English Language Education

Teacher Education and Teacher Professional Development in the COVID-19 Turn – Drajati & Putra (Eds)
© 2023 The Author(s), ISBN: 978-1-032-38970-7

Globalization and its impact on ELT in Cambodia's higher education

Vikrant Chap
Purdue University, West Lafayette, USA

ABSTRACT: This qualitative study investigates the impact of globalization on ELT curriculum design and practices in Cambodia's higher education. Adopting critical theory, this article considers cultural identity to be an inherent prestige of a nation and an important tool to ensure freedom from oppression. A purposefully selected sample of twenty participants—three management team members, five lecturers, and twelve students—participated in the semi-structured and focus-group interviews. The inquiries revolve around the perceived values of local culture, specifically its preservation and inclusion as manifest through *Global Studies* subject. The findings indicate that the presence of *Global Studies* arises from the static view of culture, and the discussion focuses on how social actors negotiate cultural identity in light of global issues through curriculum design and practices. The study paves the way for future research under the comparative educational frame to juxtapose comparative perspectives of social actors in response to globalization and local cultural identity.

Keywords: cultural identity, cultural studies, curriculum studies, globalization, global studies

1 INTRODUCTION

In the recovery stage from periods of colonization, genocide, and civil war, Cambodia's higher education institutions (HEIs) mainly focused on specialized skills. The structure of the curriculum was shaped by the pressing need for capable human resources to rebuild the economy. In 2002, the Bachelor of Education program at one university in Cambodia incorporated *Cultural Studies* subject into its TEFL major as a response to the heavy influx of expats and international agents, and in 2006, it was replaced with *Global Studies*. This transition mandated a new epistemology and methodology on the teachers' and students' parts, yet the decision was shaped primarily by external social forces, leaving behind many critical questions. To quote from Michael W. Apple (2018), critical curriculum design involves addressing a series of critical questions:

> *Whose knowledge is this? How did it become "official"? What is the relationship between this knowledge and how it is organized and taught and who has cultural, social, and economic capital in this society? Who benefits from these definitions of legitimate knowledge and who does not? What are the overt and hidden effects of educational reforms on real people and real communities? What can we do as critical educators and activists to challenge existing educational and social inequalities and to create curricula and teaching that are more socially just? (p. 3)*

Few studies have been done to critically examine these questions, and most critical curriculum studies merely play a deconstructive role (Apple 2018). Although deconstructing a curriculum facilitates its design and development, we should not lose sight of the political and cultural aspects. Sufficient considerations of these questions imply education that is multicultural and multi-perspectival (Kellner 2003). However, in a qualitative study of academic language policy in Australia, only 2 out of 10 policymakers and planners recalled involving multiple actors in curriculum construction (Fenton-Smith & Gurney 2016). The lack of constructive alignment among policymakers and policy actors was also found in a comprehensive study of English language policy in Asian countries (see Hamid & Nguyen 2016). The authors used the term *policy dumping* as opposed to *agency* to denote the absence of clear directions in language policy implementation. Hamid and Nguyen (2016) posited that critical implementation of a macro and meso curriculum policy demands the role of agency from local actors.

As a case in point, despite the significant and critical shift from *Cultural Studies* to *Global Studies* subject, there were few discussion meetings or surveys among stakeholders, and no formal teacher training was documented. The absence of critical dialogue among policy designers and implementers leaves the critical reflection questions to the individual social actors (Apple 2018). Therefore, this study attempts to draw insights from policy insiders namely management team members, lecturers, and students. As a qualitative study, it serves as a starting point of a critical conversation about curriculum policy and implementation under the force of globalization, rather than offering a comprehensive framework. Three key questions guide this study.

1) What are the perceived needs for a shift from *Cultural Studies* to *Global Studies* in the higher education curriculum in Cambodia? What is the nature of the change?
2) How are the values of local culture (dis)integrated into the syllabus of *Global Studies* subject?
3) How is the value of local culture perceived in *Global Studies* and within a broader social context?

2 LITERATURE REVIEW

Cultural responses to globalization usually rest somewhere between two extremes. Cultural monologists emphasize the preservation of local cultural identity, while postcolonial theorists advocate a critical balance between local contexts and global trends (see Matus & McCarthy 2003). With these ideologies, although globalization discourse centers on convergence and unified identity (Pacheco 2012), diversity is seen in policy responses, designs, and implementations. While diversity can be a marker of progress, the nature of diversity must be seriously examined. Some contextual variations of a working model may result from misinterpretations instead of an act of deliberate negotiation. Broadfoot's (2000) retrospective and prospective models and Vaira's (2004) convergence and divergence theses help put these matters in perspective. To put it succinctly, successful negotiation of globalization gears toward a prospective ideal and deliberate divergence, rather than just producing variations as a by-product of situational differences. In his position paper after years of working in curriculum studies, Apple (2018) argued that critical divergence in knowledge production, representation, and circulation is rare. Similarly, Smith (2003) expressed concerns over the lack of responsiveness in curriculum policy and action and outlined a three-stage model of globalization: (a) neoliberalism, (b) localization, and (c) intercultural negotiation. Although the current social landscape is at the third stage, most HEIs are still struggling with the second stage.

In developed countries such as the United States, the United Kingdom, Canada, France, Germany, Japan, and Australia, attempts to promote critical literacy to facilitate stage three are well-documented and properly understood (Zajda & Val 2021). In contrast, in certain Asian countries, the unique characteristics of higher education further complicate these matters. As formerly colonized territories, a majority of Asian HEIs have not made efforts in revising their curriculum even after achieving independence (Altbach 2005), which results in a heavy reliance on the well-established western knowledge system (Broadfoot 2000; Crossley 2012). In a study by Hamid and Nguyen (2016) focusing on language policy and planning in Asian countries—Bangladesh, India, Indonesia, Japan, Malaysia, Pakistan, and Vietnam—strong evidence of *policy dumping* was found at the macro-level, and the teachers had to assume a full responsibility to meet the changing demands of the societies and the students. This daunting placement of responsibility on the teachers' shoulders is rarely addressed in policy discourse. As Apple (2018) put it:

> *With neoliberal, neoconservative, authoritarian populist, and new managerial forces increasingly occupying the space of real policies and practices, we have little voice in the public debates over the realities of schooling and the decisions of curriculum policies and practices. The field of education deserves more. (p. 689)*

The absence of macro-level support is critical to the success of negotiating different knowledge systems. Relying on teachers to achieve this complex negotiation rests on a false assumption that teachers have all it takes. In the study by Fenton-Smith and Gurney (2016), agency in academic language policy and planning encompasses four categories: (a) people with power, (b) people with expertise, (c) people with influence, and (d) people with interest. Put in this frame, classroom teachers do not have sufficient resources to enact full agency, let alone help students respond proactively to globalization. Therefore, an effective integration requires a critical understanding of power relations, within which policy designers need to address local and global cultures right upfront. While curriculum developers typically consider "the prescriptions of the scope of content, the sequence of learning, and criteria of completion" (Waks 2003, p. 386), local cultural capital and embedded ideology often slip through their consciousness. To challenge ideological dominance, Paulo Freire critiqued the role of critical dialogue in countering the dominant or hegemonic narrative that can potentially suppress local voices (Freire 1970). The utility of such critical dialogue rests on the ability of teachers and students to appropriate knowledge, which involves "internal dispute, contestation, revision of tastes and methods, discovery or recuperative research, politics, and philosophies" (Martin 2009, p. 302). A classroom then should represent different worldviews, and the curriculum should play a more dynamic role in preparing students for an emerging complexity that is unforeseeable, fickle, and elusive (Sharmer 2009). Situated within these selected works of literature, this study attempts to look at how one Cambodian higher education institution negotiated and responded to globalization through curriculum design and implementation after years of colonization, genocide, and civil war. Operating from the lens of critical pedagogy and curriculum studies (Apple 2018; Freire 1970), the researcher considers local cultural capital as a critical resource in curriculum design and practices. Additionally, given the lack of constructive alignment in the existing curriculum, as manifest in *Global Studies*, the study explores perceptions of social actors at various levels of execution.

3 METHOD

The research study was conducted at one foreign languages institute in Cambodia. A purposive sampling strategy was used to identify the policymakers and teachers, and students were recruited through snowball sampling. Three policymakers—active players in the areas of institutional administration, department management, and curriculum development—were invited to participate in semi-structured and non-directive one-on-one interviews regarding their critical decisions to replace *Cultural Studies* with *Global Studies*. Semi-structured one-on-one interviews were conducted with five teachers with varying levels of seniority, different educational backgrounds, and distinct teaching styles. Twelve students were sorted into three groups (four participants per group) for focus-group interviews (Mertens 2010). Providing the researcher's higher social status, focus groups encouraged the students to voice their concerns more accurately than directed interviews (Mertens 2010).

This study employs a qualitative research method to study the perceptions of the policymakers, teachers, and students at the English Department. Because questions involving intercultural negotiations "are not decidable in the abstract and need to be established on an empirical case-by-case basis" (Barker 2002, p. 71), the qualitative design matches the dynamic nature of inquiry into "multiple dimensions and layers of reality" of this social phenomenon (Johnson & Christensen 2014, p. 42) and transformative/critical paradigm (Mertens 2010). Every interview was tape-recorded with participants' informed consent. Before and after each interview, the researcher used a memoing technique to perform initial "speculative" and subsequent "focused and conclusive" analyses (Johnson & Christensen 2014, p. 765). Upon the completion of data collection, the researcher transcribed the interviews verbatim and followed the process of segmenting, coding, and developing category systems (Johnson & Christensen 2014). Given the small number of participants, the researcher used manual coding and derived the themes from the memoing notes of the researcher during the research process.

4 FINDING AND DISCUSSION

Based on the coded interview data, the participants' insights on curriculum design and practices revolved around three central themes including the views of culture, cultural-global knowledge, and global-local divide.

4.1 *Static and dynamic views of culture*

Throughout the interviews, culture was more often viewed as a static construct linked to the specific ways of living of a certain region and accumulated throughout a nation's history. Consequently, the perceived need to transition from *Cultural Studies* to *Global Studies* stemmed from the need to broaden the horizon of the curriculum to capture issues that are larger than the local culture itself. The following exacts illustrate this point.

> *For Global Studies, we can talk about broader content related to our life, and the topics or themes can be more contemporary, and the students may be more interested in searching for information and presenting information to their classmates. (Lecturer 3)*
> *The term Cultural Studies means we study the traditions and customs of people in certain areas, but mostly in their past or history. And if we talk about Global Studies ... we talk about the present, these days, the lifestyles of people these days. (Focus group 1)*

Apparently, local culture seems to have lost its relevance in a globalized world. However, instead of seeking an alternative way to discuss local cultures, the participants pointed towards a new discourse about the world's issues that are considered contemporary and more relevant. Ironically, intercultural negotiations are problematic without sufficient attempts of localization (Smith 2003). From a critical pedagogy lens, education here is no longer a liberating tool that helps neutralize power relations (Freire 1970). Instead, the curriculum reinforces dominant culture and suppresses local voices. At an instructional level, without carefully considering what to adopt and what to reject, the curriculum will perpetuate a sense of dependence on the learners (Crossley 2012; Green 2003). As one lecturer mentioned, "the students seem to feel that their identity is shaking when they discuss the content written by an American author, and they pay more attention to reciting the content rather than conversing with the ideas on equal terms" (Lecturer 2).

4.2 *Inculcating cultural-global knowledge*

The participants articulated three key challenges in imparting cultural-global knowledge: (a) language vs. culture, (b) globalization vs. localization, and (c) theory vs. praxis. The following extracts highlighted their perceptions about these points:

> *We want students to master both content and language. This is difficult ... some lecturers may overemphasize content at the expense of language and vice versa. You can't learn a language without content, and the content will definitely support your language study or learning. (Lecturer 2, on language and culture)*
>
> *... teachers who teach this subject should understand broadly about their own cultures as well as foreign cultures ... the book cannot provide many materials that are related to Khmer culture. (Policymaker 3, on globalization and localization)*
>
> *We should learn something that is contemporary and that still meets social needs. There is a lot of information related to society and culture, but when it's no longer applicable, I don't want to say it's a waste of time, but it's not practical or utilitarian. (Lecturer 3, on theory and praxis)*

The tensions in setting learning objectives, selecting texts, and designing activities were visible throughout the interviews. The lack of cultural relevance implies that cultural discussions do not surface much in the classroom. Also, the responses from policymakers, lecturers, and students were not coherent with one another. This kind of divergence is a sign of misalignment in the policy design and implementation process (Crossley 2012; Fenton-Smith & Gurney 2016; Hamid & Nguyen 2016). The voices would have been more coherent had there been policy documents or meetings to exchange ideas and feedback regarding these issues.

4.3 *Global-local divide*

Interestingly, concerning globalization beyond classroom contexts, the participants had a more uniform response towards the importance of local culture, and they acknowledged the challenges in adopting western academic culture. At this point, they framed *culture* as a dynamic construct: It travels across time and space; it is not confined to a specific region; it does not lose grasp of reality when taken out of context. In a brief reference to the third research question, the participants emphasized the importance of local culture, the difficulty of reiterating its values in a global context, and the issues of integrating local culture into the global knowledge system. Even so, from the management's and lecturers' perspectives, students paid more attention to the global issues than their cultural roots.

> *Students quickly think that ... other people in more advanced countries are much better in all aspects. Because when they feel like that, I can see that there's not much national pride ... their identity is shaking. They're not confident. (Policymaker 2)*
> *I cannot find the exact reason why. But students do not use their own culture to discuss issues, maybe because first of all they are afraid to bring anything outside the book. Even though they don't understand what happens in a given example, they still try to recite everything back. (Lecturer 2)*

The students in the focus-group interviews also emphasized such frustrations in positioning their cultural identities in the discussion of global issues and expressed their struggles in relating cultural capital to the world's issues. Regarding student identities, Policymaker 1 remarked, "the textbook is usually based on the U.S. society, and if we only learn something related to the United States, it's only information." In this situation, the lecturers and students felt disconnected from the texts and ideas discussed in the class, which minimized the opportunity for liberating dialogue (Freire 1970) and culturally responsive conversations (Barker 2002). To foster healthy identity development, curriculum needs to integrate multiple cultures and perspectives (Altbach 2005; Kellner 2003; Zajda & Val 2021) and empower agency which is an indispensable element in intercultural negotiation (see Apple, 2018; Fenton-Smith & Gurney, 2016; Hamid & Nguyen, 2016; Martin, 2019).

5 CONCLUSION AND RECOMMENDATION

This research study explores the perceptions of policymakers, instructors, and students in designing and implementing a higher education curriculum through a *Global Studies* subject. The semi-structured and focus-group interviews were employed to explore their underlying perceptions surrounding the importance of *Global Studies*, the presence or absence of culturally relevant pedagogy, and the construction of culture within and beyond the classroom context. The findings revealed that the transition from *Cultural Studies* to *Global Studies* subject derived from the static paradigm whereby local culture remains in its watertight compartment and does not evolve to meet the shifting global context. Different attempts of imparting cultural-global knowledge were evident in the acts of negotiating between language and culture, globalization and localization, and theory and praxis. Nevertheless, global and local identities were seen as separate, and culturally relevant pedagogy was problematized given the lack of quality instructional materials and effective methods. These findings bear significance to the overall field of ELT. Although *Global Studies* is just a content area subject with English as a medium of instruction, an effective design and application of such a culturally complex curriculum go far beyond mere considerations of language factors. Integrating worldviews and embracing culturally relevant pedagogy call for more critical discussions and careful selections of classroom activities. With higher education as a powerhouse of knowledge production and transfer, curriculum design and implementation should address the issues of ideology before the selection of various curricular constituents such as syllabus, textbook, language, and so on. It is also worth noting that this study was done within a tight timeframe of three months as required by the university program. A qualitative study demands a diverse source of data to compensate for its lack of generalizability. In response, the researcher utilized the emergent themes from the available data to invite more critical conversations in curriculum studies under the globalization framework, rather than aiming for a comprehensive description. Further research is needed to investigate and explore how to address the issues of ideology, cultural identity, and culturally relevant pedagogy in curriculum design and implementation.

REFERENCES

Altbach, P. G. (2005). The past and future of Asian universities: Twenty-first-century challenges. In P. G. Altbach & T. Umakoshi (Eds.), *Asian universities: Historical perspectives and contemporary challenges* (pp. 13–32). The Johns Hopkins University Press.

Apple, M. W. (2018). Critical curriculum studies and the concrete problems of curriculum policy and practice. *Journal of Curriculum Studies*, *50*(6), 685-690.

Barker, C. (2002). *Making Sense of Cultural Studies: Central problems and critical debates*. SAGE Publications.

Broadfoot, P. (2000). Comparative education for the 21st century: Retrospect and prospect. *Comparative Education*, *36*(3), 357–371.

Crossley, M. (2012). Comparative education and research capacity building: Reflections on international transfer and the significance of context. *Journal of International and Comparative Education*, *1*(1), 4–12.

Fenton-Smith, B., & Gurney, L. (2016). Actors and agency in academic language policy and planning. *Current Issues in Language Planning*, *17*(1), 72–87.

Freire, P. (1970). *Pedagogy of the oppressed* (M. B. Ramos, Trans.). Continuum International Publishing Group.

Green, A. (2003). Education, globalization, and the role of comparative research. *London Review of Education*, *1*(2), 83–97.

Hamid, M. O., & Nguyen, H. T. M. (2016). Globalization, English language policy, and teacher agency: Focus on Asia. *International Education Journal: Comparative Perspectives*, *15*(1), 26–43.

Johnson, R. B., & Christensen, L. (2014). *Educational research: Quantitative, qualitative, and mixed approaches* (5th ed.). SAGE Publication.

Kellner, D. (2003). *Media culture: Cultural studies, identity and politics between the modern and the post-modern*. Routledge.

Martin, P. W. (2009). Key aspects of teaching and learning in arts, humanities and social sciences. In H. Fry, S. Ketteridge, & S. Marshall (Eds.), *A handbook for teaching and learning in higher education: Enhancing academic practice* (3rd ed., pp. 300–322). Routledge.

Matus, C., & McCarthy, C. (2003). The triumph of multiplicity and the carnival of difference: Curriculum dilemmas in the age of postcolonialism and globalization. In W. F. Pinar (Ed.), *International Handbook of Curriculum Research* (pp. 73–82). Lawrence Erlbaum Associates.

Mertens, D. M. (2010). *Research and evaluation in education and psychology: Integrating diversity with quantitative, qualitative, and mixed methods* (3rd ed.). SAGE Publication.

Pacheco, J. A. (2012). Curriculum studies: What is the field today? *Journal of the American Association for the Advancement of Curriculum Studies*, *8*, 1–18.

Scharmer, C. O. (2009). *Theory U: Leading from the future as it emerges*. The Social Technology of Presencing.

Smith, D. G. (2003). Curriculum and teaching face globalization. In W. F. Pinar (Ed.), *International Handbook of Curriculum Research* (pp. 35–54). Lawrence Erlbaum Associates.

Vaira, M. (2004). Globalization and higher education organizational change: A framework for analysis. *Higher Education*, *48*(4), 483–510.

Waks, L. J. (2003). How globalization can cause fundamental curriculum change: An American perspective. *Journal of Educational Change*, *4*(4), 383–418.

Zajda, J., & Val, R. (2021). *Globalisation and comparative education: Changing paradigms* (Vol. 24). Springer Netherlands.

Teacher Education and Teacher Professional Development in the COVID-19 Turn – Drajati & Putra (Eds)
© 2023 The Author(s), ISBN: 978-1-032-38970-7

The global orientation in Indonesian government locally endorsed secondary students' English textbooks

Sri Imelwaty
Universitas PGRI Sumatera Barat, Indonesia

Edwar Kemal & Suharni
Universiti Malaysia Kelantan, Malaysia

Lili Perpisa
Universitas PGRI Sumatera Barat, Indonesia

Ching-Yi Tien
I-Shou University, Taiwan

ABSTRACT: This paper aims to assess ELT materials from the Global Englishes perspective in the Indonesian context. Three English textbooks for secondary school students that were locally endorsed by the Indonesian government were examined whether they draw students to use English as a global language or promote traditional and unrealistic depictions of English usage. Selected data from the textbooks were analyzed in terms of its users and target interlocutors using the framework for textbook analysis (Syrbe & Rose 2018). It was found that the textbooks lack dialogues examples of international users and target interlocutors. This could limit the students' opportunities to have exposure to the use of English in the real life-worlds as English users from different cultural and linguistic upbringings communicate using their own English. It is recommended for textbook designers to include some representations of communication in outer and expanding circle contexts of English usage.

Keywords: English language teaching; English as a global language; English textbooks; teaching materials

1 INTRODUCTION

The growth of the English language from national to global language has been documented in a huge number of studies. For example, a study by Hilgendroft (2007) found a significant role of the English language for domestic use in Germany and Breiteneder (2009) reports that English also has a pivotal function for international business and relations. In educational contexts, English is considered as the de facto working language in international conferences and academic journals (Berns et al. 2007; Montgomery 2013; Seidlhofer et al. 2006). Nowadays, it is estimated that more than 3 billion people are users of English around the world (Rose & Galloway 2019). These conditions indicate that English has been used not only for intra-national functions but also international ones as a global language.

As English becomes the global language, the goal of English language teaching (ELT) should be reevaluated (Joo, Chik & Djonov 2020). In the same Vein, Sybre and Rose (2018) suggest that the students' needs for learning English as a global language have changed. This means that English language learners should not only exclusively be taught to communicate with the people from the inner-circle countries and to attain native-like proficiency but also to interact with non-native

DOI 10.1201/9781003347798-6

speakers from outer and expanding circle countries (Galloway & Rose 2015; Kachru 1992). In this study, the terms inner, outer, and expanding circle are adopted referring to Kachru's model of Concentric Circles (Kachru 1985) for identifying three types of speech communities that use English. Those countries, where English becomes the first language, are referred to as the inner circle and those countries of the former colonies where English present as a second language are referred to as the outer circle. While in the countries of the expanding circle, English has no historical and intra-national functions, but it is taught and learned at schools. It is acknowledged that Kachru's model has limitations, but it remains useful as a portrayal of the spread of English (Syrbe & Rose 2018).

In the ELT context, textbooks remain as one of the most common sources of teaching and they prompt on the way language is taught and on the students' view of the language (Matsuda 2002). However, the empirical research on the global orientation of the ELT materials evaluation has been scarcely undertaken in the Indonesian context, where different local languages and cultures coexist. Thus, this study aims to examine whether ELT textbooks that are locally endorsed by the Indonesian government draw students to use English as a global language or promote traditional and unrealistic depictions of English usage. Specifically, this paper is addressing the following research questions:

(1) Who are positioned as the owners of English in locally endorsed secondary students' English textbooks?
(2) Who are the target interlocutors in those English textbooks?

2 LITERATURE REVIEW

It is prevalent that the phenomena of using English as a global language exist in educational and working contexts due to the globalization movements. For example, English has been a working language in the European Union which results in the use of English among Europeans in daily communication (Syrbe & Rose 2018). While in educational contexts, a huge number of academic journals are published in English (Montgomery 2013). The condition of multilingualism like in Europe, Africa, and Asia also occurs in many parts of the world which need respect for diversity and cross-cultural understanding among English language users from various linguistic and cultural backgrounds.

As English becomes a global language, it is understood that ELT's goal for achieving native speaker competence is not relevant anymore. Thus, English language learners in today's world need to become intelligible speakers of English who can use English in cross-cultural exchanges to serve specific ends (Canagarajah 2007; Hu & McKay 2014; Seidlhofer, 2004). To this end, intercultural competence is the urge to enhance English language learners' awareness for promoting cross-cultural understanding in the contexts of English as a global language use (Baker 2009; Holliday 2011). English language learners need models to interact with global contexts so that they could raise their linguistic and cultural diversity, and grow their abilities in multilingual situations (Canagarajah 2007; Joo et al. 2020). However, a native-speakerism ideology keeps dominating in English textbooks as found in their discourse and visuals which exclusively link to Anglo-American countries and the privilege of English of native speakers (Hu & McKay 2014; Matsuda 2002). The discrepancy between the current ELT goals and the available materials may hinder English language learners to become English users in global contexts.

Instead of relying on the native speaker model, English language learners need to understand the English of speech communities that have certain patterns alongside other languages (Sharifian 2011). A greater number of English users in the world are not first-language English speakers (Graddol 1997). Thus, English textbooks could play a significant role in enhancing English language learners' sensitivity to translingual practices and their needs for specific goals in learning English (Hu & McKay 2014). This means that within the textbooks, various models of interactions between English users from different linguistic and cultural upbringings should be presented to develop the intercultural communication awareness of the English language learners.

A number of studies of ELT textbooks show a native speaker bias which is quite pervasive (Sardi 2002; Shin et al. 2011; Yuen 2011). Other studies show that American and British English which are the models of English from the inner-circle countries are still dominantly used in ELT materials for the educational contexts particularly in the expanding circle countries (Hu & McKay 2014; Joo et.al. 2020; Lee 2009; Song 2013; Syrbe & Rose 2018; Yamada 2010). Given the previous studies of English textbooks notably in expanding circle countries, to the best of our knowledge, the global orientation for reaching the current ELT goals has not been examined in the Indonesian context. Therefore, the ELT textbooks for secondary school students which are locally endorsed by the Indonesian government were evaluated in terms of their global orientation.

3 METHOD

This study examined the global orientation of three series of English textbooks that the Indonesian government locally endorsed in terms of the extent of the ownership of English and the target interlocutors in the textbook contents. These textbooks are extensively used within the curriculum of English language for secondary students' class 10, 11, and 12 in Indonesia. They were published in 2017 by the Indonesian ministry of education and have been distributed in hard and soft copies to schools and students. For the purpose of this study, only selected excerpts depicting dialogues among English language users were taken as data sources.

As a research design, this present study used content analysis, which is quantitative non-experimental research (Cohen et al. 2017) for examining the global orientation within the textbooks in terms of the users and target interlocutors. A framework, which was developed by Syrbe and Rose (2018) based on Global Englishes (GE) perspective (Galloway & Rose 2015), was applied for analyzing data. The framework contains curriculum components, namely target interlocutors, models of English, target cultures, and ownership of English, which are emblematic of a traditional EFL approach or a Global Englishes one (Syrbe & Rose 2018). This paper only reports on the two components, specifically the ownership of English and the target interlocutors (see table 1). The textbooks were then assessed for placing them on EFL-oriented ideology at one end, and ELF/Global Englishes-oriented ideology at the other end. All the written dialogues between two or more English users in the three textbooks were analyzed in reference to the framework. For showcasing the users of English in the real world, the frequency of their occurrences in the authentic text excerpts were counted. Regarding the represented target interlocutors in the textbooks, certain tasks were also selected across all of the textbooks 'series. Only the tasks that instruct students to think of using English in real-world scenarios were collected for data analysis. Thereafter, the target interlocutors in these communicative situations were determined based on the aforementioned framework.

Table 1. Framework for analyzing English textbooks based on GE perspective (Galloway & Rose 2015; Syrbe & Rose 2018).

	EFL-oriented ideology	Global Englishes-oriented ideology
Who are positioned as the owners of English?	Native English speakers, with an emphasis on inner circle English usage	Native and non-native English speakers, with emphasis on global English usage
Who are the target interlocutors in the materials?	Native English Speaker	Native and non-native English Speakers

4 FINDING AND DISCUSSION

4.1 *The ownership of English*

In all three textbooks, the majority of characters' nationality was Indonesians who were identified based on their names and the given explanation of the excerpts. However, there were also a number of characters who were identified as non-Indonesians based on their names and the contexts in which the communication happened. The native English-speaking characters were portrayed from the United States of America and Great Britain. The analysis results are presented in Table 2.

Table 2. Number of Excerpts depicting the English users.

English users	Indonesian and Indonesian	Indonesian and native English-speaking characters	Native English-speaking characters and native English-speaking characters	Non- native English-speaking characters from others outer/expanding circle countries	Total
Book 1	47	12	–	–	
Book 2	13	6	9	–	
Book 3	14	5	3	–	
Total excerpts	74	23	12	–	109

The three textbooks contain 109 dialogues excerpts that illustrate real-world language use between speakers of English. More than half of these (74) involved Indonesian as non-native English speaking characters that are interacting with their Indonesian counterparts. While 23 excerpts depicted communication between Indonesian with native English-speaking characters from the inner circle country. The rest (12) excerpts portrayed communication between native English-speaking characters from the inner-circle countries. There were no instances of English use between non- native English-speaking characters from others outer or expanding circle countries in any of the real-world dialogues presented in the textbooks. In addition, the depicted communications between Indonesians in English were still in the native speakers' model from the inner circle countries.

The analysis results confirm what Matsuda and Matsuda (2011, pp.173-174) pinpoint that 'an analysis of textbooks can provide only a partial picture of the classroom reality and individual teachers may make their own 'efforts to supplement what the textbooks do not provide'. This means that the sociolinguistic reality has not been represented well in those textbooks so teachers may need other sources to show the multilingual and multicultural reality of English users around the world. Thus, the findings of this study also indicate the native speaker bias in ELT textbooks (Sardi 2002; Shin et al. 2011; Yuen 2011).

These textbooks indicate for having EFL-oriented ideology that adopting native English speakers' norms. Consequently, English as the language of the inner circle countries, with some reference to expanding circle speakers (Indonesian) are presented. This has a latent meaning that English is owned by native speakers of English from the inner circle countries. Hence, the usage of English as a global language has not been reflected in the textbooks excerpts and neither the global ownership of English language.

4.2 *The target interlocutors*

In terms of the target interlocutors (see Table 3), 46 excerpts were analyzed across all three textbooks that assign students into the scenario of English usage. Almost all of them (n= 38) depicted Indonesian students' communications with other Indonesian such as their friends, and family members. In these scenarios, students were illustrated to use English for a number of reasons such as talking to their friends, obtaining information, traveling, and other spoken conversations. Only several tasks were designed to interact with native English speakers within a tourism context.

Table 3. Number of excerpts depicting target interlocutors.

Target Interlocutors	Indonesian interlocutors	Non- Indonesian (inner circle) interlocutors	Non-Indonesian (outer/expanding circle) interlocutors	Total
Book 1	22	9	–	
Book 2	10	4	–	
Book 3	6	5	–	
Total Excerpts	38	18	–	46

It was evident that these textbooks had a lack of international target interlocutors that depicted Indonesians as non-native English speakers encountering with other non-native English speakers from the outer and expanding circle countries. The textbooks represent a biased condition that Indonesians tend to use English with other Indonesian who are non-native English speakers. However, this is not the case as in the real world Indonesians do not communicate in English on a daily basis.

The current goal of ELT in the twenty century is to prepare the English language learners to be able to communicate with their counterparts in inner, outer, and expanding circles countries. Thus, secondary Students' English textbooks which were locally endorsed by the Indonesian government should represent the communication between global community English users because as students, they need to develop the ability to communicate with other English users from various linguistic backgrounds (Baker 2011; McKay 2018). A study by Hilgendorf (2007), which was undertaken in Germany, confirmed the significant role of English in several domains specifically politics and law, business and advertising, education and other domains with varying degrees. In the context of science, scientists particularly use English as their existing working language (Berns et al. 2007; Seidlhofer et al. 2006). Those studies suggest that English language has gained a crucial role in international communication in various occupational areas which should be depicted as examples in ELT textbooks.

Overall, these English textbooks have not fully fallen in Global Englishes-oriented ideology since the materials reflected sparse attention to the use of English between Indonesians with the English users from outer and expanding circle countries or the global use of English. They are still advocating traditional and atypical depictions of English usage. Ironically, these textbooks are produced for millions of Indonesian secondary students to learn English. These textbooks-driven teaching may limit their students' opportunities to have exposures to English usage in the real life-worlds. In reality, English users from different linguistic and cultural backgrounds communicate using their own English. Meanwhile, a number of English teachers are eager to increase English as a global language exposure to their students, yet the scarcity of English as a global language teaching materials hinder them to create changes in their classrooms (Syrbe & Rose 2018).

5 CONCLUSION AND RECOMMENDATION

As the aim of this study is to examine the global orientation in Indonesian locally endorse secondary students' English textbooks, it was found that the recent study findings show a lack of models in the textbooks in terms of the users of English and the target interlocutors from the outer and expanding circle countries. In addition, the textbooks have not made a clear effort to place English as a global language because of this lack of outer and expanding circle contexts of English usage coverage. While change might take time, it is suggested to English teachers in the Indonesian contexts to include some representations of communication in English referring to the outer and expanding circle contexts. In this respect, it requires a call for innovations on ELT materials to pave the ways for students toward the linguistic and cultural diversity of English. Textbooks may influence the way the English language is taught and what is being taught in the classrooms. The findings of this study may inform writers of English language textbooks to design and develop ELT materials to promote English as a global language. It is expected that they provide instructions for English language learners to communicate with other English users from various linguistic and cultural backgrounds. It is also recommended that ELT materials should be aligned to develop students' abilities in using English for future usage by adding various varieties of English in the textbooks. For example, it could provide dialogues examples and accents particularly in the listening section materials which have not been explored in this study. Further studies on other elements of global orientation in the English textbooks for secondary students in Indonesia such as models of English and target cultures need to be explored and evaluated.

ACKNOWLEDGMENT

The authors received financial support for the research and/or authorship of this article from the Directorate General of Higher Education (DIKTI), Ministry of Education, Culture, Research, and Technology Indonesia in 2021.

REFERENCES

Baker, W. (2011). From cultural awareness to intercultural awareness: Culture in ELT. *ELT Journal,* 66(1), 62–70.

Berns, M., Claes, M.T., de Bot, K., Evers, R., Hasebrink, U., Huibregetse, I., Truchot, C., & de Wijst. P. (2007). *"English in Europe." In In the Presence of English: Media and European Youth*, edited by M. Berns, K. de Bot, and Uwe Hasebrink, 15–43. Springer.

Breiteneder, A. (2009). "*English as a Lingua Franca in Europe: An Empirical Perspective*." World Englishes 28 (2), 256–269.

Canagarajah, S. (2007). Lingua franca English, multilingual communities, and language acquisition. *The Modern Language Journal*, 91, 923–939.

Cohen, L., L. Manion, and K. Morrison. (2017). *Research Methods in Education*. (7th ed). 31–30. Routledge Falmer.

Galloway, N., & Rose, H. (2015). *Introducing Global Englishes*. Abingdon: Routledge.

Graddol, D. (1997). *The Future of English*. British Council.

Hilgendorf, S. K. 2007. English in Germany: Contact, Spread and Attitudes. *World Englishes*. 26 (2), 131–148.

Hu, G., & McKay, S.L. (2014). Multilingualism as portrayed in a Chinese English textbook. In J. Conteh & G. Meier (Eds.), *The multilingual turn in language education: Opportunities and challenges* (pp. 64–88). Multilingual Matters.

Holliday, A. (2011). *Intercultural communication and ideology*. Sage.

Joo, S. J., Chik, A., & Djonov, E. (2020). The construal of English as a global language in Korean EFL textbooks for primary school children. *Asian Englishes*, *22*(1), 68–84.

Kachru, B. B. (1985). "Standards, Codification and Sociolinguistic Realism: The English Language in the Outer Circle." *In English in the World: Teaching and Learning the Language and Literatures*, edited by R. Quirk and H. Widdowson, 11–30. Cambridge University Press.

Kachru, B. B. (1992). World Englishes: Approaches, issues and resources. *Language Teaching*, 25 (1), 1–14.

Lee, I. (2009). Situated globalization and racism: An analysis of Korean high school EFL textbooks. *Language and Literacy*, 11(1), 1–1.

Matsuda, A. & Matsuda, P.K. (2011) Globalizing writing studies: The case of US technical communication textbooks. *Written Communication* 28, 172–192.

McKay, S. L. (2018). English as an international language: What it is and what it means for pedagogy. *RELC Journal*, 49(1), 9–23.

Montgomery, S. L. (2013). *Does Science Need A Global Language? English and the Future of Research*. University of Chicago Press.

Rose, H., Galloway, N. (2019). *Global Englishes for Language Teaching*. Cambridge University Press.
Syrbe, M., Rose, H. (2018). An evaluation of the global orientation of English textbooks in Germany, *Innovation in Language Learning and Teaching*, 1(2), 152–163.
Song, H. (2013). Deconstruction of cultural dominance in Korean EFL textbooks. *Intercultural Education*, 24(4), 382–390.
Seidlhofer, B., Breiteneder, A., & Pitzl, M. (2006). English as A Lingua Franca in Europe: Challenges for Applied Linguistics. *Annual Review of Applied Linguistics*, 26, 3–34.
Sharifian, F. (2011). *Cultural Conceptualization and Culture*. Amsterdam: John Benjamins.
Sardi, C. (2002). On the relationship between culture and El.T. *Kalbu Stujitos* 3, 101–107.
Shin, J., Eslami, Z.R. and Chen, W.-C. (2011). Presentation of local and international culture in current *international* English-language teaching textbooks. *Language, Culture and Curriculum*, 24(3), 253–268.
Yamada, M. (2010). English as a multicultural language: Implications from a study of Japan's junior high schools' English language textbooks. *Journal of Multilingual and Multicultural Development*, 31(5), 491–506.
Yuen, K. M. (2011). The representation of foreign cultures in English textbooks. *ELT journal*, 65(4), 458–466.

Teacher Education and Teacher Professional Development in the COVID-19 Turn – Drajati & Putra (Eds)
© 2023 The Author(s), ISBN: 978-1-032-38970-7

Technology-based English teaching: A voice from an Indonesian English teacher during COVID-19

Ma'rifatul Ilmi & Nur Arifah Drajati
Universitas Sebelas Maret, Surakarta, Indonesia

Kristian Adi Putra
Prince Sattam Bin Abdulaziz University, Al-Kharj, Saudi Arabia

Maria Hidayati
Nanyang Technological University, Singapore

ABSTRACT: This narrative inquiry aims to listen to an Indonesian English teacher of technology-based English teaching during the coronavirus (COVID-19) pandemic. This study employed the teacher's teaching diary to capture her experiences concerning teaching English with technology. As a self-guidance, the Indonesian teacher used the TPACK framework as the knowledge base to integrate the technology, pedagogy, and content knowledge. Interpretative phenomenological analysis (IPA) was adopted to analyze the data. We revealed teaching with technology as well as getting pressure from the social condition (i.e., virus spread) is not easy. It influences the teacher and students' psychological matters. It implies that there should be more training related to teaching with technology so that teachers will get more ready to execute tech-enhanced teaching as well as manage the online classroom environment so that the students can experience meaningful learning.

Keywords: COVID-19, EFL teaching, online teaching, narrative study, TPACK

1 INTRODUCTION

Online language teaching has been proven to be recommended (Moser et al. 2021). The activity of online language teaching allows teachers to build and support students through online tasks (Meskill & Anthony 2015), facilitate students' collaboration with peers and the teachers (Moore 1989), and integrate asynchronous and synchronous learning experiences – both written and oral (Meskill & Anthony 2015). The two-way interaction (oral and written) enables teachers and students to communicate at a distance (Moser, Wei & Brenner 2021). Furthermore, either oral or written interaction in a smaller group discussion may represent the experience of face-to-face interaction. Researchers have been suggesting that learning synchronously is the exact nature of language learning (Blake 2011). There are positive advantages of synchronous language learning such as reducing the sense of isolation (Gonzalez-Lloret 2020) and motivation booster (Gunes 2019). However, asynchronous learning tends to be more effective to stimulate and improve students' willingness to communicate since it allows them to think more, and produce more language without having serious anxiety about performing (Meskill & Anthony 2015).

Technology-based teaching and learning occur in all educational subject fields particularly language education, although the integration of technological aids in a language classroom is not exactly new worldwide (Blake 2011; White 2006). However, in the Indonesian context, integrating technology or digital aids into a classroom remains challenging for teachers (Drajati et al. 2018; Habibi et al. 2020; Taopan et al. 2020). A fresh finding from the study of Moser, Wei & Brenner (2021) exploring the instructional changes that emerged during COVID-19 to teachers, language, and online learning revealed that remote teachers adjusted their expectations on the learning due to

 DOI 10.1201/9781003347798-7

COVID-19. Digital experiences are always meaningful and full of insightful knowledge (Costley 2014) as well as studying individual life experiences (Clandinin & Rosiek 2007) – turning the narratives into knowledge for the individuals themselves and others. Nevertheless, teachers' lived experiences concerning their enactment of technology-based online learning are not yet examined.

To address the aforementioned gap, this present study attempts to explore an English teacher's lived experience during the online English teaching enactment in Indonesia at the age of COVID-19. This study concerned revealing the process of the teacher's teaching practice during the pandemic as well as the challenges found in the online teaching practice. The guiding research question to the study was: what does the preservice English teacher's voice in enacting technology-integrated English teaching during COVID-19 pandemic?

2 LITERATURE REVIEW

2.1 *TPACK framework in English teaching*

The massive development of information and communication technology (ICT) tools has affected instructional design worldwide. Teachers keep on learning and integrating technology to aid their instructions and obtain much more creative material resources (Tseng 2019). Considering the importance of technology as one educational aid, Mishra and Koehler (2006; 2009) extended a theoretical framework proposed by Shulman's (1986) concept of pedagogical content knowledge (PCK). Deciding to integrate technology into teaching practice, teachers should first understand how to fit the technological tools into the PCK. Technological knowledge is indeed one significant matter concerning technology-based teaching (Tseng 2019).

Scholars have been extensively investigating the TPACK enactment in various concerns such as instructional design, lesson plan, preparing the 21-century learning (Koh 2015, 2016, 2020), teachers' identity construction (Cheng & Xie 2018), framework conceptualization, and environmental enhancement (Angeli & Valinides 2009; Mishra & Koehler 2006, 2009) and various modes of English language teaching (in-person and distance class) (Tseng et al. 2016; Tseng; Cheng & Yeh 2019). The quality of online teaching is based on the teachers' presence including cognitive, teaching itself, and social (Garrison et al. 1999). In line with that, ICT tools utilized in synchronous learning are proven to be effective to establish a social presence (Bradner & Mark 2001). Among the previous studies, however, little has given much attention to an insider experience of enacting TPACK in an online learning context, particularly when people are being locked down and insisted to learn from home.

In the context of English teaching, numerous studies have focused on the adoption of the TPACK framework in the classroom. For example, Tseng, Cheng, and Yeh (2019) examined preservice English teachers' TPACK enactment in a web-conferencing context through the lens of the design thinking approach. Mixed-method was adopted in the study to measure the preservice teachers' TPACK. The results showed that two factors mainly influence the TPACK enactment in the web-conferencing context such as technical problems (i.e., background noise) and students' engagement. Later, Taopan, Drajati, and Sumardi (2020) investigated an in-service English teacher's TPACK adoption in face-to-face teaching and learning context in Indonesia. Narrative inquiry was employed to explore the teacher's experience. The study revealed that the teacher encountered common challenges of teaching with technology (e.g., problems with technical operation, internet lagging). However, teaching with the TPACK framework allows the teacher to be able to create meaningful learning.

Despite the numerous research in the field of TPACK enactment, little research has given attention to the use of autobiographical narrative study to capture the teachers' TPACK in practice in an online classroom setting. Addressing the issue, the present study attempts to explore one preservice English teacher's enactment of TPACK in an online classroom setting. Methodologically speaking, this study contributes to the growing interest in methodological consideration to explore teachers' TPACK competence in practice.

3 METHOD

This study was a kind of autobiographical study and a part of a larger master's degree thesis research project, in which the main participant was one of the authors of this study. She is a preservice English teacher who is now taking a postgraduate program majoring in English Education at one of the state universities in Indonesia. She started her teaching career in 2017 and has been teaching online courses lately starting from May 2020 to the present in a private language institution. She teaches online short courses in the TOEFL preparation class, particularly the structure and written expression section. The short course has eight meetings and is fully conducted online using several media platforms, such as WhatsApp, google classroom, Edmodo, and google meet. The researchers gathered the data in 2020 when COVID-19 pandemic had highly risen.

This study gathered the teacher's reflective journal as the primary data collection technique (Barkhuizen et al. 2014). The reflective journal mainly consisted of how she handled the class; how she facilitated students with unpredictable issues; how she utilized the technological support; how she explained the material and how she created an interesting online classroom environment to attract the students' attention so that they could be actively engaged during the learning process. Online classroom observation was the other technique of data collection to see if any critical incidents occurred during the class. To analyze the data, the Interpretative Phenomenological Analysis (IPA) approach was adopted to see the way an individual understands the lived experiences (Smith & Osborn 2008). In the process of analyzing data, we first sorted the data and did an initial noting to see the themes. Then, we searched for the general pattern(s) among the themes. In the last stage of data analysis, we made a discussion and conclusion (Smith et al. 2013). Finally, triangulation was also applied in this study to validate the results (Wilkinson & Birmingham 2003).

4 FINDING AND DISCUSSION

This study explored a preservice English teacher's voice in executing remote EFL teaching in Indonesia. To see the teacher's perception of technology-integrated English teaching, this study was guided by the TPACK framework (Mishra & Koehler 2006).

We found that the preservice English teacher confirmed her lack of technological knowledge and realized that deciding to teach online was one big decision. Online learning, in Indonesia, has been carried out lately due to the emergence of the pandemic. Though this condition was not the proper time to start a trial on online learning, remote learning was one possible solution to keep on educating while in-person education was impossible to conduct. As the preservice teacher stated "the first time I thought I used WhatsApp as the learning management system and Kahoot! application as one supporting aid to trigger my students' engagement. I had very limited knowledge of technology and was not ready to design the online English instruction at once". She then continued "it happened suddenly, and I was not ready".

Enacting the TPACK framework in classroom teaching facilitates teachers to construct the subject matter fitting into the teaching strategies and technological aids to create a more meaningful learning experience (Misirli 2016). The result of this current study showed that teaching the English language in a remote classroom environment, particularly in the age of COVID-19 was extremely challenging. Challenges come from many aspects such as the contextual factors of the online classroom environment (Tseng et al. 2019), psychological matters (MacIntyre et al. 2020; Moser et al. 2020), and technological competence (Taopan et al. 2020). Lack of technological knowledge including technophobia, technical problems, timing, instructional design, and poor engagement are the challenges that commonly emerge in the practice of technology-based teaching (Lye 2013; Ramorola 2014; Taopan et al. 2020). Therefore, training on technology integration into teaching is needed to support teachers' readiness and enhance their knowledge of technology-integrated teaching.

Furthermore, we also revealed conducting online teaching as well as being a student-teacher in COVID-19 pandemic era is a bit frustrating. The preservice teacher used to get support from the teacher educators that helped her develop her knowledge of technological, pedagogical, and content knowledge (TPACK). Such support allows the preservice teacher to take action in practicing TPACK. As she said "being a student teacher and in charge of a preservice English teacher is an extremely wonderful experience. Continuous guidance from my lecturer is another kind of motivation for me to start teaching online". Teaching online in the age of COVID-19 pandemic moreover sacrificed much energy since we need to also keep ourselves healthy. We found that feeling bored, and tired of thinking are some of the causes of the decrease in a teacher's motivation to teach. As she confirmed "motivating myself and my students all together is not easy. At the age of Covid-19, I need to think of several considerations such as my health, and my students' health. I also need to understand the situation, keep them safe, and feel comfortable during the remote learning".

This present study informed additional challenges regarding the online teaching experience during COVID-19 pandemic. Psychological matters were found to receive high confirmation of online teaching challenges. Though the common perception of "teaching is a stressful profession" (MacIntyre et al. 2020, p. 1) has been well-proven before the pandemic, the emergency of online learning in such a situation (i.e., COVID-19, lockdown) seems to give more burden to teachers (Mercer & Gregersen 2020). The results of the current study are in line with the previous claims (MacIntyre et al. 2020; Mercer & Gregersen 2020). MacIntyre, Gregersen, and Mercer (2020) maintained that psychological health and wellbeing are two main coping strategies for adapting to the "new normal" (e.g., work from home, a study from home) – increasing self-preparedness and self-motivation.

Overall, we maintain in this study that the social aspect does affect teachers' teaching performance. The adoption of TPACK in remote teaching contexts may be one of the alternatives to successfully executing technology-integrated teaching, particularly in the context of online teaching. However, teachers' knowledge itself is not sufficient to successfully implement TPACK in practice. Thus, there should be continuous training for teachers about how technology works in the classroom.

5 CONCLUSION AND RECOMMENDATION

This study offers new knowledge and consideration to execute technology-integrated teaching in the context of remote teaching through the lens of the TPACK framework. Though generalizing the study of technology-based English teaching cannot be completed with only one single case, the result of this study contributes evidence of the framework enactment in more complex teaching circumstances (i.e., distance learning). However, it does not mean that this study is without limitations. This study relies on self-experience starting from designing the online instruction until the execution of the program. Narratives from more teachers would provide a better description of the TPACK in practice. Of all that, examining whether or not the individual's experience of teaching with technology can shape teachers' identity and self-efficacy in any classroom environment is needed for future research.

REFERENCES

Angeli, C., & Valanides, N. (2013). Technology mapping: An approach for developing technological pedagogical content knowledge. *Journal of Educational Computing Research*, 48(2), 199–221.

Blake, R. (2011). Current trends in online language learning. *Annual Review of Applied Linguistics*, 31, 19–35.

Bradner, E., & Mark, G. (2001, September). Social presence with video and application sharing. In *Proceedings of the 2001 International ACM SIGGROUP Conference on Supporting Group Work* (pp. 154–161).

Cheng, S. L., & Xie, K. (2018). The relations among teacher value beliefs, personal characteristics, and TPACK in intervention and non-intervention settings. T*eaching and Teacher Education*, 74, 98–113.

Costley, K. C. (2014). The Positive Effects of Technology on Teaching and Student Learning. *Telematics and Informatics*, 19(1), 27–40.

Drajati, N. A., Tan, L., Haryati, S., Rochsantiningsih, D., & Zainnuri, H. (2018). Investigating English language teachers in developing TPACK and multimodal literacy. *Indonesian Journal of Applied Linguistics*, 575–582.
Garrison, D. R., Anderson, T., & Archer, W. (1999). Critical inquiry in a text-based environment: Computer conferencing in higher education. *The internet and higher education*, 2(2–3), 87–105.
Gonzalez-Lloret, M. (2020). Collaborative tasks for online language teaching. *Foreign Language Annals*, 53, 260–269.
Gunes, S. (2019). What are the perceptions of the students about synchronous distance learning and blended learning? *World Journal on Educational Technology: Current Issues*, 11(4), 230–237.
Habibi, A., Yusop, F. D., & Razak, R. A. (2020). The role of TPACK in affecting pre-service language teachers' ICT integration during teaching practices: Indonesian context. *Education and Information Technologies*, 25(3), 1929–1949.
Koehler, M., & Mishra, P. (2009). What is technological pedagogical content knowledge (TPACK)? *Contemporary issues in technology and teacher education*, 9(1), 60–70. Waynesville, NC USA: Society for Information Technology & Teacher Education.
Koh, J. H. L. (2020). Three approaches for supporting faculty technological pedagogical content knowledge (TPACK) creation through instructional consultation. *British Journal of Educational Technology*, 51(6), 2529–2543.
Koh, J. H. L., & Chai, C. S. (2016). Seven design frames that teachers use when considering technological pedagogical content knowledge (TPACK). *Computers & Education*, 102, 244–257.
Koh, J. H. L., Chai, C. S., Benjamin, W., & Hong, H. Y. (2015). Technological pedagogical content knowledge (TPACK) and design thinking: A framework to support ICT lesson design for 21st-century learning. *The Asia-Pacific Education Researcher,* 24(3), 535–543.
Lye, L. T. (2013). Opportunities and challenges faced by private higher education institutions using the TPACK model in Malaysia. *Procedia-Social and Behavioral Sciences*, 91, 294–305.
MacIntyre, P. D., Gregersen, T., & Mercer, S. (2020). Language teachers' coping strategies during the Covid-19 conversion to online teaching: Correlations with stress, wellbeing, and negative emotions. *System*, 94, 102352.
Mercer, S., & Gregersen, T. (2020). *Teacher Wellbeing*. Oxford University Press.
Meskill, C., & Anthony, N. (2015). *Teaching a language online*. Buffalo, NY: Multilingual Matters.
Mishra, P., & Koehler, M. J. (2006). Technological pedagogical content knowledge: A framework for teacher knowledge. *Teachers College Record*, 108(6), 1017–1054.
Mısırlı, Z. (2016). Integrating Technology into Teaching and Learning Using Variety of Models. *Ihlara Eğitim Araştırmaları Dergisi*, 1(2), 37–48.
Moore, M. (1989). Three types of interaction. *American Journal of Distance Education*, 3(2), 1–6.
Moser, K. M., Wei, T., & Brenner, D. (2020). Remote Teaching During COVID-19: Implications from a National Survey of Language Educators. *System*, 102431.
Ramorola, M. Z. (2014). Challenge of effective technology integration into teaching and learning. *Africa Education Review*, 10(4), 654–670.
Shulman, L. (1986). Those Who Understand: Knowledge Growth in Teaching, 15(2), 4–14.
Smith, J. A., & Osborn, M. (2008). *Interpretative phenomenological analysis.* In J. A. Smith (Ed.), Qualitative Psychology: A Practical Guide to Research Method (2nd ed., pp.53-80). Sage Publication.
Smith, J. A., Flowers, P., & Larkin, M. (2013). *Interpretative Phenomenological Analysis: Theory, Method and Research.* Sage.
Taopan, L. L., & Drajati, N. A. (2020). TPACK Framework: Challenges and Opportunities in EFL Classrooms. *Research and Innovation in Language Learning*, 3(1), 1–22.
Tseng, J. J., Lien, Y. J., & Chen, H. J. (2016). Using a teacher support group to develop teacher knowledge of Mandarin teaching via web conferencing technology. *Computer Assisted Language Learning*, 29(1), 127–147.
White, C. (2006). Distance learning of foreign languages. *Language Teaching*, 39, 247–264.
Wilkinson, D., & Birmingham, P. (2003). *Using research instruments: A guide for researchers*. Psychology Press.

Teacher Education and Teacher Professional Development in the COVID-19 Turn – Drajati & Putra (Eds)

Educational mobile interactive learning activities on English language teaching during the COVID-19 pandemic: Systematic literature review

Hasan Zainnuri, Ngadiso Ngadiso, Teguh Sarosa, Kristiandi Kristiandi, Muhammad Asrori, Endang Setyaningsih & Elys Putri Karismawati
Universitas Sebelas Maret, Surakarta, Indonesia

ABSTRACT: The COVID-19 outbreak now has a wide-ranging consequence on teaching and learning. Institutions and teacher educators had to react quickly to forced-shift from face-to-face to online teaching. The purpose of this study was to conduct a systematic review of the benefits of Educational Mobile Interactive Learning Activities (EMILA) in teaching EFL/ESL during COVID-19 pandemic. Prior to conducting this systematic review, 16 studies were chosen from journals indexed in the Wiley Online Library, Taylor and Francis Online, Science Direct, and SAGE Journals databases. The results show an increase in publications on the use of various media and strategies in teaching EFL/ESL online. Some benefits of using EMILA are it encourages student progress, contributes pedagogical aspects for students and teachers, and provides interaction opportunities so that students' writing and speaking skills are improved. The implications of this study for the development of online teaching and learning practices are discussed.

Keywords: COVID-19, Educational, ELT, Interactive Learning, Mobile

1 INTRODUCTION

The COVID-19 virus swiftly resulted in government instructions to follow remote learning in order to reduce the infection curve, leading to the shutdown of higher education institutions worldwide. E-learning has increased unprecedentedly in the middle of the global wave of pandemic lockdowns. Teachers and students confront a struggle to quickly adapt and attempt to find the appropriate technologies and approaches to have engaging classes even though there are several challenges and limitations. The disruption of face-to-face instruction in schools has created concerns about the outcomes for students' learning. As a result, the requirement to create education programs for student teachers undergoing teacher education preparation necessitated decisions, choices, and adaptations in order to meet not only students' needs, but also teacher education demands and the conditions where both universities and schools had to operate. To support this circumstance, academic institutions should have to be very careful to prepare teachers to provide a seamless educational process that is equipped with digital technology (Madianou 2020; Nguyen et al. 2020; Rashid & Yadav 2020). Moreover, such circumstances can accelerate schools' design throughout the 20th century in incorporating technology into classes.

Reliance on e-learning and its adaptation during a pandemic can change the way teachers take advantage of more online features (integrating interactive activities with technology). The combination of online delivery and interactive learning activities that require student-student interaction acquire them a range of skills and competencies (Bruyn 2004). In addition, the provision of interactive learning activities allows students to connect with other students on an asynchronous basis; brainstorm, elaborate and clarify the issues in relation to the learning activity, and be involved with peers' interconnection. Flipped learning has the potential to improve language instruction by providing a more richer, creative, flexible, and learner-driven setting (Arslan 2020). Researchers from 2016 to 2017 had examined various research on the advantages and disadvantages of flipped learning in EFL/ESL teaching from 2016 to 2017 (Arslan 2020) and from 2016 to 2020 (Ngo & Md Yunus 2021). Moreover, Flipped learning had a clear and significant impact on the majority of language outcomes as well (Vitta & Al-Hoorie 2020). However, considerable confusion still persists between students and educators about how well educational mobile interactive learning activities (EMILA) and online delivery can benefit them in ELT classes during the COVID-19 pandemic.

It is intended that this study will provide new insights for policymakers, curriculum designers, teachers, and researchers who are going to teach English with Educational Mobile Interactive Learning Activities in the near future. As a result, the following goal has been addressed throughout the process of the investigation: What are the benefits of Educational Mobile Interactive Learning Activities (EMILA) in the field of teaching EFL/ESL presented in the reviewed studies?

2 LITERATURE REVIEW

Technology has greatly influenced many aspects of the education sector, especially in higher education. In particular, college students who use modern technological tools under the guidance of an instructor, can learn more complex English learning materials than in previous years without such technology. Therefore, students will be more easily involved in problem-based and inquiry-based learning situations. In such circumstances, they are required to collect or analyze data, generate models, and execute complex concepts to solve problems. Recent technological advances facilitate this type of complex learning with mobile e-learning learning that integrates interactive learning activities with learning activities in the emergency remote teaching situation.

Following a previous examination of educational research on e-learning, the researcher was able to observe the changing themes of this research within the context of educational technology. Maurer and Khan (2010) did a SLR on a sample of five scientific journals and two conferences from 2003 to 2008 in order to find trends in e-learning research. They performed scientometric and content analysis using the gCLUTO software (a graphical application for clustering data sets). Learning technology, learning process, learning perspective, learning techniques, delivery systems, learning development, production variables, students, learning environment, evaluation, culture, and teachers are the categories identified. Meanwhile, Zare et al. (2016) employed five scientific published databases to identify papers on e-learning and multi-criteria decision making, which is an efficient method for analyzing several factors. As a result, the following e-learning evaluation items were identified: functionality, response time, engagement, web and course development, availability, reliability, affordability, security, consistency, trust, accuracy, flexibility, interoperability, and continuity.

Overall, the research mentioned above provides important insights into the context of educational technology. In SLR regarding e-learning, the categories of e-learning evaluation described in the preceding paragraph are advised to be examined. The approaches to e-learning evaluation adopted for this study were student progress, pedagogical contributions, and interaction (taken from Arslan (2020). This study provides an exciting opportunity to advance our knowledge of the e-learning evaluation in the context of blended-learning during the COVID-19 pandemic.

3 METHOD

To address the research questions underlying this study, a systematic review technique is adopted. Systematic reviews are an empirical strategy for identifying, selecting, and summarizing a summary of research while minimizing bias (Moher et al. 2015). By definition, a systematic review is a sort of literature review that seeks to discover all available evidence on a topic in order to minimise the impact of bias on the review conclusions. In this research, PRISMA principles (Liberati 2009) and the PRISMA extension PRISMA-P (Moher et al. 2015) were employed as a protocol-driven system to describe the preceding road map (Liberati 2009).

The selection of the literature for the current review began with a search throughout the databases of Taylor and Francis Online, Wiley Online Library, SAGE Journals, and ScienceDirect. Publications with the phrase 'online learning' in the title (or terms like 'web-based learning,' 'e-learning,' 'distance learning,' or 'remote learning') and which responded to the descriptors/topics 'English language teaching' or 'English as Second Language' or 'English for Foreign Language' were sought. Although the terms online, web-based, e-learning, distance, and remote learning are not interchangeable, they were considered relevant for the purposes of this study, which looked at any form of practice in which technology is used to facilitate the teaching and learning process in a virtual context. In terms of article source (academic journals), research methodology (empirical), and problem relevance (main emphasis on online teaching and learning activities in the context of teaching EFL/ESL), 16 papers met the inclusion criteria for this study.

4 FINDING AND DISCUSSION

As a result of the data coding procedure, the identified benefits (adopted from Arslan (2020)) of Educational Mobile Interactive Learning Activities in the field of EFL/ESL instruction were classified into three categories: student progress, pedagogical contributions, and interaction (Table 1).

Table 1. The benefits of educational mobile interactive learning activities in teaching EFL/ESL.

Categories	Sub-categories	Sample Research
Student progress	Improves language skills	(Bahari 2020; Dressman & Lee 2021; Junn 2021; Selcuk et al. 2019)
	Increases motivation	(Selcuk et al., 2019; Ucar & Kumtepe 2019; Wang & Chia 2020)
	Encourages students to be more active	(Cheung, 2021; Wang & Chia 2020)
	Provides positive attitude	(Klimova 2020; Manegre & Sabiri 2020)
	Builds self-confidence	(Klimova 2020; Selcuk et al. 2019)
	Takes on responsibility for learning	(Junn 2021; Ortogero & Ray 2021; Safdar et al., 2020)
	Increases satisfaction	(Pebriantika et al. 2021; Wang & Chia 2020)
	Improves cognitive strategies	(Escobar Fandiño et al. 2019; Klimova 2020)
Pedagogical contributions	Increases engagement level	(Escobar Fandiño et al. 2019; Wang & Chia 2020)
	Provides self-paced learning	(Ortogero & Ray 2021; Wang & Chia 2020)
	Provides immediate feedback	(Bahari 2020; Selcuk et al. 2019; Tragant et al. 2021)
	Provides collaboration opportunities	(Junn 2021; Safdar et al. 2020)
	Enables meaningful learning	(Cheung 2021; Dressman & Lee 2021; Junn 2021)
	Enables teacher to be a facilitator	(Jin et al. 2021; Manegre & Sabiri 2020; Sun et al. 2020)
	Creates a flexible learning environment	(Pebriantika et al. 2021; Tragant et al. 2021)
	Enhances personalized learning chances	(Dressman & Lee 2021; Safdar et al. 2020)
	Enhances enjoyment	(Junn 2021; Klimova 2020)
Interaction	Provides interaction opportunities (student-student)	(Cheung 2021; Junn 2021; Safdar et al. 2020; Selcuk et al. 2019; Tragant et al. 2021)
	Provides interaction opportunities (student-teacher)	(Bahari 2020; Cheung 2021; Escobar Fandiño et al. 2019; Junn 2021; Pebriantika et al. 2021; Manegre & Sabiri 2020; Tragant et al. 2021; Wang & Chia 2020)
	Provides interaction opportunities (student-material)	(Cheung 2021; Pebriantika et al. 2021; Safdar et al. 2020)

Over half of the studies explain how the usage of EMILA in the field of teaching EFL/ESL is beneficial to students' language skills. According to numerous studies, the use of flipped learning has an essential part in the advancement of students' language skills, namely vocabulary (Dressman & Lee 2021), writing (Selcuk et al. 2019), listening (Bahari 2020), reading (Ucar & Kumtepe 2019), and speaking skills (Dressman & Lee 2021), (Junn 2021). Some media, activities, and strategies which are implemented in the reviewed studies are 1) vlogging and podcasting on YouTube (Dressman & Lee 2021), 2) online synchronous computer-mediated communication (SCMC) platforms (Junn 2021), 3) web-based collaborative writing (Selcuk et al. 2019), 4) Nearpod (Wang & Chia 2020), 5) Zoom (Cheung 2021), 6) mobile app (Klimova 2020), 7) MOOC (Safdar et al. 2020), 8) online intelligent English teaching platform based on artificial intelligence (Sun et al. 2020), 9) ARCS-V-based motivational strategies (Ucar & Kumtepe 2019), and 10) Computer-mediated feedback (Bahari 2020). The above-mentioned media and strategies can be categorized as Educational Mobile Interactive Learning Activities (EMILA).

It is also worth noting that EMILA may help students become more motivated, have a more positive attitude, gain self-confidence, and be more satisfied with their language study (Klimova 2020; Pebriantika et al. 2021; Selcuk et al. 2019; Ucar & Kumtepe 2019; Wang & Chia 2020). Furthermore, according to some experts, flipped learning encourages students to apply a variety of learning techniques by allowing them to rehearse content numerous times before class, conduct research on things to learn, practice time management, and improve cognitive thinking abilities (Escobar Fandiño et al. 2019; Junn 2021; Klimova 2020; Ortogero & Ray 2021; Safdar et al. 2020).

According to a review of studies on the effectiveness of EMILA in the field of EFL/ESL instruction during COVID-19 pandemic, it helps students become more engaged in language acquisition. Immediate feedback, collaborative chances, and meaningful learning are among the benefits according to some academics (Bahari 2020; Escobar Fandiño et al. 2019; Selcuk et al. 2019; Tragant et al. 2021; Wang & Chia 2020). Furthermore, EMILA allows students to engage in meaningful language learning at their own pace and convenience, allowing them to participate with language activities at their leisure (Cheung 2021; Dressman & Lee 2021; Junn 2021; Ortogero & Ray 2021).

Evidence suggests that using EMILA can help students form meaningful experiences with their classmates both in and out of the classroom, as well as improve their language skills and knowledge (Bahari 2020; Cheung 2021; Escobar Fandiño et al. 2019; Junn 2021; Manegre & Sabiri 2020; Pebriantika et al. 2021; Selcuk et al. 2019; Safdar et al. 2020; Tragant et al. 2021; Wang & Chia 2020). The interaction students create outside of class and in-class time with peers, which strengthens and deepens their grasp of the topic, may be related to the success and satisfaction with EMILA (Cheung 2021; Pebriantika et al. 2021; Safdar et al. 2020).

This combination of findings provides some support for the conceptual premise that EMILA can boost student progress, contributes pedagogical aspects for students and teachers, and provides interaction opportunities. EMILA turns conventional language instruction into self-paced and directed learning, allowing the instructor to be a guide on the side rather than a sage on the stage, to optimize guiding in-class time and give additional possibilities for personalized teaching and evaluation on students' language learning development.

5 CONCLUSION AND RECOMMENDATION

This systematic literature review was designed to determine the benefits of Educational Mobile Interactive Learning Activities in the field of teaching EFL/ESL. This study has found out that generally there is an increase in numerous studies on the utilization of the ten above-mentioned educational mobile interactive learning media, activities, and strategies during the COVID-19 pandemic. The most obvious finding to emerge from this study is that the implementation of EMILA in EFL/ESL can encourage student progress, contribute pedagogical aspects for students and teachers, and provide interaction opportunities. An implication of this is the possibility that this study can be the recommendation for the government, universities, lecturers, and students by implementing Educational Mobile Interactive Learning Activities in EFL/ESL.

ACKNOWLEDGMENT

This research work is funded by the Centre for Research and Community Service, Sebelas Maret University, Indonesia.

REFERENCES

Arslan, A. (2020). A systematic review on flipped learning in teaching English as a foreign or second language. *Journal of Language and Linguistic Studies*, *16*(2), 775–797.

Bahari, A. (2020). Computer-mediated feedback for L2 learners: Challenges versus affordances. *Journal of Computer Assisted Learning*, *37*(1), 24–38.

Bruyn, L. L. (2004). Monitoring online communication: can the development of convergence and social presence indicate an interactive learning environment? *Distance Education*, *25*(1), 67–81.

Cheung, A. (2021). Synchronous online teaching, a blessing or a curse? Insights from EFL primary students' interaction during online English lessons. *System*, *100*, 102566.

Dressman, M., & Lee, J. S. (2021). IDLE in the classroom: Learner-driven strategies for English language learning. *The Clearing House: A Journal of Educational Strategies, Issues and Ideas*, *94*(4), 181–187.

Escobar Fandiño, F. G., Muñoz, L. D., & Silva Velandia, A. J. (2019). Motivation and e-learning English as a foreign language: A qualitative study. *Heliyon*, *5*(9), e02394.

Jin, L., Xu, Y., Deifell, E., & Angus, K. (2021). Emergency remote language teaching and U.S.-based college-level world language educators' intention to adopt online teaching in postpandemic times. *The Modern Language Journal*, *105*(2), 412–434.

Junn, H. (2021). L2 communicative competence analysis via synchronous computer-mediated communication (SCMC) as an alternative to formal classrooms. *Innovation in Language Learning and Teaching*, 1–17.

Klimova, B. (2020). Benefits of the use of mobile applications for learning a foreign language by elderly population. *Procedia Computer Science*, *176*, 2184–2191.

Liberati, A. (2009). The PRISMA statement for reporting systematic reviews and meta-analyses of studies that evaluate health care interventions: Explanation and elaboration. *Annals of Internal Medicine*, *151*(4).

Madianou, M. (2020). A second-order disaster? Digital technologies during the COVID-19 pandemic. *Social Media + Society*, *6*(3).

Manegre, M., & Sabiri, K. A. (2020). Online language learning using virtual classrooms: An analysis of teacher perceptions. *Computer Assisted Language Learning*, 1–16.

Maurer, H., & Salman Khan, M. (2010). Research trends in the field of e-learning from 2003 to 2008. *Interactive Technology and Smart Education*, *7*(1), 5–18.

Moher, D., Shamseer, L., Clarke, M., Ghersi, D., Liberati, A., Petticrew, M., Shekelle, P., & Stewart, L. A. (2015). Preferred reporting items for systematic review and meta-analysis protocols (PRISMA-P) 2015 statement. *Systematic Reviews*, *4*(1).

Ngo, H. K., & Md Yunus, M. (2021). Flipped classroom in English language teaching and learning: A systematic literature review. *International Journal of Academic Research in Business and Social Sciences*, *11*(3).

Nguyen, M. H., Gruber, J., Fuchs, J., Marler, W., Hunsaker, A., & Hargittai, E. (2020). Changes in digital communication during the COVID-19 global pandemic: Implications for digital inequality and future research. *Social Media + Society*, *6*(3), 205630512094825.

Ortogero, S. P., & Ray, A. B. (2021). Overrepresentation of English learners in special education amid the COVID-19 pandemic. *Educational Media International*, *58*(2), 161–180.

Pebriantika, L., Wibawa, B., & Paristiowati, M. (2021). Adoption of mobile learning: The influence and opportunities for learning during the Covid-19 pandemic. *International Journal of Interactive Mobile Technologies (IJIM)*, *15*(05), 222.

Rashid, S., & Yadav, S. S. (2020). Impact of Covid-19 pandemic on higher education and research. *Indian Journal of Human Development*, *14*(2), 340–343.

Safdar, M., Yasmin, M., & Anwar, B. (2020). English for specific purpose through facilitated and non-facilitated MOOCs: An analysis of the learners' perspectives. *Computer Applications in Engineering Education*, *29*(4), 786–794.

Selcuk, H., Jones, J., & Vonkova, H. (2019). The emergence and influence of group leaders in web-based collaborative writing: self-reported accounts of EFL learners. *Computer Assisted Language Learning*, 1–21.

Sun, Z., Anbarasan, M., & Praveen Kumar, D. (2020). Design of online intelligent English teaching platform based on artificial intelligence techniques. *Computational Intelligence*, *37*(3), 1166–1180.

Tragant, E., Pinyana, À., Mackay, J., & Andria, M. (2021). Extending language learning beyond the EFL classroom through WhatsApp. *Computer Assisted Language Learning*, 1–30.
Ucar, H., & Kumtepe, A. T. (2019). Effects of the ARCS-V-based motivational strategies on online learners' academic performance, motivation, volition, and course interest. *Journal of Computer Assisted Learning*, *36*(3), 335–349.
Vitta, J. P., & Al-Hoorie, A. H. (2020). The flipped classroom in second language learning: A meta-analysis. *Language Teaching Research*, 136216882098140.
Wang, J., & Chia, I. (2020). Engaging students via Nearpod in synchronous online teaching. *Management Teaching Review*, 237929812097495.
Zare, M., Pahl, C., Rahnama, H., Nilashi, M., Mardani, A., Ibrahim, O., & Ahmadi, H. (2016). Multicriteria decision making approach in E-learning: A systematic review and classification. *Applied Soft Computing*, *45*, 108–128.

Teacher Education and Teacher Professional Development in the COVID-19 Turn – Drajati & Putra (Eds)

Exploring extracurricular informal digital learning of English through Spolsky's language policy

Surya Agung Wijaya
Universitas Sebelas Maret, Surakarta, Indonesia

Melor Md Yunus
Universiti Kebangsaan Malaysia, Malaysia

ABSTRACT: This study was to explore how preservice teachers conducted extracurricular informal digital learning of English (IDLE) in the context of English as medium instruction (EMI). In the context of EMI, preservice teachers have been allowed to manage their YouTube, Instagram, and blog to promote EMI class by creating English content. Furthermore, Arabic became another language option for this EMI. The preservice teachers from three different majors (English, Arabic, and Islamic education departments) volunteered to participate in this study. Case study was chosen to approach this study for seven months. The data were generated from artifacts and photo-elicitation also known as the interview that used pictures and videos to get clarification of the phenomena. This study found that the language practices should be fostered by faculty members making training on how to use English and Arabic for social media purposes that have potential support for international class students' IDLE practices. The theoretical implication will also be discussed.

Keywords: English as a medium of instruction; extracurricular informal digital learning of English; language policy; photo-elicitation

1 INTRODUCTION

Prior studies on EMI always indicated colliding among language management, beliefs, and practices from in-class activities (see Fitriati 2015; Kirkpatrick 2014). Although EMI has been promoted as an "international program" to pursue international students to come and study, the program was dominated by local students (Hu & Lei 2014; Rahman et al. 2019). Moreover, teachers mostly showed a negative attitude on English only for comprehension purposes, therefore, EMI was practiced only with available languages (Walker et al. 2019). However, the high demand for EMI still continues due to historical, geographic, and economic purposes (Spolsky 2009).

Although previous studies on EMI were reported far from the standard of EMI, learning beyond the classroom also known as IDLE indicated different results on how students see the value of language and L2 practices. The EMI could not make students practice L2 in authentic communication (e.g., Hu & Lei 2014; Rahman et al. 2019). On the other hand, IDLE fostered students to perform L2 by having anonymity in social media that was the fundamental asset to reduce students' anxiety in L2 practices (Lee 2019a, 2019b, & 2019c; Lee & Drajati 2019; Lee & Dressman 2017; Lee & Lee 2019). Moreover, extramural IDLE was also studied by Wijaya, Asib, & Suparno (2021) to open multidisciplinary discussion. Therefore, this study is to complement the discussion on IDLE as an extracurricular program where it has the potential to be implemented by exploring language management, beliefs, and practices.

DOI 10.1201/9781003347798-9

By opening discussion extracurricular IDLE, it potentially portrays the EMI program from students' activities beyond the classroom. Therefore, when rich L2 inputs cannot fully be offered in the classroom, Benson's (2011) extracurricular IDLE as semi-structure, certification, out-of-class, self-instructed, and self-directed can substitute the inputs. Grounded up from the frameworks, two research questions are proposed.

1. How is language management developed for extracurricular IDLE?
2. How do language management and beliefs intervene in language practices in the extracurricular IDLE?

2 LITERATURE REVIEW

Spolsky (2004, 2009, & 2012) acknowledges LP into three independent components – language management, beliefs, and practices - that are interrelated to each other. On the language management, extracurricular IDLE is regulated by explicit and implicit policymakers' power because they can create programs with additional certification for students. The agenda of the program guides students to see the value of language and language use. Moreover, students' beliefs of L2 practices can be encouraged/discouraged from their own experiences through the program. As an out-of-class program, language practices rely on students' own decisions about the content and program. Authentic communication, mostly found on extramural IDLE (Lee 2019a; Lee & Drajati 2019; Lee & Lee 2019), remains a question.

Table 1. The Conceptual parameters of Language Policy and IDLE.

Language policy (Spolsky 2004, p. 5)	Extracurricular IDLE (Benson 2011)
any specific efforts to modify or influence that practice by any kind of language intervention, planning, or management.	semi-structure, certification
its language beliefs or ideology – the beliefs about language and language use	self-instructed, self-directed
its language practice – the habitual pattern of selecting among the varieties that make up its linguistic repertoire,	out-of-class

Interestingly, prior studies on LP, and social media mostly captured the language management on the Facebook group (de Bres & Belling 2014; Sherman & Švelch 2015; Tagg & Seargeant 2014). Where language management could be implicitly established in the beginning, a multilingual environment persuaded policymakers to propose explicit language choices and practices in the group after a year (de Bres & Belling 2014). Moreover, rather than seeing IDLE as a replacement, IDLE has been proposed as complementing and extending in-class L2 inputs (Dressman & Lee 2021). They proposed that extracurricular IDLE can be juxtaposed with institutional policy by having vlogging, podcasting, and watching through YouTube. The program does not give any credits to students (Lee et al. 2018). By facing various interlocutors, other scholars also utilized IDLE to promote intercultural education for EFL learners (see. Özdemir 2017; Valencia & Benavides 2018; Wu & Miller 2019).

Interweaving discussion between LP and extramural IDLE has been proposed (Wijaya et al. 2021), indicating implicit and explicit policy influencing the value of English and language choices and practices. As semi-structured activities, extramural IDLE cannot retain the power and agenda of institutions and policymakers. Therefore, policymakers and teacher educators need to create and enact an extracurricular IDLE for L2 segmentary activities.

3 METHOD

This study was conducted in an international class of one Muslim university in Indonesia. One class consisted of three different majors (English, Arabic, and Islamic education departments). EMI has been practiced alongside Arabic as an additional language. The program aims to prepare students to teach in Indonesian schools overseas. 13 pre-service teachers were recruited from the second semester who could pass the online admission tests for English and Arabic. 2 men and 11 girls have followed and joined in extracurricular IDLE. Two students joined the student body (content creator and policymaker of extracurricular IDLE) practicing under the name of institution accounts (e.g., Twitter, YouTube, Instagram, and blog), and the others followed the program.

This research was approached as a case study (Yin 2018) from March until September 2021. The data were collected from international class and student body documents for language management. Selected contents from those social media were screenshotted and used in photo-elicitation (Harper 2002) to portray language beliefs. The figures were recreated and drawn using Vector Q. The document and artifacts were analyzed through multi-modal analysis (Valencia 2015) by "studying the relationship among the multifarious modes of communication, constructing semiotic clusters to map out the multimodal design, concentrating on modes of communication as units of analysis, and establishing intersemiotic relationships among modes conjoined to generate particular meaning" (p. 4). Where the result of this analysis "management and practices" would be compared with photo-elicitation to conclude the multi-modal analysis. We classified the data based on two research questions, therefore, we drew themes considering Spolsky's (2004) framework and lexical resources.

4 FINDING AND DISCUSSION

This study portrayed extracurricular IDLE from the student body (policy input) who took the responsibility to bring the agenda of policymakers, and the others (grassroots) joined as bottom-up supporters. During the pandemic, some preservice teachers joined a teaching training program from home (e.g., using Zoom) in one Indonesian school in Malaysia. Therefore, they expected to utilize various applications by understanding management, beliefs, and practices that potentially help them in scaffolding extracurricular IDLE.

4.1 *Managing language policy "to implement and coordinate students' activities..."*

Spolsky (2004 & 2009) points out language management is encouraged by the value and ideology of each language through top-down or bottom-up models and formed to be explicit and implicit to modify the language beliefs and practices. In 2019, the student body was established under the institution where the documents directly mentioned that vision and mission must adjust the institution policies.

> *to implement and coordinate students' activities in areas of leadership, art, sport, education, student empowerment, advocacy, research and development...*

The student body captured and implemented the policy into several divisions – health, language, art, social, main committee, and media development – which were named in English. They also contextualized research and development into a dual-language debate program. From the document analysis, the researcher found that the materials given are more focused on English debate. However, during the pandemic, online debate stopped at the preliminary level. Moreover, some divisions handling IDLE activities consulted the program to the institution policymaker, and implemented with their own decisions such as receptive IDLE (e.g., watching a movie under the art division) and productive IDLE (e.g., Instagram, blogs, and YouTube under social and media development divisions) (Lee & Drajati 2019).

There is no explicit document that regulates English and Arabic to be practiced on extracurricular IDLE. Their label as international class students that can use English and Arabic create implicit power contributing beyond the management. Because the extracurricular IDLE was more varied than previous studies (see de Bres & Belling 2014; Sherman & Švelch 2015; Tagg & Seargeant 2014), the student body shared power with various students from English, Arabic, and Islamic education departments. Therefore, their value of language choices and use may be brought into the management. The top-down policy may be difficult to be negotiated by bottom-up practices (Wijaya et al. 2021), sharing and negotiating powers were found in this study where the student body came from students themselves (bottom-up practices) and had a chance to consult with faculty members.

4.2 *Language beliefs and practices "I usually see English as a big title..."*

From top-down policy and bottom-up support, language choices were made through a discussion process considering official policy and students' needs (Lenihan 2014). Blogs mostly shared information about students and alumni with overseas experiences and achievements, but only two new articles were written in English and the others were in Bahasa Indonesia. Indicating bottom-up practices bring their own agenda in the practice, or implicit power from top-down for promoting to local students affect the extracurricular IDLE. In contrast, bottom-up practices mostly believed that English should be used to promote the international class. Keles et al. (2019) point out an unbalanced power among available languages in websites.

The student body and international class students believed that English is the most valuable language for them to be practiced although Bahasa Indonesia was still argued to be used for comprehension purposes to promote to the local audiences. On the other hand, Arabic was believed to be the most difficult one. Their beliefs were captured from one picture on Instagram. Students mostly agreed and supported English placed on the top with the uppercase font "national event" targeting international viewers and resembling an international class program. Arabic was placed on the bottom with "مَنْ أَنَا لَوْلَاكُم" or meaning in context "Who I am if I am without your (teacher') guidance" (see Figure 1). The agenda of having an international profile clearly defined English as a global lingua franca and Arabic as the Islamic identity of the international class program. Spolsky (2009) points out language management at school has close relation with religion.

Student-4
I think that picture has correct composition. It uses English as the title making all people immediately engage with information in the content. Arabic is used as complement. I think is correct adding esthetic elements and introducing international class program which is popular with bilingual...I think all of this depending on the points of view. I usually see English as big title rather than Arabic as big title.

Figure 1. Celebrating teacher's day in english and arabic.

Moreover, Landry & Bourhis (1997) acknowledge "serves an important informational and symbolic function as a maker of the relative power and status of the linguistic communities inhabiting the territory" (p. 23). The student body and other students believed in English placing higher status over Arabic. Multilingual practices cannot offer equal power among available languages. In the spoken practices, mixed languages became more complicated in identifying values and powers in the practices.

(W2) I am with you... I am ki more beautiful than Zoo...
(W1) Sopo kui zoo...
(W2) Zoo ki koe…halah....
(W1) لاَ مِثْلُ مَنْ ...مِثْلُ مَنْ ...هَا...
(W2) I am ki beautiful like (name of actress)... huuu...ذَلِك

(W1) Kok...لاَ أَعْرِف... ya
(W2) أَنْتِ ...لاَ تَعْرِف Is all penonton....
(W1) تَعْرِف (name of singer) ... أَمْ لا
(Audiences) ...لا
(W1) Noh kan...لاَ تَعْرِف
(W2) Aduh ... كَيْف to... ya خَلَص ... Penonton وَ أَنْت listen to me yes...

Figure 2. Comedy-drama with on YouTube.

The student body and other students created a video drama on YouTube, *Wayang* rooted in Javanese mythologies, that practiced code-mixing and -switching to reconstruct local culture and wisdom by using English, Arabic, Bahasa Indonesia, and Javanese. The music (e.g. *Gamelan*), the backdrop (e.g., *Batik Mega Mendung*), and the costume (e.g., traditional Javanese costumes combined with veil) combined all semiotics and language practices representing acculturation of Javanese and Islamic. Moreover, English gave a sense of western resemblance. Creating their own drama script and consulting with policymakers made English that should share power with other languages for promoting the program to local students. However, some students showed disagreement with this practice because non-standard expressions destroyed meaning and form. The Arabic translator stated that there is no expression like "مِثْلُ مَنْ... هَا" in Arabic although it sounds like a repetition of the previous one. Moreover, the giving space for English is more to be performed that potentially shaped a positive attitude on culture and L2 use (see Lee 2019b; Lee & Lee 2018). The language practices cannot simply be said as 'top-down' (e.g., Bahasa Indonesia and Javanese, institutional aims of introducing their program to local students) but sometimes it can be 'bottom-up'. Students were aware of "*de facto* colliding *de jure* LP" (Keles et al. 2019) where they believed that English must be used more in extracurricular IDLE.

As an international class program, students labeled having the capacity to perform English and Arabic for in-class and out-of-class daily uses. Inconsistent policies (Fitriati 2015; Hu & Lei 2014; Kirkpatrick 2014; Rahman et al. 2019) in EMI contexts still affected extracurricular IDLE from management, beliefs, and practices. Extracurricular IDLE collided with other agendas of policymakers to promote their program to local students. The L2 practices were also less authentic because students dominated with local students. Having overseas friends as one of the programs

(Özdemir 2007; Valencia & Benavides 2018; Wu & Miller 2019) may cover up the authentic communication for L2 practices.

5 CONCLUSION AND RECOMMENDATION

This study explored extracurricular IDLE through the lens of LP. The result of this study showed that de jure policy may be different from de facto norms implemented in extracurricular IDLE where unstructured language developments and policies may appear from explicit and implicit form (Shohamy 2006; Spolsky 2004). Three stages could retain the policy including pre-, in-, and post-service IDLE design. In the beginning stage, policymakers introduce and clearly define the status of English and other languages in extracurricular IDLE explicitly, so students could reduce the use of native languages in the content. Second, giving comments in the target language from alumni and prior students may help new students to grow up in the communities. Finally, at the end of the semesters, policymakers could give feedback and reinforcement of the praxis. Indicating the theoretical implication, Spolsky's LP (2004) model can be used to evaluate extramural IDLE by seeing how language management is managed by community members that can intervene in language beliefs and practices and how language practices modify language beliefs and practices. This study was limited to one Muslim university in EFL context without any international students. Future studies may expand the discussion in the ESL context to see extracurricular IDLE designs.

ACKNOWLEDGEMENT

The researchers would like to acknowledge Lembaga Pengelola Dana Pendidikan (LPDP) Indonesia for their support and grant given in the following financial support for the researcher and/or publication of this research.

REFERENCES

Benson, P. (2011). Language learning and teaching beyond the classroom: An introduction to the field. In P. Benson & H. Reinders (Eds.), *Beyond the language classroom: The theory and practice of informed language learning and teaching* (pp. 7–16). Palgrave Macmillan.

de Bres, J., & Belling, L. (2015). Free Your Stuff Luxembourg! Language policies, practices and ideologies in a multilingual Facebook group. *Language Policy, 14*, 357–375.

Dressman, M., & Lee, J. S. (2021). IDLE in the Classroom: Learner-Driven Strategies for English Language Learning. *The Clearing House: A Journal of Educational Strategies, Issues and Ideas, 94*(4), 181–187.

Fitriati, S. W. (2015). English Bilingual Education in an Indonesian Public School. In P. Redmond, J. Lock, & P. A. Danaher (Eds.), *Educational Innovations and Contemporary Technologies: Enhancing Teaching and Learning* (pp. 87–102). Palgrave Macmillan.

Harper, D. (2002). Talking about pictures: A case for photo elicitation. *Visual Studies, 17*(1), 13–26.

Hu, G., & Lei, J. (2014). English-medium instruction in Chinese higher education: A case study. *High Educ, 67*, 551–567.

Keles, U., Yazan, B., Giles, A., & Yazan, B. (2019). Turkish-English bilingual content in the virtual linguistic landscape of a university in Turkey: exclusive de facto language policies. *International Multilingual Research Journal, 14*(1), 1–19.

Kirkpatrick, A. (2014). English as a medium of instruction in East and Southeast Asian universities. In N. Murray & A. Scarino (Eds.), *Dynamic ecologies: A relational perspective on languages education in the Asia-Pacific region* (pp. 15–29). Springer.

Landry, R., & Bourhis, R. Y. (1997). Linguistic landscape and ethnolinguistic vitality: An empirical study. *Journal of Language and Social Psychology, 16*(1), 23–49.

Lee, J. S. (2019a). EFL students' views of willingness to communicate in the extramural digital context. *Computer Assisted Language Learning, 31*(7), 692–712.

Lee, J. S. (2019b). Informal digital learning of English and strategic competence for cross-cultural communication: Perception of varieties of English as a mediator. *ReCALL. 32*(1), 47–62.
Lee, J. S. (2019c). Quantity and diversity of informal digital learning of English. *Language Learning and Technology*, *23*(1), 114–126.
Lee, J. S., & Drajati, N. A. (2019). Affective variables and informal digital learning of English: Keys to willingness to communicate in a second language. *Australasian Journal of Educational Technology*, *35*(5), 168–182.
Lee, J. S., & Dressman, M. (2017). When IDLE hands make an English workshop: Informal digital learning of English and language proficiency. *TESOL Quarterly*, *52*(2), 435–445.
Lee, J. S., & Lee, K. (2019). Informal digital learning of English and English as an international language: The path less traveled. *British Journal of Educational Technology*, *50*(3), 1447–1461.
Lee, J. S., Nakamura, Y., & Sadler, R. (2018). Effect of videoconference-embedded classrooms (VEC) on learners' perceptions toward English as an international language (EIL). *ReCall, 30*(03), 319–336.
Lenihan A. (2014). Investigating language policy in social media: translation practices on Facebook. In Seargeant P., Tagg C. (Eds), *The language of social media* (pp. 208–227). Palgrave Macmillan.
Özdemir, E. (2017). Promoting EFL learners' intercultural communication effectiveness: A focus on Facebook. *Computer Assisted Language Learning*, *30*(3), 1–19.
Rahman, M. M., Singh, M. K. M., Johan, M. & Ahmed, Z. (2020). English medium instruction ideology, management and practice: A case study of Bangladeshi private university. *English Teaching & Learning, 44*, 61–79.
Sherman, T., & Švelch, J. (2015). "Grammar Nazis never sleep": Facebook humor and the management of standard written language. *Language Policy, 14*, 315–334.
Shohamy, E. (2006). *Language policy-Hidden agendas and new approaches*. Routledge.
Spolsky, B. (2004). *Language Policy*. Cambridge University Press.
Spolsky, B. (2009). *Language Management*. Cambridge University Press.
Spolsky, B. (2012). What is language policy. In B. Spolsky (Eds.), *The Cambridge Handbook of Language Policy* (pp. 3–15). Cambridge University Press.
Tagg C., & Seargeant P. (2014) Audience design and language choice in the construction and maintenance of translocal communities on social network sites. In P. Seargeant, & C. Tagg (Eds), *The Language of Social Media* (pp. 161–185). Palgrave Macmillan.
Valencia, J. A. Á. (2015). Language views on social networking sites for language learning: the case of Busuu. *Computer Assisted Language Learning*, *29*(5), 853–867.
Valencia, J. A. Á., & Benavides, A. F. (2018). Using social networking sites for language learning to develop intercultural competence in language education programs. *Journal of International and Intercultural Communication*, *12*(1), 23–42.
Walker, T., Liyanage, I., Madya, S., & Hidayati, S. (2019). Media of Instruction in Indonesia: Implication for Bi/Multilingual Education. In I. Liyanage & T. Walke (Eds.), *Multilingual Education Yearbook 2019* (pp. 209–229). Springer.
Wijaya, S.A., Asib, A., Suparno, S. (2021). Portraying informal digital learning of English from management, beliefs, and practices. *REGISTER Journal*, 14(2).
Wu, J. G., & Miller, L. (2019). Raising native cultural awareness through WeChat: A case study with Chinese EFL students. *Computer Assisted Language Learning*, *34*(4), 552–582.
Yin. R. K. (2018). *Case study research and applications design and methods* (6th ed.). Sage.

Integrating team-based project model in higher education: Examining lecturers experiences and students' perspective

Farikah Farikah, Candradewi Wahyu Anggraeni, Moch. Malik Al Firdaus, A. Sri Haryati & Sukron Mazid
Universitas Tidar, Magelang, Indonesia

Lilik Istiqomah
Western Sydney University, Penrith, Australia

ABSTRACT: This case study research explores the design and instruction applied by lecturers in the Team-based Project Learning Model. Seeing from the students' perspective, by involving five lecturers and ten students of higher education from Indonesia, this study investigates how the lecturers accomplish the tasks of teaching to ensure that the curriculum meets the learners' needs. Virtual observation and the excerpts of the students' and lecturers' experiences were used as sources of information in collecting the data. The data were coded based on Branch (2009). The finding has shown that the ADDIE (Analysis, Design, Development, Implementation, and Evaluation) steps can be one of the learning designs in the Team-based Project learning model. Such instructional design affected developing or engaging the students' activeness in the teaching-learning process.

Keywords: ADDIE, case study, instructional design, learning model, team-based project

1 INTRODUCTION

The era of the Industrial Revolution 4.0 and the challenges of the 21st-century require higher education students to master 21st-century skills, including critical thinking and problem solving, creativity and innovation, communication, and collaboration. It is in line with the 21st-century skills that are needed. Those are technical, information, communication, collaboration, critical thinking, creativity, and problem-solving skills (Laar et al. 2020; Santos 2017). Parallel with the above concept, the Indonesian Ministry of Education and Culture issued Decree No. 3/M/2021 concerning the Key Performance Indicators (IKU) of State Universities and Higher Education Service Institutions. One of the Key Performance Indicators is the use of case methods and team-based projects in learning. One of the goals of this learning model is to improve the students' ability to solve problems, acquire new knowledge and skills in learning and make students more active in solving complex project problems with tangible results. Based on the Ministerial Decree, it is declared that the criteria for learning in the classroom must use one or a combination of problem-solving learning methods (case method) or project-based group learning (team-based projects). As one of the learning models, team-based project learning (TBL) is an interactive teaching method that is learner-centered and instructor-directed and involves self-work followed by teamwork can be an alternative solution (Gaber et al. 2020).

The Covid-19 Pandemic has forced the world community to be creative and innovative, including the world of education. Institutions around the world are currently adopting online learning to provide continuing education to students during the coronavirus-19 (COVID 19) pandemic (Yan et al. 2021). The Indonesian government has implemented a stay-at-home and school-from-home program to suppress the expansion of Covid-19. In responding to the program, the learning mode shifted to virtual classrooms. Consequently, many problems of education during the pandemic

DOI 10.1201/9781003347798-10

continue to vary. Sari (2017) stated that few teachers (lecturers) found designing and using learning media difficult. Besides, the various student characters make the classroom teachers or lecturers get in trouble. In addition, the maintenance of interactive learning techniques in the online teaching-learning process becomes a robust challenge because of the Covid-19 Pandemic. In addition, Brown et al. (2020) clarified that COVID-19 has changed institutional perspectives and required on-campus institutions to pivot to online learning. The focus of the problem is on how the best management to meet student expectations to achieve optimal results through the quality of student involvement and increase students' motivation, time, and confidence in achieving good achievement.

Accordingly, it can be inferred that one of the challenges in teaching-learning activities during a pandemic is how to develop students' critical thinking skills for problem-solving, communication skills, collaboration, and creativity. Besides, it is also a requirement of Key Performance Indicators (IKU). To cope with this learning situation, higher education institutions are expected to create a learning climate that can train 21st-century skills for students during a pandemic. To answer the challenges of lecturers during this pandemic, this study explores lecturers' experience and students' perspectives on what design and instruction are applied by lecturers in the Team-based Learning Model to accomplish the tasks of teaching to ensure that the curriculum meets the learners' needs.

2 LITERATURE REVIEW

One of the educational concerns that have become the priority is the quality of the learning process or quality of learning instruction during the pandemic. One of the ways is to make an excellent instructional design of learning. Instructional design is an activity or system that can serve as guidelines and allow users to determine the extent to which the activity has been achieved as instructional design or a systematic procedure in developing the learning program (Reiser & Dempsey 2007; Triyono 2015). All instructional design processes consist of at least five significant activities (Branch & Kopcha 2014):

1. Analysis of the setting and learner needs
2. Design a set of specifications for an effective, efficient, and relevant learner environment
3. Development of all learner and management materials
4. Implementation of instructional strategies
5. Evaluation of the development results, both formative and summative.

In teaching and learning, team-based project learning (TBL) is a relatively popular instructional approach applied in several disciplines. However, it has yet to gain popularity in education. TBL positively affects the students' learning experience and motivation. Team-based teaching media (such as Facebook) was indeed perceived as an innovative and effective tool in a student-centered learning environment that enriched students' educational experiences, increased the subject matter's relevance, and encouraged students to collaborate effectively with their peers and faculty (Rasiah 2014).

Several research findings suggest that Team-based learning has a very important role in teacher training as positive entries outnumbered negative entries. Teams with high average microteaching scores also had more positive diary entries (Samad et al. 2015). The idea of team-based learning includes grouping students into varied teams of 5-7 students who work together throughout the class (Michaelsen et al. 2002). Because it is part of the learning activities, students in TBL are strategically organized into permanent groups (for the entire learning period). Learning materials are organized into main units which are usually divided into five to seven units. Since each module begins with a readiness assurance process (RAP) before each class event, students must study the assigned material. The RAP ranges from a short test (covering the content and key concepts from reading or other activities, for example, dissection) that students first complete as individuals, then they take the same test again as a team, arriving at a consensus on each question. Immediate feedback is received by students on team tests, and then they have the opportunity to write evidence-based appeals if they feel they can make valid arguments for their answers to incorrect questions.

Following the previous steps, the lecture is the last step of the team-based learning model. This learning model allows the instructor or lecturer to provide clarification on misperceptions during the team-based teaching and learning process (Parmelee & Michaelsen 2010).

3 METHOD

This case study explores the design and instruction applied by lecturers in the Team-based Project Learning Model in accomplishing teaching tasks to ensure that the curriculum meets the learners' needs and actively engages the students. As an empirical investigation, case studies observe contemporary phenomena in real-life contexts, especially when the boundaries between phenomena and context are not clear (Yin 2003). The case study was implemented since the writer wanted to describe in detail what lecturers implemented steps and activities in the teaching-learning process by applying a team-based learning model.

It involved ten university students and five lecturers who participated voluntarily. This study involved seven female students and three male students. They are aged 18-22 years and spread from semesters 1, 3, 5, and 7. In addition, this study involved five lecturers of study programs who have teaching experience that varies between 1 year, five years to 20 years. They were recruited purposely. They were invited to participate in this study to share their experiences in joining and lecturing during virtual learning by implementing team-based learning models. Before expressing their opinions and experiences, they were informed about this study and given consent forms. This study was conducted for one semester in the academic year of 2020/2021. Virtual observation and the excerpts of the lecturers' and students' experiences were used as data sources. The data were coded based on Branch (2009). The virtual interviews were conducted casually and followed the procedures of 'narrative interviewing' (Reissman 2008). The interviews were started by asking some questions about her general experience during the teaching-learning process at university. It was followed by some specific questions requesting her to elaborate on their experience and perception in joining the teaching-learning process by implementing a team-based project learning model.

4 FINDING AND DISCUSSION

Derived from ADDIE's theory of Branch (2009), based on the interview results, it is clarified that the lectures applied five steps in designing and instructing the teaching-learning activities. Those five steps were analyzing, designing, developing, implementing, and evaluating. As a framework for the instructional planning and implementation process, the five-step ADDIE cycle evolves over time and is continuous. Each step in ADDIE has a different purpose and function in developing learning designs.

This ADDIE model is a generic learning design model that provides an organized process in developing learning materials that can be used, both in face-to-face learning and online learning (Shelton & Saltsman 2011). To get a better understanding of the implementation of the ADDIE model in the teaching-learning activities below are the excerpts of procedures and activities done by the lecturers before applying it in learning. Teachers need to know the ADDIE model procedure. This model was developed through several systematic steps as follows.

4.1 *Analysis stage*

The analysis stage is the stage done by lecturers before implementing the actions. The lectures surveyed the learners and learning environment to determine which learning problems are of high priority and should be chosen as objectives. lecturers carry out three analysis activities, including needs analysis, curriculum analysis, and student character analysis. In these analysis stages, the lecturers do classroom observation to determine the activities that occur in the classroom, especially those related to the teaching and learning process (KBM) and classroom students' characters. Below are the samples of excerpts from interviews with lecturers.

Excerpt 1. Lecturer 1

> "*I observe learning devices that have been used previously to find out the learning achievements that have been achieved.*"

Excerpt 2. Lecturer 2

> "*To know all the abilities that must be mastered by students at each stage of learning by the predetermined learning outcome of a subject (CPMK/Capaian pembelajaran Mata kuliah), then I always observe the beginning of the class before planning the learning that I will do.*"

In addition to the lecturers' data, the writer also interviewed the students about what the lecturers did before starting the class. Below is the sample of excerpts of the students' interviews

Excerpt 3. Student 1

> "*The agreement on the material that will be studied, methods of teaching, methods of assessment, media, learning, project which will be carried out, as well as hours of lectures.*"

Analysis of the learners and the learning environment is a crucial point in starting to design effective instructions. Analyzing the devices used by the lectures, the classroom setting and environment, the media, the method of teaching, and the materials implemented by the previous lecturers will determine the smoothness of following the teaching and learning process. Understanding the students' previous knowledge will also become a better start in designing the next teaching steps. The prior knowledge is a starting point for designing the following steps of teaching-learning activities (Bennett et al. 2017).

Based on the interview data above, it can be concluded that the lecturers conduct analytical activities before carrying out learning. The analysis was carried out to previous learning achievements, students' initial abilities, student conditions, class conditions, and previous learning tools. In addition, it is also to determine the components needed for the next learning stage, such as determining the characteristics of learners and analyzing the needs of students in learning. Besides, it determines the type of media to be developed and analyzes the obstacles found. These findings are in parallel with Peterson's study (2003). In the analysis stage, the lecturers' primary consideration is the target audience, in this case, the students. Analyzing the students' needs is conducted to determine the audience's needs by distinguishing between what students already know and what they need to know. The standards and competencies reviewed beforehand will assist the lecturers in the teaching-learning process.

4.2 *Design stage*

The design stage is when the lecturers translate the learning objectives into lesson plans or blueprints based on the analysis stages of observations. RPS (*Rencana Pembelajaran semester*) or other terms for the lesson plan, must be reviewed and adjusted periodically according to the development of science and technology. Before the lecturer develops learning tools, the lecturer needs to design or translate the learning objectives contained in the course achievements into a lesson plan. The creation of Semester Learning Plans (RPS) is carried out by the lecturers before carrying out classroom learning in the classroom. Below are excerpts from interviews with lecturers.

Excerpt 4. Lecturer 2

> "*Based on the learning outcomes that have been determined, I prepared the RPS first to describe the competencies that will be mastered by students.*"

Excerpt 5. Lecturer 3

> "*As lecturers, our institution requires lecturers to prepare RPS first. Besides formally fulfilling the obligations of lecturers before entering the e-learning system, it also serves to provide direction and steps for learning outcomes to be achieved.*"

Excerpt 6. Lecturer 4

"In designing the teaching-learning plan, first I determine the appropriate learning strategies with all student characteristics, and how to overcome students' problems by increasing the special attention to students who have less ability. Besides, I also plan the timing in preparing the lesson plans. In addition, I also determine learning media according to student characteristics, the students' ability, intelligence."

Translating the classroom observation results into a lesson plan is the way to arrange the teaching procedures. Since the students' competence is to be achieved, the steps and the directions for the achieved learning outcomes are described there. The lesson plans describe the teaching procedures, methods, and media of teaching and learning that will be implemented in the teaching-learning process.

From the interview results, it can be concluded that the lectures do some activities at the design stages. The activities include designing the lesson plan, designing the e-module, designing the time division, the media used, and designing the learning strategies and activities. This designing stage is a vital element of instructional design. As the designers of the teaching-learning process, the lecturers need to determine how objectives will be assessed and what forms of assessment will be used before implementation (Peterson 2003).

4.3 *Development stage*

Specific materials and procedures are created to give life to the blueprint. The development stage is the product realization stage. At this stage, the material development based on learning outcomes was carried out according to the design. Lecturers and peer team lectures will validate the materials. In the process validation, the validator uses an instrument standardized by the quality assurance institution of the university prepared previously. Below are excerpts from interviews with lecturers.

Excerpt 7. Lecturer 4

"The lesson plans I made under the provisions of the component formulation RPS as university rules."

Excerpt 8. Lecturer 5

"Besides I constructed RPS, I also plan the materials that I will implement in the classroom, besides other sources like live pictures, Youtube and others."

Excerpt 9. Lecturer 1

"After I observe the CPL, and then design the RPS, and finally I develop the materials based on CPL, and the media I used to achieve the CPL."

The development phase comes next after the teaching design (the materials and the assessment) is approved. It was designed to address the performance gap (Welty 2007). Based on the above excerpts in the development stage, the lectures develop the RPS based on CPL, the materials to achieve the CPL, and the media used to achieve the CPL. Developing the RPS and the materials are based on the previous stages; they are analyzing and the design stage. In this stage, the lecturers, as the designers of the teaching-learning process, must now refer to the results from the previous two phases and construct a product to deliver the information (Peterson 2003).

4.4 *Implementation stage*

Learners used the materials and procedures that were created. In the implementation stage, the lectures applied various activities of team-based learning models. The excerpts below are the samples of the students' interviews.

Excerpt 10. Student 2

"Class divided into six groups by the subject matter. Each team made a summary of material related to the topic to be discussed, namely one of the historical events in literature. Then each team made a presentation. The presentation is done like a talk show. Each team member shares the tasks, there are who become guides, events, and more into a resource. The students in the wild team always listened

and were required to ask questions or responses in the talk show activity. After the event, the lecturer gave an evaluation and reinforcement of the material."

Excerpt 11. Student 3

"They mostly use online platforms to conduct virtual. The lecturer shares a link to Zoom or another platform, the First meeting, the lecturer explains RPS and materials. Then, at the next meeting, the students are divided into several groups to present their results every week. Every week, the students' group presents their results, and the lecturer gives comments and adds materials if needed. Lecturer gives an assignment in ELITA."

Excerpt 12. Student 4

"In general, the steps of the lecturers in online learning are giving material directly with online conferences or indirectly with sharing sources of material and asking the student to understand it. After that, the lectures give topics for group activities. The lecturers ask the group to find and navigate the materials. The lecturers provide some quizzes or exercises to ensure that the students understand the material."

Excerpt 13. Student 5

"For the first, the lecturers give the provision of general/basic as the basis of discussion further with the group. After that, the group's division discusses issues and discusses them together in groups then, they expose results, discussion, question, and answer, give feedback and make conclusions."

From the above excerpts, it can be summarized that the lectures in implementing team-based learning apply various patterns. Each lecture has his/her way of making team activities. The roles of the lectures are as facilitators of learning, helping students navigate the materials based on CPL, and motivating them to learn. The lectures implement some patterns in the team-based project model of learning. Each of the lectures has his/her style. Among the styles, it can be inferred that the lectures do the activities such as informing the link of virtual platforms used, giving provision of the materials, grouping students, group discussion, exposing discussion results, question and answer, and reinforcement from lecturers. Referring to the above steps, it can be seen that lecturers, as designers of the teaching-learning activities, must take an active role. Their role intensifies with the advent of this phase (Peterson 2003).

4.5 *Evaluation stage*

Lecturers conduct assessments to determine the extent to which students have mastered the goals that were determined at the beginning, and are carried out to make revisions as necessary. There are various models of evaluations applied by the lecturers during team-based learning models. The sample excerpts of the students' interviews are as follows.

Excerpt 14. Student 6

"Lecturers give an evaluation on micro-teaching video directly at the time of zooming meeting later after meeting. Usually no quiz at Elita (Elearning Universitas Tidar). The UAS (Final Semester Test) and UTS (Mid Semester Test) were given following the academic calendar."

Excerpt 15. Student 7

"Quizzes, assignments, and provision of evaluation were given during the discussion of the material. In the middle of the semester, the lecturer gave UTS and UAS at the end of the semester. Immediately, every student finished performing is evaluated, explained errors, and what should be (corrected and given material) besides UTS and UAS."

Excerpt 16. Student 8

"Evaluation is given after a student writes a draft of them. After the lecturer gives feedback, students revise their writing again and send it back using ELITA media. The quiz is given in the form of a writing assignment because it is by the writing course. Lecturers also give UTS and UAS according to the academic calendar".

Based on the above excerpts, it can be inferred that the lectures implement various ways of evaluation in a team-based learning model. Besides mid-semester and final-semester tests, the lecturers gave various evaluation models to ensure that the curriculum meets the learners' needs and actively engages the students. In this stage, the lecturers as the teaching-learning designers have to determine if the problem has been solved and is relevant to training programs (Peterson 2003).

Referring to the previous findings mentioned above, in conclusion, the ADDIE (Analysis, Design, Development, Implementation, and Evaluation) model can be one of the alternative models in designing a Team-based model of teaching. The activities in each stage can be various based on the lecturers' considerations. As Peterson (2003) clarified, because of its systematic and generic structure, the ADDIE model is a useful, simple framework for instructional design. The stages can be applicable in a variety of settings as well.

5 CONCLUSION AND RECOMMENDATION

This present study offers insights about designing teaching-learning activities in a Team-based Project Learning Model. Using the ADDIE (Analysis, Design, Development, Implementation, and Evaluation) model, the lecturers show their various activities in team-based learning instruction. The lecturers translated the learning objectives into lesson plans or blueprints based on the survey results. The specific materials and procedures are created to give life to the blueprint. Then the lecturers implement various activities of team-based learning models. The last stage is evaluation. It was a stage to determine the extent to which the lecturers mastered the learning outcome. In addition, the findings show that various models or activities can be implemented in team-based learning. Each model of instruction is to ensure that the learning and the materials meet the learner's needs based on the learning outcome we determined. This study indicates that the ADDIE model can be an alternative instruction design model applied in a Team-based learning model. The activities on each stage of ADDIE will vary depending on the lecturers' consideration. It needs to elaborate more on the detailed and precise activities to achieve learning outcomes. Besides, it needs to explore deeply how the ADDIE activities inspire the students to be actively engaged in the teaching-learning activities.

REFERENCES

Bennett, S., Agostinho, S., & Lockyer, L. (2017). The process of designing for learning: understanding university teachers' design work. *Educational Technology Research and Development*, *65*(1), 125–145.

Branch, R. M. (2009). *Instructional Design-The ADDIE Approach*. New York: Springer.

Branch, R. M., & Kopcha, T. J. (2014). Instructional design models. In the Handbook *of research on educational communications and technology*. Springer, New York, NY.

Brown, A., Lawrence, J., Basson, M., & Redmond, P. (2020). A conceptual framework to enhance student online learning and engagement in higher education. *Higher Education Research & Development*, 1–16.

Dick, W., Carey, L., & Carey, J. O. (2001). *The systematic design of instruction*. 6th. New York: Longman.

Gaber, D. A., Shehata, M. H., & Amin, H. A. A. (2020). Online team-based learning sessions as interactive methodologies during the pandemic. *Medical education*, *54*(7), 666–667.

MedhatAli, M., & El-Din, H. A. K. (2015). Promoting active learning through the use of case studies: experiments in active teaching in graphic design courses.

Michaelsen L.K., Bauman-Knight A., Fink L.D. (2002). *Team-based learning: a transformative use of small groups in college teaching*. USA: Stylus Publishing.

Michaelsen, L. K., & Sweet, M. (2012). Fundamental principles and practices of team-based learning. *Team-based learning for health professions education: A guide to using small groups for improving learning*, 9–34.

Parmelee, D. X., & Michaelsen, L. K. (2010). Twelve tips for doing effective team-based learning (TBL). *Medical teacher*, *32*(2), 118–122.

Peterson, C. (2003). Bringing ADDIE to life: Instructional design at its best. *Journal of Educational Multimedia and Hypermedia*, *12*(3), 227–241.

Popil, I. (2011). Promotion of critical thinking by using case studies as a teaching method. *Nurse education today*, *31*(2), 204–207.
Reissman, C. K. (2008). *Narrative Methods for the Human Sciences*. Thousand Oaks, CA: Sage.
Samad, A. A., Husein, H., Rashid, J. M., & Rahman, S. Z. S. A. (2015). Training English language pre-service teachers using a team-based learning approach. *English Language Teaching*, *8*(1), 44–51.
Santos, J. M. (2017). 21st century learning skills: a challenge in every classroom. *International Journal of Emerging Multidisciplinary Research*, *1*(1), 31–35.
Sari, B. K. (2017). Desain pembelajaran model addie dan implementasinya dengan teknik jigsaw. *Jurnal Inovasi Pendidikan Kimia*, *9*(1), 87–102.
Seel, N. M., Lehmann, T., Blumschein, P., & Podolskiy, O. A. (2017). What is Instructional Design? In *Instructional Design for Learning*, Sense Publishers, Rotterdam.
Shelton, K., & Saltsman, G. (2011). Applying the ADDIE model to online instruction. In *Instructional design: Concepts, methodologies, tools, and applications* (566–582). IGI Global.
Solis, J. D. (2007). Robert A. Reiser and John V. Dempsey, trends and issues in instructional design and technology. *Upper Saddle River*, 336–341.
Welty, G. (2007). The 'design phase of the ADDIE model. *Journal of GXP Compliance*, *11*(4), 40–48.
Yan, L., Whitelock-Wainwright, A., Guan, Q., Wen, G., Gašević, D., & Chen, G. (2021). Students' experience of online learning during the COVID-19 pandemic: A province-wide survey study. *British Journal of Educational Technology*.
Van Laar, E., van Deursen, A. J., van Dijk, J. A., & de Haan, J. (2020). *Determinants of 21st-century skills and 21st-century digital skills for workers: A systematic literature review*. Sage Open.
Yin, R. (2003). *Case study research: Design and methods* (3rd ed.). Thousand Oaks, CA: Sage.

Math and Science Education

Teacher Education and Teacher Professional Development in the COVID-19 Turn – Drajati & Putra (Eds)
© 2023 The Author(s), ISBN: 978-1-032-38970-7

Development of biology interactive multimedia based on vegetation diversity research results for high school students

Anwari Adi Nugroho, Mahmuddah Mahmuddah & Agus Purwanto
Universitas Veteran Bangun Nusantara, Java, Indonesia

Muhammad Luthfi Hidayat
King Abdulaziz University, Jeddah, Saudi Arabia

ABSTRACT: The research objective was to determine the characteristics, feasibility, and practicality of biology interactive multimedia based on the results of research on the vegetation diversity on one of the beaches in Indonesia for grade X high school students. The focus of this research is to develop interactive multimedia for grade X high school ecology material. The research method used is research and development from Borg and Gall (1983). The product being developed is interactive multimedia biology that is packaged in a PowerPoint show so that it can be used with a laptop or device. The results of the validation assessment by multimedia experts and ecological materials show that the product is very feasible, while the assessment by the teacher shows that the product is very good so that it can be used for grade X high school students. Biology Interactive Multimedia (BIM) integrated with the research result of vegetation diversity can be used as an alternative media for biology in high school to improve the process and learning outcomes.

Keywords: Biology Learning, Interactive Multimedia, Vegetation Diversity

1 INTRODUCTION

Biology as one of the subjects in school has good potential in developing 21st-century skills. Biology has special material that is different from other sciences, namely the science that studies living things, the environment, and the relationship between the two (Sudarisman 2015). However, the biological material is often an abstract material that is difficult to reach because it contains abstract content (Brown & Schwartz 2009). In addition, the complex process and the use of complicated terms make it difficult for students to learn some biology topics (Patro 2008; Rice 2013). A teacher needs to design innovative learning to overcome these problems. One way for students to easily learn biology material is to visualize the material in the form of learning media (Çimer 2012).

The learning media used in biology learning in several schools in Indonesia for ecological material are still rarely found (Sambodo et al. 2018).. On the other hand, in learning ecology material, learning media were still in the form of illustrations, including pictures, posters, and dioramas so that makes students less able to understand the concept (Sambodo et al. 2018). Research has shown that, even after school instruction, students do not understand essential ecological concepts. This is because factors that affect learning such as the availability of information technology-based learning media are ignored in learning processes and curriculum construction (Sander et al. 2006; Wastiau et al. 2013; Yeop et al. 2019). Learning media that do not support the learning process can make it difficult for students to understand the content of the material so that student learning outcomes are low (Tabor & Minch 2013). The results of pre-study observations in one high school in the region of Indonesia show that there are still many biology teachers who are reluctant to use interactive multimedia in supporting their teaching and learning activities. This is because the biology learning media used by teachers in teaching are still in the form of visual media, posters, and pictures. Multimedia biology is needed in addition to improving the quality of the process and learning outcomes, teachers must also get used to learning based on technology and information. Mastery of technology and information is a learning need in the 21st-century. 21st-century biology learning and the COVID-19 pandemic condition demands more interactive online learning (Daniel 2020). Online learning requires the use of various instructional strategies to enhance interactive learning and provide feedback.

The development of innovative learning media is needed so that ecological learning is more interactive, supports online learning, and can increase understanding of concepts. The development of learning media by integrating technology can change the learning process in the old class to be more interactive so that students can construct their knowledge. One type of learning media and utilizing technology is multimedia (Shana 2009). Multimedia is a combination of more than one type of media such as text (alphabetical or numeric), symbols, images, images, audio, video, and animation with the help of technology to improve understanding of learning materials. Multimedia characteristics include integration, diversity, and interaction to communicate information to people for a better understanding of concepts (Guan et al. 2018).

The development of Biology Interactive Multimedia (BIM) must also pay attention to the characteristics of the material. Ecological material is the basis for studying other materials such as *Plantae, Animalia*, and other branches of biology because biology is a continuous science. Ecology learning is concrete learning, in the form of direct interaction with the environment Ecological objects consist of stratification levels including species, populations, communities, interaction patterns, material and energy flows, and climate change (Lewinsohn et al. 2015). Effective ecological teaching must involve real nature objects and employ an active data gathering in the field (Schaal et al. 2012). However, not all environments can be reached, so environments that are far from the learning space can be packaged in interactive multimedia. One of the advantages of multimedia is that it can present objects or events that are far away (Susilana & Riyana 2009). Vegetation diversity is one of the topics in ecological material. In learning the topic of vegetation diversity, students identify the various plants that exist in the vegetation. Students at high school in a district in Indonesia have difficulty in identifying the diversity of vegetation around the coast because the location of the beach is far from the school, so it is necessary to integrate local potential in the form of research data on the diversity of coastal vegetation in interactive multimedia. The integration of local wisdom in learning can clearly describe the specifics of the learning environment and learning process (Dwianto et al. 2017).

Local potential in Indonesia is very diverse and has the potential as a learning resource that can be integrated into the learning process (Hastuti et al. 2020). One of the local potentials around Indonesia is the vegetation on the coast. The coastal area has unique conditions. The specificity of environmental conditions has the opportunity to create vegetation that has certain characteristics. The coast is included in the coastal ecosystem. Along the coast is overgrown by various coastal vegetation (Purwanta & Sugiharyanto 2013). On one of the coasts in Indonesia, several pioneer plant species were found and have the characteristics of a succession of coastal vegetation preceded by the dominance of vines, namely *Ipomoea pes-caprae*. The vegetation on the beach is still not much researched.

Pre-study observation activities were carried out by recording the diversity of vegetation on one of the beaches in Indonesia. These results are used as ecological materials for the developed interactive multimedia. The results of the pre-study observation showed that there were 12 families consisting of 22 species, while the diversity of vegetation on the coast was in the moderate category. The number of plant species and their diversity in coastal vegetation has not been used for biology teaching materials in schools. Biological material in high school has the potential to be developed by integrating the results of biological research. The results of the research on the coast vegetation diversity are used as material in BIM.

The integration of the result of biological research to multimedia learning ecology has not been done before. The development of BIM that has been done previously includes interactive multimedia adobe flash biology which packs learning materials in learning media and there are no research results in it (Anggraini et al. 2017; Wulandari et al. 2019; Zairana et al. 2020), interactive multimedia based on learning models. Learning steps are a core component in learning media but the biology material presented is not based on research results(Cahyani 2017; Isnaneny et al. 2018; Topano et al. 2021), android-based biological multimedia (Kasim et al. 2021). No multimedia that has been developed previously that integrates the results of biological research. The development of interactive multimedia by integrating the results of research on coastal vegetation is an innovation of learning media that has never been developed before. Vegetation research results that are integrated in interactive multimedia have the potential to improve the quality of the process and student learning outcomes.

Biology learning with the help of interactive multimedia in the learning process can make it easier for teachers to deliver material, change abstract material into concrete, stimulate students' interest in learning, and help present a lot of complex information in a short time with little effort (Abdulrahaman et al. 2020). Well-designed interactive multimedia can help for presentation activities, classroom or laboratory learning, simulations and e-learning to enable students to process information in both verbal and image form (Alemdag & Cagiltay 2018). The use of technology in interactive multimedia changes the learning process of students who previously only heard, saw, imitated from teachers and textbooks, but with technology in learning students can receive information from the multimedia so that students can construct knowledge and understanding socially (Shana 2009). Multimedia accommodates various learning styles, such as auditory, visual, and kinesthetic learners, so that learners can choose media according to their respective learning styles (Wang et al. 2019; Weiss et al. 2006).

Development of BIM based on the research results on one of the coast in Indonesia vegetation specifically for use by high school students in grade X on ecology material. The purpose of this BIM development is to develop a BIM product based on valid and practical results so that it can be implemented in high school biology learning on ecological material. In this study, BIM products that have been made are then tested for feasibility by experts and practitioners so that BIM products can be used in schools. The results of coastal vegetation research in multimedia provide convenience for high school students in Sukoharjo in accessing information and data about coastal vegetation. The ease of students in obtaining knowledge, information and data about the material can have a positive impact on their learning outcomes. The development of interactive multimedia is expected to provide alternative biology learning media as a solution to improve the quality of biology learning processes and outcomes. Based on the background described, the formulation of the research questions is as follows:

1. What are the characteristics of BIM as a learning medium based on the results of research on plant diversity?
2. How is the feasibility of BIM as a learning medium based on the results of research on plant diversity?
3. How is the practicality of BIM as a learning medium based on the results of research on plant diversity?

2 METHOD

The research method used in this research is research & development (R&D). Is a research method used to produce products and test the effectiveness of products (Borg & Gall 1983). The procedure comprises 7 steps (Borg & Gall 1983) which have been modified, including 1) research and information collecting including the activities of observing the needs of learning media in grade X high school by observing learning in class, interviewing students and biology teachers. In addition, a literature review was also carried out to determine the type of learning media developed, namely learning multimedia. In this step, research is also carried out on the vegetation diversity in beach so that research results are obtained as material for learning multimedia. 2) Planning is a planning activity to create a multimedia draft by analyzing the basic competencies of grade X biology material. At this stage, a learning multimedia design is also planned which is developed using PowerPoint and contains interactive quizzes with macro media.

The interactive multimedia design that has been planned is then carried out by making the preliminary form of the product. 3) Develop a preliminary form of the product containing interactive multimedia creation activities with ecological materials and research results on vegetation diversity. After that, it was continued by making interactive questions using macro media in PowerPoint. 4) Preliminary field testing, at this stage expert validation of interactive multimedia products is carried out. Expert validation used in this study is the validation of material experts and learning media experts. 5) Main product revision contains activities to make product improvements based on suggestions from material experts and learning media experts. After that, an assessment of interactive multimedia products was carried out. 6) Main field testing is the stage of testing multimedia products to 10 high school students in grade X and validating practitioners to biology teachers. 7) Operational product revision contains interactive multimedia product improvement activities based on suggestions from students and biology teachers (Borg & Gall 1983).

The instruments used to collect the data needed in the development of interactive multimedia are learning observation sheets, student and biology teacher interview sheets (research and information collecting stage), validation sheets for plant ecology material experts and biology learning media experts (Preliminary field testing stage), and product readability sheet (main field testing stage). Assessment of learning media experts, material experts on plant ecology, and readability by students and teachers, the data were analyzed descriptively using the percentage feasibility technique (Riduwan 2015). The score obtained is converted to a value using the following equation.

$$K = \frac{F}{N \times I \times R} \times 100\%$$

(Riduwan 2015)

The explanation of the formula : K (Percentage of eligibility criteria), F (All respondents' answers), N (The highest score in the validation sheet), I (Number of statements), R (Number of respondents). The results of the percentage of data are then converted in the form of categories, among others: very unworthy (0% – 20%), not worth it (21% – 40%), enough (41% – 60%), worthy (61% – 80%), very worth it (81% – 100%) (Riduan 2012).

3 FINDING AND DISCUSSION

BIM was developed by integrating the results of research on the diversity of vegetation in Indonesia. BIM is used for grade X high school students, especially ecological materials. This interactive multimedia contains material about ecosystems at the high school level. The material presented in the multimedia also contains research material on the diversity of vegetation in Indonesia. Before designing interactive multimedia, research on the diversity of vegetation on one of the coast in Indonesia was conducted first. The results showed that the abundance of vegetation contained 12 families and consisted of 22 species. The highest importance index of the 22 species was *Ipomea pes-caprae, Ischaemum muticum*, and *Asystasia gangetica*. Meanwhile, the lowest significant value indexes were *Cynodon dactylon* and *Casuarina equisetifolia*. The level of diversity of coastal vegetation can be determined by using the Shannon Wiener Index (H'). The results of the analysis of the level of diversity of coastal vegetation indicate that the vegetation has a moderate diversity.

The BIM that has been made contains ecosystem material in which there are results of research on the diversity of vegetation on one of the beaches in Indonesia. BIM is validated by material ecologists and learning media experts to get a valid product. This is in accordance with research that multimedia biology needs to be validated by material experts and learning media experts (Kasim et al. 2021).

The purpose of expert validation that has been carried out on learning media products is to find errors that occur and provide suggestions that are used as consideration for revising learning media, so that learning media products are suitable for use in learning (Nurbaiti et al. 2017; Panjaitan et al. 2019; Panjaitan et al. 2016). In the validation of ecological material experts, the assessment includes aspects of learning and the content of ecological materials. The learning aspect includes indicators 1) clarity of basic competency formulation, 2) clarity of success indicators, 3) consistency between basic competency, indicators, materials, and evaluations, 4) systematic presentation of material. Aspects of the content of the ecological material include indicators 1) the suitability of the material with the content, 2) the accuracy of the order in which the material is presented, 3) the ease of the material for students to understand, 4) the depth of the material, 5) the ease with which students understand the questions, 6) the evaluation used is in accordance with the learning objectives. The results of the validation of ecological materials experts on BIM products can be seen in Table 1.

Table 1. Results of the validation of material ecologists.

No	Stage	Total score	Category
1	I	65%	Worthy
2	II	82,5%	Very worth it

The results of the validation of the material ecologist at stage I (Table 1) indicate that the interactive multimedia product has a worthy category. The results of the phase I material ecologist validation (Table 1) show that BIM has a feasible category. BIM has met the worthy learning aspects, including the formulation of basic competencies that are clearly formulated, 2) indicators of success are written clearly, 3) basic competencies, indicators, materials, and evaluations are interrelated, 4) systematic presentation of material. BIM also fulfills aspects of ecological material content, including 1) material according to content, 2) material presented in sequence, 3) material easy for students to understand, 4) material according to learning objectives, 5) questions presented are easy to understand, 6) evaluation in accordance with learning objectives. In the first stage of the validation process, the validator provides suggestions for interactive multimedia products. Suggestions from material ecologist validators can be seen in Table 2.

Table 2. Expert suggestions of plant ecology materials.

No	Suggestion	Follow-up
1	Improvements to indicators of achievement of basic competencies	Improving indicators of achievement of basic competencies
2	The depth of the material has not been reached	Added a definition of ecology
3	Less explanatory material	Schematic improvements to the concept map

Based on the suggestions from the validator (Table 2), then the BIM product was improved and validated instage II by material ecologists. The results of the validation of the material ecologist in stage II obtained a very worth it category (Table 1). Furthermore, BIM products were validated by learning media experts.

In the validation of learning media experts, the assessment includes aspects of layout and software. The layout aspect includes indicators 1) all slides on the media are easy to access, 2) display suitability, 3) suitability for menu layouts and other pages, 4) font size and proportional slide space, 5) content placement, 6) clarity of image layout and video, 7) the color composition on the media is correct, 8) attractive and proportional display design. While the software aspect includes indicators 1) supporting software for running interactive quizzes has worked well, 2) original learning media by researchers. The results of the validation of ecological materials experts on interactive multimedia products can be seen in Table 3.

Table 3. Results of the assessment of learning media experts.

No	Stage	Total Score	Category
1	I	80%	Worthy
2	II	92,5%	Very worth it

The results of the validation of learning media experts in stage I (Table 3) indicate that BIM products have a worthy category. BIM has been worthy as a learning media because it has fulfilled the good layout aspects including display, menus, pages, letters, slides, colors, image and video layouts. BIM is also supported by software that works well and the multimedia is original. In the first stage of the validation process of learning media experts, the validator provides suggestions for interactive multimedia products. Suggestions from expert validators of learning media can be seen in Table 4.

Table 4. Suggestions of learning media experts.

No	Suggestion	Follow-up
1	Fixed icon and text	Improvements to icons and clarifying writing
2	The evaluation is filled with student worksheets and interactive quizzes	Make a student worksheet menu with interactive quizzes in the evaluation menu
3	Adding a sub-menu to the material	Adding a sub-menu to the material
4	Improve navigation	Clarify navigation

Based on the suggestions from the validator (Table 4), BIM products were improved and validated in stage II by learning media experts. The results of the validation of the learning media experts in stage II obtained a very worthy category (Table 3). The results of the validation of material experts and learning media experts indicate that BIM is feasible to be continued at the next stage, namely practitioner validation. This is by the results of the study that after expert validation shows a worthy/very worthy it category, it can be continued on practitioner validation (Isnaneny et al. 2018; Nur'aini et al. 2015; Topano et al. 2021).

Practitioner's validation activities on BIM aim to determine the practicality of the product. BIM was validated by a high school biology teacher in a district in Indonesia. Practitioner validation includes an assessment of learning aspects, materials, layouts, and software. The results of the validation of the practicality of interactive multimedia by biology teachers obtained a score of 92.5% and obtained a very decent category. The feasible category by practitioners shows that the product can be practically implemented in learning. Good learning media can facilitate teachers in the learning process to achieve learning objectives (Barra et al. 2014; Mayer 2017; Zhang 2005). During the validation process to practitioners, there were several suggestions from biology teachers as improvements for interactive multimedia. Suggestions from biology teachers on interactive multimedia products can be seen in Table 5.

Table 5. Suggestions of practitioners (biology teachers).

No	Suggestion	Follow-up
1	Improvements to the writing that is still wrong	Correct the sentence
2	Paragraphs must be right and left-aligned	Arrange paragraphs to align right and left
3	Adding an image to the coastal vegetation formation	Adding an image in each formation on the coastal vegetation

After the BIM has been validated by practitioners, the next step is to conduct a product readability test by 10 students of grade X high school. In this step, BIM is simulated in the classroom, then students give suggestions for the product. Some suggestions from students include 1) the caption on the ecosystem picture needs to be enlarged to make it clearer, 2) the picture of ecosystem types needs to be enlarged, 3) the picture on the coastal ecosystem component is still black and white, needs to be replaced with colored ones to make it attractive, 4) the picture of the plant *Casuarina equisetifolia* is not clear. Furthermore, the product is improved according to the suggestions from the students so that BIM is obtained which has good readability for students. In general, suggestions from students are in the form of displaying pictures and writing on learning media. Writing and pictures on learning media are very important to support students' understanding of the material. Still, images and moving images will provide a more concrete understanding to support text abstraction (Taufiqoh & Indana 2012). The results of expert, practitioner, and legibility validation show that interactive multimedia can already be utilized or implemented as a medium for learning biology in high school. This is by research that practical (appropriate) interactive multimedia products can already be used in the classroom for learning (Kuo et al. 2015; Topano et al. 2021). The practicality of learning media is caused by students easily using learning media according to the situation and condition of students (Kuo et al. 2015; Maddison et al. 2014; Ridgway et al. 2007).

BIM based on the results of research on the coast vegetation diversity is expected to improve the quality of the process and learning outcomes of high school biology students. Interesting content and interactive activities in BIM can make teachers and students interact actively without meeting face to face in learning. Attractive learning media can increase interest, learning motivation, and learning efficiency (Tabor & Minch 2013; Tur & Marín 2015; Vagg et al. 2020). Learning materials in this BIM are presented attractively, accompanied by concrete pictures and interactive tests. The presentation of this interesting and interactive material can make it easier for students to understand the learning material (Chang 2002; Weiss et al. 2006). Multimedia is also equipped with interesting text, images, sounds, videos so that the learning material will be clearer and easier to understand. Multimedia that uses a lot of media combinations will make students use their senses more to receive and manage information so that it will have a positive impact on memory (Grimley 2007; Zheng et al. 2008), higher-order thinking skills (Almasseri & AlHojailan 2019), and student learning outcomes (Lee et al. 2014; Rasch & Schnotz 2009).

The results of research on the vegetation diversity on one of the coast in Indonesia are local potentials that can be used as teaching materials and can improve student learning processes and outcomes. Previously, biology learning media was only in the form of print media and not yet IT-based, so this BIM can be an alternative biology learning media in the 21st-century. Learning today requires multimedia that is attractive to students, easy to use and can be accessed remotely (Weiss et al. 2006). The characteristics of this multimedia based on the results of research on vegetation diversity provide new nuance for students and teachers in studying ecological materials based on local potential. Potential locations in the school environment and outside the school can foster a spirit of concern for the environment (Sahebzadeh et al. 2013), science process skills (Dwianto et al. 2017) and scientific inquiry (Cahyani 2017). In addition, the presentation of local potential in the form of biodiversity can increase students' interest in learning so that it can have a positive effect on biology learning outcomes (Fitria et al. 2017). The development of interactive biology multimedia is a simple and effective approach that motivates students and teachers to study plant science (Campbell et al. 2011), and involves the active participation of students in learning (Jacquemart et al. 2016; Subiyantoro & Nugroho 2018).

4 CONCLUSION AND RECOMMENDATION

Biology Interactive Multimedia (BIM) in this present study was developed based on the results of research on the vegetation diversity on one of the coast in Indonesia, Indonesia as the coast has sufficient diversity and abundance so that it can be used as additional ecological material for grade X high school students. The multimedia is in the form of PowerPoint which contains text, images, and graphics, videos with sound effects, and interactive quizzes using macro media. The results of the validation of ecological material experts, learning media experts, practitioners (teachers) and readability tests for high school students on multimedia products show that BIM is feasible and practical, thus it can be implemented in high school biology learning. BIM has interesting features, easy to use, so that it can make students interested and motivated to learn. BIM can also strengthen contextual learning so that it can improve the quality of the process and learning outcomes of high school biology students. In addition, BIM as a 21st-century biology learning medium contains interesting learning activities and materials so that it can increase interactions in biology learning, scientific literacy, and students' higher-order thinking skills. However, this study has limitations, that the BIM has not been integrated into specific learning models/methods so that biology teachers can use BIM in learning combined with learning models. For researchers, the results of this study can be used as a reference in developing similar multimedia or testing its effectiveness and comparing it with other learning media. This study also illustrates that local potential around the school environment or outside the school can be used as material for developing innovative learning media.

REFERENCES

Abdulrahaman, M. D., Faruk, N., Oloyede, A. A., Surajudeen-Bakinde, N. T., Olawoyin, L. A., Mejabi, O. V., Azeez, A. L. (2020). Multimedia tools in the teaching and learning processes: A systematic review. *Heliyon, 6*.

Alemdag, E., & Cagiltay, K. (2018). A systematic review of eye tracking research on multimedia learning. *Computers and Education, 125*, 413–428.

Almasseri, M., & AlHojailan, M. I. (2019). How flipped learning based on the cognitive theory of multimedia learning affects students' academic achievements. *Journal of Computer Assisted Learning, 35*(6), 769–781.

Anggraini, L., Lestari, S. R., & Handayani, N. (2017). Pengembangan multimedia interaktif biologi berbasis Adobe Flash CS6 pada materi sistem sirkulasi manusia kelas XI MIPA SMA Nasional Malang. *Jurnal Pendidikan Biologi, 10*(2).

Barra, E., Aguirre Herrera, S., Pastor Caño, J. Y., & Quemada Vives, J. (2014). Using multimedia and peer assessment to promote collaborative e-learning. *New Review of Hypermedia and Multimedia.*

Borg, W. ., & Gall, M. . (1983). *Education Research* (4th ed). New York: Longman Inc.

Brown, M. H., & Schwartz, R. S. (2009). Connecting photosynthesis and cellular respiration: Preservice teachers' conceptions. *Journal of Research in Science Teaching, 46*(7), 791–812.

Cahyani, R. (2017). Developing multimedia-assisted inquiry learning instruments for basic biology intended to foster students' scientific inquiry. *IOP Conf. Series*. IOP Publishing.

Campbell, K. R., Wilson, S. B., Christopher Wilson, P., & He, Z. (2011). Interactive online tools for teaching plant identification. *HortTechnology, 21*(4), 504–508.

Chang, C. Y. (2002). The impact of different forms of multimedia CAI on students' science achievement. *Innovations in Education and Teaching International, 39*(4), 280–288.

Çimer, A. (2012). What makes biology learning difficult and effective: Students' views. *Educational Research and Reviews, 7*(3), 61–71.

Daniel, S. J. (2020). Education and the COVID-19 pandemic. *Prospects, 49*(1–2), 91–96.

Dwianto, A., Wilujeng, I., Prasetyo, Z. K., & Suryadarma, I. G. P. (2017). The development of science domain based learning tool which is integrated with local wisdom to improve science process skill and scientific attitude. *Jurnal Pendidikan IPA Indonesia, 6*(1), 23–31.

Fitria, A. D., Mustami, M. K., & Taufiq, A. U. (2017). Pengembangan media gambar berbasis potensi lokal pada pembelajaran materi keanekaragaman development of picture media based on local potency for learning materials biodiversity. *AULADUNA: Jurnal Pendidikan Dasar Islam, 4*(2), 14–28.

Grimley, M. (2007). Learning from multimedia materials: The relative impact of individual differences. *Educational Psychology, 27*(4), 465–485.

Guan, N., Song, J., & Li, D. (2018). On the advantages of computer multimedia-aided English teaching. *Procedia Computer Science, 131*, 727–732.

Hastuti, P. W., Setianingsih, W., & Anjarsari, P. (2020). How to develop students' scientific literacy through integration of local wisdom in Yogyakarta on science learning? *Journal of Physics: Conference Series, 1440*(1).

Isnaneny, F. Y., Sajidan, S., & Masykuri, M. (2018). Pengembangan multimedia interaktif berbasis problem based learning (PBL) untuk meningkatkan kemampuan berpikir kritis mahasiswa program studi pendidikan biologi UMS pada materi sistem gerak manusia. *INKUIRI: Jurnal Pendidikan IPA*, *7*(1), 111.

Jacquemart, A. L., Lhoir, P., Binard, F., & Descamps, C. (2016). An interactive multimedia dichotomous key for teaching plant identification. *Journal of Biological Education*, *50*(4), 442–451.

Kasim, J. J. F., Haryanto, H., & Katili, F. (2021). Design of interactive multimedia mobile learning base on Android operating system for Biology subjects. In *IOP Conf. Series*. IOP Publishing.

Kuo, C. Y., Wu, H. K., Jen, T. H., & Hsu, Y. S. (2015). Development and Validation of a Multimedia-based Assessment of Scientific Inquiry Abilities. *International Journal of Science Education*, *37*(14), 2326–2357.

Lee, Y. H., Hsiao, C., & Ho, C. H. (2014). The effects of various multimedia instructional materials on students' learning responses and outcomes: A comparative experimental study. *Computers in Human Behavior*, *40*, 119–132.

Lewinsohn, T. M., Attayde, J. L., Fonseca, C. R., Ganade, G., Jorge, L. R., Kollmann, J., Weisser, W. W. (2015). Ecological literacy and beyond: Problem-based learning for future professionals. *Ambio*, *44*(2), 154–162.

Maddison, R., Foley, L. S., Olds, T. S., Ridley, K., & Jiang, Y. (2014). Validating the multimedia activity recall for children and adolescents in a large New Zealand sample. *Journal of Sports Sciences*, *32*(5), 470–478.

Mayer, R. E. (2017). Using multimedia for e-learning. *Journal of Computer Assisted Learning*.

Nur'aini, F., Chamisijatin, L., & Nurwidodo. (2015). Pengembangan media berbasis multimedia interaktif untuk meningkatkan pemahaman siswa MAN 2 batu materi kingdom animalia. *Jurnal pendidikan biologi Indonesia*, *1*(1), 35–46.

Nurbaiti, Panjaitan, R. G. P., & Titin. (2017). The properness of Adobe Flash basis interactive media for respiratory system learning material. *UNNES Science Education Journal*, *6*(3), 1662–1668.

Panjaitan, R. G. P., Lijana, & Wahyuni, E. S. (2019). The use of comics as a learning medium for ecology. *Unnes Science Education Journal*, *8*(1), 41–45.

Panjaitan, R. G. P., Savitri, E., & Titin. (2016). Pengembangan media e-comic bilingual sub materi saluran dan kelenjar pencernaan. *Unnes Science Education Journal*, *5*(3), 1379–1387.

Patro, E. T. (2008). Teaching aerobic cell respiration using the 5 Es. *The American Biology Teacher*, *70*(2), 85–87.

Purwanta, S., & Sugiharyanto, N. K. (2013). Karakteristik spasial pengembangan wilayah pesisir Daerah Istimewa Yogyakarta dalam konteks UUK DIY. *Laporan Penelitian Hibah Bersaing*. Yogyakarta.

Rasch, T., & Schnotz, W. (2009). Interactive and non-interactive pictures in multimedia learning environments: Effects on learning outcomes and learning efficiency. *Learning and Instruction*, *19*(5), 411–422.

Rice, S. C. (2013). Using interactive animations to enhance teaching, learning, and retention of respiration pathway concepts in face-to-face and online high school, undergraduate, and continuing education learning environments. *Journal of Microbiology & Biology Education*, *14*(1), 113–115.

Ridgway, P. F., Sheikh, A., Sweeney, K. J., Evoy, D., McDermott, E., Felle, P., O'Higgins, N. J. (2007). Surgical e-learning: Validation of multimedia web-based lectures. *Medical Education*, *41*(2), 168–172.

Riduan. (2012). *Metode dan teknik menyusun proposal penelitian*. Bandung: Alfabeta.

Riduwan. (2015). *Skala pengukuran variabel-variabel penelitian*. Bandung: Alfabeta.

Sahebzadeh, B., Kikha, A., Afshari, Z., & Kharadmand, Z. (2013). Effect of environmental factors for teaching of science on academic achievement and interest of students and on their teachers' job statisfaction. *International Journal on New Trends in Education and Their Implications*, *4*(April), 75–88.

Sambodo, R. A., Prayitno, B. A., Karyanto, P., & Sulistyowati, E. (2018). Developing mobile learning as ecology practical tool using three-layer observation framework. *Jurnal Pendidikan Biologi Indonesia*, *4*(3), 225–234.

Sander, E., Jelemenská, P., & Kattmann, U. (2006). Towards a better understanding of ecology. *Journal of Biological Education*, *40*(3), 119–123.

Schaal, S., Matt, M., & Grübmeyer, S. (2012). Mobile learning and biodiversity–bridging the gap between outdoor and inquiry learning in pre-service science teacher education. *Procedia – Social and Behavioral Sciences*, *46*, 2327–2333.

Shana, Z. A. (2009). Apilot study to investigate the effectiveness of multimedia CD-ROM *vis-à-vis* traditional print-based technology in teaching fourth-grade children. *International Journal on E-Learning*, *8*(3), 403–423.

Subiyantoro, S., & Nugroho, A. (2018). Android-based instructional media development procedure to enhance teaching and learning in the age of disruption 4.0. *International Conference on Applied Science and Engineering (ICASE 2018) Android-based* (Vol. 175, pp. 152–155). Atlantis Press.

Sudarisman, S. (2015). Memahami hakikat dan karakteristik pembelajaran biologi dalam upaya menjawab tantangan abad 21 serta optimalisasi implementasi kurikulum 2013. *Florea*, *2*(1), 29–35.

Susilana, R., & Riyana, C. (2009). *Media Pembelajaran*. Bandung: CV Wacana Prima.
Tabor, S. W., & Minch, R. P. (2013). Student adoption and development of digital learning media: action research and recommended practices. *Journal of Information Technology Education: Research*, *12*.
Taufiqoh, P. L., & Indana, S. (2012). Profil media pembelajaran interaktif berbasis komputer pada materi sistem peredaran darah manusia. *BioEdu*, *1*(2), 1–4.
Topano, A., Asiyah, Basinun, Walid, A., Alimi, & Febrini, D. (2021). Improving student cognitive learning outcomes through the development of interactive multimedia-based biology learning at Muhammadiyah Improving student cognitive learning outcomes through the development of interactive multimedia-based biology learning a. In *Journal of Physics: Conference Series*. IOP Publishing Ltd.
Tur, G., & Marín, V. I. (2015). Enhancing learning with the social media: student teachers' perceptions on Twitter in a debate activity. *New Approaches In Educational Research*, *4*(1), 46–53.
Vagg, T., Balta, J. Y., Bolger, A., & Lone, M. (2020). Multimedia in education: What do the students think? *Health Professions Education*, *6*(3), 325–333.
Wang, J., Mendori, T., & Hoel, T. (2019). Strategies for multimedia learning object recommendation in a language learning support system: verbal learners vs. visual learners. *International Journal of Human-Computer Interaction*, *35*(4–5), 345–355.
Wastiau, P., Blamire, R., Kearney, C., Quittre, V., Van de Gaer, E., & Monseur, C. (2013). The use of ICT in education: A survey of schools in Europe. *European Journal of Education*, *48*(1), 11–27.
Weiss, I., Kramarski, B., & Talis, S. (2006). Effects of multimedia environments on kindergarten children's mathematical achievements and style of learning. *Educational Media International*, *43*(1), 3–17.
Wulandari, T. A. J., Sibuea, A. M., & Siagian, S. (2019). Pengembangan media pembelajaran berbasis multimedia interaktif pada mata pelajaran biologi. *Jurnal Teknologi Informasi & Komunikasi Dalam Pendidikan*, *5*(1), 75–86.
Yeop, M. A., Yaakob, M. F. M., Wong, K. T., Don, Y., & Zain, F. M. (2019). Implementation of ICT policy (blended learning approach): Investigating factors of behavioral intention and use behavior. *International Journal of Instruction*, *12*(1), 767–782.
Zairana, L., Djulia, E., & Harahap, H. (2020). Pengembangan multimedia pembelajaran interaktif biologi menggunakan Adobe Flash CS 6 pada materi. *Jurnal Pelita Pendidikan*, *8*(1), 99–109.
Zhang, D. (2005). Interactive multimedia-based e-learning: A study of effectiveness. *International Journal of Phytoremediation*, *21*(1), 149–162.
Zheng, R. Z., Yang, W., Garcia, D., & McCadden, E. P. (2008). Effects of multimedia and schema induced analogical reasoning on science learning. *Journal of Computer Assisted Learning*, *24*(6), 474–482.

Hots-link mobile learning application: Instrument validity using Borich's percentage of agreement

Endang Susantini, Isnawati & Raharjo
Universitas Negeri Surabaya, Surabaya, Indonesia

Slamet Hariyadi
Universitas Jember, Jawa Timur, Indonesia

Nur Qomariyah
University of Groningen, Groningen, The Netherlands

ABSTRACT: Many HOTS-based mobile learning applications should be validated to be a reliable HOTS-learning medium. This study aimed to describe an instrument validity that could be used to assess the developed HOTS-Link mobile learning application. This study used a descriptive-qualitative research approach with an FGD research design. There were nine ratters whose expertise was Biology education chosen purposely and asked to validate the instrument. A validation sheet consisting of 63 questions with a four-point Likert scale divided into three domains namely instruction, design, and engagement. The obtained data were analyzed using Borich's inter-observer agreement to determine instrument reliability. Results depicted that 56 questions showed more than 75% of Borich's percentage of agreement, whereas, the other 7 questions were less than 75% of Borich's percentage of agreement. This study implies that the instrument consisting of 56 questions can be used as the validity instrument by not involving the extraneous 7. This study could be the reference for further research about developing valid and reliable instruments to assess HOTS-based mobile learning applications.

Keywords: Borich, HOTS, instrument, mobile learning application, validity.

1 INTRODUCTION

The development of Information and Communication Technology or ICT is advancing very rapidly in various fields including but not limited to education (del Carmen Ramírez-Rueda et al. 2021). Along with the development of technology, the role of ICT in education is indeed important (Alam & Ujjwal 2017). The use of ICT in learning has great potential to improve the quality of the learning process while, at the same time, providing opportunities for students to learn and apply technology as one of the required skills to be mastered in 21st-century education (Lawrence & Usman 2018). In addition to technical mastery, another important skill in 21st-century education is higher-order thinking skills. Higher-Order Thinking Skill (HOTS) is a thinking process that involves one's ability to apply, relate, and manipulate new information received with previous information to solve problems effectively (Schraw & Robinson 2011). HOTS is mainly characterized by the ability to think both critically and creatively (Conklin & Manfro 2012).

One form of technology integration in supporting digital learning is through the use of mobile applications. Currently, mobile learning applications have been widely available regardless of the coping fields (Arista & Kuswanto 2017). The ease of use and the freedom to access content are some

DOI 10.1201/9781003347798-12

of the advantages of using mobile learning applications (Astra et al. 2015). Many previous studies developed mobile learning applications to train HOTS, however, most of them were intended to increase students' HOTS in the field of mathematics (Dewi et al. 2020) and physics (Setiyoaji et al. 2020). To the best of the researchers' knowledge, few studies were undertaken related to developing HOTS-based mobile learning applications to train preservice teachers' skills in devising a HOTS-based lesson plan (Gunawan et al. 2021; Suwarma & Apriyani 2022). Due to its importance, the present study developed an Android-based mobile learning application called HOTS-Link. The application was available on Playstore that can be freely installed and accessed via Android-based devices. The application contained HOTS materials and their relation to 4C and Biology learning models. However, the quality of the application was still unknown. Therefore, it is necessary to develop an appropriate instrument to measure its quality as a reference in designing quality applications.

The instrument must meet the validity criteria. Widoyoko (2012) states that a valid instrument can measure what it wants to measure. In this study, the assessment items in the validation instrument referred to good application criteria by Lee & Cherner (2015). This study aimed to describe the validity of the instrument that can be used to measure the quality of the HOTS-Link mobile learning application.

2 LITERATURE REVIEW

In this ICT-based learning era, the use of Android-based mobile learning applications has been widely used by diverse education stakeholders (Nofitasari et al. 2021). Android is widely known as an operating system used by more than 85% of smartphones (Statista 2018). Several studies on the use of Android-based mobile learning applications have been carried out, including in Biology learning. Arista & Kuswanto (2017) conducted a study on an Android-based virtual lab to increase students' autonomous learning. Kocakoyun & Bicen (2017) undertook a study on the use of Android-based mobile learning applications to increase students' motivation and learning outcomes. Pahlifi and Nurcahyo (2018) also had done a study on the development of an Android-based illustrated Biology dictionary on the topic of invertebrates to increase students' learning motivation. At last, Mukarromah & Razak (2020) investigated the use of Android-based e-worksheets that made students easier to understand biological materials because of supporting interactive contents in the form of Biology pictures and videos.

In addition, Verdecchia et al. (2019) stated that producing a good quality application is indeed considerable even though it has been everyone's challenge. There are a wide variety of criteria as a measure of application quality (Nouri et al. 2018). Lee and Cherner (2015) made a comprehensive rubric to assess application quality, which consists of 24 evaluative sub-skills divided into three main aspects namely instruction, design, and engagement. The instruction aspect consists of ten sub-skills namely thoroughness, 21st-century skills, relation to future learning, error scores, feedback to teachers, level of learning material, learning cooperativeness, and accommodation of differences between individuals, media integration, and cultural sensitivity. The engagement aspect encompasses seven sub-skills assessing whether the application is likely to be in the viewer's interest. The design aspect consists of nine sub-skills measuring application design based on the ability to store user work progress, platform integration, screen design, ease of use, navigation, goal orientation, and presentation of student's information including learner control, interactivity, speed, personal preference, interest, aesthetics, and utility. In connection with a valid instrument, Pasek et al. (2015) and Creswell (2005) state that an instrument that meets the validity criteria can provide meaningful scores and allow researchers to draw good conclusions from the measured sample. This is intended so that there is a match between the assessment criteria in the instrument and aspects measured (Mohamad et al. 2015). By referring to those previous studies, the present study aimed to adapt the developed rubric to generate a validation instrument for HOTS-based mobile learning applications.

3 METHOD

This study used a descriptive-qualitative research approach with a Focus Group Discussion (FGD) research method. The present study was conducted in one of the state universities in Indonesia and focused on developing a HOTS-based mobile learning application to train preservice teachers' lesson planning skills for the HOTS learning process. The application was specialized for the Indonesian context due to different ways of thinking skills among countries. There were nine Biology education experts asked to examine the application for three days with a validation instrument. Among them, 4 experts were male and 5 others were female, of which their ages ranged from 32 to 57 years old. During the FGD process, there was an interaction of perceptions as well as interchangeable ideas and opinions, followed by an in-depth interview. The validation instrument was compiled based on good application criteria developed by Lee & Cherner (2015) containing 63 questions divided into three main aspects namely instruction, design, and engagement. The data were obtained through the completed validation instrument carried out by providing an assessment of each question item using four rating scales, namely (1) bad, (2) fair, (3) good, and (4) very good. The obtained data were then analyzed using Borich's formula of the percentage of agreement. A valid instrument was based on the provisions of the inter-rater agreement where the instrument could be used if the value of the inter-rater agreement was above 75% (Borich 1994).

4 FINDING AND DISCUSSION

The present study found that there were seven out of 63 developed items that did not meet the percentage of agreement (M < 75%). The seven items included three items of the instruction aspect, two items of the design aspect, and two aspects of the engagement aspect. The other 56 items were in accordance with Borich's percentage of agreement (M ≥ 75%) so that they were valid and could be used as a validation instrument. The following subheadings portrayed the irrelevant items including the discussion of why such phenomena occurred.

4.1 *Instruction aspect*

This subchapter mainly discussed the instructional aspects that were suggested to be excluded in the instrument due to the violence of Borich's (1994) theory. Table 1 depicts the elicited instructional aspects.

Table 1. Elicited items in instructional aspect.

Item	Constructs	Scale giving by Expert									Percentage of Agreement
		1	2	3	4	5	6	7	8	9	
	Material Sufficiency and Coverage										
13	Demonstrating a correct concept of HOTS materials and its relation to the 4C	4	4	4	2	4	4	4	4	4	67%
22	Referring to trusted and up-to-date sources	4	4	4	4	3	2	4	4	4	67%
	Student Accommodation and Discussion										
25	Providing an opportunity for discussion	4	3	4	2	4	4	4	4	4	67%

The instruction aspects comprised ten sub-skills namely the rigor of application in supporting HOTS skills, linkages to the 21st-century learning and beyond, feedback capability, material adequacy and coverage, and learner accommodation and discussion. In accordance with Table 1, there were three irrelevant items according to Borich's percentage of agreement (M < 75%), of which two of them were in the sub-skill of material adequacy and one item was in the sub-skill of coverage and learner accommodation and discussion. In connection with the sub-skill of material adequacy and coverage, the two items encompassed demonstrating a correct concept of HOTS materials and their relation to the 4C and referring to trusted and up-to-date sources (M = 67%). Whereas, in connection with the sub-skill of learner accommodation and discussion, the irrelevant item was providing an opportunity for discussion (M = 67%). Due to the minimum score of Borich's percentage of agreement, those three items were then elicited in the validation instrument.

The rationale beyond the lower Borich's percentage of agreement for the item of demonstrating a correct concept of HOTS materials and their relation to the 4C was the fact that the HOTS-Link application, previously invented and developed, did not accommodate the practice of the four Cs (e.g., communication, critical thinking, creative, and collaboration). This was acceptable since HOTS-based learning applications would less emphasize communication, creativity, and collaboration skills. Al-Zou'bi (2021) stated that general communication somehow did not require higher-order thinking skills, i.e., casual communication to approach students with lesser cognitive attainment or performance. Moreover, only specified-contextualized communication needed HOTS to speak the mind clearly and effectively (Al-Zou'bi 2021). On the other side, many scholars have decided that creativity and critical thinking skills were two un-interchangeable concepts, rather, supplementary. Even if some others argued that both skills were in a slightly closed cut-off, there were many aspects to differ such as the existence of packaging a case with an interesting language (Al-Zou'bi 2021). At last, collaborative skills were extraneous to critical thinking skills as the skills were about engaging with different people's backgrounds such as perspectives and sights (Santoso et al. 2021), and not about emphasizing how to put a claim together with relevant reason, evidence, and conclusion (Nussbaum 2021).

4.2 *Design aspect*

The design aspects that were suggested are explained below (Table 2) that show the elicited design aspect.

Table 2. Elicited items in design aspect.

Item	Constructs	Scale giving by Expert 1	2	3	4	5	6	7	8	9	Percentage of Agreement
29	Platform and Media Integration Promoting a combination of images to clarify the given concept	4	4	4	2	4	3	4	3	4	67%
36	Information Presentation Demonstrating clearly visible writing on the table	4	4	4	2	3	3	4	4	4	67%

There were five sub-skills in the design aspect covering platform and media integration, layer design, information presentation, ease of use and navigation, goal orientation, and cultural sensitivity. In accordance with Table 2, two items got scores below Borich's percentage of agreement (M < 75%). The one item in the sub-skill of platform and media integration was promoting a combination of images to clarify the given concept (M = 67%). Meanwhile, another one was in the sub-skill of information presentation, namely demonstrating visible writing on the table (M = 67%). Due to the minimum score of Borich's percentage of agreement, those three items were then elicited in the validation instrument. The item of promoting a combination of images to clarify the given concept got a low score since the HOTS-Link application did not provide illustration over the concept of higher-order thinking skills. Miller & Hartzog (2020) said that it was difficult to transform a construct into picturization. However, the application had provided the learners with detailed explanations, so that the learners might need some professional's assistance while using the application, i.e., Biology education experts. On the contrary, some scholars stated that a construct, such as critical thinking skills, actually could be picturized in a form of a mind mapping even though it was still problematic in representing good critical thinking skills (Miller & Hartzog 2020).

In most product validation sheets, the design aspect was crucial as it represented the user's attractiveness and visible application (Jomsri 2020). An attractive design of learning media might elevate learners' learning motivation (Rachmavita 2020). Rachmavita (2020) also stated that a good design should consider product appearance as well as text fonts used because such visualization did matter. To achieve an interesting design, the user's preference should be in a pre-consideration of making an Android-based application. In accordance with the present study, therefore, the validation sheet included all elements of design namely platform and media integration, layer design, information presentation, ease of use and navigation, goal orientation, and cultural sensitivity.

4.3 *Engagement aspect*

Another aspect from Borich's (1994) theory is engagement. Table 3 shows the elicited engagement aspects.

Table 3. Elicited items in engagement aspect.

Item	Constructs	Scale giving by Expert									Percentage of Agreement
		1	2	3	4	5	6	7	8	9	
	Interaction										
53	Encouraging user interaction	4	4	4	2	4	4	4	4	4	67%
	Personal Preference										
55	Promoting freedom in determining username	2	3	4	3	4	3	4	4	4	67%

The present study found that there were five sub-skills in the engagement aspect, namely learner control and discretionary access, interaction, personal preferences, interests and aesthetics, and utility. Regarding Table 3, there were two items in the engagement aspect that had low Borich's percentage of agreement, namely encouraging user interaction in the sub-skill of interaction and promoting freedom in determining username in the sub-skill of personal preference (M = 67%). The low score in encouraging user interaction implied that such an item was not necessary to be included in assessing HOTS-based mobile learning applications. This was due to the fact that not all HOTS-based applications aimed to make an interactive process with the users. Moreover, interactions were less prioritized in most Android-based applications (Ewing & Cooper 2021). Most Android-based learning applications, especially HOTS-based app, only provided specific and relevant information and examples. Hence, the item was insignificant to be included in a product validation sheet and should be elicited. The second item, namely promoting freedom in determining username, got a low Borich's percentage of agreement. This implied that users could not do profile customization, including name. Even if this was preferable in certain cases, the results of the inter-rater showed that the item should be elicited as some applications might not complicate users with registration (Mansilla et al. 2020).

In general, learning applications must have valid and appropriate criteria to be used so as to support the attractiveness of learning activities (Astra et al. 2015). On the Android platform, there were many mobile learning applications that could be accessed, so it was necessary to choose which application was good to use. Lee & Cherner (2015) stated that in order for teachers to choose good applications to use in learning, teachers needed measuring tools as the basis for selecting applications. The use of the measuring instrument must also pay attention to its validity, because the validity of the instrument was one of the important considerations when conducting an educational intervention (Lakshmi & Mohideen 2013). Choosing a good application is important.

Bhakti & Rahman (2020) stated that the use of android applications that were in accordance with user characteristics would provide a good understanding of a context. This could be measured in the HOTS-Link mobile learning application instruction domain that comprised the ease of application use in supporting learning achievement. A good application would also provide better user experience and good application performance (Hoque et al., 2020). This could be measured in the design and engagement domains developed in the HOTS-Link mobile learning application validation instrument.

5 CONCLUSION AND RECOMMENDATION

The present study aims to develop a valid instrument for assessing HOTS-based mobile learning applications. It results in a very good and valid instrument as 89% of the items of the instrument are ready-to-use and the score of the inter-rater agreement is above 75%. In this study, the items of the instrument were assessed based on the instruction, design, and engagement domains. The result of the present study is expected to be useful in the field of pedagogy in helping teachers develop similar mobile learning applications to improve students' HOTS. By inventing this valid instrument, teachers could have a reference of a valid instrument in developing an Android-based application. They know what to prepare and what to develop by referring to the instrument items. However, the present study is still limited to the Indonesian context. Due to the diverse concept of rhetoric, it is suggested for further research to investigate how to develop HOTS-based mobile learning applications according to different rhetorical communities.

ACKNOWLEDGEMENTS

The authors thank DIPA Directorate General of Higher Education, Directorate of Resources, Ministry of Education, Culture, Research, and Technology, who gave financial assistance Number SP DIPA-023.17.1.690439/2021. They also thank Universitas Negeri Surabaya that provided outstanding support and many stakeholders who were in-charge in the research completion.

REFERENCES

Alam, K., & Ujjwal, K. H. (2017). Attitude towards the use of ICT in classroom among trainee-teachers'. *Journal of Education and Development*, 7(13), 249–257.

Al-Zou'bi, R. (2021). The impact of media and information literacy on acquiring the critical thinking skill by the educational faculty's students. *Thinking Skills and Creativity*, 39, 100782.

Arista, F., S., & Kuswanto, H. (2017). Virtual physics laboratory application based on the android smartphone to improve learning independence and conceptual understanding. *International Journal of Instruction*, 11(1), 1–16.

Astra, I., M., Nasbey, H., & Nugraha, A. (2015). Development of an android application in the form of a simulation lab as learning media for senior high school students. Eurasia Journal of Mathematics, *Science & Technology Education*, 11(5), 1081–1088.

Bhakti, C. P., & Rahman, F. A. (2020). Android application development for exploration career based on Multiple Intellegence: A model hypothetical. *Journal of Physics: Conf. Ser.* 1470 012043.

Borich, G. D. (1994). *Observation Skill for Effective Teaching*. New York: Macmillan Publishing Company.

Conklin, W., & Manfro, J. (2012). *Higher Order Thinking Skill to Develop 21st Century Learners*. Huntington: Shell Education Publishing. Inc.

Creswell, J.W. (2005). *Educational Research: Planning, Conducting and Evaluating Quantitative and Qualitative Research*, 2nd Edition. New Jersey: Pearson Merrill Prentice Hall.

Carmen Ramírez-Rueda, M., Cózar-Gutiérrez, R., Colmenero, M. J. R., & González-Calero, J. A. (2021). Towards a coordinated vision of ICT in education: A comparative analysis of Preschool and Primary Education teachers' and parents' perceptions. *Teaching and Teacher Education*, 100, 103300.

Dewi, R. M., Ghofur, M. A., & Soejoto, A. (2020). Pelatihan game edukasi android berbasis hots sebagai media evaluasi pembelajaran. Abimanyu: *Journal of Community Engagement*, 1(1), 59–67.

Ewing, L. A., & Cooper, H. B. (2021). Technology-enabled remote learning during COVID-19: perspectives of Australian teachers, students and parents. *Technology, Pedagogy and Education*, 30(1), 41–57.

Gunawan, K. D. H., Liliasari, L., Kaniawati, I., & Setiawan, W. (2021). Implementation of competency enhancement program for science teachers assisted by artificial intelligence in designing HOTS-based integrated science learning. *Jurnal Penelitian dan Pembelajaran IPA*, 7(1), 55–65.

Hoque, M. A., Rao, A., & Tarkoma, S. (2020). Network and application performance measurement challenges on android devices. *ACM SIGMETRICS Performance Evaluation Review*, 48 (3): 6–11.

Jomsri, P. (2020, August). Usability evaluation for user interface design of application for recommender system to enhance the potential of community-based tourism in phatthalung, thailand. In *Journal of Physics: Conference Series*, 1627, (1). IOP Publishing.

Kocakoyun, S., & Bicen, H. (2017). Development and evaluating of educational android application. *Cypriot Journal of Educational Sciences*, 12 (2), 58–68.

Lakshmi, S., & Mohideen, M. A. (2013). Issues in reliabiliy and validity of research. *International Journal of Management Research and Review*, 3(4): 2752–2758.

Lawrence, J.E., & Usman, A. T. (2018). Factors that influence teachers' adaption and integration of ICT in technology/learing process. *Educational Media International*, 55(1), 79–105.

Lee, C-Y., & Cherner, T. S. (2015). A comprehensive evaluation rubric for assessing instructional apps. *Journal of Information Technology Education: Research*, 14, 21–53.

Mansilla, L., Milone, D. H., & Ferrante, E. (2020). Learning deformable registration of medical images with anatomical constraints. *Neural Networks*, 124, 269–279.

Miller, C. R., & Hartzog, M. (2020). "Tree thinking": the rhetoric of tree diagrams in biological thought. *Poroi: An Interdisciplinary Journal of Rhetorical Analysis & Invention*, 15(2).

Mohamad, M. M., Sulaiman, N. L., Sern, L. C., & Salleh, K. M. (2015). Measuring the validity and reliability of research instruments. *Procedia – Social and Behavioral Sciences*, 204, 164–171.

Mukarromah, N. A., & Razak, A. (2020). Problem analysis and needs electronic student worksheet based on android biology material in SMAN 3 Padang. *International Journal of Progressive Sciences and Technologies*, 24 (1), 155–162.

Nouri, R., Kalhori, S. R. N., Ghazisaeedi, Marchand, G., & Yasini, M. (2018). Criteria for assessing the quality of mHealth apps: a systematic review. *Journal of the American Medical Informatics Association*, 15(8): 1089–1098.

Nofitasari, A., Lisdiana, L., & Marianti, A. (2021). Development of my biology app learning media based on android materials of food digestion systems as student learning source at senior high school. *Journal of Innovative Science Education*, 10(1), 70–78.

Nussbaum, E. M. (2021). Critical integrative argumentation: Toward complexity in students' thinking. *Educational Psychologist*, 56(1), 1–17.

Pahlifi, D. M., & Nurcahyo, H. (2018). The effect of android-based pictorial biology dictionary on students' motivation on topic of invertebrate. *Journal of Physics: Conf. Ser.* 1241 012048.

Pasek, K. H., Zosh, J. M., Golinkoff, R. M., Gray, J. H., Robb, M. B., & Kaufman, J. (2015). Putting education in "educational" apps: lessons from the science of learning. *Psychological Science in the Public Interest*, 16 (1), 3–34.

Rachmavita, F. P. (2020, October). Interactive media-based video animation and student learning motivation in mathematics. In *Journal of Physics: Conference Series*, 1663 (1). IOP Publishing.

Schraw, G., & Robinson, D.H. (2011). *Assessment of Higher Order Thinking Skills*. New York: Information Age Publishing, Inc.

Santoso, A. M., Primandiri, P. R., Zubaidah, S., & Amin, M. (2021, March). Improving student collaboration and critical thinking skills through ASICC model learning. *Journal of Physics: Conference Series*,1806 (1), 012174. IOP Publishing.

Setiyoaji, W. T., Supriana, E., & Laksono, Y. A. (2020). Pengembangan e-book berbasis android dengan soal HOTS untuk membantu menganalisis besaran pada materi gerak lurus. *Jurnal Pendidikan Fisika dan Teknologi*, 6(1), 114.

Suwarma, I. R., & Apriyani, S. (2022). Explore teachers' skills in developing lesson plan and assessment that oriented on higher order thinking skills (HOTS). *Journal of Innovation in Educational and Cultural Research*, 3(2), 106–113.

Verdecchia, R., Malavolta, I., & Lago, P. (2019). *Guidelines for architecting android apps: a mixed-method empirical study*. 2019 IEEE International Conference on Software Architecture (ICSA), 2019, 141–150.

Widoyoko, S. E. P. (2012). *Teknik Penyusunan Instrumen Penelitian*. Yogyakarta: Pustaka Pelajar.

Teacher Education and Teacher Professional Development in the COVID-19 Turn – Drajati & Putra (Eds)
© 2023 The Author(s), ISBN: 978-1-032-38970-7

The analysis of students' metacognitive skills, learning motivation, communication skills, and their correlations

Lina Listiana, Cholifatul Chasanah, Aisyahtur Rodhiyah & Ruspeni Daesusi
Universitas Muhammadiyah Surabaya, Jawa Timur, Indonesia

ABSTRACT: This study aims to analyze the level of metacognitive skills, students' motivation, and communication skills. To interwave discussion on metacognitive skills, learning motivation, and communication skills, this study used a quantitative approach. The participants were seventy-eight senior high school students in the second grade. There were some layers that we used in this study (e.g., an essay test was used to collect data on metacognitive skills, and communication skills and motivation data were obtained through observation). The findings showed a positive relationship between metacognition and motivation with communication. The policy implication from the correlations study among metacognitive skills, learning motivation, and communication skills would also be discussed.

Keywords: communication skills, Indonesian senior high school students, learning motivation, metacognitive skills.

1 INTRODUCTION

Learning is directed towards blended and hybrid learning models in this digital era. Various efforts have been conducted so that learning remains of high quality, including the improvement of the student learning motivation factor. Learning motivation plays an important role in increasing active learning. Learning motivation is considered as one of the factors that influence learning outcomes (Burguillo, 2010). Besides that, learning motivation is also closely related to other learning activities such as communicating in assessing their work, managing learning methods, and determining learning strategies. In fact, there is a lack of students' learning motivation, which is shown in student learning activities such as discussing, communicating, questioning, and answering; even students' metacognitive skills are less developed during the COVID-19 pandemic.

Metacognitive skills are particularly important for students to improve their thinking capacity (Kozikoğlu, 2019). In the learning process, students are assisted by the empowerment of metacognitive skills to be more independent. Winne & Nesbit (2010) reviewed select findings from this voluminous literature and identify two domains of psychology where metacognitive skills have contributed to the improvement of the learning processes. Bahri & Corebima (2015) conducted research that focused on investigating the correlation between learning motivation and metacognitive skills on students' cognitive learning outcomes simultaneously in classes. Those studies concluded that learning motivation and metacognition play a role in cognitive learning outcomes. Urban et al., (2021) investigated different levels of metacognitive awareness with the differences in learning motivation and creativity among students. Besides metacognitive skills, communication skills are important for students in the learning process.

Many previous studies only describe the students' metacognitive skills without explaining the description of students' communication skills and the description of student learning motivation. Those researches do not investigate the relationship between these variables studies (metacognitive skills, communication skills, and student motivation). Therefore, this study aims to describe the level of metacognitive skills, communication skills, and student motivation, especially in Senior

 DOI 10.1201/9781003347798-13

High School of science class XI. This study also describes the correlation between these variables. This study will explore the following research questions: (1) what is the level of students' metacognitive skills, learning motivation, and communication skills? (2) Is there any correlation between metacognitive skills and communication skills? (3) Is there any correlation between learning motivation and communication skills? (4) Is there any correlation between metacognition skills and learning motivation?

2 LITERATURE REVIEW

Metacognition is a term introduced by Flavel (1979) and is interpreted as knowledge of the cognitive process. The method of measuring metacognition is assessed by giving the implementation of the assignment, which is given before the assignment, during the assignment, and after the assignment (Veenman & Spaans, 2005). On the other hand, the method of measuring metacognition can be determined according to the type of measuring instrument used to measure metacognition. Tobias & Everson (1996) explained that metacognitive measurement uses the technique of assessing knowledge monitoring, which is conducted simultaneously to evaluate the students' self-reports about declarative knowledge and procedural knowledge. Procedural knowledge related to problem-solving in mathematics or science is then assessed by giving an objectively assessed test. Metacognitive skills are classified among the higher cognitive abilities because it includes several elements such as analysis, synthesis, and evaluation. Three indicators of metacognitive skills include planning (e.g., setting goals, enabling relevant resources, choosing the right strategy), evaluation (e.g., determining the level of understanding of a person, and how to choose the right strategy), and monitoring (e.g., checking one's progress, and choose the appropriate improvement strategies when the chosen strategy does not work) (Anderson & Krathwohl, 2001). Metacognitive skills test is applied by having several questions that are integrated with metacognitive skills indicators, such as planning, monitoring, and evaluating (Krathwohl, 2002). There are seven scales (e.g., 0-7) in the metacognitive skills rubric, which include: 1) the response in his own words, 2) the order of a coherent answer, 3) the grammar or language, 4) the reason (analysis/evaluation, creation), and 5) the answer (right/less/not really/blank) (Corebima, 2009, 2011).

Keller's (1983) motivation design model has four areas that happened in sequence. They are Attention, Relevance, Confidence, and Satisfaction (ARCS). Keller & Suzuki (2004) acknowedge the use of various learning media and models that can improve students' attention so they can follow the learning well. Consequently, students can be more confident in doing their assignments based on their abilities. Satisfaction can be attained through rewards or personal attention from the lecturer. Moreover, previous research has shown that learners with more learning motivation had better learning outcomes (Burguillo, 2010). It has been supported by Daumiller et al., (2021) that students with higher achievement motivation have been shown having a more sustained commitment and more cognitive engagement with learning materials.

Communication skills, as well as metacognitive and student motivation, are important competencies that students must possess. Alias & Osman (2015) explain that communication skills are a type of learning that involves many activities that require the students to interact as a team to achieve solutions through information gained in presenting writing or presenting processes. Banks (2005) suggested four ways to implement assessment effectively: 1) focusing on the "big idea" and not depending merely on facts and certain skills; 2) involving interaction between educator and student, student and student; 3) providing multiple approaches to demonstrate the achievement of students; 4) using a scoring scheme focused on the main idea of a task. Referring to Nakamura & Valens (2001), oral measurement communication includes the following criteria: 1) presentations: content, language, eye contact; 2) interviews: comprehension, fluency, ability to explain an idea; 3) discussing and debating: times other than when asked a direct question, uses additional questions to include others in conversation, transfers skills used in dialogues to group discussions; 4) the categories used in the analysis are as follows: 1 (poor), 2 (fair), 3 (enough), and 4 (good).

Metacognitive skill is considered to play an important role in many types of cognitive activity, including comprehension by considering the correlation between metacognitive skill and cognitive learning outcome (Coutinho, 2007). Learning motivation is considered another factor that influences the learning outcome more than the learning arrangement (Paris et al., 1983). Students having high learning motivation tend to have a metacognitive strategy and tend to maintain the strategy in accomplishing the given assignment than those who have low learning motivation (Pintrich & De Groot, 1990). Furthermore, metacognition was significantly correlated with intrinsic motivation (Tas et al. (2012). Besides, communication is an important element in science education because students need to transfer their findings in the form of charts, diagrams, tables, figures, or graphs and explain them clearly (Alias & Osman, 2015). Previous research on metacognitive skills, student motivation, and communication skills has discussed that metacognitive skills helped students make a plan, follow certain progress, and monitor their learning. It is not separable from learning motivation, which has implications for the ability to communicate the results of their assignments. This study aims to reveal the correlation of three variables. Therefore, this study interweaves those theories by the multiple correlation studies, so the contribution of each factor might be revealed, where metacognitive skills potentially contribute to communication, motivation contribute to communication, and metacognitive and motivation contribute to communication.

3 METHOD

This study was correlational research conducted to reveal the contribution of metacognitive skills and learning motivation to student communication skills. There were 78 second grade students from science classes who consisted of 27 males and 51 females. There were 4 schools from 7 schools selected by purposive sampling in this study. One class was taken from each school as the research target. The school's selection as a research target was based on the availability of facilities that supported online learning during the COVID-19 pandemic. Online learning at the target school was conducted through video conference using Zoom application. The students were learning online for 2 hours and 15 minutes, which the material observed for 130 minutes (3 x 45 minutes) in each week. This research was conducted for 3 months in the odd semester of 2021.

This study used quantitative research with the correlation method where we applied an essay test with 0 to 100 scores. The scores would be clustered and interpreted into some categories (e.g., very well-developed, well-developed, beginning to develop, not so developed, and still very risky (Green, 2002) for metacognitive skills; Keller (1983) for learning motivation with some categories (e.g., good, enough, fair and poor); and Nakamura & Valens, (2001) communication skills with some categories (e.g., good, enough, fair and poor) from students' tests and activities. The students' metacognitive skills were measured by using an essay test which consisted of 8 numbers (e.g., how do hormones affect the menstrual cycle? explain your answer!) taken from pre-test and post-test. Before the test was used, a validity analysis was carried out including content validation, construct validation, and empirical validation to determine the reliability value. The concept of understanding test instruments that reflected two biology subjects was referred to as content validation. The construct validation was conducted by expert validators. Empirical validity is determined by testing the test on students who have received the subject being tested. The empirical validity used was the item validity which was calculated using the *Product Moment* correlation formula. The reliability of the learning outcomes test instrument was determined by using the *Alpha Cronbach* formula. Moreover, the students' learning motivation was measured by using a student behavior observation sheet to show developed motivation referred to as the ARCS model (Keller, 1983) which was adopted by the researcher for developing into an observation sheet grid that contained Keller's four indicators. There were two items each that became part of the learning motivation observation rubric. One student was observed by three teachers. The students in one class were divided into four groups. The observation was conducted on each student in each group three times. Moreover, each observer calculated the score from three observations. The communication skills data was collected through the observation of verbal communication aspects. Communication skills data was taken when students made presentations. The observation sheet contained the observation rubric guide with a numeral rating scale, which consisted of 6 indicators each with 2 items being observed (Nakamura & Valens, 2001). The observation assessment from the online work presentation was conducted twice by each student in the group. During the presentation, each group member took turns delivering the results of their work. The presentation was held twice, with different subjects including the reproductive hormone system and family planning program. The data of this study were analyzed by using regression analysis to examine the correlation through SPSS 27.

This research hypothesized that there was a positive correlation between metacognitive skills and learning motivation with students' communication skills.

4 RESULT AND DISCUSSION

The results of the analysis of students' metacognitive skills, communication skills, and learning motivation were presented in Table 1 to Table 3. Meanwhile, the results of the correlational tests between these variables are presented in Table 4 to Table 6. The results of this study revealed more about the relationship between these variables and their contributions. The contribution of metacognitive skills and learning motivation to communication skills, both were good predictors. The analysis of metacognitive skills, communication skills, and learning motivation had implications for further research. They were for the development of learning strategies that were applied to different populations and academic abilities.

4.1 *Level of students' metacognitive, motivation and communication skills*

The level of metacognitive skills varied, according to the results of a test essay designed to measure students' metacognitive skills. Table 1 shows the level of metacognitive awareness.

Table 1. The categories of students' metacognitive skills.

Scores	Categories	Number of students	Percentage (%)
$81 \leq X - 100$	Very well-developed	26	33,33
$61 \leq X < 80$	Well-developed	44	56,41
$41 \leq X < 60$	Begins to develop	8	10,26
$21 \leq X < 40$	Not so developed	0	0,00
$X < 20$	Still very risky	0	0,00
Total		78	100,00

Table 1 showed that the students' metacognitive skills categories of very well-developed were amount 33,33%, well-developed was 56,41%, and begins to developed was 10,26%. Generally, the students' metacognitive skills have developed well. In the planning stage, it showed that students could solve problems (e.g., how to regulate the population growth rate control?). Task orientation could assist students to understand their tasks. As a result, students could plan actions to solve the problem. The students worked in groups to plan task completion strategies and share assignments. When students worked systematically, it was possible to monitor the results that they were working on. Likewise, monitoring showed that students were able to monitor their learning progress and correct errors by finding relevant information from other sources. Moreover, in the evaluation stage, it showed that students were able to provide an assessment of themselves and others, such as when students in a group could assess their answers and assess the result of assignments that were presented by other groups. Vuorre & Metcalfe (2021) explained that students could know the material that has been understood and not understood by giving the assignment so that they could control their cognitive process. It was connected to metacognitive assessment that there was a relationship with the accuracy of the answers to the given task. Consequently, metacognitive skills led to the ability of high-order thinking involving active control of a particular cognitive process in learning, so metacognitive skills assisted students to make a plan, follow certain progress, and monitor their learning process (Hacker et al., 2009; Imel, 2002). Therefore, metacognitive activities, such as analyzing tasks, planning, monitoring, checking, and reflecting on the learning process, were required to be taught to students through complex tasks.

Based on the findings of the student learning motivation observation sheet, it was discovered that learning motivation categories were diverse. Table 2 shows many types of learning motivation.

Table 2. The result of students' learning motivation.

Scores	Categories	Number of students	Percentage (%)
$81{,}25 < T \leq 100$	Good	10	12,82
$62{,}5 < T \leq 81{,}25$	Enough	31	39,74
$43{,}75 < T \leq 62{,}5$	Fair	18	23,08
$25 < T \leq 43{,}75$	Poor	19	24,36
Total		78	100,00

The categories of the students' learning motivation were good (12,82%), enough (39,74%), fair (23,08) and poor (24,36). The students actively paid attention to the teacher's explanations and did the assignments. They were quite tenacious in facing difficulties and preferring independent learning. Four primary components encouraged student learning motivation based on Keller's (1983) ARCS model. Students' attention; indicates that students' activities in asking, listening, and answering were conducted well. From a relevant perspective, students still lacked the courage to express their opinions in class discussion forums, but they were quite able to defend their opinions and reasons in front of other friends. Moreover, students tried to do assignments according to their abilities, but their confidence was still lacking. Furthermore, students were not easily discouraged from doing something in class and were not ashamed if they failed and they were able to rise again to be better. Therefore, their satisfaction indicated that they felt pleased with the test result, and the result of their efforts was in line with their expectations. Keller & Suzuki (2004), point out that it was necessary to develop a focus on the relevance dimension of motivation (intrinsic) to improve student motivation and performance in class. We can suggest that students' motivation is quite good. The remaining result was 47,44% of students who had fair and poor motivation.

Some of the students' learning motivation was still lacking. It was shown that students were still less confident in doing their assignments because they lacked the understanding of the learning materials and felt embarrassed to ask about incomprehension. A teacher could make students feel confident by convincing them that they could complete the assignments if they were facilitated and help students to establish positive expectancies for success (Driscoll, 1994; Keller & Suzuki, 2004). Students could get satisfaction through a motivational strategy such as verbal reinforcement, reward, personal attention, and relevant feedback (Keller & Suzuki, 2004). Therefore, it indicated that motivation was an important factor that could improve learning achievement. Motivation was an internal factor that needed an external stimulus, and the teacher's role was very helpful in improving students' learning motivation, especially in interest (attention), relevance, confidence, and satisfaction.

The results of the communication skills observation sheet on the presentation element revealed that communication skill categories were diverse. Table 3 shows the results of the communication skills categories.

Table 3 shows that the categories of students' communication skills were enough (61,54%) and 35,90% in the poor category. The results from the observation of students' oral communication skills through presentations were good enough. Students' oral communication skills have shown fluency when presentations during discussions and debates in answering questions directly. Active

Table 3. The categories of students' communication skills.

Scores	Categories	Number of students	Percentage (%)
$81{,}25 < T \leq 100$	Good	0	0,00
$62{,}5 < T \leq 81{,}25$	Enough	48	61,54
$43{,}75 < T \leq 62{,}5$	Fair	2	2,56
$25 < T \leq 43{,}75$	Poor	28	35,90
Total		78	100,00

participation of students in communicative activities could be seen when they were doing assignments. Additionally, high student metacognition about oral communication could be shown by more students raised their hands to ask questions. There was interactive discussion between student to student and students to lecturer, and the presentation of the content of the material presented was also quite good. However, the students' still need to improve their communication skills during the learning process. Huang et al. (2010), acknowledge that communication skills need to be trained, especially when the students work in teams. Students need to increase the quality and quantity of writing skills and oral communication skills that were developed in the curriculum (Tucker & Mccarthy, 2001). Therefore, the use of appropriate learning strategies would facilitate students' improvement in their communication and argumentation skills. The use of information searching strategies would improve students' experience in argumentation activities (Reisoğlu et al., 2020). Therefore, students' communication skills were required continuously to be trained through interactive communication, which empowered active involvement in communicating between students and students, and students and teachers. The use of certain learning models and strategies also increased communication activities.

4.2 *The correlations between metacognitive skills, motivation, and communication skills*

The results of the correlational test between these variables are presented in Table 4 to Table 6. The summary of the regression analysis of the correlation between metacognitive skills and communication skills was presented in Table 4.

Table 4. The regression summary of the correlation between metacognitive skills and communication skills.

Model Summary				
Model	R	R Square	Adjusted R Square	Std. Error of the Estimate
1	.331[a]	.110	.098	16.04384

a. Predictors: (Constant), METACOGNITIVE

Table 4 showed that metacognitive contribution was tested simultaneously with communication by 11% (R square = 0.110), while the remaining 89% was influenced by other factors. There is a positive correlation between students' metacognitive skills and communication skills (r = 0.331).

Based on Table 5 showed that the contribution of motivation was tested simultaneously with communication of 71.1% (R square = 0.711), while the remaining 38.9% was influenced by other factors. There is a very strong positive correlation between students' learning motivation and communication skills (r = 0.843).

Table 6 showed that the motivation and metacognitive contributions were tested simultaneously with communication of 71.4% (R square = 0.714), while the remaining 29.6% was influenced by

Table 5. The regression summary of the correlation between students' learning motivation and communication skills.

Model Summary				
Model	R	R Square	Adjusted R Square	Std. Error of the Estimate
1	.843[a]	.711	.707	9.14683

a. Predictors: (Constant), MOTIVATION

Table 6. The regression summary of the correlation between metacognitive skills and learning motivation with communication skills.

Model Summary				
Model	R	R Square	Adjusted R Square	Std. Error of the Estimate
1	.845[a]	.714	.706	9.16110

a. Predictors: (Constant), METACOGNITIVE, MOTIVATION

other factors. There is a very strong positive correlation between students' metacognitive skills and learning motivation with communication skills ($r = 0.845$).

The finding of this study revealed that the contribution of metacognitive skills to communication skills was low. It happened because the ability of students to plan, organize, and determine strategies for solving complex tasks was still underdeveloped. As a result, the understanding was not good, which has implications for communication skills and the ability to argue in discussions. The students of senior high school in metacognitive skills contribute to improve the learning process which has the implications for cognitive learning outcomes (Listiana et al., 2016). Besides the metacognitive skill, learning motivation is seen as a factor that extremely influences students' final cognitive learning outcomes. Students with higher achievement motivation have been shown to have more sustained commitment to learning and more cognitive engagement with learning materials (e.g., more attention to learning materials) (Daumiller et al., 2021; Daumiller & Dresel, 2020; Yang et al., 2021). Liu's research (2020) revealed that students can increase their learning activities and participation with strong intrinsic and extrinsic motivation in the learning process. According to Sato & Lam (2021) metacognitive skills, such as the ability to plan, monitor, regulate, and examine their cognitive skills were actually under the influence of student motivation. It was supported by Marulis & Nelson (2020), that students learning motivation strongly correlated to metacognitive skills. Reisoğlu et al. (2020) said that argumentation is an effective tool in developing metacognitive skills and information searching strategies. Cavagnetto (2010) explained that the students' participation in arguing can develop communication skills and push metacognition.

5 CONCLUSION AND RECOMMENDATION

This study aims to reveal the correlation of metacognitive skills and learning motivation to student communication skills. Relationship between motivation and metacognition with communication showed a positive result. It is identified that the contribution of motivation to communication is higher than the contribution of metacognitive to communication. Students with high motivation have the potential to improve their communication skills. The contributions of motivation to metacognitive are also tested simultaneously with communication and are very high. It shows that students with high motivation and metacognitive skills can encourage communication skills. From a policy perspective, it can consider to integrate the curriculum about metacognitive skills and communication skills into learning objectives or design curriculum by supporting the development of goals during learning. Further research needs to reveal more about the relationship between metacognitive skills, learning motivation, and communication skills by using another approach, the Group Investigation Integrated Think Talk Write (GITTW).

ACKNOWLEDGEMENTS

This project was supported and funded by DRPM Ministry of Research, Technology, and High education, Republic of Indonesia. We would like to thank everyone who has contributed to the completion of this paper.

REFERENCES

Alias, A., & Osman, K. (2015). Assessing oral communication skills in science: a rubric development. *Asia Pacific Journal of Educators and Education, 30*, 107–122.

Anderson, L. W., & Krathwohl, D. R. (2001). *A taxonomy for learning, teaching and assessing (A revision of Bloom's taxonomy of educational objectives)*. New York: Addison Wesley longman, Inc.

Bahri, A., & Corebima, A. D. (2015). The contribution of learning motivation and metacognitive skills on cognitive learning outcomes of students within different learning strategies. *Journal of Baltic Science Education, 14*(4), 487–500.

Banks, S. R. (2005). *Classroom assessment: Issues & practices*. Boston, MA: Pearson Education Inc.

Burguillo, J. C. (2010). Using game theory and competition-based learning to stimulate student motivation and performance. *Computers and Education, 55*(2), 566–575.

Cavagnetto, A. R. (2010). Argument to foster scientific literacy: A Review of argument interventions in K-12 science contexts. *Review of Educational Research* (Vol. 80, Issue 3).

Corebima, A. (2009). Metacognitive skill measurement integrated in achievement test. In *Paper was presented on 3rd International Conference on Science and Mathematics Education*. Penang: 10-12 November 2009.

Corebima, A. (2011). The validation of metacognitive skill measurement integrated in Achievement Test. *The 4th Redesigning Pedagogy International Conference*, 1–10.

Coutinho, S. A. (2007). The relationship between goals, metacognition, and academic success. *Educate, 7*(1), 39–47.

Daumiller, M., & Dresel, M. (2020). Researchers' achievement goals: Prevalence, structure, and associations with job burnout/engagement and professional learning. *Contemporary Educational Psychology, 61*, 1–17.

Daumiller, M., Rinas, R., Olden, D., & Dresel, M. (2021). Academics' motivations in professional training courses: effects on learning engagement and learning gains. *International Journal for Academic Development, 26*(1), 7–23.

Driscoll, M. P. (1994). *Psychology of learning for instruction*. MA: Allyn & Bacon.

Flavel, J. H. (1979). Metacognition and cognitive monitoring: A new area of cognitive and developmental inquiry. *American Psychologist, 34*(10).

Green, R. (2002). *Better thinking learning an introduction to cognitive Education*. Western Cape Education Department.

Hacker, D. J., Keener, M. C., & Kircher, J. C. (2009). Writing is applied metacognition. In Hacker D. J., Dunlosky J., Graesser A. C. (Eds.). In *Handbook of metacognition in education*. New York: Routledge.

Huang, D., Leon, S., Hodson, C., Torre, D. La, Obregon, N., & Rivera, G. (2010). *Preparing students for the 21St century: Exploring the effect of afterschool participation on student's collaboration skills, communication skills, and self-efficacy*. Los Angeles, CA: University of California, National Center for Research on Evaluation, Standards, and Student Testing (CRESST).

Imel, S. (2002). Metacognition skills for adult learning. Trends and issues alert. *ERIC Clearinghouse on Adult, Career, and Vocational Education, Columbus, OH., 39*, 1–4.

Keller, J. M. (1983). *Motivational design of instruction (Instructional design theories and models: An Overview of their current status)*. Hillsdale, NJ: Lawrence Erlbaum.

Keller, J., & Suzuki, K. (2004). Learner motivation and e-learning design: A multinationally validated process. *Journal of Educational Media, 29*(3), 229–239.

Kozikoğlu, İ. (2019). Investigating critical thinking in prospective teachers: Metacognitive skills, problem solving skills and academic self-efficacy.*Journal of Social Studies Education Research, 10*(2), 111–130.

Listiana, L., Susilo, H., Suwono, H., & Suarsini, E. (2016). Empowering students' metacognitive skills through new teaching strategy (group investigation integrated with think talk write) in biology classroom. *Journal of Baltic Science Education*, *15*(3), 391–400.

Liu, I. F. (2020). The impact of extrinsic motivation, intrinsic motivation, and social self-efficacy on English competition participation intentions of pre-college learners: Differences between high school and vocational students in Taiwan. *Learning and Motivation*, *72*, 1–10.

Marulis, L. M., & Nelson, L. J. (2020). Metacognitive processes and associations to executive function and motivation during a problem-solving task in 3–5 year olds. *Metacognition and Learning*, *16*(1), 207–231.

Nakamura, Y., & Valens, M. (2001). Teaching and testing oral communication skills. *Journal of Humanities and Natural Science*, 43–53.

Paris, S. G., Lipson, M. Y., & Wixson, K. K. (1983). Becoming a strategic reader. *Contemporary Educational Psychology*, *8*(3), 293–316.

Pintrich, P. R., & De Groot, E. V. (1990). Motivational and self-regulated learning components of classroom academic performance. *Journal of Educational Psychology*, *82*(1), 33–40.

Reisoğlu, İ., Eryılmaz Toksoy, S., & Erenler, S. (2020). An analysis of the online information searching strategies and metacognitive skills exhibited by university students during argumentation activities. *Library and Information Science Research*, *42*(3).

Sato, M., & Lam, C. D. (2021). Metacognitive instruction with young learners: A case of willingness to communicate, L2 use, and metacognition of oral communication. *Language Teaching Research*, *25*(6), 899–921.

Tas, C., Brown, E. C., Esen-Danaci, A., Lysaker, P. H., & Brüne, M. (2012). Intrinsic motivation and metacognition as predictors of learning potential in patients with remitted schizophrenia. *Journal of Psychiatric Research*, *46*(8), 1086–1092.

Tobias, S., & Everson, H. T. (1996). *Assessing Metacognitive Knowledge Monitoring*. New York: College Enterence Examination Board.

Tucker, M. L., & Mccarthy, A. M. (2001). Presentation self-efficacy: Increasing communication skills through service-learning. *Journal of Managerial Issues*, *13*(2), 227–244.

Urban, K., Pesout, O., Kombrza, J., & Urban, M. (2021). Metacognitively aware university students exhibit higher creativity and motivation to learn. *Thinking Skills and Creativity*, *42*.

Veenman, M. V. J., & Spaans, M. A. (2005). Relation between intellectual and metacognitive skills: Age and task differences. *Learning and Individual Differences*, *15*(2), 159–176. https://doi.org/1

Vuorre, M., & Metcalfe, J. (2021). Measures of relative metacognitive accuracy are confounded with task performance in tasks that permit guessing. *Metacognition and Learning*.

Winne, P. H., & Nesbit, J. C. (2010). The psychology of academic achievement. *Annual Review of Psychology*, *61*, 653–678.

Yang, J., Zhang, Y., Pi, Z., & Xie, Y. (2021). Students' achievement motivation moderates the effects of interpolated pre-questions on attention and learning from video lectures. *Learning and Individual Differences*, *91*(152), 102055.

Social Science Education

Teacher Education and Teacher Professional Development in the COVID-19 Turn – Drajati & Putra (Eds)
© 2023 The Author(s), ISBN: 978-1-032-38970-7

Exploring the effectiveness of PBL-GIS toward students' spatial thinking skills during the pandemic

M. Sigit Permadi, Sarwono Sarwono & Yasin Yusup
Universitas Sebelas Maret, Surakarta, Indonesia

ABSTRACT: This study aims to determine the effectiveness of spatial thinking with PBL-GIS compared to PBL without GIS on students' learning outcomes during the COVID-19 pandemic in a senior high school in Indonesia. This research was conducted in grade ten of two social science classes with 21 students, wherein the class was selected through simple random sampling consisting of 21 students. This is an experimental study using a post-test control group design. The data were collected using Spatial Thinking Ability Test, questionnaires, and documentation, while data analysis was carried out employing the Mann Whitney test. The result showed that the formula was the Mann Whitney test of 0.19 > 0.05, which implies that PBL-GIS was equally effective as PBL without GIS during the COVID-19 pandemic. Thus, this study implies that both models have the potential to improve the students' spatial thinking during COVID-19 era, giving a chance to the teacher to integrate one model wisely according to the classroom and students' conditions.

Keywords: COVID-19, PBL-GIS, Spatial thinking, STAT

1 INTRODUCTION

The coronavirus (COVID-19) has spread and affected many fields worldwide. It also affected schools globally; when the WHO declared it a pandemic, many schools had to conduct online classes. Throughout Indonesia, the government has implemented large-scale social restrictions and temporarily stopped social activities to reduce the risk of new cases of COVID-19. Thus, Indonesia's learning process changes from face-to-face (offline) learning to distance learning (Kemendikbud 2020). However, not all schools in Indonesia are ready for online learning because of several problems, such as connectivity, infrastructure, and the unpreparedness of teachers to use online tools (Azhari & Fajri 2021). In addition, the problem faced by students in the learning of flood geography and river flows is the lack of application of contextual material and the application of GIS media that can explain abstract material becomes real during the pandemic, causing low spatial thinking skills such as mentioning location names, choosing the right location based on spatial phenomena, imagining spatial data representation, and selecting river flow patterns based on image appearance (Imaniar et al. 2021).

Previous research has suggested the use of geo-technological devices, online or offline, such as Global Positioning System, cell phones, personal digital assistants, or computers to facilitate students' spatial thinking in learning geography (Kerski et al. 2013; Zhu et al. 2016; González & Torres 2020). Furthermore, in teaching geography, teachers need to teach not only physical and non-physical phenomena but also Geographic Information Systems (GIS) to facilitate students' spatial thinking skills and accomplish 21st-century skills. This is in line with (Lee & Bednarz 2009) who stated that teachers are encouraged to be creative in teaching and learning activities, for example, using GIS applications, which can improve students' spatial thinking through the implementation of problem-based learning.

Several studies have shown the use of PBL-GIS and PBL without GIS to facilitate students in learning geography. King (2008) argued that online PBL-GIS is more interesting and increases

DOI 10.1201/9781003347798-14

students' learning motivation than the traditional method. On the other hand, research by Amin, Sumarmi, Bachri, Susilo, and Basith (2020), showed that PBL without GIS could also improve spatial thinking skills with the help of hybrid media. The research corresponds to some studies, which stated that GIS could be replaced with some simple media. Conventional maps or Google Earth Android applications assisted by e-books and videos suited for any situation can provide a learning experience to improve students' spatial thinking (Kerski et al. 2013; Yusup et al. 2018; Zhu et al. 2016). However, some studies above show that both models contribute to students' learning under normal conditions without a pandemic. A study of PBL-GIS and PBL without GIS during the COVID-19 pandemic is still underexplored. This study fills the gap by showing the effectiveness of both models for students' learning during the pandemic with a proposed question: to what extent does the use of the two PBL learning models above improve students' spatial thinking ability during this COVID-19 pandemic? Through this study, teachers can design creative learning activities to enhance students' spatial thinking abilities by implementing PBL-GIS or PBL without GIS.

2 LITERATURE REVIEW

Howarth and Sinton (2011) explain that PBL-GIS is a conceptual learning model technique that integrates GIS operating skills in real-world problems by fostering analytical and critical thinking skills.Kerski et al. (2013), Gis and Group (2018), Hintermann, Bergmeister, and Kessel (2020) explain that PBL-GIS is effective and helps introduce GIS media in schools, and supports the achievement of the implementation of 21st-century geography learning paradigm, namely creativity, innovation, critical thinking, and problem solving.

Moreover, Kerski (2003) and Oberle (2020) state that PBL-GIS integrates five geographic investigation skills to solve problems comprising asking the geographic questions, acquiring geographic data, exploring geographic data, analyzing geographic data, and answering geographic data. Several studies showed that PBL without GIS could improve students' self-directed learning skills (Golightly 2018), spatial thinking (Amin et al. 2020; Wijayanto et al. 2020), and spatial intelligence with conventional maps (Marzuqi 2019). Liu, Bui, Chang & Lossman (2010) conducted an experimental study in secondary high school in Singapore, which compared PBL-GIS (experimental class) and PBL without GIS (control class). The results showed both models specifically. In the experimental class, students can demonstrate analytical and evaluation skills well, while the control class is stronger with the ability to remember. Yusup et al. (2018) conducted a research and development study in a senior high school that showed that PBL-GIS could effectively improve the students' spatial thinking ability by 30% with the help of e-book learning media. Those previous studies mentioned above compared the use of PBL-GIS and PBL without GIS in synchronous learning. However, the study about it in online learning is still underexplored and needs to be addressed in this proposed study.

3 METHOD

The research was conducted in grade ten of social science classes with 21 students with 21 students divided into two classes, 8 students in the experimental class and 13 students in the control class in Surakarta City, Indonesia during the COVID-19 pandemic. This is an experimental study using a post-test control group design with simple random sampling data collection techniques. The topic of geography being taught is watershed and flood areas adapted to school conditions that are prone to flood disasters. Data collection was done using documentation, interviews, and spatial thinking tests, totaling 7 post-test questions. Data collection was carried out using WhatsApp and Zoom applications, the data collection procedure are: 1) Documentation or collection of image and video recordings on smartphone or computer. 2) Online interviews conducted before and after the learning session was conducted. 3) Tests to find out students' spatial thinking abilities, which are carried out after learning sessions. The data was analyzed using the SPSS 20 application. The dependent variable of this research is (X1) PBL-GIS and (X2) PBL without GIS. The independent variable is (Y) spatial thinking ability. The experimental class and control class data were compared using non-parametric statistical analysis, namely Mann Whintey test, because the total sample in the class was less than 20 students.

4 FINDING AND DISCUSSION

This section begins with a description of the measurement of the results of spatial thinking of two learning models, namely PBL-GIS and PBL without GIS learning geography for watersheds and floods. The spatial thinking skills addressed in this topic lead the students to be able to mention the name of the location, choose the right location based on spatial phenomena, imagine spatial data representations, and choose river flow patterns based on image appearance. PBL GIS activities are carried out by teaching materials using GIS practices. Meanwhile, PBL without GIS is not practiced. These results of the measurements are the basis for the decision whether the two models can be used during a pandemic or not. According to Sahin (2009), PBL-GIS and PBL without GIS correspond to the 21st-century learning paradigm highlighting students' ability to seek references from various learning sources (literacy), formulate problems, think analytically, cooperate in solving problems, and utilize technology tools in their learning. Table 1 shows the difference in the implementation of PBL-GIS and PBL without GIS.

Table 1. PBL-GIS and PBL without GIS implementation in this study.

PBL-GIS	PBL-without GIS
GIS practice activities	No GIS practice
Assignment through GIS practice report and presentation.	Assignment via poster and presentation
Use 5 geographic inquiry process	Use 5 geographic inquiry process
Lesson materials on watershed and floods assisted with modules and video	Lesson materials on watershed and floods assisted with modules and video

The GIS applied to the PBL model during the pandemic is a mobile GIS. GIS mobile SW maps are one of the GIS tools that is easily obtained by downloading it on the play store or app store for free. If teachers cannot use mobile GIS, they can use conventional GIS, such as maps, drawings, or diagrams in modules using PBL models without GIS.

The indicator in this study to measure spatial thinking is one indicator "Analyzing space changes caused by humans and nature as an effort to preserve watersheds against floods" consisting of seven questions. Lee and Bednarz (2009) explain the spatial thinking indicator refers to the taxonomy of spatial thinking consisting of three components, namely the concept of space, tools of representation, and processes of reasoning.

4.1 *Average value PBL-GIS and PBL without GIS*

The average value of PBL-GIS and PBL without GIS was carried out after the learning model was applied at the end of the lesson using seven essay questions related to watershed and flood material adapted to contextual problems in the student's environment. The average value of spatial thinking results can be observed in Figure 1.

The average value in Figure 1 indicates the mean of post-test experiment class 82 and control group 78, which did not signify much difference. This means that each group gets good spatial thinking test results and meets the minimum completeness criteria of geography learning in senior high school above 72.

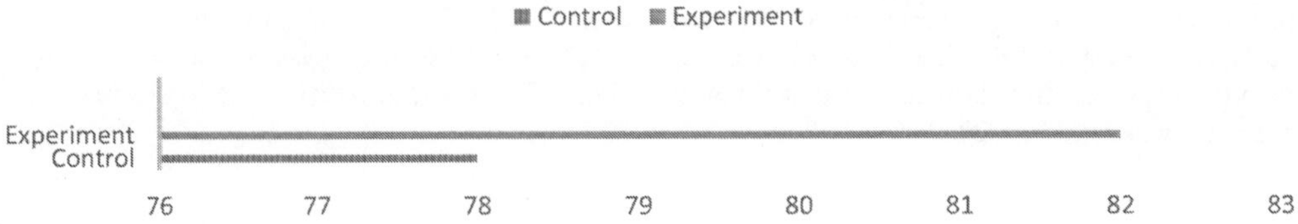

Figure 1. Average value PBL-GIS and PBL-without GIS.

4.2 *Mann Whitney test in PBL-GIS and PBL without GIS Spatial Thinking*

PBL-GIS and PBL without-GIS were compared to determine whether there were significant differences in the results of the spatial thinking post-test or not. The results of the Mann Whitney test are displayed in Table 2.

If the value of asymp Sig > 0.05 then the H0 hypothesis is accepted

If the value of asymp Sig < 0.05 the the H1 hypothesis is accepted

Hypothesis

H0: PBL-GIS is equally effective than PBL without GIS during the COVID-19 pandemic

H1: PBL-GIS is more effective than PBL without GIS during the COVID-19 pandemic

Table 2. Mann Whitney test in PBL-GIS and PBL without-GIS spatial thinking.

Mann Whitney test
PBL GIS Class-PBL without GIS
Post-test Experiment – Post-test Control
asymp. Sig (2 tailed)
0.19

According to the results of the Mann Whitney test shown in Table 2, the result was 0.19 > 0.05 H0 is accepted, which indicates that there is not much difference between PBL-GIS and PBL without-GIS during the COVID-19 pandemic. Therefore, PBL-GIS is equally effective as PBL without-GIS during the COVID-19 pandemic. Teachers can implement these models in their classrooms. Teachers can choose one of these models. The results are relevant to other research. King (2008) compared PBL-GIS and without PBL online with the conclusion that PBL-GIS model was more effective in improving learning outcomes. Ratinen & Keinonen (2011) and Schlemper, Athreya & Czajkowski (2018) also found that PBL improved student-teachers' geographical thinking skills, community awareness, spatial thinking, and geospatial technology. A study by Wahyuningtyas (2021) found that PBL-GIS plays an important role in creating and improving students' spatial thinking skills; students can better recognize the condition of the surrounding environment from various aspects of both threats, vulnerability, and risk of disaster.

In conclusion, the implementation of PBL-GIS or PBL without GIS learning models in this study showed that both models are equally effective in improving students' spatial thinking skills in geography learning. Teachers could choose a suitable learning model for their students by adjusting their abilities and resources (time and energy efficiency). For a good result in learning, teachers can collaborate with their colleagues and parents to provide direct supervision in this online learning era.

5 CONCLUSION AND RECOMMENDATION

This study addresses the effectiveness of using PBL-GIS and PBL models without GIS for students' spatial thinking skills in learning geography during the COVID-19 pandemic. According to the test result and discussion, it is concluded that both learning models are equally effective in improving students' spatial thinking. The chosen topics for the learning and test about watersheds and floods indirectly suggest the students to be more aware, sensitive, and care about their surroundings, especially flood disasters in a pandemic situation. Through this study, teachers are encouraged to design creative lessons to facilitate students' learning, especially for their spatial thinking skills during this pandemic. Different learning models to improve students' spatial thinking in online learning are suggested for future research.

REFERENCES

Amin, S., Sumarmi, Bachri, S., Susilo, S., & Bashith, A. (2020). The effect of problem-based hybrid learning (PBHL) models on spatial thinking ability and geography learning outcomes. *International Journal of Emerging Technologies in Learning*, *15*(19), 83–94.

Azhari, B., & Fajri, I. (2021). Distance learning during the COVID-19 pandemic: School closure in Indonesia. *International Journal of Mathematical Education in Science and Technology*.

de Miguel González, R., & de Lázaro Torres, M. L. (2020). WebGIS implementation and effectiveness in secondary education using the digital atlas for schools. *Journal of Geography*, *119*(2), 74–85.

Gis, A. P., & Group, T. S. (2018). Bridging high school and introductory undergraduate courses in geographic information science and technology. *Journal of Geography*, *117*(4), 165–173.

Golightly, A. (2018). The influence of an integrated PBL format on geography students' perceptions of their self-directedness in learning. *Journal of Geography in Higher Education*, *42*(3), 460–478.

Hintermann, C., Bergmeister, F. M., & Kessel, V. A. (2020). Critical geographic media literacy in geography education: Findings from the MiDENTITY project in Austria. *Journal of Geography*, *119*(4), 115–126.

Howarth, J. T., & Sinton, D. (2011). Sequencing spatial concepts in problem-based GIS instruction. *Procedia – Social and Behavioral Sciences*, *21*, 253–259.

Imaniar, N., Karyanto, P., & Yusup, Y. (2021). Spatial thinking profile for 2017 Surakarta senior high school. *IOP Conference Series: Earth and Environmental Science*, *683*(1).

Kemendikbud. (2020). *Surat Edaran Mendikbud No 4 Tahun 2020.*

Kerski, J. J. (2003). The implementation and effectiveness of geographic information systems technology and methods in secondary education. *Journal of Geography*, *102*(3), 128–137.

Kerski, J. J., Demirci, A., & Milson, A. J. (2013). The global landscape of gis in secondary education. *Journal of Geography*, *112*(6), 232–247.

King, E. (2008). Can PBL-GIS work online? *Journal of Geography*, *107*(2), 43–51.

Lee, J., & Bednarz, R. (2009). Effect of GIS learning on spatial thinking. *Journal of Geography in Higher Education*, *33*(2), 183–198.

Liu, Y., Bui, E. N., Chang, C. H., & Lossman, H. G. (2010). PBL GIS in secondary geography education: Does it result in higher-order learning outcomes? *Journal of Geography*, *109*(4), 150–158.

Marzuqi, M. I. (2019). The Effect of problem-based learning models through map media on spatial intelligence and problem solving skills. *The Indonesian Journal of Social Studies*, *2*(2), 77.

Oberle, A. (2020). Advancing students' abilities through the geo-inquiry process. *Journal of Geography*, *119*(2), 43–54.

Ratinen, I., & Keinonen, T. (2011). Student-teachers' use of Google Earth in problem-based geology learning. *International Research in Geographical and Environmental Education*, *20*(4), 345–358.

Sahin, M. C. (2009). Instructional design principles for 21st century learning skills. *Procedia – Social and Behavioral Sciences*, *1*(1), 1464–1468.

Schlemper, M. B., Athreya, B., Czajkowski, K., Stewart, V. C., Athreya, B., Czajkowski, K., & Stewart, V. C. (2018). Teaching spatial thinking and geospatial technologies through citizen mapping and problem-based inquiry in grades 7-12 teaching spatial thinking and geospatial technologies through citizen mapping and problem-based inquiry in grades 7-12. *Journal of Geography*, *0*(0), 1-.

Yusup, Y., Santoso, S., & Istifarida, B. (2018). Pengembangan e-book berbasis problem based learning-gis untuk meningkatkan kecakapan berfikir keruangan pada siswa kelas X SMA N 1. *Paedagogia*, *20*(2), 134.

Zhu, L., Pan, X., & Gao, G. (2016). Assessing place location knowledge using a virtual globe. *Journal of Geography*, *115*(2), 72–80.

Identifying students learning style: Effort to anticipate degradation in geography online learning

Hikari Dwi Saputro, Chatarina Muryani & Sarwono Sarwono
Universitas Sebelas Maret, Surakarta, Indonesia

ABSTRACT: This paper presents a quantitative study that seeks to identify students' learning styles to create appropriate nature-based geography learning techniques in online learning. The data were garnered by distributing an online questionnaire containing sixteen questions version 8.01 adopting the parameters developed by Fleming (2006). The VARK (Visual-Aural-Read-Kinesthetic) learning style test questionnaire was administered to 277 high school students in Indonesia. The results showed that the single learning style most favored by students is the aural learning style with a total of 103 students (37.18%), while the combination learning style most favored is the AK (Aural-Kinesthetic) learning style which is a combination of aural and kinesthetic learning styles. The implication of this research will be a pivotal basis for educators in designing online learning media to maximize geography online learning practices.

Keywords: Geography Online Learning; Learning Style; VARK

1 INTRODUCTION

The Covid-19 pandemic has disrupted the sustainability of all areas of life including the educational curriculum of various levels. The sudden transition from face-to-face learning to distance learning carries a considerable risk of learning loss for students. In this pandemic era, parents tend to be worried if their children could not follow any online learning instructions from home. A shift from face-to-face learning to online learning during the pandemic requires the adaptation and innovation of teachers to the use of technology in their classroom (Iglesias-Pradas et al. 2021). Looking at this issue, teacher readiness is an essential step in understanding the way to support the transition to online learning (Scherer et al. 2021). Optimal instructional instruction requires a diagnosis of student learning styles that reveal students' preferences to adjust material presentation (Pashler 2009). Chen (2014) revealed that learning styles could be considered valid predictors of success in an online or hybrid learning environment.

The study of learning styles helps facilitate the design of teacher learning activities to meet the different needs of students (Coffield et al. 2004). Variations in student learning styles affect how students think in receiving external stimuli, thinking, and solving problems (Wong & Hsu 2016). Hsu (2016) found a significant relationship between learning styles and attitudes in convincing students to continue using a particular learning system. Another study conducted by Chui, Molesworth, and Bromfield (2021) reported that the relationship between learning styles and learning outcomes showed a beneficial effect when the delivery of feedback through the media was following the learning style of the students. Pérez-Marín, Paredes-Velasco & Pizarro (2022) investigated student learning styles during a pandemic by applying multi-mode teaching covering all VARK preferences. The digital multi-mode framework using the VARK preference has proven to improve learning outcomes and students feel satisfied with learning. Thus, teachers need to be

DOI 10.1201/9781003347798-15

aware of each student's different learning styles and be able to effectively adapt learning strategies and methods to meet student learning needs (Hassan etal. 2019).

Knowing learning styles allows students to identify their learning tendencies. Learning styles influence learning outcomes through their relationship with learning strategies (Jie & Xiaoqing 2006). The VARK system adopted in the prior studies so far has focused on investigating the student learning style preferences which lead to a correlation of learning outcomes. This study examines explicitly the learning style preferences of students' VARK model which will be implemented in geography learning. The research was conducted to explore the question: what are the preferences of students' learning styles in geography class during online learning situations?

2 LITERATURE REVIEW

Learning style is considered a valid predictor of success in a web-based learning environment (Chen & Chiou 2014). Studying learning style preferences allows educators to accommodate students in learning (Chui, et. al. 2021). There are several learning style models, including the Felder & Silverman model, the Kolb model, the Honey and Mumford model, and the VARK model by Fleming. Leite, Svinicki, & Shi (2009) argue that Fleming's VARK model allows feedback to be adjusted and presented in an appropriate modality. VARK's learning style is an assessment tool that equips students to maximize their learning abilities. VARK has four different categories of the abbreviation "VARK" itself, namely visual, read, audio, and kinesthetic (Fleming 2006). A visual learning style with a "V" code means that students can receive the best information from what they see. The Aural learning style "A" is related to the sense of the ear, which means that students can process information well through sound. Learning style "R" stands for reading/writing which means students tend to like the information in the form of written text. Kinesthetic learning style with code "K" means that students more easily understand information through experience, practice, and things related to reality. There is a "multimodal" category, namely students who fall into more than one sensory category from visual, aural, reading, or kinesthetic (Fleming 2012).

The study of learning styles provides a perspective on the future implications of designing learning environments. Identification of the various dimensions of learning styles provides greater awareness of the unique characteristics of students. Troussas, Chrysafiadi, & Virvou (2021) conducted a study of personalized learning style preferences in learning systems and applications. Identification of students' learning style preferences is obtained through an account which is then entered into the model according to their learning style. This study indicates that a good learning acceptance rate is a hybrid learning style model by students and underlines its pedagogical potential. A study conducted by Pérez-Marín et al. (2022) investigated student learning styles in digital learning conducted for five months. They found that students were quicker to understand the learning that combined various learning resources using the VARK multi-mode digital approach. However, geography learning practice requires reconsideration in implementing distance education during the pandemic to adjust student tendencies.

Each class consists of students with diverse learning styles. In implementing learning, teachers need to ensure those teaching methods, learning opportunities, and assessment methods are varied in geography classes to develop students evenly (Healey, et. al 2005). Teachers can maximize learning by designing teaching frameworks that incorporate different learning styles. The application of various teaching and learning strategies by accommodating all learning style preferences by geography educators can help in developing students' self-directed learning (SDL) skills (Golightly 2019). Project-based learning strategies are proven to be able to attract students' interest and prepare students for success in the next stage of learning geography (Ives-Dewey 2008). Studying geography professionally needs to engage students towards a real-world experiential perspective. This research focuses on studying student learning style preferences which are used as the basis for developing online learning media to maximize the practice of learning geography.

3 METHOD

This research is quantitative research using a questionnaire. This study on learning styles involved 277 students from three high schools in Indonesia. The distribution of respondents based on gender in this study was male n = 47 or 17% and female n = 230 or 83%. Data was collected through a self-administered online questionnaire. Questionnaire links are distributed to respondents (students) accompanied by instructions for filling out. Students gave a good response and sent a questionnaire during online school time. The questionnaire consists of 16 online questions (Fleming 2006). Instructions direct students to choose answers according to their preferences. Each question in the VARK questionnaire has four answer choices: visual, aural, read, and kinesthetic. The pattern of answer choices for the VARK questionnaire in this study is as follows:

Table 1. VARK Questionnaire assessment.

Question	A	B	C	D
Question 1	K	A	R	V
Question 2	V	A	R	K
Question 3	K	V	R	A
Question 4	K	A	V	R
Question 5	A	V	K	R
Question 6	K	R	V	A
Question 7	K	A	V	R
Question 8	R	K	A	V
Question 9	R	A	K	V
Question 10	K	V	R	A
Question 11	V	R	A	K
Question 12	A	R	V	K
Question 13	K	A	R	V
Question 14	K	R	A	V
Question 15	K	A	R	V
Question 16	V	A	R	K

Each question can have single, multiple, or zero answers. Calculation of learning style profile refers to the definition (Vasileva-Stojanovska et al. 2015), which gives a score of 0 to 6 on each learning style variable. A score of 6 indicates very strong, a score of 5 indicates strong, and a score of 4 indicates weak for single modalities. Scores 3, 2, and 1 each indicate a bimodal, trimodal, and multimodal learning style, while a score of 0 means that there is no particular modality preference. The VARK system shows a strong tendency for student learning styles to refer to the following learning styles (Chang-Tik 2018):

Table 2. Learning VARK styles tendency.

V	Visual (very strong, strong, mild)
A	Aural (very strong, strong, mild)
R	Read/Write (very strong, strong, mild)
K	Kinesthetic (very strong, strong, mild)
Bimodal	VA, VR, VK, AR, AK, RK
Trimodal	VAR, VAK, ARK, RKV
Multimodal	VARK

Learning style scores can serve as predictors of outcomes for evaluating learning methods (Leite et al. 2010). The data results are presented based on the central tendency of each respondent who fills out the questionnaire. The mode is used to indicate the preferences that respondents often choose in each question in the questionnaire. Each student has identified their learning style preferences which will fit into one of the modalities (Visual, Audio, Read, or Kinesthetic), bimodal, trimodal, or multimodal.

4 FINDING AND DISCUSSION

From the overall responses to the learning style questionnaire result, it is known that the students have almost all categories in the VARK learning style. Through the sixteen questions in the learning style questionnaire, the following results were found in the following table:

Table 3. Students' learning style (VARK).

Learning Style	Very Strong	Strong	Mild	Total
Visual	29	6		36
Aural	94	7		103
Read	37	10		48
Kinesthetic	33	3		34
Bimodal				51
Trimodal				5
Multimodal				2

A total of 219 students (79.06%) have chosen unimodal including, V (visual), A (aural), R (read), and K (Kinesthetic) as their learning style, while the remaining 58 students (20.94%) chose a bimodal, trimodal, and multimodal as their learning styles. Students who choose the bimodal learning style are 51 students (18.41%) with details of the VA learning style for as many as 9 students (3.25%), VR learning style for as many as 8 students (2.89%), VK learning style as many as 5 students (1.81%), learning style AR numbered 11 students (3.97%), AK learning style was 13 students (4.69%), and RK learning style was 5 students (1.81%). Students who choose the trimodal learning style are 5 students (1.81%) with details, the VAR and VAK learning styles are chosen by 1 student each, and the ARK learning style is chosen by 3 students. There are 2 students who choose the VARK multimodal learning style.

The single learning style most favored by students is the aural learning style with a total of 103 students (37.18%), while the combination learning style most favored by 13 students is the AK learning style which is a combination of aural and kinesthetic learning styles. Some dominant learning styles need to be accepted based on professionalism about how students can learn and obtain information in certain ways. The teacher must have no specific learning style, but it is a consideration in varying teaching methods and strategies.

During online learning in the pandemic era, student engagement and understanding of geography learning materials tends to decrease. The teacher varies learning based on the dominant learning style, including aural learning style followed by reading, visual and kinesthetic learning styles. The implementation of hybrid geography learning that is carried out effectively has been proven to impact a more comprehensive student network compared to the traditional face-to-face format (Herrick 2010). In an attempt to create an engaging and attractive atmosphere in the classroom, learning style preferences become a basis for teachers in designing and presenting the materials. In line with Santana-Mancilla et al., (2019), game-based teaching methods with multi-mode interactions can equip students with important skills, such as user engagement, task-centered system design, behavior models, creativity, metaphors, and graphic screen design. Online learning currently requires students to learn from home, so teachers need to formulate geography learning practices as optimally as possible. An excellent strategy to support the practice of learning geography is designed according to the students' single and multiple modality tendencies.

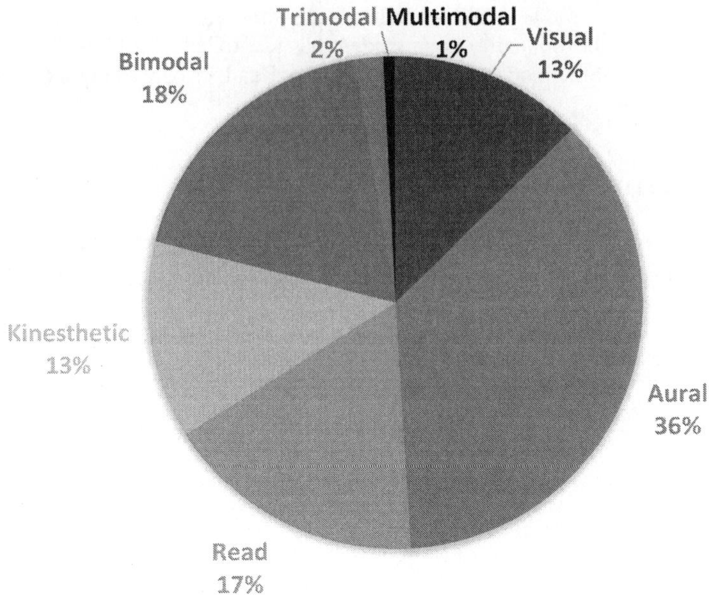

Figure 1. Percentage of students' learning style.

5 CONCLUSION AND RECOMMENDATION

This study aims to identify the student learning styles preferences in online learning. This study found that the aural learning style was the most dominant learning style, followed by read, visual, and kinesthetic learning styles. Drawing from the result, it gives insight for teachers to take into account these four major learning styles preferences to be applied in their own virtual classroom. Adjusting the material and its deliverance in online geography learning will contribute to increasing students' motivation and positive outcomes in online learning situations. Implementation in the practice of online geography learning potentially shows positive changes, but this still cannot exceed the performance of actual practice in the field. Taking into account the results from the study of learning style preferences is needed to prepare online learning media to be implemented in hybrid and online classes. Further studies might investigate the learning styles and performances of different situations by including variables that can maximize the practice of geography learning.

REFERENCES

Chang-Tik, C. (2018). Impact of learning styles on the community of inquiry presences in multi-disciplinary blended learning environments. *Interactive Learning Environments*, *26*(6), 827–838.

Chen, B. H., & Chiou, H.-H. (2014). Learning style, sense of community and learning effectiveness in hybrid learning environment. *Interactive Learning Environments*, *22*(4), 485–496.

Chui, T. K. S., Molesworth, B. R. C., & Bromfield, M. A. (2021). Feedback and Student Learning: Matching Learning and Teaching Style to Improve Student Pilot Performance. *International Journal of Aerospace Psychology*, *31*(2), 71–86.

Coffield, F., Moseley, D., Hall, E., & Ecclestone, K. (2004). *Learning styles and pedagogy in post-16 learning: A systematic and critical review*. Learning and Skills Research Centre & Department for Education and Skills.

Fleming, Neil D. (2006). *V.A.R.K. Visual, Aural/Auditory, Read/Write, Kinesthetic*. Bonwell Green Mountain Falls.

Fleming, Neil D. (2012). *Facts, Fallacies and Myths: VARK and Learning Preferences*. VARK Learn Limited. Retrieved from: http://vark-learn.com/wp-content/uploads/2014/08/Some-Facts-About-VARK.pdf.

Golightly, A. (2019). Do Learning Style Preferences of Preservice Geography Teachers Matter in Self-Directed Learning? *Journal of Geography*, *118*(4), 143–156.

Hassan, M. A., Habiba, U., Majeed, F., Shoaib, M., & Hassan, M. A. (2019). Adaptive gamification in e-learning based on students' learning styles. *Interactive Learning Environments*, *29*(4), 545–565.

Healey, M., Kneale, P., & Bradbeer, J. (2005). Learning styles among geography undergraduates: An international comparison. *Area*, *37*(1), 30–42.

Herrick, C. (2010). Lost in the field: Ensuring student learning in the 'threatened' geography fieldtrip. *Area, 42*(1), 108–116.

Hsu, L. (2016). An empirical examination of EFL learners' perceptual learning styles and acceptance of ASR-based computer-assisted pronunciation training. *Computer Assisted Language Learning*, *29*(5), 881–900.

Iglesias-Pradas, S., Hernández-García, Á., Chaparro-Peláez, J., & Prieto, J. L. (2021). Emergency remote teaching and students' academic performance in higher education during the COVID-19 pandemic: A case study. *Computers in Human Behavior*, *119*.

Ives-Dewey, D. (2008). Teaching experiential learning in geography: Lessons from planning. *Journal of Geography*, *107*(4–5), 167–174.

Jie, L., & Xiaoqing, Q. (2006). Language Learning Styles and Learning Strategies of Tertiary-Level English Learners in China. *RELC Journal, 37*(1), 67–90.

Leite, W. L., Svinicki, M., & Shi, Y. (2010). Attempted validation of the scores of the VARK: Learning styles inventory with multitrait-multimethod confirmatory factor analysis models. *Educational and Psychological Measurement*, *70*(2), 323–339.

Pashler, H., McDaniel, M., Rohrer, D., Bjork, R. (2009). Learning styles: Concepts and evidence. *Psychological Science in the Public Interest, 9*(3), 105–119.

Pérez-Marín, D., Paredes-Velasco, M., & Pizarro, C. (2022). Multi-mode Digital Teaching and Learning of Human-Computer Interaction (HCI) using the VARK Model during COVID-19. *Educational Technology & Society, 25*(1), 78–91.

Santana-Mancilla, P. C., Rodríguez-Ortiz, M. A., García-Ruiz, M. A., Gaytan-Lugo, L. S., Fajardo-Flores, S. B., & Contreras-Castillo, J. (2019). Teaching HCI skills in higher education through game design: A Study of students' perceptions. *Informatics, 6*(2), 22.

Scherer, R., Howard, S. K., Tondeur, J., & Siddiq, F. (2021). Profiling teachers' readiness for online teaching and learning in higher education: Who's ready? *Computers in Human Behavior*, *118*(December 2020), 106675.

Troussas, C., Chrysafiadi, K. & Virvou, M. (2021). Personalized tutoring through a stereotype student model incorporating a hybrid learning style instrument. *Education and Information Technologies, 26*, 2295–2307.

Vasileva-Stojanovska, T., Malinovski, T., Vasileva, M., Jovevski, D., & Trajkovik, V. (2015). Impact of satisfaction, personality and learning style on educational outcomes in a blended learning environment. *Learning and Individual Differences*, *38*, 127–135.

Wong, Lung-Hsiang., & Hsu, Ching-Kun. (2016). Effects of learning styles on learners' collaborative patterns in a mobile-assisted, Chinese character-forming game based on a flexible grouping approach, *Technology, Pedagogy and Education*, *25*(1), 61–77.

Appendix: The VARK Questionnaire

1. When I am learning I:
   a. see patterns in things.
   b. like to talk things through.
   c. read books, articles and handouts.
   d. use examples and applications.
2. I want to learn about a new project. I would ask for:
   a. examples where the project has been used successfully.
   b. a written report describing the main features of the project.
   c. an opportunity to discuss the project.
   d. diagrams to show the project stages with charts of benefits and costs.
3. A website has a video showing how to make a special graph or chart. There is a person speaking, some lists and words describing what to do and some diagrams. I would learn most from:
   a. reading the words.
   b. watching the actions.
   c. seeing the diagrams.
   d. listening.
4. I want to learn to do something new on a computer. I would:
   a. start using it and learn by trial and error.
   b. read the written instructions that came with the program.
   c. talk with people who know about the program.
   d. follow the diagrams in a book.
5. I need to find the way to a shop that a friend has recommended. I would:
   a. use a map.
   b. ask my friend to tell me the directions.
   c. find out where the shop is in relation to somewhere I know.
   d. write down the street directions I need to remember.
6. I have a problem with my heart. I would prefer that the doctor:
   a. gave me something to read to explain what was wrong.
   b. showed me a diagram of what was wrong.
   c. described what was wrong.
   d. used a plastic model to show me what was wrong.
7. I want to find out about a house or an apartment. Before visiting it I would want:
   a. a plan showing the rooms and a map of the area.
   b. to view a video of the property.
   c. a printed description of the rooms and features.
   d. a discussion with the owner.
8. I want to learn how to take better photos. I would:
   a. use the written instructions about what to do.
   b. ask questions and talk about the camera and its features.
   c. use diagrams showing the camera and what each part does.
   d. use examples of good and poor photos showing how to improve them.
9. I prefer a presenter or a teacher who uses:
   a. question and answer, talk, group discussion, or guest speakers.
   b. diagrams, charts, maps or graphs.
   c. handouts, books, or readings.
   d. demonstrations, models or practical sessions.
10. When choosing a career or area of study, these are important for me:
    a. applying my knowledge in real situations.
    b. communicating with others through discussion.
    c. working with designs, maps or charts.
    d. using words well in written communications.

11. I want to find out more about a tour that I am going on. I would:
    a. use a map and see where the places are.
    b. talk with the person who planned the tour or others who are going on the tour.
    c. look at details about the highlights and activities on the tour.
    d. read about the tour on the itinerary.
12. I have finished a competition or test and I would like some feedback. I would like to have feedback:
    a. using graphs showing what I achieved.
    b. using examples from what I have done.
    c. using a written description of my results.
    d. from somebody who talks it through with me.
13. I want to learn how to play a new board game or card game. I would:
    a. read the instructions.
    b. use the diagrams that explain the various stages, moves and strategies in the game.
    c. listen to somebody explaining it and ask questions.
    d. watch others play the game before joining in.
14. I want to save more money and to decide between a range of options. I would:
    a. consider examples of each option using my financial information.
    b. read a print brochure that describes the options in detail.
    c. use graphs showing different options for different time periods.
    d. talk with an expert about the options.
15. I want to assemble a wooden table that came in parts (kitset). I would learn best from:
    a. watching a video of a person assembling a similar table.
    b. advice from someone who has done it before.
    c. diagrams showing each stage of the assembly.
    d. written instructions that came with the parts for the table.
16. When learning from the Internet I like:
    a. interesting design and visual features.
    b. audio channels where I can listen to podcasts or interviews.
    c. interesting written descriptions, lists and explanations.
    d. videos showing how to do or make things.

 ISBN: 978-1-032-38970-7

Exploring students' motivation in geography online learning

Siti Dwi Lestari, Ahmad Ahmad & Nunuk Suryani
Universitas Sebelas Maret, Surakarta, Indonesia

ABSTRACT: Learning loss is increasingly prone to occur in Indonesia so students' learning motivation must be increased. This study aims to determine students' perceptions of their online learning motivation in geography and test students' online learning motivation when using an android-based learning application. This study employed a quantitative research design and experimental research methodology. The data was collected using an online 5-point Likert scale questionnaire that involved 108 senior high school students. The measurement of the mean difference was carried out using t-test analysis. In addition, to support the result of the questionnaire, the researchers also shared an open-ended question to get a deeper understanding of students' online learning motivation. The results showed that intrinsic motivation has a strong influence on triggering students' enthusiasm for learning. Intrinsic motivation contains a sense of ego-sense in students that encourages their enthusiasm to learn. Nevertheless, this study found that the role of parents became the "core" motivation of students to study online from home professionally. Independent sample t-test statistics revealed a statistically significant difference (0.05 significance) between the means of the two groups. Students who use the Android-based learning application have better motivation than those who do not. The use of Android-based learning applications in learning geography has succeeded in creating stronger student motivation.

Keywords: Geography Online Learning; Perception; Motivation to Learn; Teacher Innovation.

1 INTRODUCTION

The transformation of online learning in the era of the COVID-19 pandemic poses complex problems. Teachers cannot monitor students as they do during face-to-face learning. The transition period causes an average learning loss of 66%, and the gap is widening due to a lack of support for learning resources (Sabates et al. 2021). All subjects feel the impact of distance learning, including geography. Geography is related to nature which requires the use of geospatial technology to support high school geography classes in Indonesia (Schlemper et al. 2019). Learning geography will be more potential if it involves the surrounding environment in teaching and learning activities. The COVID-19 pandemic limits the scope of learning geography enhanced by nature (Brookfield 2021). Logical thinking is needed to find digital learning innovations that are low-cost and easy to use according to needs in Indonesia.

Students' learning motivation is related to their learning achievement (Chang et al. 2014). Students who have learning motivation tend to be more actively involved in learning which can increase the effectiveness of their learning (Law et al. 2019). However, this pandemic has resulted in learning patterns that decrease learning motivation. The condition of students during distance learning is very heterogeneous both from a social and economic perspective. The pandemic is life-changing and causes some parents to be unable to provide technological facilities to study from home (Maqableh & Alia 2021). The use of innovative GIS (Geographic Information System) teaching methods in geography learning for underachieving students can stimulate and increase their intrinsic motivation (Bikar et al. 2022). Learning using mobile applications can increase students' motivation in studying geospatial technology (Ridha & Kamil 2021).

 DOI 10.1201/9781003347798-16

Innovations in learning that have been carried out tend to direct their contribution to students' cognition. Researchers are trying to find out students' perceptions of factors that can encourage learning motivation with the conditions they experience. It is useful to know the condition of students' motivation to learn from home as a basis for whether they can absorb learning well. The learning application innovations provided can be accessed offline to overcome signal limitations and internet quotas. This study aims to determine student perceptions and the effect of innovation on student learning motivation. The purpose of this study raises the following questions: 1) What are students' perceptions of the factors that drive their online learning motivation?; 2) How are the differences in students' online learning motivation using android-based learning applications and without using the application?.

2 LITERATURE REVIEW

Student learning motivation could be positive or negative (Harlen & Crick 2003). Motivation is classified into two types, namely intrinsic motivation and extrinsic motivation. Intrinsic motivation refers to factors within students, whereas extrinsic motivation refers to external factors that encourage students to be more active. Tüzün (2006) classified categories as motivational elements including identity presentation, social relations, playing, learning, achievement, reward, immersive context, fantasy, uniqueness, creativity, curiosity, control and ownership, and support context. Motivation is shown through students' actions, efforts, and dedication to achieving learning goals (Brophy 2013; Hanus & Fox 2015).

Mobile devices have great potential to transform traditional classrooms into more interactive and interesting ones (Shen et al. 2008). Mobile learning allows educators to teach without being limited by time and place (Huang et al. 2010). Research conducted by Buckley & Doyle (2014) found that gamification intervention in learning effectively stimulated students' intrinsic motivation. This gamification design is very suitable to be implemented in a mobile application. This is reinforced by the findings in the research of Su & Cheng (2014) that mobile technology and gamification can achieve higher motivational learning performance. The use of mobile learning media by paying attention to the flow experience is proven to be able to encourage students to develop more excellent learning (Hung et al. 2015). Fletcher, Kickbusch, & Huijser (2020) added that the development of mobile geography applications creates an impression for students which is believed to increase motivation and geospatial skills. The educational utility of the gaming tablet can sustain attention and involving students in the learning process creates greater learning motivation (Hung et al. 2015). Law, Geng, & Li (2019) conveyed that the positive motivation of students can increase the effectiveness of learning. Studies of learning motivation will always be needed to find novelty following a dynamic learning environment.

For this reason, the teacher's role in creating a learning climate and environment is important to encourage students' learning motivation. There has been no development of geography learning applications that can be accessed offline to face the online learning transition, especially in Indonesia. The researcher tries to provide various geography learning content packaged in mobile applications that can be accessed offline to anticipate students with limited quotas. This application requires one-time installation online, and then it can be accessed offline. The results of this study can be used as a basis for developing geography learning resources during the pandemic following the needs in Indonesia.

3 METHOD

This study uses quantitative methods. The research methodology of this study is experimental research. The data collection of students' learning motivation was done using a questionnaire, and was filled out by 108 high school students in class 10. The learning motivation questionnaire is in the form of a 5-point Likert scale that is divided into two, namely intrinsic and extrinsic motivation with 20 questions. The questionnaire was compiled using modifications that included categories: learning, product content, social relations, and rewards (Tüzün 2006). The selection of the learning motivation category in the questionnaire was based on consideration of the effects arising from the developed learning application. Researchers also shared an open-ended question to understand students' online learning motivation better. Questionnaires were distributed to each control and experimental class to find out the comparison. Data analysis is based on the average Likert scale scores to distinguish students' learning motivation. Quantitative data were analyzed using t-test analysis to determine the significance of students' learning motivation in online learning.

4 FINDING AND DISCUSSION

The product development has gone through validation and testing stages before being widely used. Students in the Experimental class in this study were using an Android-based geography learning application, meanwhile, the Control class as the sample includes all students using textbooks in pdf format and interactive PowerPoint. The application contains material features and evaluation questions that students can access offline to reduce costs while also avoiding signal limitations and internet quotas. The content in the Android-based geography learning application is designed to encourage students' online learning motivation through the integration of material visualization in the form of images, maps, audio-video, simple educational games, and animations.

Figure 1. Lithosphere material screenshot of the android-based geography learning application.

4.1 *Motivational factors for students' online learning*

Researchers designed a questionnaire containing the open-ended question, namely "What can motivate you to be more active in learning geography online?". This question aims to explore what factors can encourage students' online learning motivation. Through the answers to these open-ended questions, researchers could find out the perceptions of students' self-study motivation that have not been revealed. Students gave various answers according to their respective perceptions.

In general, students' online learning motivation can be categorized into intrinsic and extrinsic motivation. Based on the questionnaire data, it was found that 61 students preferred the former as a motivation to make them more active in learning geography online, with 42 students preferred the latter. Beyond prediction, 5 students did not know what things could motivate them to study geography online.

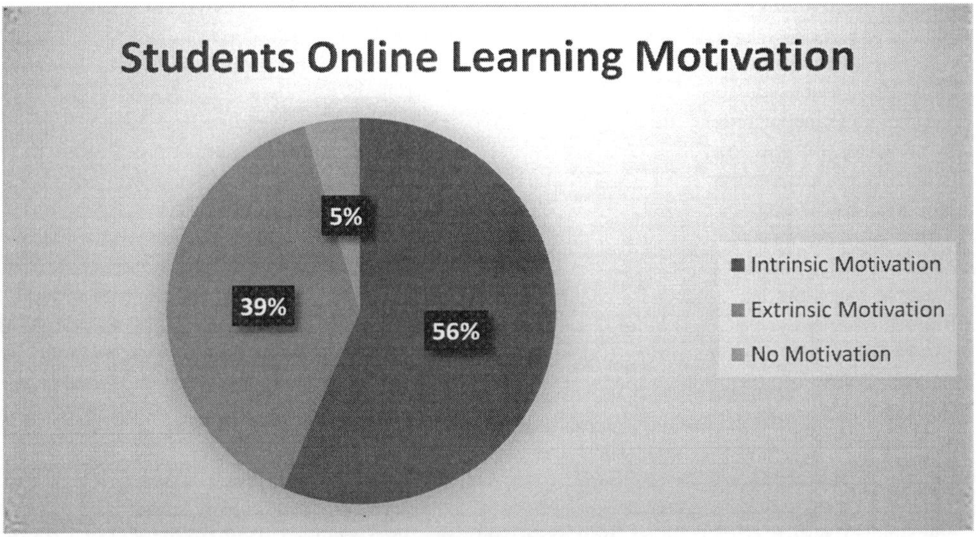

Figure 2. Percentage of students' online learning motivation.

This study revealed three main factors of students' self-intrinsic motivation in online learning, namely results-oriented, future-oriented, and self-interest to always increase knowledge. Results-oriented motivation is formed within the students with the goal of getting the best learning outcomes in school. Examples responses:

"My motivation to study is to get the best lesson value." – AIK
"I want to get good grades." – YNK
"Much better lesson value than what we usually get." – FNH
"To get maximum lesson value." – PEA

Most of the future-oriented student responses are related to the students' expectations of being able to take the next level of education:

"My motivation to study geography is to get good grades so I can enter one of the best universities." – ANH
"Study hard so you can realize your dreams." – FK
"I will accumulate a lot of knowledge for the future and move up to a higher level." – IKN
"Study hard to reach your goals." – L
"The future begins with studying hard and hard work will not betray the results." – RS

From the excerpts above, we can see that some students are planning to continue their studies at the university of their dreams. Therefore, teachers can encourage students to study harder by providing insights into future opportunities.

Students also have self-interest, one of which is the craving for more knowledge. This relates to the ambition that already exists within themselves. Positive thinking is important for every individual at school as a basis for carrying out learning. Examples responses:

"I want to master or better understand geography material." – AYL
"Increasing insight and knowledge." – DA
"Discovering new things." – NBA

"So that I can better understand all the learning materials." – LP
"I want to know more about geography." – LANF
"To be smarter." – AFE

Intrinsic motivation refers to the attraction to an activity to obtain the desired result (Rheinberg & Engeser 2018). It is closely related to interest and involvement. In this research, intrinsic motivation is summed up in the three main factors that have been described. However, there is something that has not been discussed, the "ego-centric" motivation. Ego is not always a negative thing. The researcher can clearly understand that intrinsic motivation contains self-ego as seen in the students' answers above. Furthermore, teachers as mentors in online classes must set out to assist students in their self-management as well. This is a challenge for the education system, especially in Indonesia, to face the future.

The results showed that extrinsic motivation in online learning comprises three key factors: internet network, learning strategy, and parent-related factors. Factors related to the internet network are crucial in online learning. While face-to-face learning does not require an internet network, online learning is currently all done through an internet network. Schools, teachers, and students depend on it to carry out learning. And with how crucial internet network is, not all of them can fulfill that condition. Example responses:

"I will study hard if the network connection is better." – SF
"I wish my internet connection is good." – MWNH
"If only I have a lot of quota." – SNF

The majority of extrinsic motivation is oriented to the strategies used by teachers in online learning. The use of learning strategies that lead to deep understanding can excite students. This orientation is the main factor that determines whether or not students are motivated to study hard. Example responses:

"Interesting learning topics and simple explanations." – AKL
"Video learning or via Zoom or Google Meet conference." – M
"Fun learning strategy." – IA
"Interesting and fun material presentation." – DO
"The learning media and point system." – KN
"On point learning method." – FMA

A few students responded that their learning motivation was based on parental factors. This was revealed by 3 students from a total of 108 respondents in this study. About 2.7% of the total participants. Researchers think that although the amount is low, this cannot be taken lightly.

This statement relates to students' online learning activities carried out from home. So the role of parents to provide a safe and comfortable learning environment from home needs to be emphasized. Example responses:

"I study hard because of my parents." – PAF
"Parents." – TSO
"Parents and comfortable environment." – FRA

We often categorize high school students in a particular age category in which they are looking for an identity. More efforts are needed by teachers in realizing the online learning that students want. Students' extrinsic and intrinsic motivation is equally important. These main motivation factors can be utilized as a reference for what elements need to be focused on in online learning. Learning is carried out according to the students' characteristics for optimal learning outcomes.

Schools are essential in bringing out the best of the students, whether it's knowledge or morals. The majority of a child's life is spent in school, which allows opportunities to learn character values and better pedagogy (Brannon 2008). It should always be emphasized that the first and foremost educators of a person are from the family. Individual characters are formed from childhood, so it is critical for every household to provide the best parenting for their children. Especially in online learning, students spend more time at home. Comprehensive cooperation between schools and families is needed to ensure students' motivation to learn online.

4.2 *Students' online learning motivation using android-based learning application*

Researchers designed a questionnaire containing questions that lead to students' self-motivation during online learning. It includes intrinsic and extrinsic motivation containing the categories of learning, product content, social relations, and rewards the first ten questions lead to an assessment of students' self-intrinsic motivation and the next ten questions lead to an assessment of students' extrinsic motivation.

Assessment of students' intrinsic and extrinsic motivation varies. The average results in the Experimental class and Control class showed different results. The development of android-based geography learning application media used in the experimental class can increase students' learning motivation in each of the following categories:

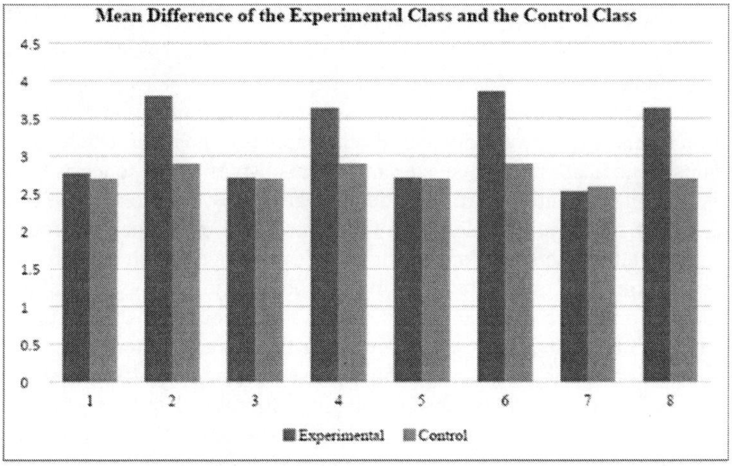

Figure 3. Mean difference between the experimental class and the control class.

The preliminary study's interviews with teachers revealed students had low motivation to study geography. The results in this study were surprising because the geography learning application showed excellent performance in enhancing students' online learning motivation. Some students' learning motivation is still stagnant, but it's only a minority. Implementing geography learning applications in the online learning era is still relevant to encourage student learning motivation. Through enhancing students' online learning motivation, we hope that it will also be in line with improving student learning outcomes.

The results of the differences (t-test) of students' learning motivation are presented in Table 1. Table 1 shows that the Experimental and Control groups' learning motivation has a statistically significant difference. These results indicate that the Experimental group's learning motivation is better than the control group in all categories. The development of innovative products in online learning in the form of an Android-based geography learning application is proven to be able to develop students' learning motivation. Careful content preparation to optimally target student motivation categories can support this success. The product is supported by the visualization of images, videos, maps, and animations. In line with Elaish, Ghani, Shuib, Al-Haiqi (2019) research verdict that students enjoyed and expressed pleasure with the mobile game application even when they use it during their spare time at home. Students tend to be more active in previous learning. Some students also took the initiative to ask about the part of the material that they did not understand in this application.

Table 1. Student online learning motivation for each category.

		Pre-questionnaire		Post-questionnaire			
Category	Research Group	Mean (1 to 5)	SD	Mean (1 to 5)	SD	t-Stat	t-Critical
Learning	Experimental	2.77	0.90	3.80	1.11	4.45	1.70
	Control	2.70	0.90	2.90	0.90		
Product Content	Experimental	2.71	0.91	3.64	0.80	3.53	1.70
	Control	2.70	0.91	2.90	0.91		
Social relations	Experimental	2.71	0.72	3.86	0.72	4.80	1.70
	Control	2.70	0.90	2.90	0.90		
Rewards	Experimental	2.53	0.81	3.64	0.81	4.13	1.70
	Control	2.59	0.93	2.70	0.99		

5 CONCLUSION AND RECOMMENDATION

This study was conducted to determine students' perceptions of online learning motivation and test their learning motivation when using an android-based learning application. Students' perceptions of self-motivation in learning are very diverse. The research results in this article are consistent and extend to previous identifications. Ego-sense can foster self-motivation to learn and compete. The teacher's task is to ignite the enthusiasm of students' learning by interpreting the ego into positive energy. The results of the t-test conducted on the control and experimental groups showed that students were motivated by the use of applications in learning geography. Further results showed that most of the experimental group students understood learning using applications (M = 3.80), application product content received attention (M = 3.64), social relations supported material understanding (M = 3.86), students were happy with the reward system provided (M = 3.64). This geography learning application that can be accessed online and offline can increase student learning motivation. This is a good response to the given stimulus. However, students' strong self-motivation will determine their academic success. The limitations of this study are the limited time and scope of research subjects. Research can be continued more deeply by involving the perspective of students' guardians regarding children's learning motivation which raises new questions related to the context of the role of parents in motivating online learning from home.

REFERENCES

Bikar, S. S., Rabe, B. R. Z., Mahat, H., Sharif, S., & Talin, R. (2022). The impact of geography information system integrated teaching on underachieving students' intrinsic motivation. *International Research in Geographical and Environmental Education*, 1–16.

Brannon, D. (2008). Character Education: It's a Joint Responsibility. *Kappa Delta Pi Record, 44*(2), 62–65.

Brookfield, K. (2021). 'Nature-enhanced learning' and geography education. *Journal of Geography in Higher Education*, 1–16.

Brophy, J. E. (2013). *Motivating students to learn*. New York: Routledge.

Buckley, P., & Doyle, E. (2016). Gamification and student motivation. *Interactive Learning Environments, 24*(6), 1162–1175.

Chang, C. S., Liu, E. Z. F., Sung, H. Y., Lin, C. H., Chen, N. S., & Cheng, S. S. (2014). Effects of online college student's Internet self-efficacy on learning motivation and performance. *Innovations in Education and Teaching International, 51*(4), 366–377.

Elaish, M. M., Ghani, N. A., Shuib, L., & Al-Haiqi, A. (2019). Development of a Mobile Game Application to Boost Students' Motivation in Learning English Vocabulary. *IEEE Access, 7*, 13326–13337.

Fletcher, A., Kickbusch, S., & Huijser, H. (2020). Authentic learning using mobile applications and contemporary geospatial information requirements related to Environmental Science. *Journal of Geography in Higher Education*.

Hanus, M. D., & Fox, J. (2015). Assessing the effects of gamification in the classroom: A longitudinal study on intrinsic motivation, social comparison, satisfaction, effort, and academic performance. *Computers & Education, 80*, 152–161.
Harlen, W, & Crick, R. D. (2003). Testing and Motivation for Learning. Assessment in Education: Principles, *Policy & Practice, 10*(2), 169–207.
Huang, Y. M., Lin, Y. T., & Cheng, S.-C. (2010). Effectiveness of a mobile plant learning system in a science curriculum in Taiwanese elementary education. *Computers & Education, 54*(1), 47–58.
Hung, C-Y., Sun, J. C-Y., & Yu, P. T. (2015). The benefits of a challenge: student motivation and flow experience in tablet-PC-game-based learning. *Interactive Learning Environments, 23*(2), 172–190.
Law, K. M. Y., Geng, S., & Li, T. (2019). Student enrollment, motivation and learning performance in a blended learning environment: The mediating effects of social, teaching, and cognitive presence. *Computers and Education, 136*, 1–12.
Maqableh, M., & Alia, M. (2021) Evaluation online learning of undergraduate students under lockdown amidst COVID-19 Pandemic: The online learning experience and students' satisfaction. *Children and Youth Services Review, 128*, 106160.
Rheinberg, F., & Engeser, S. (2018). *Intrinsic Motivation and Flow. In Motivation and Action, Third Edition* (pp. 1–909). Springer
Ridha, S., & Kamil, P. A. (2021) The Problems of Teaching Geospatial Technology in Developing Countries: Concepts, Curriculum, and Implementation in Indonesia. *Journal of Geography, 120*(2), 72–82
Schlemper, M. B., B. Athreya, K. Czajkowski, V. C. Stewart, and S. Shetty. 2019. Teaching spatial thinking and geospatial technologies through citizen mapping and problem-based inquiry in grades 7-12. *Journal of Geography, 118* (1), 21–34.
Shen, R., Wang, M., & Pan, X. (2008). Increasing interactivity in blended classrooms through a cutting-edge mobile learning system. *British Journal of Educational Technology, 39*(6), 1073–1086.
Su, C-H, & Cheng, C-H. (2015). A mobile gamification learning system for improving the learning motivation and achievements. *Journal of Computer Assisted Learning, 31*(3), 268–286.
Tüzün, H. (2006). *Multiple Motivations Framework. In Affective and emotional aspects of human-computer interaction: game-based and innovative learning approaches* (pp. 59–92). IOS Press.

Teacher Education and Teacher Professional Development in the COVID-19 Turn – Drajati & Putra (Eds)
© 2023 The Author(s), ISBN: 978-1-032-38970-7

School counselors' perception of online guidance and counseling amid the Covid-19 pandemic

Riswani Riswani & Raden Decue Berlian Purnama
Universitas Islam Negeri Sultan Syarif Kasim Riau, Riau, Indonesia

Solihah Hayeesama-ae
Fatoni University, Khao Tum, Thailand

Sukma Erni, Nurhayati Nurhayati & Meimunah Syofariatun Moenada
Universitas Islam Negeri Sultan Syarif Kasim Riau, Riau, Indonesia

ABSTRACT: This article examines the school counselors' perception of online guidance and counseling. The participants were two school counselors in junior high school who designed and implemented technology-integrated guidance and counseling amid the COVID-19 pandemic. Data were collected from their reflective journals and interviews and analyzed thematically through the lens of positive and negative dimensions of emotion to explore participants' perceptions of online guidance and counseling. The results showed that the pandemic situation was not a barrier to conducting guidance and counseling. One of the positive effects of integrating technology in guidance and counseling was saving time. It was also found that despite poor internet connectivity in their respective areas, the participants had limited skills in integrating technology in guidance and counseling. This indicates that participants should maintain positivity in the new normal. The implications of the study are discussed.

Keywords: COVID-19, narrative study, online service guidance and counseling, school counselor

1 INTRODUCTION

Over the past few decades, the use of technology in school guidance and counseling has been considered an important component to assist school counselors. School counselors must involve technology as a work team so that they are involved in the technological revolution (Sabella et al. 2010). Counselors not only used technology for helping with administrative tasks but also for the counseling process to improve students' social skills. Technology creates a sense of comfort for counselors, and technology does not reduce the quality of counseling. The effect of using technology can help counselors who conduct online counseling look different from traditional counselors (Osborn et al. 2018).

The emergence of the COVID-19 pandemic has brought changes in various aspects of human life (WHO 2020). This encourages the integration of technology in guidance and counseling. However, integrating technology into guidance and counseling is not easy because not all school counselors have the knowledge to use technology and adequate technological facilities. This pressure creates a dilemma for school counselors. On the one hand, they have to face a challenge that requires them to adopt technology, and face the constraint of the lack of available facilities . This triggers the emergence of negative emotions in the counselors. Changes in academic activities to online learning modes are unavoidable during the pandemic (Corbera et al. 2020).

Researchers have investigated the school counselor and the integration of technology in guidance and counseling. Osborn, Peterson, and Hele (2018), for example, highlighted that the job of virtual school counselor is not regulated by time and place, and requires self-discipline to do

 DOI 10.1201/9781003347798-17

well. Wakimoto & Lewis (2019) found that lack of time and lack of perceived value by school administrators were the most cited reasons for not continuing to use ePortfolios as professionals or integrating ePortfolios as a tool in a comprehensive school counseling program. More recently, Kosir et al. (2020) investigated the predictors of work stress in elementary and upper-secondary school teachers and school counselors in the initial period of online education in schools during the COVID-19 pandemic. The results indicated that school professionals who reported higher ICT self-efficacy had more positive attitudes toward distance education.

Despite the existing research on school counselors and technology, little is known about school counselors' perception of online guidance and counseling amid the COVID-19 pandemic. This research aims to address this issue by exploring the narrative story of the school counselors. The findings in this study are expected to contribute to the emotional mapping of the school counselors to adapt to the integration of guidance counseling into the new normal. This study's guiding research question was: how is the experience of school counselors who integrated technology in online guidance and counseling during the COVID-19 pandemic?

2 LITERATURE REVIEW

The perception of school counselors adopting technology in guidance and counseling has been widely discussed among researchers. Beidoglu, Dinçyüre, and Akıntug (2015) used a descriptive approach to identify school counselors' opinions concerning the use of ICT in school counseling services. The results of their study revealed that school counselors had overall positive opinions about the use of ICT in school counseling. Cronin et al. (2018) investigated school counselor technology use and school-family-community partnerships, and it was found that school counselors who used technology more frequently had higher self-ratings of developing school-family-community partnerships. Kosir et al. (2021) involved 964 school professionals as participants to investigate the predictors of work stress. School counselors found that school professionals with higher ICT self-efficacy had more positive attitude toward distance education. Different research designs were used by Puhy et al (2021). They examined counselor and student perceptions of an mHealth technology platform used in a school counseling setting. They found that school counselors endorsed the use of the platform to enhance treatment engagement and reduce stigma. These studies show a positive perception of school counselors on the integration of technology in guidance and counseling.

On the other hand, the study also found the reluctance of school counselors to integrate technology into guidance and counseling. Ngarajan and Yuvaraj (2019) explored the self-reported knowledge of counselors about technology in counseling. They recruited 11 practicing counselors through snowball sampling. The analysis revealed that the counselors preferred face-to-face counseling and were not using technology for their mainstream practice, but all were quite aware of the process, the benefits, and costs of using different forms of technology. Foon et al (2020) found that fear of miscommunication due to unfamiliar online language, lack of professional development technical skill training, and lack of technical resources were key deterrents that deter their intentions to offer e-counseling. Code, Santos, and de-Prado (2021) analyzed the scientific literature about online educational guidance, with a focus on the COVID-19 pandemic. A total of 42 articles were studied. The main results show that many counselors were at a disadvantage and in a very stressful situation, as they did not have the necessary digital training. These studies show a negative perception of school counselors on the integration of technology in guidance and counseling.

While the perception of school counselors adopting technology in guidance and counseling is gaining traction in guidance and counseling research, there is a lack of studies on the perception of school counselors on the integration of technology amid the COVID-19 pandemic as seen from the positive and negative emotions. Emotions influence one's resources when planning or deciding on actions (Russell 2003). Blanchette and Richards (2010) stated that the behavioral consequences of emotions result from the general principle of congruence: pleasant emotions facilitate attention to and accessibility of a repertoire of positive actions, whereas negative emotions facilitate attention to and accessibility of a repertoire of negative actions. For example, anxious individuals identify more negativity in emotionally ambiguous facial expressions, which indicates that emotions influence perceptual attentional processes. Therefore, research on the perception of school counselors on integrating technology in guidance and counseling services that explores their positive and negative emotions needs to be done. As postulated by Kember (1997) and Akerlind (2003, 2008), a broader understanding of practitioners' conceptions is first needed before planning training interventions.

3 METHOD

The researchers used qualitative research methods, which adapted narrative inquiry (Clandinin 2006) to examine school counselors' stories describing the reflective practice of integrating technology in guidance and counseling. Pseudonyms are used for the confidentiality of the participants. The first school counselor, Abdul (male), is an experienced school counselor with 5 years of counseling experience and has been enthusiastic about applying technology from the early years since he entered the counseling profession. The second school counselor, Fikriah (male) has 15 years of counseling experience. Both school counselors have a BA degree in guidance and counseling in education.

The data were collected through online semi-structured interviews and reflective journals. The interview was conducted twice. Procedurally, we interviewed the participants through the Zoom meeting application. After they were interviewed, they were asked to write reflective journals about their experiences of conducting online guidance and counseling services in schools. As there was no clear description of their emotions, we finally conducted a second interview to confirm their emotions.

For data analysis, the researchers conducted a thematic analysis (Braun & Clarke 2006) following the 12 dimensions of positive emotion (Yih et al. 2019) and 6 dimensions of negative emotions (Koenen et al. 2018). The thematic analysis has several steps of analyzing. First, we transcribed the interviews. Then, we read the data from the reflective journals and interviews repeatedly while coding the data. After the data were coded, we then generated the themes concerning positive and negative emotions during online counseling. Finally, we interpreted the data to report the results of the present study. Furthermore, we used triangulation and member checking to validate the study results.

4 FINDING AND DISCUSSION

The present study investigated the school counselors' perceptions of online guidance and counseling. Through thematic analysis, we revealed two general themes, such as (1) the positive emotions of the participants through the advantages of online counseling and (2) the negative emotions that participants shared through the obstacles faced in conducting online guidance and counseling.

4.1 *Positive emotions: The benefits of online counseling amid the COVID-19 pandemic*

Positive emotions are the coherence of appraisal, emotion, motivation, and coping expressed by participants through words of excited speech, challenge, happy, and hope to witness the benefits of online counseling. According to Abdul, he is very happy with the existence of an online counseling guidance policy during the COVID-19 pandemic because the implementation of guidance and counseling is not bound by time and place, and this saves Abdul's time, as seen in the interview excerpt:

> *I'm so excited about online guidance and counseling policies during this pandemic, besides avoiding COVID-19, it also saves my time. In my school, the schedule for individual counseling depends on the school counselor and it can be carried out synchronously and asynchronously. I can counsel students anywhere and anytime, the important thing is that there is an agreement on time and place between me and my student (Zoom interview, June 10, 2021).*

Fikriah also feels that online counseling saves time when conducting a need assessment to compile an annual program for counseling and guidance. This triggers the emergence of Fikriyah's positive emotions toward online guidance and counseling. Fikriyah's positive emotions as seen in the interview excerpt:

> *Before the online guidance and counseling policy, I distributed the instrument for need assessment to set an annual program of guidance and counseling to students in the classroom. Some students are slow to return the instruments they have filled in, and some even do not return the instruments. It's different with online distribution. Students return the instrument immediately. I am so happy and excited to compile an annual guidance and counseling program (Reflection Journal, June 20, 2021)*

Both school counselors agreed that collaboration with parents to solve their children's problems through online guidance and counseling would be easier, as seen in the excerpt from Abdul:

> *The presence of parents at school to help solve the problems faced by their children is sometimes needed, but not all parents have time to come to school. Now, the counselor will make a video call to the parents or the counselor will do an online meeting. I am happy because technology really helped my work as a counselor to be more effective and efficient.*

This excerpt indicates Abdul's positive emotions toward online guidance and counseling. Adopting technology in guidance and counseling not only saves his time but also creates ease of collaboration between school counselors and parents in solving problems faced by students. According to Abdul, not all parents have time to come to school to help solve problems faced by their children. Through online parents can be invited to discuss their children and they do not need to come to school.

4.2 *Negative emotions: Online counseling constraints during the COVID-19 pandemic*

Although the adoption of technology in guidance and counseling services is not new and the development of technology that can be adopted in counseling services is growing rapidly, some school counselors still face obstacles in integration of technology in professional practice. These obstacles affect their emotions and become negative. It is described through some negative activating emotions, e.g., anger, irritation, and tension and some negative deactivating emotions, e.g., sadness, guilt, and helpless (Koenan et al. 2018).

One of the obstacles faced by participants was the limited internet network provided by the school. Abdul shared that

> *There is no Wi-Fi in the counseling room. If I want to be online, I have to go to the second floor in the IT room. I am tired of having to go up and down. That is the only thing I can do instead of having to pay for my quota. Sometimes, I think it is better to do face to face counseling than online counseling. (Abdul's Reflective journal, June 20, 2021).*

This excerpt shows that Abdul felt negative emotions. The school does not provide Wi-Fi in the counseling room but it provides it in the IT room. It led Abdul to climb to the IT room on the second floor every time he wanted to seek online guidance and counseling. He feels tired and thinks that conventional guidance and counseling is better than online counseling. Different from Abdul, the network limitation experienced by Fikriyah is an unstable and disconnected network. It makes Fikriyah feel very disturbed. She illustrated it by the following interview excerpts

> *I'm very helpless when I'm online and suddenly the network goes out, or the power goes out, and this happens a lot. Then, I lose my concentration and motivation to do online guidance and counseling services for my students. (Zoom interview, 10 June 2021).*

Another finding that triggers the emergence of participants' negative emotions is that students do not have devices that can be used for online guidance and counseling such as gadgets, computers, or laptops. It is expressed by Fikriyah as follows:

> *Some students are absent in my online guidance and counseling class. I call them to know the reason why they do not attend my class, but they do not reply. I found out the problem when I did a home visit, they do not have the device, or they have, but must take turns using it with their siblings. It makes me helpless and sad (Reflective journal, June 20, 2021).*

Fikriyah mentioned when some students did not attend her online guidance and counseling class, she did *home visits*. She felt negative emotions because she found her student did not have a device for online learning. There was also a student who had devices but had to take turns using it with siblings for online learning.

Obstacles in carrying out online guidance and counseling are not only outside factors but also some inside factors. Fikriah said that earlier she was not much aware about technology integration into her work. As a result, she relied heavily on others for providing online guidance and counseling. It triggers the emergence of negative activating and deactivating emotions, eg, tension and less confidence simultaneously when other people did not have time to help her. This is expressed by Fikriah as follows:

> *I have been assisted by my son to do my online guidance and counseling service because I have limited skill in adopting technology for it. I design the contents while he determines the appropriate media and edits it. I feel tension and less confidence in front of my students when my son does not have time to help me. (Zoom interview, July 1, 2021).*

Unlike Abdul, Fikriyah faced difficulty observing and identifying nonverbal processes that her client showed such as client body movements when individual counseling online in this reflective journals:

> *It is difficult for me to know whether my clients are just pretending about their feelings or not. I couldn't observe their nonverbal even though I had asked my clients to face the camera when counseling online. I'm annoyed with myself for not being able to fully help students (Zoom interview, July 1, 2021)*

Fikriyah exhibits negative emotions in the form of feeling irritated by herself because of her limitations in identifying the non-verbal responses shown by students in the individual counseling process. As a result, it is difficult to identify and evaluate client and psychological problems experienced by clients.

5 DISCUSSION

This study examined two school counselors' perceptions of online counseling during the COVID-19 pandemic. From an analysis of narrative research, we identified two emotional patterns, namely positive and negative, that reflect the perception of school counselors toward online counseling:

1. Positive emotions are associated with the comfort that school counselors get from online counseling (e.g., save time, ease school counselor-parents collaboration)
2. Negative emotions are obstacles that counselors encounter when carrying out online counseling, (e.g., limited device network facilities, and limited mastery of online counseling techniques by counselors)

While both school counselors have the same positive emotional tendencies toward online counseling that they reflect in the feeling of pleasure (Yih et al 2019), there is little difference in terms of time-saving. Abdul mentioned online counseling saves his time in terms of carrying out individual counseling because counseling can be held anytime and anywhere depending on the agreement with the student. Fikriah stated that it saves her time on the need assessment program because the results of the need assessment are used to support the planning of the annual guidance and counseling program, which can be done more faster and accurately than traditional ones. It seems that the integration of technology in guidance and counseling is a necessity for both participants. Lazarus (1991) views that when one is committed to pursuing an important goal, one will experience positive emotions from assessing smooth goal progress or goal achievement, and negative emotions from assessing goal failure or delay.

Concerning the negative emotion, Fooh et al (2020) stated that school counselors were receptive in offering e-counseling in school with the conditions that they were given adequate and appropriate professional development, technical training, and e-counseling facilities. This study suggests that the limited internet network in schools and the unavailability of devices for students and the lack of skills of integrated technology in guidance and counseling as obstacles faced by school counselors lead to the emergence of negative emotions in participants toward integrating technology in guidance and counseling in schools. Limited facilities and skills can encourage participants to prefer traditional face-to-face counseling rather than online counseling. As per the study conducted by Nagarajan (2019), the counselors preferred face-to-face counseling and were not using technology for their mainstream practice because of the existence of limited facilities provided by the school.

6 CONCLUSION AND RECOMMENDATION

This study examined the school counselors' perception of online guidance and counseling during the COVID-19 pandemic. The result of the study indicated that school counselors showed positive and negative emotions during the enactment of online counseling activities. The positive emotions referred to the counselors' convenience while the negative emotions referred to the challenges encountered during the online counseling. Furthermore, school counselors had the desire to adapt to online guidance and counseling. It implied that schools and institutions should provide training and workshops to increase the school counselors' knowledge of guidance and counseling in the online environment. Due to the number of participants, the findings of the present study could not be generalized. Future studies should consider more participants to explore the school counselors' perception of online counseling.

REFERENCES

Akerlind, G. S. (2003). Growing and developing as a university teacher: Variation in meaning. *Studies in Higher Education, 28,* 375–390.

Akerlind, G. S. (2008). A phenomenographic approach to developing academics' understanding of the nature of teaching and learning. *Teaching in Higher Education, 13*, 633–644.

Beidoglu, M.; Dinçyürek, S.; Akıntug, Y. (2015). The opinions of school counselors on the use of Information and communication technologies in school counseling practices: North Cyprus schools. *Comput. Hum. Behav. 52*, 466–471

Blanchette, I., and A. Richards. 2010. The Influence of Affect on Higher Level Cognition: A Review of Research on Interpretation, Judgement, Decision Making and Reasoning. *Cognition & Emotion 24*: 561–595.

Braun, V., & Clarke, V. (2006). Using thematic analysis in psychology. *Qualitative Research In Psychology, 3(2)*, 77–101

Cronin, S., Marguerite, O., Emily, C., Brita, C., Jessica, D., Camille, M., and Megan, R. (2020). School Counselor Technology Use and School-Family-Community Partnerships. *Journal of School Counseling*, 6(16).

Corbera, E., Anguelovski, I., Honey-Roseìs, J., & Ruiz-Malleìn, I. (2020). Academia in the time of COVID-19: Towards an ethics of care. *Planning Theory & Practice, 21(2)*, 191–199.

Clandinin, D. J. (2013). *Engaging narrative inquiry.* Routledge.

Foon, L. W., Zainudin, Z.N., Yusop, Y.M., Othman, W.N.W. (2020). E-counseling : The intention,motivation and deterrent among school counselors. *Univ. J. Educ. Res. 8*, 44–51

Kettunen, J., Vuorinen, R., & Sampson, J. P. (2013). Career practitioners' conceptions of social media end career services. *British Journal of Guidance and Counseling, 41(3)*, 302–317.

Košir, K et al. (2020). Predictors of perceived teachers' and school counselors' work stress in the transition period of online education in schools during the COVID-19 pandemic. *Educ. Stud.*

Koenen, A.-K., Vervoort, E., Kelchtermans, G.,et al. (2018). Teacher sensitivity in interaction with individual students: the role of teachers' daily negative emotions. *European Journal of Special Needs Education, 1–16.*

Kember, D. (1997). A reconceptualisation of the research into university academics' conceptions of teaching. *Learning and Instruction, 7,* 255–275.

Lazarus, R. S. (1991). *Emotion and adaptation*. New York, NY: Oxford University Press.

Nagarajan, M., & S, Y. (2019). Mental health counselors perceptions on use of technology in counseling. *Current Psychology.*

Puhy, E.C. (2021). Counselor and student perceptions of an mHealth technology platform used in a school counseling setting. *Psychology in the Schools.* 1–15.

Osborn, S., W., Peterson, R., Hele (2018). Virtual school counseling. *Journal of Professional School Counseling*, Volume 18.

Puhy, C.E., Litke, S.G., Silverstein. M.J., Kiely. J.R., Pardes. A., McGeoch. E., and Daly. B.P. (2021) Counselor and student perceptions of an mHealth technology platform used in a school counseling setting. *Psychology in the Schools.* 1–15.

Sabella, R. A., Ponyton. T., Isaacs,M.I . (2010). School counselors perceived importance of counseling technology competencies. *Computers in Human Behavior 26 (4)*:609–617

Russell, J. A. (2003). *Core Affect and the Psychological Construction of Emotion. Psychological Review 110*:145–172

Yih, J., Kirby, L. D., & Smith, C. A. (2019). Profiles of appraisal, motivation, and coping for positive emotions. *Cognition and Emotion, 34*(3), 1–17.

Wakimoto, D. K., Lewis, R.,E.(2019). School counselors' changing perceptions of ePortfolios: From graduate students to professional. *The Internet and Higher Education, 41.* 45–50

World Health Organization (WHO). *Coronavirus disease 2019 (COVID-19), situation report*- 51. 2020.

Teacher Education and Teacher Professional Development in the COVID-19 Turn – Drajati & Putra (Eds)

Puberty and sexuality matters: Teachers' experiences in dealing with autistic adolescents

Mahardika Supratiwi
Universitas Sebelas Maret, Surakarta, Indonesia

Faaizah Shahbodin
Universiti Teknikal Malaysia Melaka, Melaka, Malaysia

Gunarhadi Gunarhadi & Sugini Sugini
Universitas Sebelas Maret, Surakarta, Indonesia

ABSTRACT: Autistic adolescents have communication barriers, so the issue of puberty, sexuality, and relation with the opposite sex becomes more challenging for them when entering puberty. Teachers also face varied challenges in dealing with these autistic adolescents, especially in school. The aim of this study is to explore the experience of teachers in dealing with their adolescents who are autistic. This study used a phenomenology approach and involved 16 special school teachers in a city, Indonesia. Focus group discussion is used to collect research data. The research data were analyzed using thematic analysis proposed by Braun & Clarke (2006). The results produced several themes, among others: the sexual behavior of autistic adolescents and the experience of teachers when dealing with sexually aroused autistic adolescents in school. This research suggests that teachers are expected to improve their understanding of puberty conditions in autistic adolescents and provide media to explain sexuality for autistic adolescents.

Keywords: Autistic Adolescents, Puberty, Sexuality, Special Education Teacher

1 INTRODUCTION

The complex world of sex and appropriate sexual behavior can be very challenging for people with autism spectrum disorders (ASD) and, without guidance, many of those with ASD face vulnerable situations (Tony et al. 2014). ASD is at great risk of becoming a sexual victim and/or engaging in sexual behavior. Furthermore, Parchomiuk (2019) stated that individuals with ASD reported experiencing sexual violence in various forms. The results of this study also showed that low knowledge about sexuality in people with autism spectrum disorders is one of the factors that make them being victims of sexual violence.

Koller (2000) mentioned that individuals with developmental disabilities, including autism and/or mental retardation, experience barriers to expressing their sexual needs, this can be seen in the social skills of autistic adolescents in expressing their biological needs at puberty. The barriers can trigger inappropriate behavior such as masturbating in inappropriate places. The most frequently reported inappropriate sexual behavior was masturbating in public (Beddows & Brooks 2016), displaying indecent behavior or activities (Pecora et al. 2020), and inappropriate behavior towards the opposite sex (Kellaher 2015). There are no reports of inappropriate homosexual behavior or sexual interest in much younger children (Fernandes et al. 2016).

Inappropriate sexual behavior occurs because of sexual urges that arise in adolescence not supported by socialization and understanding of social norms that regulate sexual behavior in autistic children as other children receive (Sullivan & Caterino 2008), negative emotional and physical experiences, emotional and sensory problems, lack of stimulation (in people with lower intellectual function) (Parchomiuk 2019). Based on MacKenzie (2018), the sexual behavior of individuals with ASD is problematic. The stigma is reinforced by their characteristics such as dislike of being touched and self-mutism so it is considered not interested in romantic relationships. These conditions make parents and teachers face challenging situations.

After parents, teachers are important agents involved in the education of autistic children. Some teachers who have autistic students portray the sensory impacts, lack of understanding, mental health problems, and complexity associated with their autistic students' behavior (Wood & Happé 2021). The challenging situation faced by teachers is increasing during the puberty of their students. Typically, individuals with ASD exhibit sexual behavior such as showing their genitals in public, hugging, kissing, smelling, masturbating, undressing, stroking, and touching the genitals of others (Bekirogullari et al. 2011). These sexual behaviors are often carried out in schools and are known by special education teachers. In another study, teachers of autistic adolescents highlighted that autistic adolescent girls are facing struggles experienced during puberty, such as pre-menstrual tension (PMT) and a lot of behaviors a week before the period (Cummins et al. 2020). Individuals with ASD need to receive an adequate understanding of sexuality, from the anatomy of their sexual organs to directing sexual behavior to be accepted by society. In this case, special education teachers have an important role in providing adequate understanding for individuals with ASD. Based on the discussion related to this issue, this study aims to answer the research question below:

"What experiences do teachers have at school when dealing with autistic teenagers who are going through puberty?"

2 LITERATURE REVIEW

Autism spectrum disorder (ASD) is a neurodevelopmental disorder characterized by deficits in social communication and the presence of restricted interests and repetitive behaviors (American Psychiatric Association 2013). Neurodevelopmental disorders in ASD result in an individual's ability to communicate, build relationships with others and respond well to the environment (Gense & Gense 2003). Signs of the disorders may emerge from birth or at a very early age in the developmental stage and have a lifelong influence on how children behave. All individuals with ASD have deficits in 1) social interaction, 2) verbal and nonverbal communication, and 3) repetitive behaviors or interest (American Psychiatric Association 2013; Lewis 2003). Skills in the domains of social interaction, communication, and behavior are very important for sexual development and sexual health (Dewinter et al. 2013). Individuals with ASD exhibit discomfort, anger, aggressiveness, and depressive behavior because they do not know how to react to stimuli in the event of sexual arousal (Bekirogullari et al. 2011). Children with Low Function of Autism perceive masturbation as a behavior that can be done repeatedly and therefore, they do it more often. Simultaneously, it has also been confirmed that boys with autism who have mental retardation can exhibit themselves, masturbate in public, and touch other's genital area (Kalyva 2010).

Kalyva (2010) reported that, children with Low Function Autism (LFA) exhibit more problematic behavior than children with High Function Autism (HFA) or Autistic Syndrome. Although the problematic behavior of HFA is less, it is more worrying because autism with average intelligence is more likely to be part of social interactions in families or communities that provide opportunities for sexual experiences. The increased interaction innate in classroom settings between students with and without ASD have consequence in negative interactions and consecutive social isolation from peers unless students are transfered together to learn about and know each other well, without prejudice to stigmatize them. So it is necessary to educate non-autistics about autism. Sex education is aimed at teaching appropriate socio-sexual behavior and promoting positive peer interactions. When teached about sex education, it should not only focus on 'adjusting the problem', which, in most cases, is aimed at reducing 'problematic' sexual behavior; but it also must be fixed to the level of student development.

Most teachers, regardless of their experience working with children with ASD, as well as teachers with or without postgraduate studies, have an optimistic or even overly optimistic view of ASD's ability to cope with their own problems (Syriopoulou-Delli et al. 2016). Aggressive behavior, anger, in ASD in the face of sexual desire as a result of limited knowledge of how to react to ASD, neither educational psychologists nor special education teachers have information about this, and certainly poses a challenge for teachers, who may not know how to react when faced with such behavior. Especially the challenge on what to do if an ASD child masturbates (Bekirogullari et al. 2011).

Teachers argue that ASD lacks awareness of privacy-related rules and that its knowledge is more limited to how to respond and behave sexually. Individuals with ASD are distant from social interaction in their own world, the only source of sexual information is sex education to be delivered (Bekirogullari et al. 2011). Some of the leading sexual behaviors in schools have been proclaimed by teachers, perhaps invisible to parents or caregivers, when compared to children with HFA or AS. This is an important source of previous information about children's sexual behavior with the LFA. (Kalyva 2010).

Individuals with ASD should be helped to recognize the sexual terms of their bodies. Children know sexual differences and similarities are the goal of biological recognition. Recognizing his own body is important because it allows the child to know concepts such as menstruation, erection and also to provide an appropriate response to stimuli (Bekirogullari et al. 2011). For that reason, special education teachers and parents must assist individuals with ASD especially to provide an understanding of sexuality and communicate effectively through interactive learning media.

3 METHOD

This qualitative research uses a phenomenology approach which is conducted on 16 special school teachers in Surakarta, Indonesia who handled students with autism as research participants. Focus group discussions were used to collect research data. Teachers were interviewed using face-to-face and semi-structured interviews by the researchers. Researchers had ensured that the interviews were audio-recorded and subsequently transcribed verbatim with the teachers' consent. The transcribed verbatim was analyzed using thematic analysis using Braun & Clarke (2006), which allowed identification, analysis, and description of the themes contained in the data set to inform the report. Researchers conducted interviews and made transcription. Furthermore, the coding of the transcription was repeated until the main themes and sub-themes were obtained from the results of interviews with respondents.

4 FINDING AND DISCUSSION

The two major themes were identified in this study, including the sexual behavior of autistic adolescents that appeared in school and teacher experiences when dealing with autistic adolescents at school. Two major themes and sub-themes can be seen in Table 1.

Table 1. Main themes and sub-themes identified from transcript data.

Main themes	Sub-themes
Sexual behavior	Masturbation, rubbing genitals, holding breasts, kissing objects (i.e. dolls)
Experiences faced by teachers	angry, diverting other activities, consulting with parents

4.1 *Sexual behaviors of autistic adolescents that appeared in school*

Research participants stated about sexual behaviors carried out by autistic students in their schools. Sexual behaviors that often appear in schools include: masturbating in public places, rubbing genitals in front of the teacher, holding breasts (for autistic girls), and kissing dolls in the gym. One participant also mentioned that there were students who had experienced injuries to their genitals due to the high frequency of masturbating in a day.

> *And for the other children, there was also a case, where the genitals were injured because it was like that because he (the student) kept masturbating continuously, until one day we (the teacher) found that he was wearing diapers. So we asked the parents why it was bleeding. The doctor did a total screening so that the blood was not from the inside, but because of the wound in the area of his penis due to excessive masturbation to such injuries. (Participant 2)*

Most teens (boys and girls) with autism will yell or come at the opposite sex to attract attention spontaneously. Sometimes they burst upon the opposite sex and then hugged and/or kissed them. Participants also stated that the person who caught their attention was their teacher or classmate.

> *The problems experienced by our children who are going through puberty are that most of our children are attracted to the opposite sex. So whether it's from the same teacher or friends, sometimes they come to his/her crush, and then hugged him/her spontaneously. (Participant 5)*

The sexual behaviors of autistic adolescents mentioned by the study participants are in accordance with the research by Beddows and Brooks (2016) which explains that adolescents with ASD engage in inappropriate sexual behavior. Masturbation in public places, exhibitionism, touching private parts of the body in public, the use of inappropriate objects, and having an obsession with certain objects are some examples mentioned in Beddows and Brooks (2016). Adolescents with ASD experience physical changes and reproductive functions like adolescents in general, but the age at which puberty occurs varies from individual to individual (Marieb & Hoehn 2007).

4.2 *Teachers' experiences when dealing with autistic adolescents at school*

The results of the study showed that the participants experienced challenges in dealing with puberty in adolescent students' with ASD. These challenges include when facing explosive anger from students.

> *So maybe it's because their understanding about social norms is not good enough so that they show their anger explosively in front of the public. Normal children will not do such things because they understand not to explode their anger in public. But for our students (the students with ASD), we still can't control their emotions in public places. (Participant 8)*

When faced with students who show or rub their genitals at school, teachers do some action to solve this problem, such as warn them to not do it, provide other interesting activities like drawing to distract their concern on sexual behavior, and give advice to parents to teach their children at home related to this behavior. Other learning media provided by the teacher include picture cards about what children can and cannot do, using motion and songs as interactive learning, and visual learning such as coloring for children.

> *During our learning activities, when we received our students, there was a kind of MoU with parents which basically had to have any kind of cooperation and openness because sometimes, most of the parents felt taboo about this puberty issue. So, indeed we usually talk together. We, the teachers and parents, talk (sharing) from heart to heart through a routine visitation program. We told the parents that they can be more relaxed and then they convey things that it is taboo. (Participant 12)*

Previous research has found that communication and relationships between teachers and parents are essential for autistic children (Cheung & Pomerantz 2012). Teachers explain that frequent communication gives parents the opportunity to share their experiences (Harper & Pelletier 2010). Parents can share information about the child's barriers, strengths, weaknesses, and needs so that teachers will increase their awareness of student barriers and needs (Walker et al. 2012). The regular and effective use of communication to address and identify a child's level of ability appropriately expands opportunities to stimulate students with rigor, facilitate them to continually achieve more (Josilowski & Morris 2019).

Special education teachers deal with autistic students' sexual behavior that appears in schools, including masturbation, rubbing genitals, holding breasts, and kissing objects. In addition, teachers also face the anger of autistic students. The teacher tries to overcome this by distracting students with other activities and consulting with parents.

5 CONCLUSION AND RECOMMENDATION

Puberty is a challenging time not only for adolescents with ASD, but also for parents, caregivers, and teachers. This study explored the experiences of special education teachers that taught adolescents with ASD. In dealing with adolescents with ASD, teachers faced some challenges such as found inappropriate behavior (masturbation, exhibitionism, offenses, etc) and bad-tempered expressions shown by the students. Teachers' solutions related to the problems were providing interesting activities to distract the students' inappropriate behavior and assisting the parents to teach the proper sexual education for the children. Besides, the teachers can cooperate with the parents by encouraging them to build better communication about sexual behavior with their ASD child. In addition, teachers can work with parents to build more open sexual communication with their children. Furthermore, this research can be used as a recommendation for teachers to create educational media for teaching puberty for adolescents with ASD. These educational media include picture cards, using motion and songs as interactive learning, and drawing and coloring for children. For adolescents, the teacher can give social stories, videos, puzzles, or other physical activities which can distract them from their sexual desires. The recommendation for further research is that researchers can consider the types of teaching strategies that have been given to teach sex education. In addition, researchers can also explore the problems faced by parents of autistic teenagers who are going through puberty and how to overcome them.

REFERENCES

American Psychiatric Association. (2013). Diagnostic and Statistical Manual of Mental Disorders, (5th-ed). In *The 5-Minute Clinical Consult Standard 2016: Twenty Fourth Edition*. American Psychiatric Association.

Beddows, N., & Brooks, R. (2016). Inappropriate sexual behaviour in adolescents with autism spectrum disorder: what education is recommended and why. *Early Intervention in Psychiatry*, *10*(4), 282–289.

Bekirogullari, Z., Gulsen, C., & Soyturk, K. (2011). The information and attitude levels of the educational psychologists and special education teachers in the process of sex education for the adolescents with autism. *Procedia - Social and Behavioral Sciences*, *12*, 638–653.

Braun, V., & Clarke, V. (2006). Using thematic analysis in psychology. *Qualitative Research in Psychology*, *3*(2), 77–101.

Charlop, M. H., Lang, R., & Rispoli, M. (2018). *Evidence-Based Practices in Behavioral Health.* Cham: Springer Nature.

Cheung, C. S.-S., & Pomerantz, E. M. (2012). Why does parents' involvement enhance children's achievement? The role of parent-oriented motivation. *Journal of Educational Psychology*, *104*(3), 820–832.

Cummins, C., Pellicano, E., & Crane, L. (2020). Supporting minimally verbal autistic girls with intellectual disabilities through puberty: Perspectives of parents and educators. *Journal of Autism and Developmental Disorders, 50,* 2439–2448.

Dewinter, J., Vermeiren, R., Vanwesenbeeck, I., & van Nieuwenhuizen, C. (2013). Autism and normative sexual development: A narrative review. *Journal of Clinical Nursing*, *22*(23–24), 3467–3483.

Fernandes, L. C., Gillberg, C. I., Cederlund, M., Hagberg, B., Gillberg, C., & Billstedt, E. (2016). Aspects of Sexuality in Adolescents and Adults Diagnosed with Autism Spectrum Disorders in Childhood. *Journal of Autism and Developmental Disorders*, *46*(9), 3155–3165.

Gense, D. J., & Gense, M. H. (2003). *Autism Spectrum Disorders and Visual Impairment: Meeting Students' Learning Needs*. AFB Press.

Harper, S. N., & Pelletier, J. (2010). Parent involvement in early childhood: A comparison of English language learners and English first language families. *International Journal of Early Years Education*, *18*(2), 123–141.
Hodges, H., Fealko, C., & Soares, N. (2020). Autism spectrum disorder: Definition, epidemiology, causes, and clinical evaluation. *Translational Pediatrics*, *9*(8), S55–S65.
Josilowski, C. S., & Morris, W. (2019). A Qualitative Exploration of Teachers' Experiences with Students with Autism Spectrum Disorder Transitioning and Adjusting to Inclusion: Impacts of the Home and School Collaboration. *The Qualitative Report, 24(6)*, 1275–1286.
Kalyva, E. (2010). Teachers' perspectives of the sexuality of children with autism spectrum disorders. *Research in Autism Spectrum Disorders*, *4*(3), 433–437.
Kellaher, D. C. (2015). Sexual Behavior and Autism Spectrum Disorders: an Update and Discussion. *Current Psychiatry Reports, 17 (4)*, 25–32.
Koller, R. (2000). Sexuality and adolescents with autism. *Sexuality and Disability*, *18*(2), 125–135.
Lewis, V. (2003). *Development and disability, 2nd Edition*. Blackwell Publishing.
Lord, C., & McGee, J. P. (2001). *Educating Children with Autism*. Committee on Educational Interventions for Children with Autism. Washington, DC: National Academy Press.
MacKenzie, A. (2018). Prejudicial stereotypes and testimonial injustice: Autism, sexuality and sex education. *International Journal of Educational Research*, *89*(October), 110–118.
Marieb, E., & Hoehn, K. (2007). *Human Anatomy & Physiology, 7th Edition*. Pearson Benjamin Cummings.
Ninci, J., et.al. (2015). Meta-analysis of Single-Case Research on Teaching Functional Living Skills to Individuals with ASD. *Review Journal of Autism and Developmental Disorders*, *2*(2), 184–198.
Parchomiuk, M. (2019). Sexuality of Persons with Autistic Spectrum Disorders (ASD). *Sexuality and Disability*, *37*(2), 259–274.
Pecora, L. A., Hooley, M., Sperry, L., Mesibov, G. B., & Stokes, M. A. (2020). *Sexuality and Gender Issues in Individuals with Autism Spectrum Disorder*.
Postorino, V., Siracusano, M., Giovagnoli, G., & Mazzone, L. (2018). *Aspects of Sexuality During Development in Autism Spectrum Disorder*. Sexual Dysfunctions in Mentally Ill Patients. 63–73.
Sevlever, M., Roth, M. E., & Gillis, J. M. (2013). Sexual abuse and offending in autism spectrum disorders. *Sexuality and Disability*, *31*(2), 189–200.
Sullivan, A., & Caterino, L. C. (2008). Addressing the Sexuality and Sex Education of Individuals with Autism Spectrum Disorders. *Education and Treatment of Children, 31*(3), 381-394.
Syriopoulou-Delli, C. K., Cassimos, D. C., & Polychronopoulou, S. A. (2016). Collaboration between teachers and parents of children with ASD on issues of education. *Research in Developmental Disabilities*, *55*, 330–345.
Tony, A., Isabelle, H., & Nick, D. (2014). *Autism Spectrum, Sexuality and the Law: What Every Parent and Professional Needs to Know*. Jessica Kingsley Publishers.
Walker, S., Dunbar, S., Meldrum, K., Whiteford, C., Berthelsen, D., Carrington, S., Hand, K., & Nicholson, J. (2012). The transition to school of children with developmental disabilities: Views of parents and teachers. *Australasian Journal of Early Childhood*, *37*(3), 22–29.
Wood, R., & Happé, F. (2021). What are the views and experiences of autistic teachers? Findings from an online survey in the UK. *Disability & Society, 1–26.*
Yapko, D. (2003). *Understanding Autism Spectrum Disorders: Frequently Asked Questions*. Jessica Kingsley Publishers.

Teacher Education and Teacher Professional Development in the COVID-19 Turn – Drajati & Putra (Eds)
© 2023 The Author(s), ISBN: 978-1-032-38970-7

Design of deaf instruction during the COVID-19 pandemic

Alfina Fadilatul Mabruroh, Gunarhadi Gunarhadi & Herry Widyastono
Universitas Sebelas Maret, Surakarta, Indonesia

Mohd Hanafi Mohd Yasin
Universiti Kebangsaan Malaysia, Bangi, Malaysia

ABSTRACT: The value of this study is to draw teaching presence aspect analysis of the Community of Inquiry for the design of deaf instructions during COVID-19. This study applied a narrative qualitative research approach. Data were collected through deep interviews and document studies by two teachers of a special school in Surakarta, one of these was an experienced teacher engaged in teaching profession since 1991 and a novice teacher who practices teaching since 2004. The data was analyzed using the design and organization instruction under the teaching presence concept of the COI framework, which consists of creating categories of coding the data, reducing the data, presenting the data, and withdrawing conclusions. Based on the design and organization of deaf instruction, teaching presence aspect of the COI framework gives a design of asynchronous learning. Findings revealed that teachers have different ways to provide instructional materials and conduct assessments. The experienced teacher made a module and deaf students were expected to go to school for the module collection every week, meanwhile, the novice teacher prepared the module and sent it via WhatsApp. Indirectly, the design and organization of deaf instruction under the teaching presence aspect of the COI framework gives deaf students new experiences using technology in learning and provides emotional understanding and communication practices through discussion. The implication is that the government provides professional development to the teachers providing special education.

Keywords: COVID-19, Designing instruction for students, Special Education, Teacher Change

1 INTRODUCTION

The emergence of the COVID-19 pandemic changed numerous aspects of life. For instance, social distancing was encouraged to prevent the spread of this pandemic. All infected countries were made aware of the spread of the pandemic, all educational institutions were closed, and all stakeholders were expected to stay at home (Madhesh 2021). The education system is one of the first institutions affected by the emergence of the COVID-19 pandemic; numerous schools were closed (Zhou et al. 2020). This pandemic forced the quick adaptation of teachers and students to practice distance learning (Alshawabkeh et al. 2021; Reimers et al. 2020). Teachers sought a practical way to complete tasks during remote learning (Wang 2021).

In the context of special education learning, the COVID-19 pronounced the need for deaf students to learn remotely. Deaf children had a risk of experiencing language delays as these students lack the hearing ability (Hall et al. 2019). Their ability to communicate was very lacking (Meinzen-Derr et al. 2017). Thus, it was difficult for the students with disabilities to adapt to COVID-19 learning (Baroni & Lazzari 2020). Online learning was more popular and widespread than ever before, which offered great potential to reach students, on the other hand, barriers exist for deaf students (McKeown & McKeown 2019). Previous literature showed COVID-19 engendered “emergency remote teaching" which had difficulties associated with either online teaching infrastructure, the inexperience of teachers, the information gap, (Bozkurt & Sharma 2020), lack of mentoring and support, or complex environment at home (Carrillo & Flores 2020). According to Maddrell, Morrison, and Watson, applying a community of inquiry (CoI) framework has an impact on the optimal design of online learning environments (Maddrell et al. 2017). This study will explain the design and organization which consist of setting curriculum and methods of deaf instruction under the teaching presence aspect of the COI framework. Issues on deaf learning and CoI are very rarely analyzed because of past injustices based predominantly on race and gender and enough attention has not been paid to the issue of disabilities (Satar 2019). In this study, the researchers examine the following research question: How do the teachers create the design of deaf instruction under the teaching presence of the CoI framework?

2 LITERATURE REVIEW

This study investigated how the teacher provides new instruction learning to adapt to remote learning. This was assessed under the teaching presence of the CoI framework by Garrison, Anderson, and Archer (2000). The CoI model is an in-depth collaborative learning process in an online inquiry community involving interrelated components of presence (Garrison et al. 2000). The CoI framework suggested three essential elements for fostering student engagement and learning, which include social presence (SP), cognitive presence (CP), and teaching presence (TP) (Micsky & Foels 2019). The SP is defined as the skill to connect members of the community for instance building trust, building purposeful environment, making personal relationships with other community members (Garrison & Akyol 2009; Pifarré, Guijosa & Argelagós 2014), and supporting critical thinking in a community of members (Garrison & Anderson 2003; Pifarré, Guijosa & Argelagós 2014). SP needed online participants to build meaningful relationships that involve effective expression, open communication, and group cohesion (Liu & Yang 2014). The CP delineates sustained communication between participants of a community which is aimed at meaning to meaning-making (Garrison Anderson & Archer 2000; Pifarré et al. 2014). CP triggered participants to explore ideas then structure and initiate resolutions and confirmations using cooperation and reflection (Liu & Yang 2014). Meanwhile, TP refers to the course design process, discussion facilitation, and direct instruction (Liu & Yang 2014) for leading personally meaningful and educationally worthwhile learning outcomes (Anderson et al. 2001; Pifarré et al. 2014).

The CoI model framework is based on John Dewey's Learning Theory, which believes that learning experiences need to integrate individual and group interests through collaboration so that learners will be responsible for actively constructing and confirming meaning (Pratiwi et al. 2016). The CoI occurs when a group of individuals engages in a process of empirical or conceptual inquiry into problematic conditions for instance involving the formation of knowledge, the meaning, ethics, and value formation (Haynes 2018). Recently, a study using the CoI framework for teacher learning by designing and investigating online participatory practice showed feedback and provided observation of each other's practices in the teaching presence, which allowed the teachers to recognize problems in feedback practice and develop principles of effective written feedback through collaborative efforts (Yang 2016). The factors that underpin students' satisfaction with their e-learning experience with a CoI, concluded cognitive presence and social presence affected e-learning satisfaction directly (Lee et al. 2021). Then, a study investigated K-12 online learning success using a CoI framework and self-regulated learning (SRL) theory through 696 online surveys of students from 8th to 12th grade that show that mentor presence is positively affecting than social presence and cognitive presence (Zhang & Lin 2021). Cognitive presence, social presence, and mentor presence have non-statistically significant differences in males and females, and the network speed is important to strengthen the potential of students to create an in-depth and purposeful learning experience (Lau et al. 2021). Furthermore, the CoI framework involved college students, who revealed that the components of effectiveness and the level of exploration accounted for the majority of students' social and cognitive presences (Guo et al. 2021).

This study portrayed CoI in deaf learning whose members of learning had disabilities in communication. A number of the CoI studies including deaf as a subject are very rare. However, deaf students require a communication ability that involves the social presence of the CoI model. With analysis design and organization of instruction learning, which is an indicator of teaching presence aspect, communication and process learning of CoI will occur. As Shea & Bidjerano (2013) state that SP is predicted by TP and a portion of the variance in SP is affected by the design of the course. The design and organization of deaf instruction under the teaching presence of the COI framework will be analyzed and described narratively.

3 METHOD

This study applied a qualitative research approach and involved two teachers of a special school in Surakarta, Indonesia. Data were collected through deep interviews and document studies from two teachers. The participants consist of Mrs. C (56 years old) who has teaching experience of 30 years as she is practicing since 1991 and Mrs. M (36 years old) who has teaching experience of 17 years as she is practicing since 2004. The data were collected through interviews to find out information related to how the teacher developed the design of deaf instruction in the COVID-19 pandemic. The teachers were interviewed by chatting via WhatsApp to dig into the facts of the learning process. Participants prefer chatting by text to find out data rather than interviewing by call or video call because teachers need to remember the past event. Study documents are used to provide concrete evidence of how instructional designs and assessments were arranged. Similarly, the study of the narrative inquiry approach portrayed a three-dimensional narrative inquiry space including temporality, people/social interaction, and places (Li & Craig 2019). Clandinin (2006) asserted narrative which provided the possibility to understand how personal and social interaction are intertwined in teachers' lives and how these experiences are shaped by the larger social and institutional narratives in which they live (Mendieta & Barkhuizen 2019). The data were analyzed by Miles and Huberman (1994) which consists of creating categories of coding the data, reducing the data, presenting the data, and withdrawing conclusions. Coding, reducing, and presenting the data, which reflected design and organization instruction analyzed by teaching presence concept of CoI framework, thus conclusions drawn from the number of facts of the learning process.

4 FINDING AND DISCUSSION

The emergence of COVID-19 changed face-to-face learning to remote learning, consequently, teachers changed methods, strategies, and tools to maintain effective teaching and learning process. Mrs. C stated that studying from home enhances her creativity in teaching. As a result, she said that she was very busy and tried to adapt to distance learning, as shown in the transcript below:

> *I am not a young teacher anymore. I tried to hold effective learning during COVID-19 although I has limited access to learning technology. Whereas, my students did not have distance learning experiences and use technology in learning. I had to provide simple learning to the deaf to avoid learning loss. It made me in a busy condition, hence deaf students as visual learners needed to gave visual material learning to assist their comprehension. In early distance learning, I made design learning involving deaf students' parents in Whatsapp to inform schedule and school assignments. Besides I made and printed modules and media learning so that my students can take it at school. That's made me so tired.*

She had to learn technology usage, but at the same time, her students were not skilled enough and did not own a smartphone for distance learning. Therefore, she decided to make a group WhatsApp for their parents first, made the strategy for her learning based on the meeting with stakeholders in the school, and tried to distribute the information via WhatsApp. However, it did not enable the students to increase their learning performance and created inconvenience for their parents' activities, so the school made a policy to make a paper-based module. Mrs. C explained that distance learning builds knowledge to improve learning and enhance technical skills. Furthermore, distance learning has a positive impact that makes teachers, deaf students, and their parents used to enriched experience of learning, knowing natural learning motivation of deaf students, and learn more independently (time management). Parents involved in early remote learning helped in study adaptation during distance learning.

Meanwhile, Mrs. M asserted that distance learning forced teachers to adapt to the lack of infrastructure for deaf students, her inexperience using technology, and limited information for accessing pedagogical knowledge (model learning to hold online learning). Mrs. M used the WhatsApp group for remote learning. She and deaf students attended online classes from 08.00 until 13.00. This policy was made by Mrs. M and students in March 2020 through discussion in the WhatsApp group. In early COVID-19, Mrs. M had trouble adapting to online learning and engaging her students because the cost of the internet quota was more extravagant. It forced Mrs. M to find an alternative to hold the online learning by asynchronous and synchronous learning. Asynchronous learning introduced the WhatsApp groups as online classes. Distributing and communicating by WhatsApp group helped in getting habituated for synchronous learning. After 3 weeks, Mrs. M took 15 minutes meeting only once a week. According to Hrastinski which divides two types of online learning, namely asynchronous and synchronous online learning (Hrastinski 2008; Adedoyin & Soykan 2020), Mrs. M tried to address the challenges faced by the deaf students. She tried to create an opportunity with WhatsApp group to habituate deaf students learning with technology.

Different from Mrs. M, Mrs. C created and gave a paper-based module once a week with weekly practice questions. Mrs. C did not use WhatsApp every day for distance learning. She argued that deaf students need to go to school because many students did not have a device for online learning, they used their parents' smartphones to communicate during remote learning. Mrs. C occasionally attempted to interact with the deaf students through WhatsApp groups and Google Meetings for assessment. Mrs. C and Mrs. M stated the WhatsApp group not only was effective to stimulate and record deaf students' communication skills but also distributing images to illustrate materials in modules. Either Mrs. C or Mrs. M designed instruction during the COVID-19 pandemic by creating paper-based modules every week from The Government's Core Competence and Basic Competence. They were required to adjust lesson plans before COVID-19 and transform it into simple lesson plans. The lesson plan during COVID-19 was adjusted from lesson plans before COVID-19, followed by recompilation and condensation from book materials. After this, they used to compile moderate material learning based on a government-published social textbook through resume materials in the book and made a quiz to provide practice questions. The process of recompiling the module during COVID-19 and condensing the material is shown in Figures 1 to 4.

A number of findings show that teachers arrange and provide CoI framework directly. *First,* the teacher made the design instruction dependent on their lack of pedagogy knowledge, skill in operating technology, and the condition of students who did not have a device for remote learning. Teachers choose to maintain simple learning such as designing paper-based modules or create modules that are distributed via WhatsApp. It seems like they introduced synchronous learning through WhatsApp group as an online class for promoting deaf students to access new learning experiences using technology in learning. According to the CoI model, the design and organization of instruction learning is a part of teaching presence. It seeks to generate metacognitive awareness among the students, in addition to guiding the teaching process (Garrison & Arbaugh 2007; Keles 2019). In the context of deaf learning, learning using technology was something new for deaf students. They usually learn in a traditional model with face-to-face learning. During the COVID-19 pandemic, they have the opportunity to access technology in learning. That reinforced Dewey's thought that learning experiences need to integrate individual and group interests through collaboration so that learners will be responsible for actively constructing and confirming meaning.

Second, Mrs. C printed the modules and the quiz and distributed them to deaf students through school, Mrs. M distributed through WhatsApp once a month, Mrs. C uses Google Meets and

- 713 -

K. KOMPETENSI INTI DAN KOMPETENSI DASAR ILMU PENGETAHUAN SOSIAL SMALB TUNARUNGU

KELAS: X

Tujuan kurikulum mencakup empat kompetensi, yaitu (1) sikap spiritual, (2) sikap sosial, (3) pengetahuan, dan (4) keterampilan. Kompetensi tersebut dicapai melalui proses pembelajaran intrakurikuler, kokurikuler, dan/atau ekstrakurikuler.

Rumusan Kompetensi Sikap Spiritual adalah "Menerima dan menjalankan ajaran agama yang dianutnya". Adapun rumusan Kompetensi Sikap Sosial adalah "Menunjukkan perilaku jujur, disiplin, tanggung jawab, santun, peduli, dan percaya diri dalam berinteraksi dengan keluarga, teman, guru, dan tetangga serta cinta tanah air". Kedua kompetensi tersebut dicapai melalui pembelajaran tidak langsung (*indirect teaching*), yaitu keteladanan, pembiasaan, dan budaya sekolah dengan memperhatikan karakteristik mata pelajaran, serta kebutuhan dan kondisi siswa.

Penumbuhan dan pengembangan kompetensi sikap dilakukan sepanjang proses pembelajaran berlangsung dan dapat digunakan sebagai pertimbangan guru dalam mengembangkan karakter siswa lebih lanjut.

Kompetensi pengetahuan dan kompetensi keterampilan dirumuskan sebagai berikut ini.

KOMPETENSI INTI 3 (PENGETAHUAN)	KOMPETENSI INTI 4 (KETERAMPILAN)
3. Memahami pengetahuan faktual dan konseptual dengan cara mengamati, menanya dan mencoba berdasarkan rasa ingin tahu tentang dirinya, makhluk ciptaan Tuhan dan kegiatannya, dan benda-benda yang dijumpainya di rumah, di sekolah dan tempat bermain	4. Menyajikan pengetahuan faktual dan konseptual dalam bahasa yang jelas, sistematis, logis dan kritis, dalam karya yang estetis, dalam gerakan yang mencerminkan anak sehat, dan dalam tindakan yang mencerminkan perilaku anak beriman dan berakhlak mulia

Figure 1. KI and KD from the government.

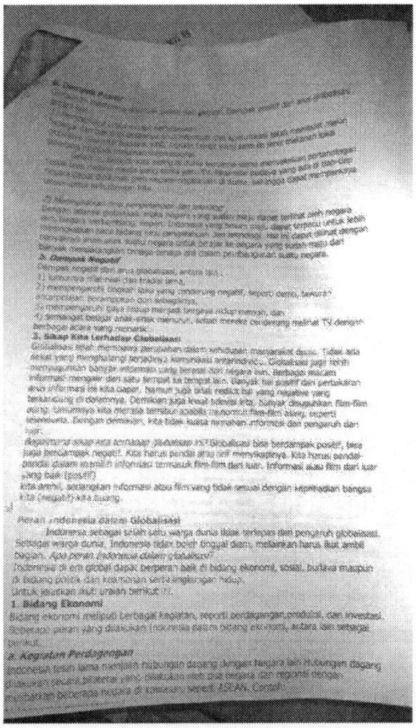

Figure 2. Module made by Mrs. C.

Figure 3. Area to compile the task and take the modules by Mrs. C.

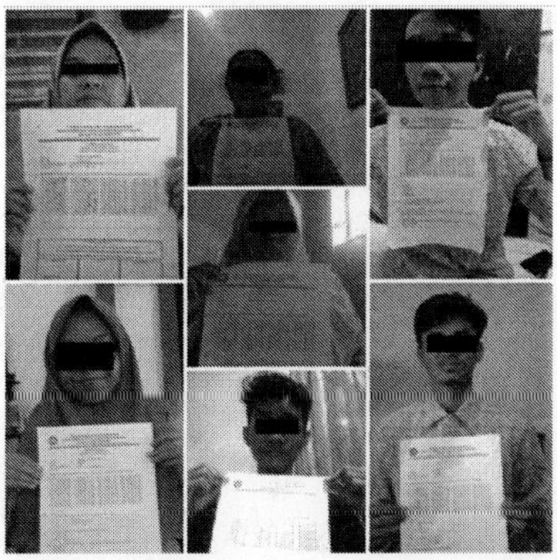

Figure 4. Synchronous learning by Mrs. M.

WhatsApp to control the final exam and enable the students to express their feeling, problem, and challenges during remote learning. It implies that teachers obtain teaching presence aspect through planning and managing the teaching process. Based on the teaching presence aspect, it leads to the development of the social learning environment such that it enables active and successful learning, and perform as an expert in given area of study (Anderson et al. 2001; Keles 2019). Teachers arrange and provide a CoI framework indirectly through designing instruction that implicates technology and discussion session in online learning. Indirectly, deaf students need to improve their communication ability through discussion sessions using face-to-face meetings and chatting on WhatsApp. Majeski, Stover & Valais (2018) revealed that emotional perception and emotional understanding are a part of teaching presence that can be demonstrated by using reflective listening with students when they express their feelings toward assignments, course material, the instructor, and other learners.

5 CONCLUSION AND RECOMMENDATION

This study aimed to investigate how the teacher created the design of deaf instruction under the teaching presence of the CoI framework. The module required to stimulate metacognitive awareness and avoid learning loss during COVID-19. Module consists of not only material learning but also matters to the discussion in the WhatsApp group. How the deaf student composed a sentence and delivered their idea in the chatting room is the evidence of improvement in their experience in language delays and communication. The result showed that the determining design and organization of deaf instruction under the teaching presence aspect of the COI framework gives a design of asynchronous learning that provides deaf students new experiences for using technology in learning and give emotional understanding and communication practices through discussion. Thus, the findings of this study could be the recommendation for the stakeholders in the school and the government to provide teacher professional development for special education teachers. It makes teachers able to imply the best design and organization of instructional learning for achieving personally meaningful and educationally worthwhile learning outcomes. A vital suggestion for future research is researchers are required to investigate CoI theoretical framework in special education to promote optimal online learning quality for disabled students.

REFERENCES

Adedoyin, O. B., & Soykan, E. (2020). Covid-19 pandemic and online learning: the challenges and opportunities. *Interactive Learning Environments*, 1–13.

Alshawabkeh, A. A., Woolsey, M. L., & Kharbat, F. F. (2021). Using online information technology for deaf students during COVID-19: A closer look from experience. *Heliyon*, 7(5).

Anderson, T., Rourke, L., Garrison, D. R. & Archer, W. (2001). Assessing teaching presence in a computer conferencing context. *Journal of Asynchronous Learning Networks*, 5(2), 1–17.

Baroni, F., & Lazzari, M. (2020). Remote teaching for deaf pupils during the COVID-19 emergency. *Proceedings of the 14th IADIS International Conference E-Learning 2020, EL 2020 – Part of the 14th Multi-Conference on Computer Science and Information Systems, MCCSIS 2020*, 170–174.

Bozkurt, A., & Sharma, R. (2020). Emergency remote teaching in a time of global crisis due to CoronaVirus pandemic. *Asian Journal of Distance Education*, *15*(1).

Clandinin, D. J. (2006). Narrative inquiry: A methodology for studying lived experience. *Research Studies in Music Education*, 27(1), 44–54.

Carrillo, C., & Flores, M. A. (2020). Covid-19 and teacher education: a literature review of online teaching and learning practices. *European Journal of Teacher Education*, *43*(4), 466–487.

Dubnewick, M., Clandinin, D. J., Lessard, S., & McHugh, T. L. (2017). *The centrality of reflexivity through narrative beginnings: towards living reconciliation:*, *24*(6), 413–420.

Garrison, D. & Akyol, Z. (2009) Role of Instructional Technology in the Transformation of Higher Education, *Journal of Computing in Higher Education*, 21(1), 19–30.

Garrison, D. R., & Arbaugh, J. B. (2007). Researching the community of inquiry framework: Review, issues, and future directions. *Internet and Higher Education*, 10(3), 157–172.

Garrison, D.R. & Anderson, T. (2003) *E-learning in the 21st Century: a framework for research and practice*. London: RoutledgeFalmer.

Garrison, R., & Vaughan, N. D. (2008). *Blended Learning in Higher Education: Framework, Principles, and Guidelines*. Willey.

Garrison, R., Anderson, T., & Archer, W. (2000). Critical inquiry in a text-based environment. *The Internet and Higher Education*, *2*(2), 87–105.

Guo, P., Saab, N., Wu, L., & Admiraal, W. (2021). The community of inquiry perspective on students' social presence, cognitive presence, and academic performance in online project-based learning. *Journal of Computer Assisted Learning*, *37*(5), 1479–1493.

Hall, M. L., Hall, W. C., & Caselli, N. K. (2019). Deaf children need language, not (just) speech. *First Language*, 39(4), 367–395.
Haynes, F. (2018). Trust and the community of inquiry. *Educational Philosophy and Theory*, *50*(2), 144–151.
Hrastinski, S. (2008). Asynchronous and synchronous e-learning. *Educause Quarterly*, 31(4), 51–55.
Keles, E. (2019). Use of Facebook for the community services practices course community of inquiry as a theoretical framework. *Computers & Education*, 116, 203–224.
Lau, Y. Y., Tang, Y. M., Chau, K. Y., Vyas, L., Sandoval-Hernandez, A., & Wong, S. (2021). Covid-19 crisis: exploring community of inquiry in online learning for sub-degree students. In *Frontiers in Psychology* (Vol. 12).
Lee, R., Hoe Looi, K., Faulkner, M., & Neale, L. (2021). The moderating influence of environment factors in an extended community of inquiry model of e-learning. *Asia Pacific Journal of Education*, *41*(1), 1–15.
Li, J., & Craig, C. J. (2019). A narrative inquiry into a rural teacher's emotions and identities in China: through a teacher knowledge community lens. *Teachers and Teaching: Theory and Practice*, *25*(8), 918–936.
Liu, C.-J., & Yang, S. C. (2014). Using the community of inquiry model to investigate students' knowledge construction in asynchronous online discussions. *Journal of Educational Computing Research*, 51(3), 327–354.
Maddrell, J. A., Morrison, G. R., & Watson, G. S. (2017). Presence and learning in a community of inquiry. *Distance Education*, 38(2), 245–258.
Madhesh, A. (2021). Full exclusion during COVID-19: Saudi deaf education is an example. *Heliyon*, *7*(3), e06536.
Majeski, R. A., Stover, M., & Valais, T. (2018). The community of inquiry and emotional presence. *Adult Learning*, 29(2), 53–61.
McKeown, C., & McKeown, J. (2019). Accessibility in online courses: understanding the deaf learner. *TechTrends*, *63*(5), 506–513.
Meinzen-Derr, J., Wiley, S., McAuley, R., Smith, L., & Grether, S. (2017). Technology-assisted language intervention for children who are deaf or hard-of-hearing; a pilot study of augmentative and alternative communication for enhancing language development. *Disability and Rehabilitation: Assistive Technology*, *12*(8), 808–815.
Mendieta, J., & Barkhuizen, G. (2019). Blended language learning in the Colombian context: a narrative inquiry of teacher ownership of curriculum change. *Computer Assisted Language Learning*, 1–21.
Micsky, T., & Foels, L. (2019). Community of Inquiry (CoI): a framework for social work distance educators. *Journal of Teaching in Social Work*, *39*(4–5), 293–307.
Miles, M., & Huberman, A. (1994). *Qualitative data analysis: An expanded sourcebook* (second edition). Thousand Oaks, CA: Sage Publications.
Pifarré, M., Guijosa, A., & Argelagós, E. (2014). Using a blog to create and support a community of inquiry in secondary education. *E-Learning and Digital Media*, 11(1), 72-87.
Pratiwi, E. R., Wonorahardjo, S., & Arief, M. (2016). *Partisipasi Online Dalam Pembelajaran Community Of Inquiry (Coi) Pada Materi Distilasi. 1*, 1410–1416.
Reimers, F., Schleicher, A., Saavedra, J., & Tuominen, S. (2020). *Supporting the continuation of teaching and learning during the COVID-19 pandemic*. 1–38.
Satar, A. A. (2019). Innovative strategies for creating inclusive spaces for hearing-impaired and visually impaired students in an open distance and e-learning (ODeL) Environment A Case Study of the University
Shea, P., & Bidjerano, T. (2013). Understanding distinctions in learning in hybrid, and online environments an empirical investigation of the community of inquiry framework. *Interactive Learning Environments*, 21(4),
Wang, C. X. (2021). CAFE: an instructional design model to assist k-12 teachers to teach remotely during and beyond the covid-19 pandemic. *TechTrends*, *65*(1), 8–16.
Yang, S. H. (2016). Conceptualizing effective feedback practice through an online community of inquiry. *Computers and Education*, *94*, 162–177.
Zhang, Y., & Lin, C. H. (2021). Effects of community of inquiry, learning presence and mentor presence on K-12 online learning outcomes. *Journal of Computer Assisted*
Zhou, L., Wu, S., Zhou, M., & Li, F. (2020). 'School's out, but class' on', the largest online education in the world today: taking China's practical exploration during the covid-19 epidemic prevention and control as an example. *SSRN Electronic Journal*.

Teacher Education and Teacher Professional Development in the COVID-19 Turn – Drajati & Putra (Eds)

Promoting innovative teaching practices: Teachers' voices in creating E-comic based local history lesson

Ema Agustina
Universitas PGRI South Sumatera, Palembang, Indonesia

Theresia Manalu
Universitas Pendidikan Indonesia, Bandung, Indonesia

ABSTRACT: The current study unveils teacher's insights on creating new teaching media with local history immersion in a high school setting. It highlights teachers' beliefs, motivations, considerations, and challenges in providing E-comic-based local history immersion experience to improve students' knowledge and engagement. The study employed a descriptive qualitative design. The verbal data were collected through in-depth interviews with one experienced history teacher in Indonesia. To interpret the data, open coding, axial coding, and selective coding were employed. The results revealed that E-comic-based local history does stimulate students' curiosity by moving beyond ordinary to extraordinary sensation, especially in the very beginning of teaching and learning activities. The philosophical belief also impacted teachers' motivation in creating e-comic. To conclude, innovative touch in teaching makes teaching and learning activities alive. Based on the findings, it is proven that there is still a history teacher who can meet the challenges of 21st-century learning by creating innovative teaching, E-comic for example. Implications for history teachers are fully discussed.

Keywords: E-comic, Innovative teaching, and learning, Teachers' professional development

1 INTRODUCTION

Of the many teachers' pedagogical responsibilities, creating innovative teaching and learning environments is one key pivotal element. Today, creating innovative teaching is the current idea and possibility for teachers to deal with the current virtual learning environment (VLE) and digital platforms (Rienties et al. 2016). This phenomenon turns attention to teachers in relation to the teaching function and style (Spiteri et al. 2018). Since this phenomenon offers new demand for educational practices, it requires teachers to create and to imply new teaching practices in regard to innovative teaching media (Faulkner & Latham 2016). E-comic, as one of the exemplary media, is considered as powerful in promoting students' skills to achieve particular learning goals (Lazarinis & Panagiotakopoulos 2015; Melliou et al. 2014; Murti et al. 2020; Vassilikopoulou et al. 2011). The involvement of E-comic on any subject matter in this sense leads to a new way of teaching that goes beyond a virtual representation of the traditional physical classroom.

Studies on local history in history teaching materials have led to the improvement of many aspects in teaching. Previously mentioned, Ma'unah et al. (2018) did a study to investigate the attractiveness and efficacy of the interaction between local history and interactive learning materials. The findings of the study revealed that a local history learning package was appealing and effective for use in history classes. Visual forms of learning are incredibly crucial for today's youngsters who grew up with television, video, and the Internet, therefore technology is altering the way we live and learn (Magro et al. 2014)

 DOI 10.1201/9781003347798-20

In parallel to what previously has been stated, it is indispensable to learn from others' experiences for good things they have done before. There have been many studies researching teachers' experiences and teachers' variances. But, it is essential to look deeper into the teacher's media that infuses local history into a digital comic. To bridge this notion, the present study attempted to unveil history teachers' receipts in creating publicly marketed E-comic-based local history teaching media. It also captured challenges on the practical use of E-comic during a virtual learning environment. Finally, it answered the question: how a history teacher created an E-comic-based local history to promote innovative teaching.

2 LITERATURE REVIEW

Throughout history, the teaching and learning approach is dynamically changing based on the emerging trends in society. This alteration entails an innovative way of teaching competencies in the classroom to answer the problematic issues that happen due to the dynamic way of people in learning and understanding (Redecker 2009). Innovative teaching is considered as the answer to recent education demands to meet the educational needs and goals of students. Providing new methods, strategies, and techniques are associated with innovative teaching that can stimulate students' creativity (Ferrari & Schoolnet 2009). The issue of authenticity, student-centered, collaborative, and real learning situations also have close relationship to today's innovative teaching in practices (Brandon 2004).

In this context, the use of E-comic with local historical content as teaching media is one of exemplary innovative teaching. The infusion of digital transformation and authenticity to teaching media is an excellent result of innovative teaching that teachers can create and develop, especially history teachers. The use and development of ICT in education as the impact of digitalization has brought significant changes in the field of education. Recent research on ICT illustrates that teachers commonly have a positive attitude towards ICT use in their classroom regardless of age, teaching experience, gender, ICT skills, ICT training, and ICT experience (Semerci & Aydin 2018). In it, local history is one of authentic content that is appropriate to students' prior knowledge and history learning functions in history subject matter. Previously, authentic materials were materials that were used and existed in society for other purposes instead of education including anything that is used to communicate in the real environment (Gebhard 2009).

Next, studies on local history in history teaching materials have led to the improvement of many aspects in teaching. Ma'unah et al. (2018) conducted a study to look deeper at the interaction between local history and interactive learning materials and its attractiveness and effectiveness. Research results showed that the learning package of local history is considered attractive and effective to be utilized in history learning. Then, research and studies say that today's children, who grew up with television, video, internet, and visual modes of learning are extremely important, so that the technologies are changing the way people live and learn (Magro et al. 2014).

In general, this whole framework issues a construct that fits with the conceptual method of this research. Further, this study seeks the Indonesian history teacher's conceptualization of beliefs, attitudes, motivations, perceptions, and challenges of practical use of E-comic especially during virtual learning environment (VLE). Above all, this study confirmed the questions: How do teachers design and integrate local history E-comic based teaching media in teaching history during virtual learning environment (VLE) and what are teacher's knowledge, perceptions, attitudes, and challenges towards the integration of local history E-comic based teaching media.

3 METHOD

This current study employed a descriptive qualitative design as it focuses on exploring teacher's voices in incorporating as well as in creating local history based into E-comic as one of dynamic teaching style (Faulkner & Latham 2016; Kereluik et al. 2013; Griffin 2015; Rudy 2017). To seek rich data, online interviews were intensively conducted to one male experienced history teacher in Indonesia, who made his own commercial E-comic for history subjects. Interestingly, the full infusion of local history based, specifically the history of Surabaya as one of historical cities in Indonesia and the E-comic became the focus of this study. Given the qualitative data of interview protocols, developed open-ended questions were issued at the motivation for innovating E-comic based local history, practical use in the classroom, perceptions of factual benefits, and challenges. In addition to the data analysis, the interview results were comprehensively explained by transcribing and describing his responses to the interview questions.

To interpret the data, open coding, axial coding, and selective coding were employed (Creswell 2017; Straus & Corbin 2015). First, in open coding, the data were labeled, categorized, and conceptualized based on the findings. Then, it was continued to the axial coding by identifying central phenomena to the categorization previously conceptualized. Finally, selective coding was attempted as the final step in analyzing the data to discover the cases that were uncovered in the two previous coding.

4 FINDING AND DISCUSSION

This study explored teacher's insights on creating E-comic based local history lessons in a high school that highlights teachers' beliefs, motivations, considerations, and challenges to improve students' knowledge and engagement with literature or other subject matter. Based on the content analysis, we found three major themes in the study: (1) Personal and professional identity, (2) Evaluation of E-comic based local history lesson, (3) Beliefs and knowledge about innovative teaching, and (4) Challenges in designing innovative teaching media, to get the history teacher's insights in relation to what his personal experience for his classroom.

4.1 *Personal and professional identity*

Prior to the participant's education background, the selected teacher in this study was an experienced teacher who commenced his first career as a history teacher since he finished his bachelor degree in 2005 and continued his master degree in 2007. Brief information is presented in the following table.

Table 1. Teacher's voice on *personal and professional identity.*

Indicators	Teacher's description
1. Personal	Having technological skill to create and design E-comic
2. Professional	Experiencing on creating E-comic, following teacher training to Busan and Melbourne, founder SAMISANOV (One week for one innovation), a teaching movement to force teaching for creating new learning environment and instructions.

Regarding his professional identity, in 2013 and 2017 he visited Busan and Melbourne to learn about innovative teaching and learning in the educational context. It is also worth mentioning that he was a founder of an educational movement in 2018 that inspires and asks for teachers in Indonesia to make an innovative and creative change every week. Beside providing weekly innovative teaching methods, this movement also debuted a commercial e-comic that pops up the local history as its plot. All content and design are the creative ideas of him based on his experiences as a history teacher. From this, in the same year, he was rewarded as one of the innovative and creative teachers by the Indonesian Ministry of Education.

Moreover, according to Pennington and Richards (2016), the concept of teacher identity emphasizes a teacher's individual characteristics and how they are integrated with the possibilities and potentials provided by the teacher's institutional identity, as well as the content and methods of a specific field, as they are realized in specific teaching contexts. These identities then bring evidence to the teacher's voices on deeper understanding of educational experiences in innovative teaching and learning. This notion resonates with some previous studies (Beijaard 2004; Izadinia 2016; Körkkö 2016) that argued how teacher reflective skills on identities can positively impact professional development. Therefore, regarding this result, seeking the pattern of experienced and professional teacher's belief and knowledge in innovative teaching-learning might be evidence of a teacher's professional development.

4.2 *Beliefs and knowledge about Innovative Teaching*

There is no doubt that the participant knows well what and how innovative teaching-learning is conducted inside and outside the classroom. It is clearly seen from his professional achievement and what he has done for better educational practices. Completely, the brief landscape of the result is presented as follows.

Table 2. Teacher's voice on *beliefs and knowledge about innovative teaching.*

Indicators	Teacher's description
1. Definition	Teaching with an innovative touch that invents and creates new methods.
2. Aims	To get the best learning experiences
3.Teaching activities	Demonstrating new learning environment
4.Teaching instructions	Using and creating new instructions that concerns to the learning demand

When questioned about innovative teaching regarding definition, aims, and activities, he defined himself as a determined person who always tried to deliver materials with 'innovative touch' while experimenting with new teaching models beforehand. Prior to this result, Kim et al. (2013) argued that teachers' beliefs influence fundamentally on teachers' knowledge and practices to their adoption of new technology. Simply, if the teacher finds it is essential to provide a new tool then it will be provided (Ajzen & Madden 1986; Kim et al. 2013). In fact, beliefs of someone, cover all perceptions, assumptions, opinions which play role as the basic information then react to the glance judgment due to the brief consideration and analysis as if it supports the belief which then concluded to be knowledge (Vongkulluksn et al. 2018).

Philosophically, the participant believes that a good teacher is a teacher who eagerly keeps innovating and creating something new to teaching and makes sure that teaching activities run as planned. As much research has highlighted teachers' innovative and creative beliefs, it supported the teacher's notion that people are dynamic and keep changing. There is no reason to stick around to only one particular patron (Bereczki & Kárpáti 2021) although they are still struggling with the appropriate technology for their courses (Voogt & McKenney 2016).

In relation to the use of the participant's E-comic based local history, the participant believes that emerging ICT and teaching materials could possibly answer the 21st-century learning especially for history subject matter. In line with Taimalu and Luik (2018) who identified the impact of beliefs and professional knowledge of teachers on the infusion of technology. The identification resulted in teachers' professional knowledge and beliefs both directly and indirectly giving impact to technology integration and pedagogical knowledge. Thus, the participant designed and developed this e-comic due to the characteristics of history itself in which actors, events, and plot exist to make a story chronologically. Thus, folklore became the reason for immersing local history in his e-comic. The immersion of this local history component to e-comic form is the part of his innovative touch in his classroom.

4.3 *Evaluation of E-comic based local history*

In the context of practical use, the participant claimed that this E-comic based local history has gone much further in educational practices in Indonesia. It has been commercially sold to the public. It is true that e-comic is one of the media in forcing innovative teaching and learning in his classroom. This media had answered the 21st-century learning demand due to its digital networking and authentic content immersion (Anagün 2018; Trilling & Fadel 2009). For today, no more and no less, issues of these both things take serious part especially in the virtual learning environment. The teacher's description can be seen as follows.

Table 3. Teacher's voice on *beliefs and knowledge about innovative teaching.*

Indicators	Teacher's description
1. Innovation	Local history learning moves beyond to digital platform
2. Appropriateness	Digital form in digital era
3. Educational value	Local history goes to wider landscape of learning

Accordingly, the comic content has various advantages, including boosting interest in reading and desire to study because it contains many illustrations appropriate for their developmental stage. Prior to this study, earlier, this E-comic based local history was created to only boost students' interest in learning history. But it all changes. Today, however, students are disposed to be acquainted with their gadget (Norton 2003). For that reason, the participant designed and created a digital material in the form of comics that is immersed with local history. Several studies have proven the effectiveness of 21st-century skills improvement such as critical thinking and character even in another subject (Lestari et al. 2021; Hosler 2011). Regarding the E-comic based local history subject in this study, the e-comic is only used firstly in his classroom but along with high demand for digitalization, it is published commercially.

As a result, it is undeniable that digital revolution brings direct impact on teaching practices as demonstrated in e-learning. It also supported that today's children, who grew up with television, video, internet, and visual modes of learning are extremely important, so the technologies are changing the way we live and learn (Magro et al. 2014). The borderless time and spaces of e-comic give more access and chance to students to learn. All these matters, of course, do support innovative teaching in the classroom.

4.4 *Challenges in designing innovative teaching media*

Regarding many issues found in designing a E-comic based local history, these are several challenges the participant found. First, by changing standardized written textbook to be a history comic, the content material infused is not as complete as textbook. However, it was, for him, only a matter of media used to get students' interest. He madeE-comic based on local history as a center of interest in the classroom to boost students' interest and motivation to learn history (Norton 2003).

Second, the issue of processing time in relation to book publishers. Previously, the book-keeping process was not easy to do. The participant personally admitted that his e-comic is not easily accepted by major publishers that support good printing. It is even possible that the e-comic is not publicly printed because it is considered unprofitable. Therefore, he chose a minor publisher so that his e-comic would be printed immediately. But, unfortunately, the printed result is not good enough according to him. There are so many significant things missed such as the quality of color and paper, editing, design, and indexing. Next, for the second time, he tried to start with his own editing process. On the basis of his curiosity and last experience, the participant tried to cover the minus things from the previous try by doing all the editing and design process on his own and cooperated with a legal publisher. With that piece of knowledge, he argued that processing time, especially editing and designing stages, would always bring out. If this also happened to another teacher, he suggested that we keep on trying to find a stepping-stone to all of these problems.

Last, it is a matter of language. It has been known that the language of comics consists of both word and image. On the one hand, visual and written language are paired to deliver the meaning of a particular set (Kim et al. 2016). The process of deciphering meaning in a limited number of words formally and metaphorically is a new challenge in creating a comic. Specifically, his e-comic has been written and designed to meet students' needs of history subject mastery in the classroom. This aim and context pushes him to create ordinary but formal language that students can understand easily. Dictions and images in every intertext and pages have to be rethought many times to make it acceptable to the educational context, especially for students. As such it is essential to assist students in developing their self-contextual knowledge about creativity and innovation to optimize the advantages of innovative teaching and learning (Kaufman & Beghetto 2013).

5 CONCLUSION AND RECOMMENDATION

This study unveils teacher's insights on creating new media with local history immersion in a high school setting. It tries to highlight teachers' beliefs, motivations, considerations, and challenges in providing students a E-comic based local history immersion experience to improve students' knowledge and engagement with literature or other subject matter. Thus, regarding the study result, a new medium with local history immersion does stimulate students' curiosity by moving beyond ordinary to extraordinary sensation in the very beginning of teaching and learning activities. The philosophical belief also impacts teacher's motivation in creating new media. Innovative touch in teaching is also considered makes teaching and learning activities alive. Based on the findings on the use of E-comic with local historical content as a learning medium, it is theoretically proven if there is still a history teacher who can meet the challenges of 21st-century learning by innovating through E-comic as a medium in the classroom. Although there are some obstacles in the process of actualizing, the use of E-comic with local history content can improve students' interest and motivation in learning.

These factual experiences suggest that it is essential to create new approaches on media to teaching that can create a new effective learning environment to prepare students for 21st-century learning. Finally, with regard to the limitations of this study, it should be noted that a more experienced teacher is desirable and could occupy the future work. It is also possible to conduct a study deeper at the dawn of innovative teaching with other crucial variables to make wider and richer results. Looking ahead, more research is needed to investigate deeper to media development and innovative teaching practices to a different level of students' grade. Among other things, it is recommended to investigate teacher's actual practices when creating innovative media. An investigation into the notion of teachers' made media is guaranteed to offer significant new perspectives on what teachers really know and do when creating an innovative media which is digitally produced globally.

REFERENCES

Ajzen, I., & Madden, T. J. (1986). Prediction of goal-directed behavior: Attitudes, intentions, and perceived behavioral control. *Journal of Experimental Social Psychology*, 22, 453–474.

Anagün, Şengül S. (2018). Teachers' perceptions about the relationship between 21st century skills and managing constructivist learning environments. *International Journal of Instruction, 11*(4), 825–840.

Bereczki, E. O., & Kárpáti, A. (2021). Technology-enhanced creativity: A multiple case study of digital technology-integration expert teachers' beliefs and practices. *Thinking Skills and Creativity,* 39(2021).

Brandon, B. Y. B. (2004). Applying instructional systems processes to constructivist learning environments. *The E-Learning Developers' Journal.*

Douwe Beijaard; Paulien C. Meijer; Nico Verloop (2004). Reconsidering research on teachers' professional identity. *Teaching and Teacher Education, 20(2),* 107–128.

Faulkner, J., & Latham, G. (2016). Adventurous lives: Teacher qualities for 21st century learners. *Australian Journal of Teacher Education, 41*(4), 137–150.

Ferrari, A., & Schoolnet, E. (2009). *Innovation and creativity in education and training in the EU member states: Fostering creative learning and supporting innovative teaching (Seville, JRC-IPTS).*

Hosler, J., & Boomer, K. B. (2011). *Are comic books an effective way to engage nonmajors in learning and appreciating science?* CBE Life Sciences Education, 10(3), 309–317.

Izadinia, Mahsa (2016). Preservice teachers' professional identity development and the role of mentor teachers. *International Journal of Mentoring and Coaching in Education*, *5*(2).
Kaufman, J. C., & Beghetto, R. A. (2013). In praise of Clark Kent: Creative metacognition and the importance of teaching kids when (not) to be creative. *Roeper Review, 35*, 155–165.
Kereluik, K., Mishra, P., Fahnoe, C., & Terry, L. (2013). What knowledge is of most worth: Teacher Knowledge for 21st century learning. *Journal of Digital Learning in Teacher Education*, *29*(4), 127–140.
Kim, C., Kim, M. K., Lee, C., Spector, J. M., & DeMeester, K. (2013). Teacher beliefs and technology integration. *Teaching and Teacher Education*, 29, 76–85.
Körkkö, Minna; Kyrö-Ämmälä, Outi; Turunen, Tuija (2016). Professional development through reflection in teacher education. *Teaching and Teacher Education, 55, 198–206.*
Lazarinis, F., & Panagiotakopoulos, C. (2015). E-comics in teaching: evaluating and using comic strip creator tools for educational purposes. *2015 10th International Conference on Computer Science & Education (ICCSE), Iccse*, 305–309.
Lestari, P F., Ahmadi, F., Rochmand, R. (2021). The implementation of Mathematics comic through contextual teaching and learning to improve critical thinking ability and character. *European Journal of Educational Research*, *10*(1), 497-508.
Ma'unah, S., Umamah, N., Sumardi, S., & Afita Surya, R. (2018). The enhancement of attractiveness and effectiveness of history learning using local history interactive teaching material. *American Journal of Educational Research*, *6*(11), 1531–1538.
Magro, G., De Carvalho, J. R., & Marcelino, M. J. (2014). Improving history learning through cultural heritage, local history and technology. *Proceedings of the 10th International Conference on Mobile Learning 2014, ML 2014*, 34–40.
Melliou, K., Moutafidou, A., & Bratitsis, T. (2014). *Digital comics use to develop thinking dispositions in early childhood education*. 10–12.
Murti, D. K., Gunahardi, G., & Winarno, W. (2020). Development of educational comic with local wisdom to foster morality of elementary school students: A need analysis. *International Journal of Educational Methodology*, *6*(2), 337–343.
Norton, B. (2003). The motivating power of comic books: Insights from Archie comic readers. *The Reading Teacher*, 57(2), 140–147.
Patrick Griffin and Esther Care. (2015). Educational Assessment in an Information Age. In *Assessment and Teaching of 21st Century Skills*.
Pennington, M. C.; Richards, J. C. (2016). Teacher identity in language teaching: Integrating personal, contextual, and professional factors. *RELC Journal*.
Redecker, C. (2009). Review of learning 2.0 practices: Study on the impact of Web 2.0 innovations on education and training in Europe.*European Commission Joint Research Centre: Institute for Prospective Technological Studies*.
Rienties, B., Giesbers, B., Lygo-Baker, S., Ma, H. W. S., & Rees, R. (2016). Why some teachers easily learn to use a new virtual learning environment: A technology acceptance perspective. *Interactive Learning Environments*, *24*(3), 539–552.
Rudy, R. I. (2017). The dynamics of EFL teaching in Indonesia: Be innovative teachers through social media. *English Language Teaching and Research*, *1*(1), 29–37.
Semerci, A., & Aydin, M. K. (2018). Examining high school teachers' attitudes towards ICT use in education. *International Journal of Progressive Education*, *14*(2), 93–105.
Spiteri, M., Nu, S., & Rundgren, C. (2018). Literature review on the factors affecting primary teachers' use of digital technology. *Technology, Knowledge and Learning*, *25*(1), 115–128.
Taimalu, Merle., & Piret, Luik. (2018). The impact of believes and knowledge on the integration of technology among teacher educators: A path analysis. *Teaching and Teacher Education*, *79*, 101–110.
Trilling, B., & Fadel, C. (2009). 21st century skills: Learning for life in our times. San Francisco: Jossey-Bass.
Vassilikopoulou, M., Retalis, S., Nezi, M., & Boloudakis, M. (2011). Pilot use of digital educational comics in language teaching. *Educational Media International*, *48*(2), 115–126.
Vongkulluksn, Vanessa W.; Xie, Kui; Bowman, Margaret A. (2017). The role of value on teachers' internalization of external barriers and externalization of personal beliefs for classroom technology integration. *Computers & Education*.
Voogt, Joke., & Susan McKenney. (2016). TPACK in teacher education: Are we preparing teachers to use technology for early literacy. *Technology, Pedagogy, and Education,* 1–15.

Teacher Education and Teacher Professional Development in the COVID-19 Turn – Drajati & Putra (Eds)
© 2023 The Author(s), ISBN: 978-1-032-38970-7

Entrepreneurship learning at vocational secondary school in Indonesia: A systematic literature review

Susantiningrum Susantiningrum, Siswandari Siswandari, Soetarno Joyoatmojo & Izza Mafruhah
Universitas Sebelas Maret, Surakarta, Indonesia

Budi Wahyono
Yamaguchi University, Yamaguchi, Japan

ABSTRACT: This paper aims to systematically review the problem of entrepreneurship learning and the strategy to handle it in Vocational Secondary School (VSS). The primary source of data was research articles published in Indonesia from 2000–2020. Three stages of filtering were done to find out the articles that fit the research objectives. A total of 22 related articles were identified. Data were analyzed using a qualitative approach to the meta-synthesis method. The finding revealed that the most discussed problem is the entrepreneurship learning model and evaluation. Meanwhile, the strategy to overcome problems in the learning model is by formulating a new learning model that emphasizes practice instead of theory, and the strategy to overcome problems about learning evaluation is by increasing the competency of entrepreneurship teachers. This implies that learning entrepreneurship in Indonesia still needs improvement in order to change the mindset of students from job seekers to job creators.

Keywords: Entrepreneurship learning, learning model, systematic literature review, vocational secondary school.

1 INTRODUCTION

Entrepreneurship has been proven to be one of the programs that can improve the quality of human resources. In fact, entrepreneurship is the biggest job vacancy in the 21st-century that anyone can enter (Hameed & Irfan 2019; Valliere & Peterson 2009). However, those who will succeed in achieving success in the field of entrepreneurship are people who are creative and innovative. In order to produce creative and innovative human resources, entrepreneurship education is embedded in the curriculum at every level of education in Indonesia, starting from the elementary, primary, secondary, and higher education.

Secondary Schools in Indonesia are divided into two kinds of school, namely Senior Secondary School (SSS) and Vocational Secondary School (VSS). The learning outcome of VSS is to prepare skilled workers to fill existing job opportunities. So far, the development of job opportunities has always been less than the number of VSS graduates; as a result, many of them are unemployed. The number of unemployed from VSS in August 2020 reached more than 1.3 million or 13.55% of the total number of unemployed in Indonesia (BPS-Statistics Indonesia 2020). For this reason, VSS students need to be provided with more provisions on entrepreneurship skills so that they are able to create jobs when they graduate.

Entrepreneurship learning must be able to provide opportunities for students to carry out entrepreneurial activities to produce products (Kim et al. 2020). In the end, it is hoped that alumni can become job creators through entrepreneurship learning, or self-employed (Killingberg et al. 2020). The role of educational institutions in influencing the interest in entrepreneurship for their students is still very high (Bazan et al. 2020), however, there are very few research articles that discuss specifically the problem of entrepreneurship learning and strategies in dealing with it. Publications in Indonesia mostly only explain the implementation of a particular model or method of entrepreneurship learning in the classroom action research format. This issue is opposed to the very dynamic development of entrepreneurship learning.

Entrepreneurship learning at VSS in Indonesia has undergone several changes. The latest curriculum states that learning entrepreneurship at VSS must be able to provide students with the provisions to become entrepreneurs as indicated by the significant increase in the number of learning hours. One program that has been launched is the VSS Revitalization Program which emphasizes increased interest and entrepreneurial skills. Starting in 2019, it's raising the slogan "VSS is ready to work, ready to be an entrepreneur". However, there is no general description of the literature on entrepreneurship learning in VSS. Referring to the need for entrepreneurship learning development, this literature review was objective to investigate the track record of published research on entrepreneurship learning, which is focused on the problem and the strategy to handle entrepreneurship learning in VSS, especially in Indonesia.

2 LITERATURE REVIEW

2.1 *Entrepreneurship learning*

Entrepreneurship builds a concept that was developed in the sociological and psychological traditions. In the early 18th century, Richard Cantillon described entrepreneurship as pervasive and he stated the entrepreneur has a pivotal role in the economy (Brown & Thornton 2013). Entrepreneurship is essentially the nature, characteristics and character of a person who has the will to bring innovative ideas into the real world creatively. In terms, the definition of entrepreneurship can be interpreted as "the backbone of the economy", which is the nerve center of the economy for controlling the economy of a nation (Bruton et al. 2018; Salmivaara & Kibler 2020). Epistemologically, entrepreneurship is a value needed to start a business or a process of doing something new and different. Entrepreneurship is the application of creativity and innovation to solve problems and efforts to take advantage of opportunities faced daily.

Entrepreneurship is a skill that can be developed through education. Reorientation learning as the first step is needed to deal with entrepreneurship as the embedded curriculum so it can be grown programmatically. Programmed entrepreneurship learning in groups and accompanied by facilitators supports the achievement of mastery of students' entrepreneurial skills (Cavallini et al. 2019). Entrepreneurship learning that is considered successful is when students master sufficient and high entrepreneurial skills, this can be achieved through the learning process in the classroom and special training (Laguna-Sánchez et al. 2020). It would be even better if entrepreneurship learning became a hidden curriculum that was applied to all lessons inside and outside the classroom so as to shape the school culture.

In line with the development of school culture, entrepreneurship learning needs to be linked to attitudes and behaviors such as discipline, responsibility, hard work, enthusiasm for learning, creativity, environmental care, and other positive attitudes that are included in the soft skills category. The role of soft skills here is very crucial because they affect 90% of individual success. The right educational strategy will be able to equip students with reliable entrepreneurial skills, especially to create students who are creative and innovative (Kakouris 2021). This is important because, in any career choice, entrepreneurial skills become a positive character that supports individual achievement (Susilaningsih 2015).

2.2 *Vocational secondary school*

VSS as secondary education has a big role in planning and creating professional and productive human resources to prepare them to enter the world of work. This is in line with Lauglo and Maclean

(2005): "that the vital role of vocational education in creating jobs has a role that will fill the need for employment in the business and industrial world". The existence of VSS is the spearhead of improving the quality of Indonesian human resources. Currently, the number of VSS in Indonesia is 14 thousand units, consisting of 25% State VSS and 75% Private VSS. Meanwhile, the number of VSS students is more than nine million students who are divided into 9 areas of expertise, 49 skill programs, and 146 skill competencies (Ministry of Education and Culture of Indonesia 2018).

In terms of policy, the government has issued Presidential Instruction Number 9/2016 concerning VSS Revitalization which is intended to improve the quality and competitiveness of human resources. It aims to increase the competitiveness and independence of Indonesian human resources so that alumni of VSS are expected to be ready to face the second demographic bonus era. It is estimated that this will occur in 2030-2040. At that time, the productive age population was predicted to reach 64 percent of the total projected population of 297 million people (Ministry of National Development Planning 2017). VSS alumni are expected to have dual abilities, as job seekers and job creators, especially in the field of entrepreneurship. For this reason, a more focused strategy is needed, especially with improving entrepreneurship learning in VSS that leads to strengthening employability skills and entrepreneurial skills.

The development of employability skills and entrepreneurial skills is relevant to the achievement of Sustainable Development Goal 4 (SDG4), namely "Quality Education" set by the United Nations and must be achieved by 2030. UNESCO (2017) recommends that education for sustainable development should be included in all formal education curricula: primary, secondary, and higher education. A specific target as a result of learning in quality education is that students should be able to "recognize the importance of their own skills to improve their lives, especially for work and entrepreneurship". Entrepreneurship learning is an important point in developing entrepreneurial skills and job skills. For this reason, an in-depth study of entrepreneurial learning in Indonesia is needed to be able to formulate an entrepreneurial learning model that is able to equip students with employability and entrepreneurial skills (Hayes et al. 2020). On the other hand, the role of institutions is also important for the quality and quantity of entrepreneurial learning. However, not all institutions play a similar role; rather, there is a dynamic relationship between institutions and economic development (Chowdhury et al. 2019).

3 METHOD

The method used in this study is a systematic literature review (SLR). It is a specific method to identify, evaluate, and interpret the result of research in accordance with the question, topic, or phenomenon of the research (Kitchenham 2004; Noordzij et al. 2009). The study consisted of an SLR of research on entrepreneurship learning in Indonesia, especially in VSS, seeking thereby to identify the problem and strategies to handle it.

Searching for literature was done by using the keywords "entrepreneurship learning at vocational school" and "*Pembelajaran Kewirausahaan di SMK*" (*Bahasa*) in Google Scholar search engine. The finding of articles limits the publication date to 2000 and 2020. For each manuscript, preliminary relevance was determined by the title. Then, if the content seemed to discuss entrepreneurship learning in VSS. Because all articles published in the past twenty years, so that we can build the review on the recent literature considering information. The result of finding, screening, and quality assessment is drawn in a flow diagram as shown in Figure 1.

The two main things that form the basis of filtering are articles published in journals that already have a DOI and cited by a minimum of 10 citations. These things are done at the same time to ensure that the articles have been scrutinized by the academic community. Thus the articles analyzed are articles that have good quality from an academic point of view.

Data analysis in this study was carried out using the meta-aggregation method, which focused on answering research questions by summarizing various research results, especially those related to a problem in the academic field (Lewin 2008).

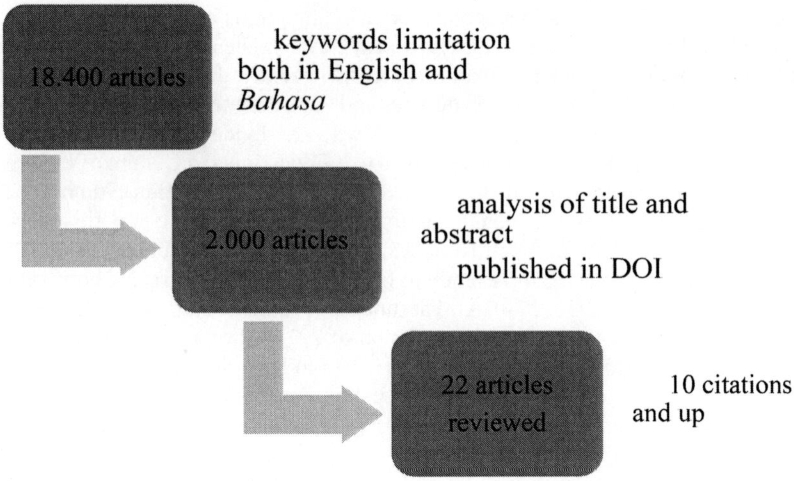

Figure 1. The flow of filtering and screening.

4 FINDING AND DISCUSSION

This study analyzed 22 articles to reach the research objectives, exploring problems and strategies to solve entrepreneurial problems in VSS. In general, it was found that entrepreneurship learning has been implemented in VSS but still experiencing some problems. In addition, the learning outcomes of entrepreneurship learning have not focused on changing the mindset of students to become entrepreneurs. However, it only provided knowledge about alternative activities that can be carried out if they do not get jobs that match their fields of expertise.

All articles that have been filtered and screened are then continued for in-depth analysis. The result is that there are five problematics in entrepreneurship learning researched as well as five problem-solving strategies suggested by the authors. The problems of entrepreneurship learning consist of problems in the field of curriculum, teaching media, learning evaluation, learning models, and learning activities (see more details in Table 1). Meanwhile, the suggested strategies are the development of entrepreneurship centre/laboratory, strengthening collaboration with industry, development of teacher competencies, development of entrepreneurship learning model, and strengthening practice than theory in entrepreneurship learning.

Most of the authors or as many as 64% have an interest in researching learning models and learning evaluation, followed by 29% in the field of curriculum and learning activities and the rest in the field of teaching media. Details of research results are presented in the two tables below.

The two problems of entrepreneurship learning that are most often encountered are about learning models and evaluation. Basically, these two things happen because the ability of entrepreneurship teachers is not maximized, such as lack of experience in entrepreneurship or have never had business activities.

Referring to the result of the research, the problems that are most interesting in being used as research themes for entrepreneurship learning researchers at VSS in Indonesia are learning evaluation and learning model. Meanwhile, the suggested entrepreneurship learning development strategy is the development of teacher competencies and the development of the entrepreneurship learning model. It was further conveyed that it was necessary to improve the quality of learning related to entrepreneurship at VSS so that it has relevance to the character of VSS graduates who should be ready to work and be entrepreneurs because students think that the practice of entrepreneurship learning benefits students in entrepreneurial practice. It is even stated that the higher the practice activity, the higher the readiness for entrepreneurship (Aprillianty 2012; Anggraeni &

Table 1. Entrepreneurship learning problems.

Author (year)	Curriculum	Teaching media	Learning Evaluation	Learning model	Learning activities
Mulyani (2011);	√				
Marini, Hamidah (2014);					
Winarno (2016)					
Lastariwati (2012).		√			
Anggarini, Sukardi. (2016);			√		
Sabatari, Hariyanto, (2013);					
Diyanti (2013); Indriyani, Margunani (2019);					
Vemmy (2012); Iswandari (2017);					
Mahfud, Pardjono. (2012).					
Nurbudiyani, (2013); Murtini, (2016);				√	
Komara, Adiraharja, (2020);					
Jusmin, (2012); Trisnawati (2014);					
Anggraeni, Hamanik.(2015);					
Laa, Winata, & Meilani (2017).					
Aprilianty (2012);					√
Prabandari, Rasyid (2015); Rahayu (2012).					
Total	3	1	7	7	3

Table 2. Strategy for handling entrepreneurship learning problems.

Author (year)	Development of entrepreneurship centre/laboratory	Strengthening Collaboration with Industrial	Development of teacher competences	Development of entrepreneurship learning model	Strengthening Practice than theory
Lastariwati (2012);	√				
Marini, Hamidah, (2014);					
Rahayu (2012);					
Sabatari, Hariyanto (2013)					
Aprillianty (2012);		√			
Vemmy, (2012);					
Anggarini, Sukardi (2016);			√		
Diyanti (2013);					
Indriyani, Margunani (2019);					
Komara, Adiraharja, (2020);					
Trisnawati (2014);					
Winarno (2016)					
Iswandari (2017);				√	
Laa, Winata, Meilani (2017);					
Murtini (2016);					
Mulyani, (2011);					
Nurbudiyani (2013)					
Anggraeni, Hamanik (2015);					√
Jusmin, (2012);					
Mahfud, Pardjono (2012);					
Prabandari, Rasyid (2015);					
Total	4	2	6	5	4

Hamanik 2015; Jusmin 2012). Thus, entrepreneurship subjects are needed by students, as a form of motivational training and entrepreneurial character.

As stated by the global entrepreneurship index, an entrepreneur is someone who has abilities, aspirations, and attitudes, which can be achieved through entrepreneurial trial and error dynamics. For this reason, entrepreneurship learning must apply entrepreneurial framework conditions consisting of the education system, market structure, infrastructure, R&D systems, financial sectors, corporate sector, and government, as illustrated in Figure 2.

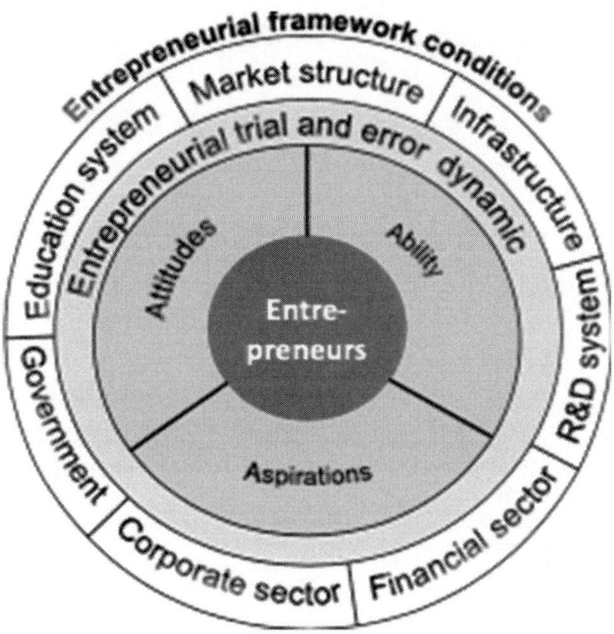

Figure 2. The entrepreneurial ecosystem configuration (Acs et al. 2019).

The implementation of education that supports entrepreneurship learning produces academic entrepreneurship that has an impact on success in business activities (Guo et al. 2019), where the output is known as an academic entrepreneur because it is able to bridge business with scientific work (Jain et al. 2009). This opens up new methods for collaboration between academia and business, to advance entrepreneurial learning that combines insights from research and experience to increase relevance and effectiveness in practical applications (Wright & Wallis 2015).

Entrepreneurship learning that provides more practical activities than theory will arouse students' interest in starting a career in entrepreneurship (Dheer & Lenartowisz 2019). Based on this, it is necessary to develop an entrepreneurial learning model that is able to equip students with entrepreneurial skills and work skills. For example, an innovative learning model (Serdyukov 2017; Tang & Li 2021) is proven to be able to make students more interested and willing to learn sincerely (Ghina 2014). In addition, the most important factors in the development of entrepreneurship learning are the competence of the teacher himself, the more creative and innovative the teacher will be, the more able he will be to organize meaningful entrepreneurial learning (Seikkula-Leino et al. 2010). VSS in Indonesia can integrate entrepreneurship learning from the values of local wisdom, namely the emotional closeness factor between teachers and students (Komara & Adiraharja 2020). Thus, meaningful entrepreneurial learning will come true, and VSS will become a school that already has an entrepreneurial framework as a learning atmosphere that enables its students to develop their entrepreneurship attitudes through real experiences.

5 CONCLUSION AND RECOMMENDATION

This research found an effective correlation between the problems of entrepreneurship learning and the solutions offered. The choice of the type of model and the evaluation of entrepreneurship learning can be pursued by developing a learning model and increasing the competence of the entrepreneurship teacher itself. With the increase in entrepreneurial competencies consisting of knowledge, skills, and attitudes, in principle, teachers will be able to overcome problems in preparing, managing, and evaluating the learning process. Meanwhile, by developing an entrepreneurial learning model that prioritizes practice and is in accordance with the characteristics of VSS students, it is expected to provide a meaningful learning experience so that interest and attitude as an entrepreneur grows. The limitation of the research is only using Google Scholar to find articles. Further research can work with other search engines to find more suitable articles.

REFERENCES

Acs, Z.J., Szerb, L., Lafuente, E., & Markus, G. (2019). *The Global Entrepreneurship Index 2019*. The Global Entrepreneurship and Development Institute, Washington, D.C., USA.

Bazan, C., Gaultois, H., Shaikh, A., Gillespie, K., Frederick, S., et al. (2020). A Systematic literature review of the influence of the university's environment and support system on the precursors of social entrepreneurial intention of students. *Journal of Innovation and Entrepreneurship*. 9:4. 1–28.

Brown, C., & Thornton, M. (2013). How entrepreneurship theory created economics. *Quarterly Journal of Austrian Economics*. 16 (4), 401–420.

Bruton, G.D., Zahra, S.A., & Cai, L. (2018). Examining Entrepreneurship Through Indigenous Lenses. *Entrepreneurship Theory and Practice*. 42 (3), 351–361.

BPS – Statistics Indonesia. (2020, February). Tingkat pengangguran terbuka Februari 2020. *Statistics Official News*. Retrieved from: https://www.bps.go.id/pressrelease/2020/05/05/1672/februari-2020–tingkat-pengangguran-terbuka–tpt–sebesar-4-99-persen.html

Cavallini, F., Carpitelli, A., Corsano, P., & Cihon, T. M. (2019). Entrepreneurship education: The case of the ABA master at the university of studies of Parma. *Journal of Entrepreneurship Education*, *22*(1), 1–13.

Chowdhury, F., Audretsch, D., & Belitski, M. (2019). Institutions and entrepreneurship quality. *Entrepreneurship Theory and Practice*. 43 (I), 51–81.

Dheer, R.J.S., & Lenartowicsz, T. (2019). Cognitive flexibility: Impact on entrepreneurial intentions. *Journal of Vocational Behavior*, 115 (2019) 103339.

Ghina, A. (2014). Effectiveness of Entrepreneurship Education in Higher Education Institutions. *Procedia - Social and Behavioral Sciences*, *115*, 332–345.

Guo, F., Restubog, S.L.D., Cui, L., Zou, B., & Choi, Y. (2019). What determines the entrepreneurial success of academics? Navigating multiple social identities in the hybrid career of academic entrepreneurs. *Journal of Vocational Behavior*. 112 (6), 241–254.

Hameed, I. & Irfan, Z. (2019) Entrepreneurship education: a review of challenges, characteristics, and opportunities. *Journal of Entrepreneurship Education*. 2 (3–4), 135–148.

Hayes, D., Subhan, Z., & Herzog, L. (2020). Assessing and understanding entrepreneurial profiles of undergraduate students: implications of heterogeneity for entrepreneurship education. *Journal of Entrepreneurship Education*. 3, 151–195.

Jain, S., George, G., & Maltarich, M. (2009). Academics or entrepreneurs? Investigating role identity modification of university scientists involved in commercialization activity. *Research Policy*, 38 (6), 922–935.

Kakouris, A. (2021). Teaching creativity in entrepreneurship: Embolden or discourage? *Industry and Higher Education*. 35 (4), 465–470.

Killingberg, N. M., Kubbered, E., & Blenker, P. (2020). Preparing for a future career through entrepreneurship education: Towards a research agenda. *Industry and Higher Education*. XX (X), 1–12.

Kim, G., Kim, D., Lee, W.,J., & Joung, S. (2020). *The effect of youth entrepreneurship education program: Two large-scale experimental studies*. Sage Open, July-September: 1–21.

Kitchenham, B. (2004). *Procedures for Performing Systematic Reviews*. Eversleigh: Keele University.

Laguna-Sánchez, P., Abad, P., de la Fuente-Cabrero, C., & Calero, R. (2020). A university training programme for acquiring entrepreneurial and transversal employability skills, a students' assessment. *Sustainability*, *12*(3), 796.

Lauglo J. and Maclean R. (2005). *Vocationalisation of Secondary Education*. The World Bank, Africa Region Human Development Department, Netherlands.
Lewin, S. (2008). *Methods to Synthesis Qualitative Evidence Alongside a Cochrane Intervention Review*. London: London School of Hygiene and Tropical Medicine.
Noordzij, M., Hooft, L., Dekker, F. W., Zoccali, C., & Jager, K. J. (2009). "Systematic Reviews and Meta-analyses: When They Are Useful and When to Be Careful." *Kidney International*. 76 (11), 1130–36.
Ministry of Education and Culture. 2018. *Statistik Pendidikan*. Retrieved from: http://statistik.data.kemdikbud.go.id/index.php/page/smk.
Ministry of National Development Planning/Bappenas. 2017. Bonus Demografi 2030-2040: Strategi Indonesia terkait ketenagakerjaan dan Pendidikan. *Siaran Pers*. Retrieved from: http://www.bappenas.go.id.
Salmivaara, V., Kibler, E. (2020). "Rhetoric Mix" of Argumentations: How policy rhetoric conveys meaning of entrepreneurship for sustainable development. *Entrepreneurship Theory and Practice*, 44 (4), 700–732.
Seikkula-Leino, J., Ruskovaara, E., Ikavalko, M., Mattila, J., & Rytkola, T. (2010), "Promoting entrepreneurship education: the role of the teacher?", *Education + Training*, 52 (2), 117–127.
Serdyukov, P. (2017). Innovation in education: what works, what doesn't, and what to do about it? *Journal of Research in Innovative Teaching & Learning*, *10*(1), 4–33.
Susilaningsih. (2015). Pendidikan Kewirausahaan di Perguruan Tinggi: Pentingkah untuk Semua Profesi? *Jurnal Economia*, *11*(1), 1–9.
Tang, D. & Li, X. (2021). Thoughts on innovation and entrepreneurship mode reform of college students in the context of COVID-19. *International Journal of Electrical Engineering Education*. 0(0) 1–10.
Valliere, D., Peterson, R. (2009). Entrepreneurship and economic growth: Evidence from emerging and developed countries. *Entrepreneurship and Regional Development*, 21 (5–6), 459–480.
Wright, B. & Wallis, S.E. (2015). *Using integrative propositional analysis for evaluating entreprenenurship theories*. Sage Open. July-September: 1–9.

Reviewed articles:
Anggarini, F., Sukardi, S. (2016). Pengembangan modul pembelajaran kewirausahaan model student company di SMK Negeri 1 Godean. *Jurnal Pendidikan Vokasi*, 6 (1), 24–30.
Anggraeni, B., Hamanik, H. (2015). Pengaruh pengetahuan kewirausahaan dan lingkungan keluarga terhadap minat berwirausaha siswa kelas XI SMK Islam Nusantara Comal Kabupaten Pemalang. *Jurnal Dinamika Pendidikan*, 10 (1).
Aprillianty, E. (2012). The effect of entrepreneur personality, entrepreneurship knowledge, and environment on entrepreneurial interest of vocational high school students. *Jurnal Pendidikan Vokasi*, 2 (1), 311–324.
Diyanti, S. I. (2013). Pengaruh hasil belajar mata pelajaran kewirausahaan dan lingkungan keluarga terhadap minat berwirausaha siswa SMK Gema 45 Surabaya. *Jurnal Pendidikan Ekonomi, 1*(3).
Indriyani, L., Margunani, M. (2019). Pengaruh kepribadian, pendidikan kewirausahaan, dan lingkungan keluarga terhadap minat berwirausaha. *Economic Education Analysis Journal*, *7*(3), 848–862.
Iswandari, A. (2017). Pengaruh motivasi intrinsik, pengetahuan kewirausahaan, dan kepribadian terhadap minat berwirausaha pada siswa SMKN 12 Surabaya. *Jurnal Ekonomi Pendidikan dan Kewirausahaan*, 1 (2).
Jusmin, E., (2012). Pengaruh latar belakang keluarga, kegiatan praktik, dan pelaksanaan pembelajaran kewirausahaan terhadap kesiapan berwirausaha siswa. *Jurnal Kependidikan*. 42 (2), 144–151.
Komara, E., Adiraharja, M.I. (2020). Integrasi Nilai-nilai Kearifan Lokal dalam Pembelajaran Kewirausahaan di SMK Negeri 10 Kota Bandung. *Mimbar Pendidikan*, 5 (2), 117–130.
Laa, N., Winata, H., & Meilani, R.I. (2017). Pengaruh model pembelajaran kooperatif tipe student teams achievement division terhadap minat belajar siswa. *Jurnal Pendidikan Manajemen Perkantoran*, 2 (2), 139–148.
Lastariwati, B. (2012). The roles of enterpreneurship class for vocational high school of tourism. *Jurnal Pendidikan Vokasi,* 2 (1), 71–80.
Mahfud, T., Pardjono. (2012). Praksis Pembelajaran kewirausahaan pada unit produksi jasa boga. *Jurnal Pendidikan Vokasi,* 2 (1), 27–40.
Marini, C.K., Hamidah, S. (2014). The effects of self-efficacy, family environment, and school environment on the entrepreneurial interest of the culinary service department students at VHSS. *Jurnal Pendidikan Vokasi,* 4 (2), 195–207.
Mulyani, E. (2011). Model Pendidikan Kewirausahaan di Pendidikan Dasar dan Menengah. *Jurnal Ekonomi dan Pendidikan*, 8 (1), 1–18.
Murtini, W. (2016). "GEPPRAK" model implementation in entrepreneurial learning to increase entrepreneurship intention in vocational high schools. *Jurnal Pendidikan Vokasi,* 6 (3), 335–345.

Nurbudiyani, I. (2013). A school cooperative-based entrepreneurship learning model in the vocational high school of business and management group. *Jurnal Pendidikan Vokasi*, 3 (1), 53–67.
Prabandari, E.T., Rasyid, A.A. (2015). The effect of entrepreneurship teaching through a business centre, pratical work, and family background on the entrepreneurship competence. *Jurnal Pendidikan Vokasi,* 5 (1), 1–14.
Rahayu, W.P. (2012). Sikap kewirausahaan siswa sekolah menengah kejujuan. *Jurnal Ilmu Pendidikan*, 18 (1), 98–103.
Sabatari, W., Hariyanto, V.L. (2013). Upaya Pembelajaran Kewirausahaan di SMK Potret Komitmen Terhadap Standar Nasional Proses Pendidikan dan Pembelajaran. *Journal of Technological and Vocational Education.* 21 (3), 285–293.
Trisnawati, N. (2014). Pengaruh pengetahuan kewirausahaan dan dukungan sosial keluarga pada minat berwirausaha siswa SMK Negeri 1 Pamekasan. *Jurnal Ekonomi Pendidikan dan Kewirausahaan*, 2 (1), 57–71.
Winarno, A. (2016). Entrepreneurship Education in Vocational School: Characteristics of teachers, schools, and risk implementation of the curriculum 2013 in Indonesia. *Journal of Education and Practice*, 7 (9), 122–127.
Vemmy, C. (2012). Factors affecting entrepreneurial intention of engineering vocational school students. *Jurnal Pendidikan Vokasi,* 4 (2), 117–125.

In-Service Teacher Professional Development

Teacher Education and Teacher Professional Development in the COVID-19 Turn – Drajati & Putra (Eds)
 ISBN: 978-1-032-38970-7

Teachers' continuous professional development: Current implementation, problems, and needs in Indonesia

Tri Murwaningsih & Nur Rahmi Akbarini
Universitas Sebelas Maret, Indonesia

Muna Fauziah
Institut Agama Islam Nahdlatul Ulama Kebumen, Indonesia

Subroto Rapih
Yamaguchi University, Japan

ABSTRACT: This research aims to scrutinize the problems and the teachers' needs in the continuous professional development implementation. Qualitative research was used as this study's design. This study involved 120 vocational high school teachers in the Surakarta Residency. To collect data, in-depth interviews, documentation, and open questionnaires were conducted. The interactive analysis model was then employed as a data analysis technique. The study results revealed several findings: (a) the teachers continuous professional development program has been implemented quite well; (b) some problems experienced by teachers comprised limited material, lack of qualified trainers and teachers' interest in scientific work, training on weekdays, the administrative burden, inadequate facilities; and (c) teachers need systematic and continuous training, instructors according to the field of science, and training on non-effective days. These findings are beneficial for schools to pay more attention to the problems and the teachers' needs about continuous professional development.

Keywords: Continuous professional development, teachers, vocational high school.

1 INTRODUCTION

The teacher's professional commitment is the force that binds teachers to carry out actions according to the goals that have been set (Ayele, 2014). This is reflected in the level of concern and commitment to the school as a social unit, committed to unique individual students, and committed to quality learning. To face global challenges in the digital era, teachers must be professional and have a global perspective (Malik, 2018). However, the fact remains that there are still quite serious problems related to the existence of teachers as professional educators. Problems that often arise related to teacher commitment in the current digital era are the lack of teacher readiness and professionalism as one of the main components in the learning process (Avidov-Ungar & Forkosh-Baruch, 2018). Teachers must have the learning, personality, social, and professional problem-solving competencies needed by schools (Darling-Hammond et al., 2019). Therefore, teachers need to recognize and take a role in Continuous Professional Development (CPD) (Malm, 2009) to have self-improvement and competency development. The introduction and taking of these roles are in line with the standard of the learning process. The standard of the learning process is a minimum criterion regarding the implementation of learning to obtain graduate learning outcomes (Senan, 2018). The standard of the learning process can be achieved if it is supported by teachers who have the competence and have professional commitment. Professional teachers are needed, especially in the era of technological development.

DOI 10.1201/9781003347798-22

In addition, Tacconi et al. (2021), argued that teachers encountered challenges to develop pedagogical competencies including working conditions, unsatisfactory wages, poor career opportunities, and lack of teacher training. Another fact that describes the professionalism of teachers can be seen from the results of the teacher competency test. In 2019, teacher competency test results in Indonesia only reached an average score of 53.02, from the expected 55 (Ministry of Education and Culture of Indonesia, 2019). It means that the competence of vocational high school teachers in Indonesia is still below the standard. Hence, teachers need to carry out continuous professional development to improve their competency.

Numerous studies have investigated the implementation of continuous professional development. For instance, Fedosejeva et al. (2018) investigated the implementation of Education for Sustainable Development (ESD). The results show that the implementation of ESD involves teachers fully. Niemi (2015) investigated the implementation of teacher continuous professional development. His research study identified a constraint in the form of a lack of early teacher preparation that affected the unpreparedness of professional learning programs. Hence, teacher preparation and development have become an increasing focus of study. So far, teacher professional development has been done through mentoring, coaching, modeling, and providing feedback to teachers (Darling-Hammond et al., 2019). Salite (2008) said that teachers can become agents of educational change when they are willing to expose themselves to innovations and teaching challenges. On the other hand, Gholami & Qurbanzada (2016) found that courses on sustainable professionalism could not be unsuccessful so that CPD activities must be oriented in the field (class).

From various previous studies, it is implied that teacher professional development programs need to be carried out. Obstacles experienced by previous researchers need to be considered and provide opportunities for the success of CPD activities. This opportunity becomes an international project to make schools a center for CPD. This is because schools are places for planning, implementing, and evaluating CPD programs (implementation, barriers, and teacher needs). This study aims to scrutinize the problems and the teachers' needs in the continuous professional development implementation. This study answers the question: what are the problems and needs of teachers in the implementation of continuous professional development?

2 LITERATURE REVIEW

CPD is approved by many parties as a component that can assist teachers in meeting the demands of the current school system and bureaucracy (Tomasik et al., 2021). This development is also alleged to be an effective technique to improve the teachers' abilities. Heller, Daehler, Wong, Shinohara, and Miratrix (2012) explain that professional development is an effort, method, and perspective on practicing independently. Rodman (2019) added that the CPD program provided opportunities for teachers to learn, interact, and collaborate in groups. CPD is implemented by identifying development needs and planning how to fulfill them (Chuang, 2019). From this understanding, it is found that CPD is carried out by each individual who is aware of the need for competency development. This is also explained by Ward and van der Mars (2020) that CPD is an effort to update knowledge and increase professional competence that is carried out consciously. This competency improvement can be done either in formal or informal form.

Teachers are required to pay attention to the principles of implementing CPD (Minister for Empowerment of State Apparatus and Beaurocracy Reform, 2009). First, CPD must focus on the child's success. Second, teachers have the right to develop themselves regularly, systematically, and sustainably, with the CPD preparation process starting from the school. Third, schools are obliged to provide opportunities for teachers to take part in the CPD program. Fourth, teachers are subject to sanctions if they cannot show improvement after participating in the CPD program. Fifth, the CPD material scope should concentrate on student learning, current educational research, technology, and the arts. Sixth, the CPD process must start with the teachers themselves. Seventh, CPD must contribute to the realization of the school's vision, mission, and goals. Eighth, CPD activities are carried out in schools or their surroundings to maintain the relevance of activities. Ninth, CPD must encourage the teaching profession recognition.

This continuous professional development can be put into practice at every level of education. Vocational high schools are places of learning for students who want to work right away (Gasskov, 2000). In line with this opinion, Clarke & Winch (2007) also contended that students who want to work directly will choose vocational education as a place to study. Vocational education is also a place to start an independent business. It is reinforced by Pavlova (2009) that vocational schools aim to prepare students directly for work and as a place to provide productive special training. From this review, it is known that vocational high school teachers are expected to provide work readiness (skills, insights, and student behavior) needed in work. Thus, teachers need to know the problems of implementing CPD and the needs of teachers in the field. This study answers the question: what are the problems and needs of teachers in the implementation of CPD?

3 METHOD

This qualitative study was conducted in nine vocational high schools in Indonesia. The participants of this research are 120 vocational high school teachers in the field of administration and offices who are selected purposely. For data collection, in-depth interview techniques, documentation, and open questionnaires (with a Likert scale) were utilized. The researcher started by distributing the questionnaire instrument first to the respondents in collecting data. The questionnaire was addressed to all teachers as the sample of this study. Then, the researcher also completed the research data by interviewing nine teachers at random in each school. Interview activities were carried out with interview guidelines. Interview activities are only a complement to the distribution of the questionnaire. Furthermore, the researcher also conducted a documentation study by making a checklist sheet. Besides, the data validity technique employed method and source triangulation. The researchers adopted the interactive data analysis model from Miles and Huberman (Miles & Huberman, 1994) to analyze the data. First, we collected data through the distribution of questionnaires, interviews, and documentation. Second, we make data reduction by selecting relevant and meaningful data. We focus on data that solve a problem or answer all research questions. Third, we present the data in the form of narrative writing to describe the situation that occurred. Fourth, we make tentative conclusions after the data has been collected. During the conclusion-drawing process, the researcher also verified the data to get a final/fixed conclusion.

4 FINDING AND DISCUSSION

This result and discussion are presented into three sections. The first section presents the implementation of teacher CPD. The second section presents the explanation of the problems experienced by teachers during implementing CPD. The third section presents a description of the needs of teachers in implementing CPD.

4.1 *Implementation of teacher CPD activities*

In this study, data collection on implementation CPD refers to three aspects, namely personal development, scientific publications, and innovative works. The results can be seen in Table 1 below.

Based on the questionnaire results, it can be described that the teacher has carried out self-development and collective activities. This activity is evidenced by the participation of teachers in functional training twice, collective workshops in teacher working groups once, and teacher participation in scientific activities twice. Referring to the results of Kabadayi (2016), which concluded that in-service training is important for teachers in schools as a way to develop professionalism and improve knowledge, quality of teaching, and learning for continuing education in the community.

Table 1. The results of implementation of teacher CPD.

Indicator	Implementation of Teacher CPD	Mean per item	Percentage (%)
Personal Development	1. Training	3.72	18.11
	2. Scientific Meeting	2.95	14.39
Scientific Publication	1. Scientific Presentation	2.17	10.57
	2. Publication of Research Results	1.60	7.79
	3. Publication of Books, etc.	2.79	13.59
Innovative Works	1. Discovering Art/Technology	2.60	12.66
	2. Making Media/Teaching Aids	1.82	8.87
	3. Development of Tests, etc.	2.88	14.02

Furthermore, on the research results concerning the scientific publication implementation in the last three years, data were obtained that the teachers on average made presentations at scientific forums as presenters once. A small number of teachers made popular scientific articles in the field of formal education. They also produced textbooks and modules. A small number of teachers could also write papers in the form of reports on research results in the field of education in their schools. They also made a paper in the form of a scientific review in the field of education.

Moreover, the innovative work that a small number of teachers had done included making simple class teaching tools once, making complex and simple teaching aids twice, and participating in preparing standards, guidelines, and questions at the provincial level once. Based on data, then teachers need to combine technology and media so that students' interest in learning is higher. Amin and Greenwood (2018) proved that media-making activities bridge the teacher to be more professional, support the transfer of knowledge, and advance the intellectuals of the teacher in education.

Therefore, it could be concluded that the CPD implementation has helped teachers to develop themselves, produce scientific publications and innovative work. However, the results obtained were still not optimal, mainly in the innovative work development. In this development, teachers have not been productive in producing innovative work because of some busyness at school.

4.2 *Teacher problems in CPD activities*

The results of the interview resulted in six problems faced by teachers in carrying out CPD activities. First, teachers have different characteristics. The impact is that the implementation of CPD is less than optimal because each teacher has different achievement targets and levels of satisfaction. Second, there is no socialization of CPD. Third, the teacher gains a partial understanding. As a result, teacher performance is not optimal. As stated by Madiclum & Mistades (2021) that a complete understanding will help achieve one's skills and bring about change for the better (achievement).

Fourth, teachers are not ready to participate in CPD activities. This is because teachers still have many administrative responsibilities and non-academic programs at school. The unpreparedness of teachers will greatly affect the implementation of school programs (Tomasik et al., 2021), just like CPD. Fifth, teachers are less motivated to carry out CPD activities. Motivation is the basis of a teacher in self-development. As explained by Riggio (2014); Estrada et al. (2019) that motivation can stimulate people to change their behavior and maintain themselves in achieving goals.

The last problem is that the training and workshop methods cannot be applied in the field. In the implementation of the CLA, the training method also greatly influences the results of the training of the participants. This finding explains that the method used is not suitable. After conducting further interviews, it turned out that the trainer only used the room to deliver the material. Craft (2000) identified that training weaknesses usually occur because the trainer dominates indoor activities. Finally, the training cannot satisfy the trainees.

4.3 *Teacher needs in CPD activities*

The third research objective is to describe the needs of teachers in the implementation of CPD. To obtain this data, the researchers conducted interviews with teachers. Based on the results of the interviews, the following data were obtained.

Table 2. List of teacher needs in continuous professional development.

Aspect	Teacher Needs
Self-development	Productive subject training, school revitalization, curriculum development, e-filing archives, entrepreneurship, application of learning models, online business and digital communication, vocational learning, e-book writing, and preparation of teaching materials, modern office technology training, protocol, master of ceremonies, competency assessors, e-learning
Scientific Publications	CAR preparation training, writing scientific papers, writing books, making questions
Innovative Works	Making media, teaching materials, modules, digital learning media, teaching aids, effective educational videos, virtual class training, and graphic design

From table 2 above, it is known that the needs of teachers consist of aspects of self-development, scientific publications, and innovative work. The main needs that are not seen in education and training regarding independent development and school focus, such as education and training on curriculum development, assessor competence, and e-learning. Currently, the need for self-development is indeed a priority for teachers. According to the teachers, the fulfillment of self-development is easier to obtain than other aspects of CPD. Likewise, teachers who have attended the training have proven successful in improving the quality of learning in their classrooms (Sumantri et al., 2018). The teaching profession must focus on productivity to acquire knowledge of curriculum development, teaching methods, assessment, or evaluation through education and training (Petrovska et al., 2020; Rutherford & Lovorn, 2018). On the other hand, the need for scientific publications is still limited to CAR training. Teachers are encouraged to develop themselves through classroom action research as teacher professional development (Meesuk et al., 2020).

From these findings, it can be concluded that CPD activities must adapt to the needs of teachers (self-development, scientific publications, and innovative work). The goal is to achieve standards and improve professional competence, especially about learning services (Furjanic & Trotman, 2000). Not only that, but previous studies also support the need to meet the needs of teachers in the CPD program. Asiyah et al. (2021) found data that professional development that considers the needs of teachers has a positive effect on teacher work commitment and teacher learning quality. Thus, the implementation of CPD will be maximized by meeting the needs of teachers, such as training needs, scientific discussion forums, preparation of CAR, and making innovative works.

5 CONCLUSION AND RECOMMENDATION

This study has a purpose to analyze the problems and the teachers' needs in the continuous professional development implementation. The implementation of the CPD program was carried out by self-development, scientific publications, and innovative works. Self-development is carried out by participating in workshops, training, and mentoring activities. Scientific publications are focused on participating in national/international seminars and conducting research/services. The innovative works are manifested in training in the manufacture of learning media/teaching aids. However, it encountered many obstacles such as materials, administration, facilities, time, instructors/trainers, costs/financial, teacher readiness, and others. The limitation of this research is that this research was only conducted in certain schools. Further research is needed to demonstrate the barriers and needs of primary/junior secondary school teachers in sustainable professional development.

REFERENCES

Amin, M. Al, & Greenwood, J. (2018). The UN sustainable development goals and teacher development for effective english teaching in Bangladesh: A gap that needs bridging. *Journal of Teacher Education for Sustainability*, *20*(2), 118–138. https://doi.org/10.2478/jtes-2018-0019

Asiyah, S., Budi, W. B., Hidayah, N., & Supriyanto, A. (2021). The effect of professional development, innovative work and work commitment on quality of teacher learning in elementary schools of Indonesia. *Eurasian Journal of Educational Research*, *2021*(95), 227–246.

Avidov-Ungar, O., & Forkosh-Baruch, A. (2018). Professional identity of teacher educators in the digital era in light of demands of pedagogical innovation. *Teaching and Teacher Education*, *73*, 183–191.

Ayele, D. (2014). *Teachers' job satisfaction and commitment in general secondary schools of hadiya zone, in southern nation nationality and people of regional state*. Jimma University.

Chuang, S. (2019). Exploring women-only training program for gender equality and women's continuous professional development in the workplace. *Higher Education, Skills and Work-Based Learning*, *9*(3), 359–373.

Clarke, L., & Winch, C. (2007). *Approach, development, and system*. Routledge.

Craft, A. (2000). *Creativity across the primary curriculum: Framing and developing practice*. Routledge.

Darling-Hammond, L., Flook, L., Cook-Harvey, C., Barron, B., & Osher, D. (2019). Implications for educational practice of the science of learning and development. *Applied Developmental Science*, *24*(3), 97–140.

Day, C. (2002). Schools as learning communities: Building capacity through network learning. *Education*, *30*(3), 19–22.

Estrada, Á. C. M., Vera, J. G., Ruiz, G. R., & Arrebola, I. A. (2019). Flipped classroom to improve university student centered learning and academic performance. *Social Sciences*, *8*(11).

Fedosejeva, J., Boče, A., Romanova, M., Iliško, D., & Ivanova, O. (2018). Education for sustainable development: The choice of pedagogical approaches and methods for the implementation of pedagogical tasks in the anthropocene age. *Journal of Teacher Education for Sustainability*, *20*(1), 157–179.

Furjanic, S. W., & Trotman, L. A. (2000). *Turning training into learning*. AMACOM.

Gasskov, V. (2000). Managing vocational training systems: A handbook for senior administrators. *Economics of Education Review*, *22*(6), 57–75.

Gholami, J., & Qurbanzada, I. (2016). Key stakeholders Attitudes towards teacher education programs in TEFL: A case study of Farhangian University in Iran. *Journal of Teacher Education for Sustainability*, *18*(2), 5–20.

Heller, J. I., Daehler, K. R., Wong, N., Shinohara, M., & Miratrix, L. W. (2012). Differential effects of three professional development models on teacher knowledge and student achievement in elementary science. *Journal of Research in Science Teaching*, *49*(3), 333–362.

Kabadayi, A. (2016). A suggested in-service training model based on Turkish preschool teachersí conceptions for sustainable development. *Journal of Teacher Education for Sustainability*, *18*(1), 5–15.

Madiclum, P. C., & Mistades, V. M. (2021). Self-directed learning assisted online module in teaching kinematics: Implication to students' level of understanding. *3rd International Conference on Modern Educational Technology*, 58–62.

Malik, R. S. (2018). Educational Challenges in 21St Century and Sustainable Development. *Journal of Sustainable Development Education and Research*, *2*(1), 9.

Malm, B. (2009). Towards a new professionalism: enhancing personal and professional development in teacher education. *Journal of Education for Teaching*, *35*(1), 77–91.

Meesuk, P., Sramoon, B., & Wongrugsa, A. (2020). Classroom action research-based instruction: The sustainable teacher professional development strategy. *Journal of Teacher Education for Sustainability*, *22*(1), 98–110.

Miles, M. B., & Huberman, A. M. (1994). *An expanded sourcebook: Qualitative data analysis*. Sage Publications.

Minister for Empowerment of State Apparatus and Beaurocracy Reform, (2009).

Ministry of Education and Culture of Indonesia. (2019). *Regional education balance*. Ministry of Education and Culture of Indonesia.

Niemi, H. (2015). Teacher professional development in Finland: Towards a more holistic approach. *Psychology, Society, & Education*, *7*(3), 279–294.
Pavlova, M. (2009). *Technology and vocational education for assustainable development empowering individuals for the future*. Spinger Science Business Media.
Petrovska, S., Sivevska, D., Popeska, B., & Runcheva, J. (2020). Mentoring in teaching profession. *International Journal of Cognitive Research in Science, Engineering and Education*, *6*(2), 47–56.
Riggio, R. E. (2014). *Endustri ve orgut psikolojisine giris [Introduction to industrial and organizational psychology]*. Nobel Yayıncılık.
Rodman, A. (2019). *Personalized professional learning: A job-embedded pathway for elevating teacher voice*. ASCD.
Rutherford, D. J., & Lovorn, C. (2018). Research-based program development: Refining the service model for a geographic alliance. *Journal of Geography*, *117*(2), 88–98.
Salite, I. (2008). Educational action research for sustainability: Constructing a vision for the future in teacher education. *Journal of Teacher Education for Sustainability*, *10*, 5–17. https://doi.org/10.2478/v10099-009-0021-6
Senan, N. A. M. (2018). Developmental review program impact on enhancing the effectiveness of "Teaching and Learning" in accounting program: A case study in a Saudi University. *Entrepreneurship and Sustainability Issues*, *6*(2), 1001.
Sumantri, M. S., Prayuningtyas, A. W Rachmadtullah, R., & Magdalena, I. (2018). The roles of teacher-training programs and student teachers' self-regulation in developing competence in teaching science. *Advanced Science Letters*, *24*(10), 7077–7081.
Tacconi, G., Tūtlys, V., Perini, M., & Gedvilienė, G. (2021). Development of pedagogical competencies of the vocational teachers in Italy and Lithuania: Implications of competence-based VET curriculum reforms. *European Journal of Training and Development*, *45*(6/7), 526–546.
Tomasik, M. J., Helbling, L. A., & Moser, U. (2021). Educational gains of in-person vs. distance learning in primary and secondary schools: A natural experiment during the COVID-19 pandemic school closures in Switzerland. *International Journal of Psychology*, *56*(4), 566–576.
Ward, P., & van der Mars, H. (2020). Confronting the challenge of continuous professional development for physical education teacher educators. *Journal of Physical Education, Recreation & Dance*, *91*(1), 7–13.

Teacher Education and Teacher Professional Development in the COVID-19 Turn – Drajati & Putra (Eds)

Discovering in-service teachers' perceived TPACK competence in EFL classroom: A self-assessment during emergency remote learning

Sri Haryati & Issy Yuliasri
Universitas Negeri Semarang, Semarang, Indonesia

Joko Nurkamto
Universitas Sebelas Maret, Surakarta, Indonesia

Sri Wuli Fitriati
Universitas Negeri Semarang, Semarang, Indonesia

ABSTRACT: This study investigated in-service teachers' perceptions of TPACK competency, with a focus on the integration of technological, pedagogical, and content knowledge in instructional practice. It examined the perceived TPACK competencies of 156 English teachers from rural and urban locations across the Indonesian archipelago at the Senior High School level. The information for this survey came from a questionnaire about EFL teachers' understanding and competency with TPACK. Teachers' Technology Knowledge (TK), Content Knowledge (CK), Pedagogical Knowledge (PK), Pedagogical Content Knowledge (PCK), Technological Content Knowledge (TCK), Technological Pedagogical Knowledge (TPK), and TPACK are moderate to good, according to the study. However, a few teachers are still having difficulty integrating technology into their classrooms and responding to modern technologies, indicating that they lack TK. This implies the establishment of a technology-based Professional Development program for EFL teachers in order to improve their TPACK competency in the face of an emergency remote learning situation.

Keywords: EFL teachers, emergency remote learning, TPACK competence

1 INTRODUCTION

The importance of TPACK integration in the instructional process is highlighted by the development of technology in 21st-century learning (21CL). Furthermore, the widespread development of the Covid-19 pandemic necessitates a shift in learning mode from direct engagement to online learning, emphasizing the significance of technology in the teaching and learning process. Koehler and Mishra (2005) established the TPACK framework, which stresses the knowledge convergence of technology, pedagogy, and content. It requires teachers to be able to synthesize these three categories of information, as well as the relevant knowledge formed when they collide (Koehler & Mishra 2009; Voogt et al. 2013). The framework instructs teachers on how to use design thinking to transform existing knowledge into a new, more realistic form of knowledge that can aid learners in learning through technology (Angeli & Valanides 2009; Chai et al. 2019). The study found that in-service secondary school teachers profited greatly from the TPACK-oriented teaching practice course, and that they successfully produced instructional designs and implemented teaching practices using the TPACK framework (Cahyono et al. 2016).

Regarding the issue on the manifestation of TPACK in English language learning in secondary level, some empirical researches on TPACK in myriad point of interest has been conducted including TPACK in related to multimodal literacy (Drajati et al. 2018), teachers' belief in TPACK (Abitt 2011;

 DOI 10.1201/9781003347798-23

Kim 2017; Hsu et al. 2020; Lai & Lin 2018), teachers' perception and attitude about TPACK (Cheng 2017; Hsu et al. 2020; Koh et al. 2014; Roussinos & Imoyiannis 2019; Tokmak 2014; Turgut 2017), the development of teachers' TPACK (Cabero & Barroso 2016; Foulger et al. 2015; Gill & Dalgarno 2017; Rienties et al. 2020), teaching approach in integrating TPACK (Al-Harthi et al. 2018; Bueno-Alastuey et al. 2018), teachers' TPACK assessment (Bilici et al. 2016; Bas, G., & Senturk, C. 2018; Cavanagh & Koehler 2013; Cheng 2017; Nazari et al. 2019; Kozikğlu & Babacan 2019), the implementation of TPACK (Clausen et al. 2019; Gómez 2015; Phillips 2016; Tseng 2019). In bottom line, the teachers stated that they can bring out students' initial notions about the topic of inquiry utilizing technology. It is found that almost every respondent expressed enthusiasm about adopting technology for a variety of goals (Drajati et al. 2018).

The implementation of TPACK in teaching practice is also being looked into. Teachers' pedagogical approaches for implementing TPACK in the classroom were discovered to be quite varied. It varies based on their educational choices, which are impacted by personal preferences, topic cultures, and school environments. The patterns reflected various forms of technological pedagogical and content knowledge development in teaching (Szeto. E. & Cheng, A.Y.N. 2016). Considering the diverse pattern in TPACK implementation, the emphasis in recent years has been placed on assessing teachers' self-efficacy in integrating information and communication technology (ICT) into teachers' instructional practices (Almerich et al. 2016; Chai et al. 2013). It is proven that teachers' TPACK and attitude both have a direct correlation with technology integration thus the teachers' self-assessment in their TPACK competence becomes the essential aspect to be investigated since no study discusses it further. This study examines in-service teachers' TPACK competence in Indonesian EFL classrooms as a self-assessment in transformative learning during the Covid-19 world pandemic. The global pandemic massively alters the teachers' attitude which further compels them to develop their competence related to technology integration during emergency remote learning. Thus, specifically, this study attempted to answer this research question: How do in-service teachers perceive TPACK competence in EFL classroom?

2 LITERATURE REVIEW

2.1 *TPACK framework*

TPACK is a popular term that refers to the successful use of technology in the classroom (Schmidt et al. 2009). It serves as a framework for examining what teachers need to know in order to integrate technology into their classrooms and how they might obtain that knowledge. The concept of TPACK was firstly initiated by Shulman who introduced the concept of Pedagogical and Content Knowledge (PCK) which intersect each other. The notion of TPACK is further expanded by Koehler and Mishra in light of the fast integration of technology in many sectors (2006). TPACK is a useful framework for thinking about the knowledge that teachers need to bring technology into their classrooms and how they might get that knowledge (Koehler 2010; Schmidt et al. 2009). TPACK consists of three bodies of core knowledge (technology, pedagogy, and content) and their intersections such as Technology Knowledge (TK), Content Knowledge (CK), Pedagogical Knowledge (PK), Pedagogical Content Knowledge (PCK), Technological Content Knowledge (TCK), Technological Pedagogical Knowledge (TPK), and Technological Pedagogical Content Knowledge (TPACK). Each component has specific characteristics namely (1) TK is a knowledge related to any kind of technologies; (2) PK is a knowledge related to methods and instructional process covering classroom management, assessment, planning and students' learning; (3) CK is knowledge related to the content or subject matter to be taught; (4) TPK is a knowledge on how technology can integrate and assist teaching; (5) TCK is a knowledge on how technology can represent the content of certain subject matters; (6) PCK is a knowledge on applying teaching strategies to deliver the teaching content; and (7) TPACK that is the knowledge in integrating technology in teaching to deliver certain subject matter. The intersection of TPACK components is visualized in Figure 1.

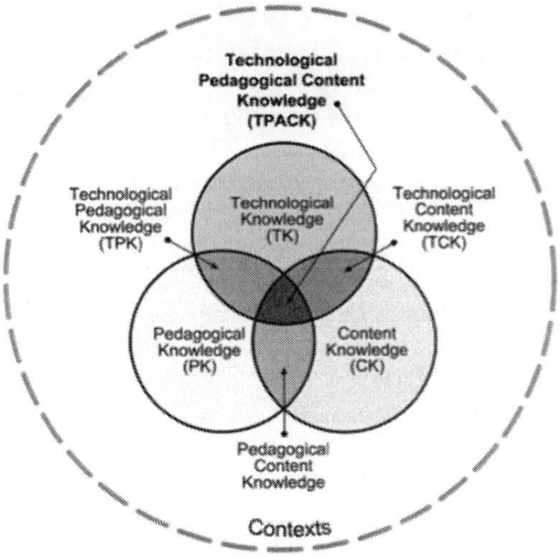

Figure 1. Mishra and Kohler TPACK framework (2006).

2.2 *TPACK competence viewed from the perspective of gender, age and teaching experience*

Gender, age, and teaching experience all have an impact on teachers' TPACK competence. According to study, female teachers have lower self-confidence in technology knowledge (TK) than male teachers (Koh et al. 2014; Roig-Vila et al. 2015), but better self-confidence in PK (Lin et al. 2013). Lin et al. (2013) also discovered that older female teachers lack confidence in their comprehension of technological applications like TK, TPK, TCK, and TPACK. There were no significant gender disparities in instructors' perceptions of total TPACK in PK and CK. Men were more confident in subject matter knowledge (CK) but less proficient in pedagogical representation and techniques (PK) at the same time (Jang & Chang 2016; Liu et al. 2015). In TPACK, the gender factor is linked to the chronological age factor. Several studies have revealed that older teachers had a worse feeling of self-efficacy for TK (Hsu et al. 2017; Kazu & Erten 2014) and TPACK, especially when it comes to instructional use of the Internet (Lee & Tsai, 2010). Older instructors, on the other hand, may have a higher level of PCK self-efficacy (Kazu & Erten 2014). The fact proves that age variable significantly improves teachers' competence in perceiving TPACK since their technology sensitivity differs in a range of time.

Experienced teachers are likely to have a lower sense of self-efficacy in their total TPACK, including game-based PCK (Hsu et al. 2017), understanding of the Internet for educational purposes (Lee & Tsai 2010), and constructivist-oriented TPACK (Koh et al. 2014). Additionally, it was discovered that teaching experience had a negative correlation with teachers' perceived understanding of technology such as TK, TPK, TCK, and TPACK (Roig-Vila et al. 2015). In comparison to experienced teachers, novice teachers tended to be less confident in their subject content knowledge (CK) and instructional techniques (PK) (Jang & Chang 2016; Jang & Tsai 2012). Additionally, other research discovered that as teachers' years of teaching experience rose, their PCK self-efficacy improved, but their perceived competence in technology usage (TK) dropped (e.g., Kazu & Erten 2014). Although experienced instructors lack confidence in their TK, they may be more enthusiastic about using technology in their education than less experienced teachers (Saudelli & Ciampa 2016).

3 METHOD

This study was designed to investigate in-service EFL teachers' TPACK competence as a self-assessment in transformative learning during the world pandemic of Covid-19. This study applied quantitative research, especially survey study. By applying descriptive research, this study used an online platform to describe the overall EFL teachers' TPACK competence. The participants of the study are 156 in-service EFL teachers at the Senior High School level from both rural and urban areas in Indonesia. The demographic information of the participants is presented in Table 1.

Table 1. The demographic information of the participants.

No	Variables	Number of Respondent	Percentage
1		Gender	
	Male	129	80.6%
	Female	31	19.4%
2		Age	
	20–25 years	3	1.9%
	26–30 years	26	16.3%
	31–35 years	18	11.3%
	36–40 years	19	11.9%
	41–45 years	36	22.5%
	46–50 years	23	14.4%
	51–55 years	22	13.7%
	56–60 years	13	8.1%
3		Teaching Experience	
	Less than 5 years	17	10.6%
	5–10 years	22	13.8%
	11–15 years	39	24.4%
	16–20 years	38	23.8%
	21–25 years	19	11.9%
	More than 25 years	25	15.6%

This survey study included data from a questionnaire developed by Ellas, Majid, and Narasuman (2019) in the form of an online survey to explore EFL teachers' TPACK competence they possessed and understood. The questionnaire consists of 35 questions focusing on teachers Technology Knowledge (TK), Content Knowledge (CK), Pedagogical Knowledge (PK), Pedagogical Content Knowledge (PCK), Technological Content Knowledge (TCK), Technological Pedagogical Knowledge (TPK), and Technological Pedagogical Content Knowledge (TPACK). Each item of this survey was rated on a 5-point Likert scale (ranging from 1, 'strongly disagree,' to 5, 'strongly agree'). The descriptive quantitative data is further being analyzed using the Mishra and Kohler (2006) TPACK framework as the main theory used in this study. The result of the quantitative data in Likert scale is further being interpreted based on the theory of TPACK framework.

4 FINDING AND DISCUSSION

The survey on teachers' TPACK competence highlights their different competence in integrating TPACK. The survey on in-service teachers' TPACK competence in EFL classroom as a self-assessment during the Covid-19 pandemic transformative learning reveals their competence in seven components, namely TK, CK, PK, PCK, TCK, TPK, and TPACK. The survey result for each component is presented in Table 2.

Further, the more obvious teachers' TPACK competence is depicted in Figure 2 which shows the gradation of teachers' competence in technological, pedagogical and content knowledge.

The study reveals the demography of teachers' competence in integrating technology for educational purposes considering the pedagogical and content issues viewed from the TPACK competence

Table 2. The survey on teacher's TPACK competence.

TPACK Components	SDA	DA	N	A	SA
TK	13	127	340	480	136
CK	2	18	133	392	88
PK	1	18	181	687	209
PCK	1	3	57	194	61
TCK	0	9	69	190	48
TPK	3	33	226	415	107
TPACK	7	71	414	636	124

Note: SDA= Strongly Disagree; DA = Disagree; N = Neutral; A = Agree; SA = Strongly Agree.

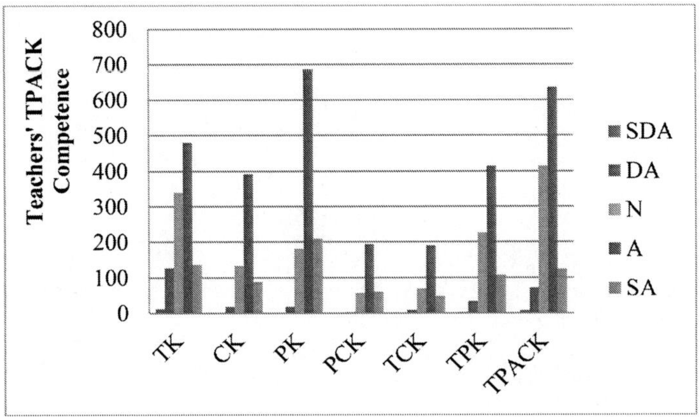

Figure 2. Teachers' TPACK Competence.
Note: SDA= Strongly Disagree; DA = Disagree; N = Neutral; A = Agree; SA = Strongly Agree

in seven components as stated by Koehler et al. (2014). The descriptive result shows that teachers have a moderate to good TK, CK, PK, PCK, TCK, TPK, and TPACK. The survey indicates that teachers possess a good competence in assessing students' performance in a classroom as well as excellent classroom management to adapt to the classroom situation. Moreover, they are proficient in selecting as well as applying teaching approaches. It shows teachers have the high competence in PK among other TPACK components. The data also attests that EFL teachers were satisfied with their TPACK, TK, and PCK competence. However, compared to other aspects, their confidence in mastering subject knowledge (CK), applying teaching strategies to deliver subject matters (PCK), and adopting instructional techniques with the help of technology (TPK) was comparatively low. The findings go along with Cheng (2017) who found out that teachers' mostly challenging in mastering the CK, TK and TPK. due to the gender differences in case that females demonstrated less competence in their subject CK and TK than males did, and females who teach need to improve their CK and TK more than males. This study also reveals that teachers are mostly challenged in technology aspect where gender becomes the prominent aspect in this issue. It is due to the number of the respondents which is dominated by females than males where female teachers have lower self-confidence in knowledge related to technology (TK) than male teachers (Koh et al. 2014; Roig-Vila et al. 2015), yet higher self-confidence in knowledge related to pedagogical aspect (PK) (Lin et al. 2013). The finding is also supported by the teachers' age and teaching experience as the affected variables. Senior teachers who had great teaching had a negative correlation with teachers' perceived understanding of technology such as TK, TPK, TCK, and TPACK (Roig-Vila et al. 2015). It means that they have a barrier in applying the technology for learning.

5 CONCLUSION AND RECOMMENDATION

The Covid-19 world pandemic has changed the mode of learning into blended, even online learning that demands the teachers' TPACK competence in integrating TPACK in their classroom practice. It assumes that after several times having emergency remote teaching, teachers' TPACK competence is improved. This study focus examines in-service teachers' perceived TPACK competence in EFL classrooms as a self-assessment during emergency remote learning. It reveals that in-service teachers have moderate to good TPACK competence covering the seven components of TPACK namely Technology Knowledge (TK), Content Knowledge (CK), Pedagogical Knowledge (PK), Pedagogical Content Knowledge (PCK), Technological Content Knowledge (TCK), Technological Pedagogical Knowledge (TPK), and Technological Pedagogical Content Knowledge (TPACK). Teachers are mostly competent in PK yet challenging in mastering the technology for educational purposes. Gender, age and teaching experience are affected variables that influence teachers' TPACK competence. The current study implies that the teachers' continuing professional development specifically in penetrating technology knowledge in classroom practice is needed to strengthen their TPACK competence. The investigation of teachers' beliefs and perspectives in using technology in the EFL classrooms is further recommended to be investigated to sharpen and depict the complete strand of teachers' TPACK competence related to this study.

REFERENCES

Anderson, H., & Gehart, D. (2007). *Collaborative therapy: Relationships and conversations that make a difference*. Routledge.

Angeli, C., & Valanides, N. (2009). Epistemological and methodological issues for the conceptualization, development, and assessment of ICT-TPCK: Advances in Technlogical Pedagogical Content Knowledge (TPCK). *Computers & Education*, 52(1), 154–168.

Cahyono, B., Kurnianti, D., & Mutiaraningrum, I. Indonesian EFL teachers' application ofTPACK in in-service education teaching practices. *International Journal of English Language Teaching*, 4(5), 16–30.

Drajati, N.A., Tan, L., Haryati, S., Rochsantiningsih, D., & Zainnuri, H. (2018). Investigating English language teachers in developing TPACK and multimodal literacy. *Indonesian Journal of Applied Linguistics, 2*(3), 575–582.

Hsu, C.Y., Tsai, M.J., Chang, Y.H., & Liang, J.C. (2017). Surveying in-service teachers' beliefs about game-based learning and perceptions of technological pedagogical and content knowledge of games. *Educational Technology & Society, 20*(1), 134–143.

Jang, S.J., & Chang, Y. (2016). Exploring the technological pedagogical and content knowledge (TPACK) of Taiwanese university physics instructors. *Australian Journal of Educational Technology, 32*(1), 107–122.

Kazu, I.Y., & Erten, P. (2014). Teachers' technological pedagogical content knowledge selfefficacies. *Journal of Education and Training Studies, 2*(2), 126–144.

Koehler, M. J. (2010). *The history of TPACK 1999-Present*. Michigan State University Press

Koehler, M. J., & P. Mishra. (2009). What is technological pedagogical content knowledge? *Contemporary Issues in Technology and Teacher Education*, 9(1), 60–70.

Koh, J.H.L., Chai, C.S., & Tsai, C.C. (2014). Demographic factors, TPACK constructs, and teachers' perceptions of constructivist-oriented TPACK. *Educational Technology & Society, 17*(1), 185–196.

Lee, M.H., & Tsai, C.C. (2010). Exploring teachers' perceived self-efficacy and technological pedagogical content knowledge with respect to educational use of the World Wide Web. *Instructional Science, 38*, 1–21.

Lin, T.C., Tsai, C.C., Chai, C.S., & Lee, M.H. (2013). Identifying science teachers' perceptions of technological pedagogical and content knowledge (TPACK). *Journal of Science Education and Technology, 22*(3), 325–336.

Liu, Q., Zhang, S., & Wang, Q. (2015). Surveying Chinese in-service K12 teachers' technology, pedagogy, and content knowledge. *Journal of Educational Computing Research, 53*(1), 55–74.

Roig-Vila, R., Mengual-Andres, S., & Quinto-Medrano, P. (2015). Primary teachers' technological, pedagogical and content knowledge. *Comunicar 45*, 151–15.

Szeto. E. & Cheng, A.Y.N. (2016). Pedagogies across subjects: What are pre-service teachers' TPACK patterns of integrating technology in practice? *Journal of Educational Computing Research, 0*(0).

Teacher Education and Teacher Professional Development in the COVID-19 Turn – Drajati & Putra (Eds)

Digital literacy enhancement in new normal era: Teachers' voice of sustainable professional development

Fitrotul Mufaridah, Ilza Mayuni & Romdani Romdani
Universitas Negeri Jakarta, East Jakarta, Indonesia

ABSTRACT: Digital literacy became one skill that teachers developed during the Covid-19 pandemic. This skill is essential for teachers to be better ready in actualizing 21st-century learning in classroom practices. Therefore, this study aims to describe the teachers' views and practices on digital literacy, and how the teachers' digital literacy sustainably developed in the new normal era. This survey study collected the data through questionnaires to 53 junior high school teachers. This study found teachers' agreement on the importance of digital literacy in online teaching. The online practices give teachers meaningful experiences in digital literacy. It also provided the teachers' need for digital literacy enhancement, however only 35.8% of them have planned programs to answer their needs. The result conveys teachers' inability to prove sustainability on digital literacy development. The implication is for policymakers to design the teacher's digital literacy development.

Keywords: 21st-century learning, digital literacy, teacher professional development.

1 INTRODUCTION

During the covid-19 pandemic, there is a shifting in the learning experience of teachers and students from face-to-face activities into virtual activities. Shifting learning from offline to online provides a problematic situation for teachers and students in most schools to carry out online or blended learning (Sá & Serpa 2020; Tejedor et al. 2020). Various studies review the implementation of media, platforms, and strategies in learning during the pandemic (Millar et al. 2021), and the challenges in digital transformation such as computer literacy, the opportunity of migrating methods, and success in handling efficient online learning processes (Adedoyin & Soykan 2020; Almazova et al. 2020). The implementation of online or blended learning, in addition to the availability of facilities and access factors, the digital competence of teachers to integrate ICT in their teaching is certainly a very important factor to pay attention to (Falloon 2020; Fraile et al. 2018; Instefjord & Munthe 2016; Tomczyk 2020).

In addition to problems with facilities and networks, the lack of digital competence of the teachers and students is also part of the obstacles in implementing online or blended learning during the covid-19 pandemic (Frolova et al. 2021; Sánchez-Cruzado et al. 2021). Likewise, students' digital competencies are significant for participating in online learning, so their competencies need to be developed. Their technology practice is not for academic purposes (Dashtestani & Hojatpanah 2020), but their digital competency should contribute to enhancing learning quality (Rolf et al. 2019; Thorell et al. 2015). The competence of teachers and at what level of their digital literacy competence has been investigated (Akayoğlu et al. 2020; Fraile et al. 2018), and teacher necessity of ICT literacy has also been scrutinized (Parra 2021). Integrating ICT into teaching provides a real learning space for teachers to hone their digital skills.

Those researches mentioned above have provided findings on the issue of ICT integration on blended or online learning, and teachers' competency on ICT. Regarding digital literacy, the teachers' view on the importance of digital literacy practices on the teaching and learning process, and their development program on it during and post the pandemic are not yet explored. Otherwise,

 DOI 10.1201/9781003347798-24

although blended or online teaching was implemented since the covid-19 pandemic, teachers' digital literacy development is not totally well-planned yet. As it is part of teacher professional development, a program to facilitate the development is essential. Therefore, this study is to explore: 1) teachers' views on digital literacy in English teaching; 2) teachers' digital literacy practice in English teaching; 3) teachers' digital literacy development program. Concerning to the objectives, the research questions of this study are:

1. What are teachers' voices on digital literacy during the pandemic?
2. How are teachers' digital literacy practices in English teaching during the pandemic?
3. How do teachers need and plan for digital literacy development?

2 LITERATURE REVIEW

Digital literacy is an essential skill to face the digitalization era and is encouraged at all educational levels. The term comes to specific ones such as information literacy, media literacy (Christ & Potter 1998), computer literacy (Tsai 2002), internet literacy (Harrison 2018), and currently multimodal literacy (Heydon 2007; Yeh & Tseng 2020). All have been connected with the proper use of digital methods in teaching and learning as part of insight into digital literacy (Gruszczynska et al. 2013). Further, a term of digital inclusion is clearly defined as different strategies planned to guard that each person has the same chance to access, and skills to take advantage of digital technologies and systems (Alhassan & Adam 2021; Aziz 2020; Reisdorf & Rhinesmith 2020). It demands the volume of utilizing digital technology to stimulate responsive participation in several dimensions during the covid-19 pandemic (Sá & Serpa 2020; Santos & Serpa 2017). The term digital literacy first came out around 1997, when Paul Gilster introduced it in his book as:

> *a set of skills to access the internet, find, manage and edit digital information; join in communications, and otherwise engage with an online information and communication network. Digital literacy is the ability to use and evaluate digital sources, tools and services properly, and apply them to lifelong learning processes (Gilster 1997).*

Teachers' digital competency is part of teacher professional development that is also termed as continuous professional development (CPD). It can be formulated as formal or informal support and activities which are mapped out to encourage teachers to develop as professionals through courses and in-school training such as coaching, mentoring, self-study, and action research (Coldwell 2017). Teachers are responsible for activating their programs or planning to find their progress in digital competencies to support and fulfill their need in processing online teaching. Coldwell focuses on active teachers' professional development, while Fraillon et al. (2014) differentiates pre-service from in-service teacher professional development. In this perspective, conventional approaches to enhancing digital abilities in teacher education have pivoted on encouraging students' digital literacy (Borthwick & Hansen 2017). It strengthens the issue that in-service teachers are active teachers who have enabled themselves to advance in digital competency.

In facing a pandemic, where variety and complex problems come to the teaching and learning process, teachers with school support need to do priority programs to address digital competence as an emergency problem. Here Ala-Mutka et al. (2008) have stated that educational institutions do not only need to utilize a particular platform for learning digital skills but should also be embedded in the teaching and learning process of all subjects. In a different aspect, Omotayo and Haliru (2020) suggested that learners should be conditioned to be motivated to get digital competency to be involved successfully in their learning goal.

Concerning the context of this study, the focus is on in-service professional development. The focus is more on the character of professional development activities, both internal and external development. Teachers are competent in accomplishing relevant learning material themselves and do not need external support, but a combination of both external and internal professional development to be most effective (Goodyear 2017). The teacher is possible to access both, internal and external support, to acquire their digital competency development for the integration of Computer-Assisted Language Learning (CALL) in the English as a Foreign Language context.

3 METHOD

This survey study uses a mix-method design with implementing both qualitative and quantitative approaches to get the validity of the findings and help to detect the variety of teachers' views and experiences (Schoonenboom & Johnson 2017). This study is to answer three research questions based on a survey of individual perspectives and experiences on digital literacy during the covid-19 pandemic. This survey study used a questionnaire and interview designed to understand English teachers' voices and experiences in implementing online or blended learning related to the integration of ICT in their teaching activities and their digital competency development need during the covid-19 pandemic.

It collected data from questionnaires to 53 junior high school teachers and interviewed 15 teachers of them. 250 teachers of private schools in East Java have answered the questionnaires, but only 53 of them fulfill the criteria. The participants of this study are selected from junior high school teachers who formulated digital utilities in their teaching and learning process as the operation of digital literacy through the activities during the pandemic. They got their understanding and experiences individually on integrating ICT in their teaching activities during the pandemic. Therefore, only data from 53 teachers were analyzed to answer the three research questions. The data analysis was about their views, practices, needs, and planning for digital literacy development. It was displayed in the percentage distribution and analyzed qualitatively based on thematic analysis (Braun & Clarke 2006).

4 FINDING AND DISCUSSION

4.1 *Teachers' voices on digital literacy during the pandemic*

Technology can be used as a learning tool to overcome learning problems during a pandemic. Some teachers have no experience conducting online teaching activities. However, facing the issue of learning during the pandemic pushes them to immigrate from face-to-face activity to online learning activity. For that reason, teachers agree that technology is an important tool to facilitate learning activities. Since technology can support learning, digital literacy proceeds in it. The teachers viewed the importance of the digital literacy process in English learning.

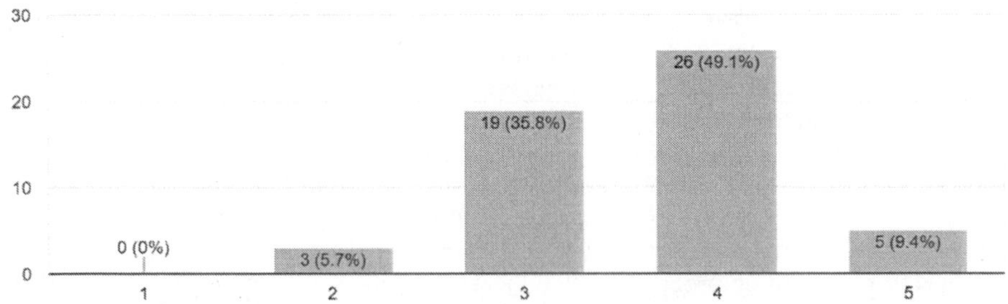

Figure 1. Teachers' view on digital literacy.

Based on the graphic, 58.5 % of teachers perceived the importance of digital literacy for supporting teaching and learning activities with technology integration. Although it gives a positive view on it, there are 5.7% of them considered different from the most number. Related to the teachers' voices in the interview, teachers declared their views on the importance of digital literacy in teaching and learning. So, this study confirmed finding on the advance of digital literacy through ICT integration for the online teaching process during the pandemic, while a study done by Alhassan focused on the significant influence of ICT usage on Quality of Life (QoL) (Alhassan & Adam 2021). Here digital literacy on access to technology, information, and communication facilities has become one of the significant aspects of 21st-century learning (Churchill 2020; Hazaea & Alqahtani 2020; Martzoukou et al. 2020; Rajagopal et al. 2020).

4.2 *Teachers' practices of digital literacy during the pandemic*

Teachers realize that online teaching is different from face-to-face teaching. Online learning provides a comfortable chance for students to be innovative by experiencing computer technology, but in terms of social presence, social interaction, and student's satisfaction, online learning is viewed as lacking compared to face-to-face learning (Bali & Liu 2018). Most teachers have been pressured to implement online teaching practices during the covid-19 pandemic, but it conveys them to learn by doing hard. Although teachers do not have enough preparation in designing lesson planning for online teaching and learning, they have tried to utilize technology to support the online learning processes. The demand to implement online learning activity directed the teachers to digital literacy practices. It was stated in the interview that the practices of integrating ICT on online learning have given them digital literacy development, from not knowing any kind of virtual platform and other digital communication, they are finally able to use and operate them step by step better. They acquired experiences in digital literacy development through the practices. The teachers do not only learn the digital ICT device but also implement them in their teaching process. This finding shows the practices from the teachers' side only, without any focus on the educational institution (Ala-Mutka et al. 2008).

The experiences of teachers' digital literacy practices in English teaching and learning from the questionnaire are shown in the figure below (Figure 2).

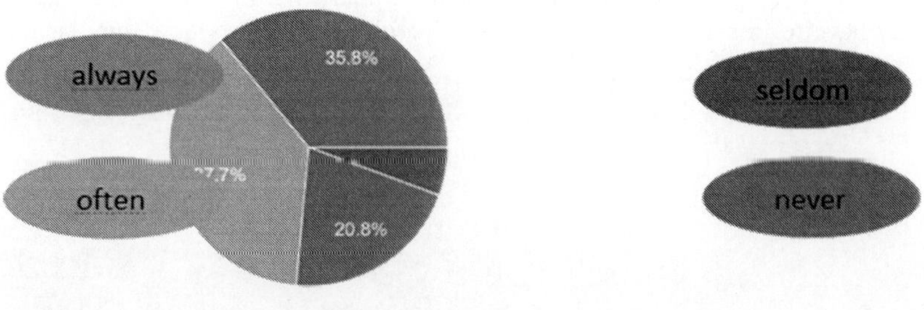

Figure 2. Teachers' practices of digital literacy in English teaching and learning.

The percentage showed that 37.7% of teachers often get digital literacy experiences by utilizing technology tools for facilitating online learning during the pandemic. There are 35.8% of teachers who declare always to find the digital literacy experience during the online learning practices, and 20.8% seldom say in getting the digital literacy experience. Only 5.7% of teachers stated that they do online learning during the pandemic but they do not get any meaningful experience in digital literacy through integrating technology in their teaching activity. The result recognizes the meaning of practice. Teachers who lead themselves to integrate the ICT utilities in their teaching activities have brought them to get the meaningful experience on it, but those who depend on some others' help in utilizing the ICT tools have got less experience (Guggemos & Seufert 2021). So, teaching with technology is not only about utilizing the ICT tools but also about how to independently utilize the ICT in their teaching practices in order to get significant experiences.

4.3 *Teachers' need and planning for digital literacy development*

After implementing online or blended learning during the pandemic, the experiences make the teachers recognize that they need to develop their digital ability better and much better. Challenges or difficulties in operating and taking advantage of implementing certain technology utilities have taught them the importance of mastering a new competency through the learning or training process. They realize their need for being able to provide a digital literacy process to their students, to motivate the students to participate actively using digital competence for bringing them to reach their learning goal (Omotayo & Haliru 2020). Teachers' competencies in handling online learning and their positive attitude toward digital literacy have strengthened students' motivation to access digital ICT. While this study focuses on teachers' efforts to motivate the students in online learning participation by utilizing digital ICT, Omotayo and Haliru's study had focused on digital libraries. The further situation on teachers' need for digital enhancement is showed in the graphic below.

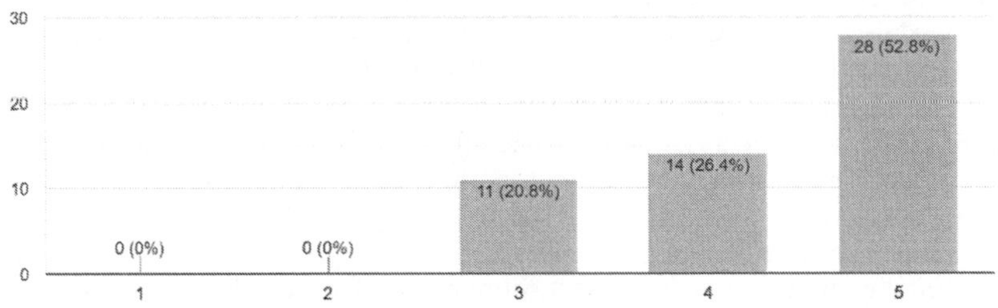

Figure 3. Teachers' need for digital literacy enhancement.

Different choice levels of confession showed 53 teachers or 100 % stated their need to develop their digital ability. They realize their flaws and weaknesses when integrating technology in learning, so they need to continue mastering ICT to improve the quality of their teaching in order to motivate the students much better. From the interview data, their needs are not actualized in good participation in such a digital literacy development program. Their awareness of the importance of digital literacy development could not lead them to plan what they have to do next, and what program they have to join intently. In fact, the following data shows how they have any digital enhancement program yet.

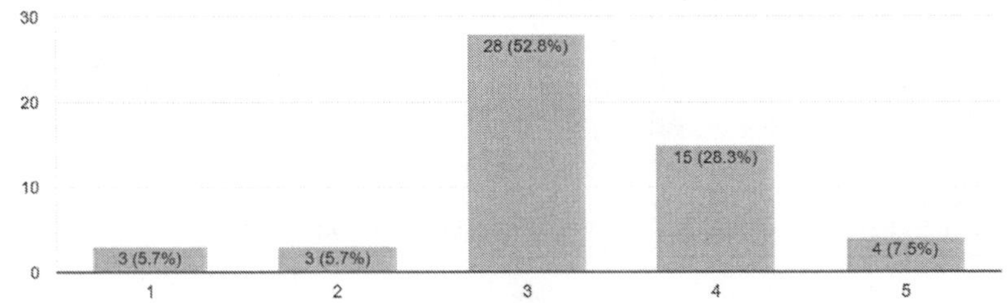

Figure 4. Teachers' plan on digital literacy improvement.

It displayed only 19 or 35.8% of teachers who have programs to develop their digital competency. Many of them declare in the interview that they are confused about how to follow up on their weaknesses in digital literacy development. They could not manage proper planning to enhance their competency. They are not independent yet in handling their personal development needs in their professional field and solving the problem. The teachers' internal motivation is not enough to formulate the proper planning to develop their digital competence. The hard situation of the teachers reflected their expectation on school attention and help for reaching the development access. It reflected that teachers need external support to enhance their digital literacy (Morris et al. 2003). Here what Goodyear explained about internal and external support becomes right. Planning action in teacher digital literacy enhancement in this pandemic era will be more effective by having a combination of both external and internal support (Goodyear 2017).

5 CONCLUSION AND RECOMMENDATION

This study investigated the teachers' voice on the importance of digital literacy, teachers' practices on the digital literacy process, online teaching and learning activities, and teachers' professional development in digital competency. First, only 5.7% of teachers disagreed on the importance of the digital literacy process and their teaching. Their teaching, however, was not integrating proper technology to support their teaching and learning practices. Second, teachers acquire digital literacy experiences by utilizing technology in their online teaching activities. The practices give them benefits in solving digital challenges and difficulties. Third, teachers realize their need for digital literacy development. However, their awareness was not followed by their participation in programs about digital competency development. It figures out the teachers' inability to formulate their digital literacy development planning. The implication of these findings could be pointed to teachers themselves and policymakers in an educational institution. It becomes a crucial reflection for teachers regarding their professional development. It provides recommendations for policymakers to design coaching programs for teacher digital literacy development. This study does not cover interviews for such a type of digital literacy development program, and how effective they plan the program, so it can be suggested to further research to investigate it.

REFERENCES

Adedoyin, O. B., & Soykan, E. (2020). Covid-19 pandemic and online learning: the challenges and opportunities. In *Interactive Learning Environments*.

Akayoğlu, S., Satar, H. M., Dikilitş, K., Cirit, N. C., & Korkmazgil, S. (2020). Digital literacy practices of Turkish pre-service EFL teachers. *Australasian Journal of Educational Technology*.

Ala-Mutka, K., Punie, Y., & Redecker, C. (2008). Digital competence for lifelong learning. Institute for Prospective Technological Studies (IPTS), European Commission, *Joint Research Centre. Technical Note: JRC, 48708*, 271–282.

Alhassan, M. D., & Adam, I. O. (2021). The effects of digital inclusion and ICT access on the quality of life: A global perspective. *Technology in Society*.

Almazova, N., Krylova, E., Rubtsova, A., & Odinokaya, M. (2020). Challenges and opportunities for Russian higher education amid covid-19: Teachers' perspective. *Education Sciences, 10*(12), 1–11.

Aziz, A. (2020). Digital inclusion challenges in Bangladesh: the case of the National ICT Policy. *Contemporary South Asia*.

Bali, S., & Liu, M. C. (2018). Students' perceptions toward online learning and face-to-face learning courses. *Journal of Physics: Conference Series*.

Borthwick, A. C., & Hansen, R. (2017). Digital Literacy in Teacher Education: Are Teacher Educators Competent? In *Journal of Digital Learning in Teacher Education*.

Braun, V., & Clarke, V. (2006). Using thematic analysis in psychology. *Qualitative Research in Psychology*.

Christ, W. G., & Potter, W. J. (1998). Media literacy, media education, and the academy. *Journal of Communication*.

Churchill, N. (2020). Development of students' digital literacy skills through digital storytelling with mobile devices. *Educational Media International*.

Coldwell, M. (2017). Exploring the influence of professional development on teacher careers: A path model approach. *Teaching and Teacher Education*.

Dashtestani, R., & Hojatpanah, S. (2020). Digital literacy of EFL students in a junior high school in Iran: voices of teachers, students and Ministry Directors. *Computer Assisted Language Learning*.

Falloon, G. (2020). From digital literacy to digital competence: the teacher digital competency (TDC) framework. In *Educational Technology Research and Development*. Springer.

Fraile, M. N., Peñalva-Vélez, A., & Lacambra, A. M. M. (2018). Development of digital competence in secondary education teachers' training. *Education Sciences*.

Fraillon, J., Ainley, J., Schultz, W., Friedman, T., & Gebhardt, E. (2014). Preparing for life in a digital age: The IEA international computer and information literacy study international report. In *International Association for the Evaluation of Educational Achievement (IEA)*.
Frolova, E. V., Rogach, O. V., Tyurikov, A. G., & Razov, P. V. (2021). Online Student Education in a Pandemic: New Challenges and Risks. *European Journal of Contemporary Education*.
Gilster, P. (1997). *Digital Literacy*. Wiley.
Goodyear, V. A. (2017). Sustained Professional Development on Cooperative Learning: Impact on Six Teachers' Practices and Students' Learning. *Research Quarterly for Exercise and Sport*.
Gruszczynska, A., Merchant, G., & Pountney, R. (2013). "Digital futures in teacher education": Exploring open approaches towards digital literacy. *Electronic Journal of E-Learning*.
Guggemos, J., & Seufert, S. (2021). Teaching with and teaching about technology – Evidence for professional development of in-service teachers. *Computers in Human Behavior*.
Harrison, C. (2018). Critical Internet Literacy: What Is It, and How Should We Teach It? *Journal of Adolescent and Adult Literacy*. https://doi.org/10.1002/jaal.713
Hazaea, A. N., & Alqahtani, A. A. J. (2020). Competences in digital online media literacy: Towards convergence with emergency remote EFL learning. *International Journal of Media and Information Literacy*, *5*(2), 164–175.
Heydon, R. M. (2007). Making meaning together: Multi-modal literacy learning opportunities in an inter-generational art programme. *Journal of Curriculum Studies*.
Instefjord, E., & Munthe, E. (2016). Preparing pre-service teachers to integrate technology: an analysis of the emphasis on digital competence in teacher education curricula. *European Journal of Teacher Education*.
Martzoukou, K., Fulton, C., Kostagiolas, P., & Lavranos, C. (2020). A study of higher education students' self-perceived digital competences for learning and everyday life online participation. *Journal of Documentation*, *76*(6), 1413–1458.
Millar, S.-K., Spencer, K., Stewart, T., & Dong, M. (2021). Learning Curves in COVID-19: Student Strategies in the 'new normal'? *Frontiers in Education*, *6*.
Morris, M., Chrispeels, J., & Burke, P. (2003). The Power of Two: Linking External with Internal Teachers' Professional Development. *Phi Delta Kappan*.
Omotayo, F. O., & Haliru, A. R. (2020). Perception of task-technology fit of digital library among undergraduates in selected universities in Nigeria. *Journal of Academic Librarianship*.
Parra, G. (2021). Educators' ICT Literacy: A Prerequisite for the Future of Education Post-COVID-19. *Asian Journal of Education and E-Learning*.
Rajagopal, K., Firssova, O., de Beeck, I. O., Van der Stappen, E., Stoyanov, S., Henderikx, P., & Buchem, I. (2020). Learner skills in open virtual mobility. *Research in Learning Technology*, *28*, 1–18.
Reisdorf, B., & Rhinesmith, C. (2020). Digital inclusion as a core component of social inclusion. In *Social Inclusion*.
Rolf, E., Knutsson, O., & Ramberg, R. (2019). An analysis of digital competence as expressed in design patterns for technology use in teaching. *British Journal of Educational Technology*.
Sá, M. J., & Serpa, S. (2020). COVID-19 and the promotion of digital competences in education. *Universal Journal of Educational Research*.
Sánchez-Cruzado, C., Santiago Campión, R., & Sánchez-Compaña, M. T. (2021). Teacher digital literacy: The indisputable challenge after covid-19. *Sustainability (Switzerland)*, *13*(4), 1–29.
Santos, A. I., & Serpa, S. (2017). The Importance of Promoting Digital Literacy in Higher Education. *International Journal of Social Science Studies*.
Schoonenboom, J., & Johnson, R. B. (2017). How to Construct a Mixed Methods Research Design. *KZfSS Kölner Zeitschrift Für Soziologie Und Sozialpsychologie*.
Tejedor, S., Cervi, L., Pérez-Escoda, A., & Jumbo, F. T. (2020). Digital literacy and higher education during COVID-19 lockdown: Spain, Italy, and Ecuador. *Publications*.
Thorell, M., Fridorff-Jens, P. K., Lassen, P., Lange, T., & Kayser, L. (2015). Transforming students into digital academics: A challenge at both the individual and the institutional level Approaches to teaching and learning. *BMC Medical Education*, *15*(1).
Tomczyk, Ł. (2020). Skills in the area of digital safety as a key component of digital literacy among teachers. *Education and Information Technologies*.
Tsai, M. J. (2002). Do male students often perform better than female students when learning computers?: A study of Taiwanese eighth graders' computer education through strategic and cooperative learning. *Journal of Educational Computing Research*.
Yeh, H. C., & Tseng, S. S. (2020). Enhancing multimodal literacy using augmented reality. *Language Learning and Technology*.

Teacher Education and Teacher Professional Development in the COVID-19 Turn – Drajati & Putra (Eds)
© 2023 The Author(s), ISBN: 978-1-032-38970-7

Reinforcing student character: Ethnopedagogy approach to support teacher change

Kenfitria Diah Wijayanti, Djatmika Djatmika, Sumarlam Sumarlam & Wakit Abdullah Rais
Universitas Sebelas Maret, Surakarta, Indonesia

Dewi Pangestu Said
Yamaguchi University, Yamaguchi, Japan

ABSTRACT: The study aims at describing the application of the ethnopedagogical approach as a step to reinforce students' character to support teacher change. It is a qualitative study with a pedagogical collaborative ethnographic approach research with teachers and students as the participants. Data collection was carried out using observation, interview, and documentation techniques. The data analysis techniques used include domain analysis, taxonomic, component analysis, and cultural themes. The results of this study proved that applying the ethnopedagogical approach with media based on the local wisdom *Sapa Aruh* tradition is practical because it can increase a sense of nationalism, and amplify students' positive character. Therefore students are motivated to preserve the local culture. The theoretical implication of this research is that it can provide scientific contributions through the collaboration of ethnography and pedagogy. Meanwhile, the practical implication is expected that it can reinforce students' character and support teacher change.

Keywords: Character education, ethnopedagogy, local wisdom, teacher change, tradition of sapa aruh.

1 INTRODUCTION

Ethnopedagogy collaborates the local wisdom of a culture with the world of education. The ethnopedagogy places the noble values of the culture as the foundation of learning. Local culture is a social potential that can configure the character of students. Javanese culture has a philosophy of character education, namely *Tri Rahayu*. It contains *memayu hayuning salira* as the meaning of self-quality improvement, *memayu hayuning bangsa* to fight for the progress of the nation, and *memayu hayuning bawana* to struggle for world progress. *Memayu hayuning salira* is a form of how men improve their quality. It can be measured from the material side related to science (mundane) and the non-material side related to morality, religiosity, and ethics. The principle of 'You reap what you sow' will apply in life. Consequently, men must be conscientious in competing in internalizing goodness. Meanwhile, *memayu hayuning bangsa* is related to men who fight for their country and nation. A good citizen should have social control, obey the law, and try to avoid all the prohibitions of the applicable law. At the same time, *memayu hayuning bawana* is connected to how humans improve their quality to keep the world alive so that life continues in peace and comfort for all its inhabitants. Hence, internalizing the values of wisdom in local culture in education will bring many positive impacts. Apart from strengthening positive character, ethnopedagogy can also support the preservation of local culture.

Several studies related to ethnopedagogy and educational values have been conducted previously. Ethnopedagogical research was conducted by Betaubun (2019) and character education research was carried out by Samani (2013). Meanwhile, Wanhar and Hasibuan (2021) described the differences in the character development of students taught by ethnopedagogical methods compared

DOI 10.1201/9781003347798-25

to conventional learning models. The results showed that the character development of students taught by the ethnopedagogic learning model was higher than those taught by the conventional learning model. Meanwhile, Agboola and Tsai (2012) explained that character education is a discipline that develops with a deliberate effort to optimize students' ethical behavior. The outcomes of character education are used to prepare future leaders. The progress of character education is a shared responsibility by parents, teachers, and also administrators as stakeholders. Ethnopedagogy and character education are like two sides of a coin that complement each other. Consequently, the collaboration of the two becomes a suitable composition in amplifying students' character.

Those previous studies of ethnopedagogy have been conducted using various methods. However, only a few seem to integrate teaching materials in the form of oral traditions. The study then comes to answer the research question below: How does the application of the ethnopedagogical approach as a step to strengthen the character of students to support teacher change?

2 LITERATURE REVIEW

Betaubun (2019) stated that ethnopedagogy is a learning approach by incorporating elements of local wisdom into learning so that learning becomes more meaningful. Silitonga, et al (2021) explained that ethnopedagogy is the knowledge that reflects the results of the idea of practicing skills. The ethnopedagogical approach becomes the implementation of teacher change in order to reinforce the character of students. The tradition of *Sapa Aruh* is used as a learning medium that can confirm students' speaking and ethical skills. Ethnopedagogy develops on the basis of understanding the ethnocultural of an ethnic group by raising the values of local wisdom. Placement of local values as the basis for learning has the aim of fostering cultural awareness in the younger generation. Ethnopedagogy is related to character education which has three main pillars, those are family, school, and community. Samani (2013) stated that habituation is created in situations and conditions that allow students anywhere to get used to behaving according to their values and has become their character. Since they have been internalized and personified through intervention, such as through an atmosphere of learning interaction designed to achieve learning outcomes and the purpose of character building by implementing a structured learning experience. Eighteen Character Values were formulated by the Indonesian Ministry of National Education with the hope of supporting the development of the nation's character. These character values are responsibility, creativity, honesty, discipline, hard work, independence, religion, curiosity, love for the homeland, communicative, independence, nationalism, tolerance, democratic, love to read, love peace, social care, and environmental care. The success of learning is determined by the way the teacher transfers knowledge. In addition, teachers must be able to adjust the characteristics of the students they teach. Teachers are required to be creative when conveying information related to learning materials.

The teachers' role is very important in responding to changing times. The presence of teachers is expected to lead education in a better direction. Therefore, teachers must adapt to the development of the world of education and the needs of students. The competencies possessed by teachers as professionals, educators, social, and personality figures are needed in helping to realize noble ideals in the intellectual life of the nation. Teachers must be able to make changes that start from within the classroom. The curriculum that has been determined is not a patent matter, but it can be conditioned according to the students' circumstances. Changes in teaching methods, media, and techniques are the keys to success in order to produce quality, innovative, and creative learning. Teacher change is not a personal problem for the teacher. However, it is also the responsibility of the leadership and the institutions that oversee it (Whitworth & Chiu 2015). Teacher change included five dimensions: curriculum, teaching, learning, teacher roles, and teaching and learning (Tam 2015).

Kaasila and Lauriala (2010) suggested that collaborative and interactionist teacher change can integrate sociocultural and constructivist perspectives focusing on cultural and situational factors. Meanwhile, Rodgers (2002) added that the process of social interaction and development at the individual level also has an important role in supporting teacher change. In addition, Guskey

(2002) stated that professional and sustainable individual development will support teacher change. Hubber, Tytler, & Haslam (2010) emphasized the elements of language use and the importance of the personal aspect for teachers to recognize the development of students' representational resources. Penlington (2008) reinforced if dialogue serves as an effective tool for teacher change and becomes a central activity in professional learning programs. Smith and Gillespie (2007) argued that the involvement of collaborative cultural elements will support the learning effectively.

The research gap of this article is the application of the Javanese oral tradition culture of *Sapa Aruh* to support teacher change. Ethnopedagogy which is closely related to cultural elements is the right composition in the teacher change design. New innovations in learning patterns can create character students by internalizing local cultural values. Learning patterns that are full of character education and cultural values can reinforce character education to students. Teachers as agents of change can combine aspects of local cultural wisdom with conventional curricula. The use of proper teaching material and exact methods can support the character changes on students, and eventually will impact to the teacher change. Hence, the specific objective to be achieved in this research is to explore the implementation of an ethnopedagogical approach based on the Javanese culture of the oral tradition of *Sapa Aruh* to support teacher change.

3 METHOD

This research is qualitative research using a collaborative pedagogical ethnographic approach. Ethnography is a verbal description of a cultural situation that is manifested by expressive symbols in language (Spradley 2007). Etnographic research aims at describing cultural situations identified through conversations in certain events where the data as the reality are the focus of the research. It was conducted at one of the elementary schools in Surakarta city Indonesia. The participants involved in this study were three teachers and thirty students. Data in this study are in the form of events and documents. The data were in the form of events related to teaching and learning activities that were carried out in online learning. While the document data in the form of syllabus, lesson plans, and field notes from observations.

Data in the study were collected using observation techniques and interviews with an informant that was one of Kasunanan Palace dignitaries who acts as an expert on Javanese language and culture. Interviews were conducted to get the validity of the data test as they were related to the oral tradition of *sapa aruh* and the application of learning media as a form of teacher change.The basic technique is the tapping technique, which is used to obtain data by listening comprehension. It was conducted to collect data in the form of conversations that contain elements of *Sapa Aruh* on the Javanese ethnic community in Indonesia. Data analysis used were Spradley model analysis techniques including domain, taxonomy, componential, and cultural themes analysis. Several stages were taken as a start, the researchers explained the domain analysis which aims to get a general picture of the research object. Next, by using taxonomic analysis, they were analyzed in more detailed data that have been obtained based on the domains known or defined in the previous analysis stage. After that, in the form of componential analysis, they analyze the domains found or that have been determined in the previous analysis stage. Finally, they get an overview and be able to deduce the pattern of cultural themes on the data findings.

4 FINDING AND DISCUSSION

There are several findings made based on the observations. The first is that the innovations made by the teacher by collaborating the conventional curriculum with the local wisdom of the *Sapa Aruh* oral tradition can strengthen students' character. Second, the project-based learning model with role-playing makes it easier for students to understand the material. The third finding is to make scenarios that are prioritized using the oral tradition of *Sapa Aruh* dialogue can have a big effect on changing students' attitudes. The courage to innovate in learning activities will support teacher change programs. In a school environment, the teachers become a facilitator for students, especially at the basic education level. As a model and the main figure in student education, the teachers direct and become examples for their students. Teachers act as agents of change. They are also students' influencers to their success or failure. To support the change of teachers to become more professional, innovation is needed in the models, methods, and media used in classroom learning.

The innovation carried out in Javanese language learning is to use a combination of ethnopedagogical approaches that are presented in a project-based learning model. Project-based learning is applied in a classroom in the form of role-playing. The purpose of students playing a role in strengthening character education is to practice solving problems through direct demonstrations, being able to identify problems, conduct analysis, perform roles, and discuss with colleagues. The basic competence of speaking skills by applying *unggah-ungguh* using *Sapa Aruh* oral tradition media is one of the more challenging learning alternatives for students. Through the characters played, the students can provide space to interact with other students with themes related to character education. Zuhriyah (2009) stated that the essence of role-playing in character education lies in the emotional involvement of actors and observers in real problem situations. Through role-playing in the character education, it is hoped that students can (1) explore their feelings, (2) gain insight into their attitudes, values, and perceptions, (3) develop skills and attitudes in solving the problems they face, and (4) explore the core of the problem that is played in various ways (Mulyasa 2012).

Figure 1. Synchronous online learning atmosphere.

Several stories with themes related to character education were provided for the students. Nevertheless, the stories presented were packed in a simple and applicable form according to their daily lives. Character instilling points were inserted in the message of the stories. After that, students were guided to discuss the plot, theme, and message of the stories that had been presented in a simple but communicative delivery. The students then responded to the display of the story to explore their feelings about the content of the stories. Indirectly, students were trained to solve problems that occurred in the storylines. After the discussion process ended, the students were invited to create a scenario at the end of the story by directly involving them in role-playing. The characters were played by several students who then finished the story.

The dialogues contained in the scenario emphasize the strengthening of the students' characters. As stated by Penlington (2008) that dialogue is a central activity in professional learning programs, then the closing dialogue carried out by students is directed by the teacher to stay focused on the theme, namely character reinforcing. An ethnopedagogical approach with the theme of local cultural wisdom will be firmly entrenched in students' minds. Character reinforcement that is done slowly and deeply has a big effect on changing students' attitudes. Kaasila & Lauriala (2010) argued that teachers as collaborative and interactionist change agents are able to integrate sociocultural and constructivist perspectives that focus on cultural and situational factors. Therefore, teachers must be able to innovate in collaborating cultural elements in each lesson.

The teacher's role is very important in implementing the Javanese language education curriculum. Consequently, teachers must be able to integrate local wisdom into teaching materials based on the mapping of basic competencies that have been determined. These innovations can support teacher change.

Silitonga, et al (2021) argued that the approach that links local culture with learning activities in the classroom encourages the younger generation to dominate and maintain local cultural values indirectly. Local culture can change students' perspectives on macroscopic and microscopic language content that is extraneous to students. Teachers' role is very important in the implementation of the Javanese language education curriculum. Teachers must be able to integrate local wisdom into teaching materials based on the mapping of predetermined basic competencies. Therefore, these innovations can support teacher change. Teachers are not only in charge of transferring knowledge to students, but also responsible for managing, directing, and planning in learning

Changes in teachers can be observed by their actions in the classroom as a facilitator that can design more effective learning. As a result, the learning objectives can be achieved optimally. An active, creative, and interactive class can be created when teachers can read the needs of their students. Teachers who are able to think critically and adjust the times can be classified as agents of change. They must have a strong determination to continue to learn and work so that they can produce lifelong learners who will later contribute to the teachers themselves, students, and society in general. Small-scale teacher changes can start from the individual and the classroom. These small changes will over time bring big changes to the world of education in Indonesia.

5 CONCLUSION AND RECOMMENDATION

The purpose of the study is to express the application of the ethnopedagogical approach as a step to strengthen students' character to support teacher change. Based on the results of the findings and discussions, it can be concluded that this study proves that the application of an ethnopedagogical approach with the media of the *Sapa Aruh* oral tradition based on local wisdom can have a major effect on changing student attitudes. Furthermore, the purpose of applying the ethnopedagogical approach can increase a sense of nationalism, strengthen positive character in students, and motivate students to preserve local culture. Changes in teachers can be seen in the teachers' ability to innovate and raise cultural issues into a supporting curriculum so that learning objectives can be achieved optimally. Teachers must be able to integrate local wisdom into teaching materials based on the mapping of basic competencies that have been determined. Moreover, these innovations are expected to support teacher change. Teachers are not only assigned to transfer knowledge to students but are also responsible for managing, directing, and planning in learning. The research has two implications, theoretically and practically. The theoretical implication of this research is that it can provide scientific contributions through the collaboration of ethnography and pedagogy. Meanwhile, the practical implication of this research is that it can contribute to strengthening students' character and supporting teacher change. The recommendation for further research is to explore more about ethnopedagogical studies by supporting local cultural changes in the context of teachers.

ACKNOWLEDGEMENT

This research was funded by the RKAT of PTNBH Universitas Sebelas Maret for the 2021 Fiscal Year through a doctoral dissertation research scheme (pdd-uns) with Contract Number: 260/UN27.22/HK.07.00/2021.

REFERENCES

Agboola, A., & Tsai, K.C. (2012). Bring character education into the classroom. *European Journal of Educational Research, 1*(2), 163–170.

Betaubun, P. (2019). Papua contextual science curriculum contains with indigenous science (Ethnopedagogy study at Malind Tribe Merauke). *International Journal of Civil Engineering and Technology*, *10*(02), 1994–2000.

Boody, R. M. (2008). Teacher Reflection as Teacher Change, and Teacher Change as Moral Response. *Education*, *128*(3).

Ertmer, P. A., & Ottenbreit-Leftwich, A. T. (2010). Teacher technology change: How knowledge, confidence, beliefs, and culture intersect. *Journal of research on Technology in Education*, *42*(3), 255–284.

Guskey, T. R. (2002). Professional development and teacher change. *Teachers and teaching*, *8*(3), 381–391.

Hubber, P., Tytler, R., & Haslam, F. (2010). Teaching and learning about force with a representational focus: Pedagogy and teacher change. *Research in Science Education*, *40*(1), 5–28.

Kaasila, R., & Lauriala, A. (2010). Towards a collaborative, interactionist model of teacher change. *Teaching and teacher education*, *26*(4), 854–862.

Mulyasa. (2012). *Manajemen pendidikan karakter.* Jakarta: PT. *Bumi Aksara*.

Nathan, M. J., & Knuth, E. J. (2003). A study of whole classroom mathematical discourse and teacher change. *Cognition and instruction*, *21*(2), 175–207.

Penlington, C. (2008). Dialogue as a catalyst for teacher change: A conceptual analysis. *Teaching and teacher education*, *24*(5), 1304–1316.

Rodgers, C. R. (2002). Seeing student learning: Teacher change and the role of reflection. *Harvard educational review*, *72*(2), 230.

Samani, M. (2013). *Konsep dan Model Pendidikan Karakter*, Cet. 3. Bandung: Remaja Rosdakarya.

Silitonga, R. W., Cahyana, U., Purwanto, A., & Rahmawati, Y. (2021). The effectiveness and learning outcomes of mobile learning integrated with ethnopedagogy on metacognitive ability of students in North Tapanuli. In *AIP Conference Proceedings, 2331(1), 040011*. AIP Publishing LLC.

Smith, C., & Gillespie, M. (2007). Research on professional development and teacher change: Implications for adult basic education. *Review of adult learning and literacy*, *7*(7), 205–244.

Spillane, J. (2002). Local theories of teacher change: The pedagogy of district policies and programs. *Teachers college record*, *104*(3), 377–420.

Spradley, J. P. (2007). Metode Etnografi (Rev-ad). Tiara Wacana, Yogyakarta. Diterjemahkan Misbah Zulfa Elizaberth. Buku asli diterbitkan tahun 1997: *The Etnographic interview*.

Tam, A. C. F. (2015). The role of a professional learning community in teacher change: A perspective from beliefs and practices. *Teachers and Teaching*, *21*(1), 22–43.

Wanhar, F. A., & Hasibuan, A. M. (2021). The Effect of Etnopedagogy Learning Model on Student Character Development in the Pandemic Era. *Budapest International Research and Critics Institute (BIRCI-Journal): Humanities and Social Sciences*, *4*(4), 8240–8247.

Whitworth, B. A., & Chiu, J. L. (2015). Professional development and teacher change: The missing leadership link. *Journal of science teacher education*, *26*(2), 121–137.

Zuhriyah, I. A. (2009). Aplikasi Model Role Playing Sebagai Upaya Pembelajaran Nilai dalam Pendidikan IPS Sd/mi. *Madrasah: Jurnal Pendidikan dan Pembelajaran Dasar*, *2*(1).

Teacher Education and Teacher Professional Development in the COVID-19 Turn – Drajati & Putra (Eds)
 ISBN: 978-1-032-38970-7

Engaging teachers in classroom action research: Lesson learned from teachers professional development program in Indonesia

Isnanik Juni Fitriyah, Munzil Munzil, Yessi Affriyenni, Erti Hamimi, Jihan Roidah Affifah & Cica Adelia Permatasari
Universitas Negeri Malang, Java, Indonesia

Erni Yulianti
Universiti Teknologi Malaysia, Johor Bahru, Malaysia

ABSTRACT: This study describes the effect of workshops on improving teachers' skills in carrying out classroom action research (CAR). This ethnographic study uses interview sessions and questionnaires distributed to 65 junior high school teachers who teach different subjects. The qualitative data were analyzed using thematic analysis. The stages of research include need analysis, data collection, and evaluation. CAR training activities for junior high school teachers focused on guiding teachers on knowledge about CAR, assisting in planning and implementing CAR, and creating CAR articles. The results indicate a clear progression in teachers' classroom assessment skills. This research concludes that CAR is essential for junior high school teachers and can be improved through training activities.

Keywords: Class Action Research, Teacher Professionalism, Training, Workshop

1 INTRODUCTION

Learning is a relatively permanent change in behavioral tendency. Learning is a process to achieve the learning objectives in each subject (Joung & Anderman 2020). Factors that influence learning success include cleanliness of the heart, perseverance, understanding, and relationships between educators and students. In this case, the teacher plays a significant role in teaching (Peker et al. 2021). The role of the teacher starts from designing lesson planning until the learning evaluation (Colognesi et al. 2020). The learning process in the field does not run smoothly because the learning process has several obstacles such as passive students, lack of students' motivation, and understanding. In practice, teachers in Indonesia are members of the Association of Teachers of the Republic of Indonesia (ATRI) and the Subject Teachers' Conference (STC). The demand for learning in the 21st-century is for the teachers to act as active learners who support literacy programs for students. Thus, teachers can further improve their skills to create active, innovative, and interesting learning for students who are constantly changing (Chang et al. 2020).

A professional teacher is a teacher who knows the character of his/her students well and can adapt his/her learning model to the student (Noermanzah 2020). In this case, the teacher can identify learning problems and formulate appropriate solutions for the next lesson. Efforts that teachers can do are to carry out classroom action research (CAR) (Niemi 2018). CAR activities are carried out by analyzing problems in a class and then providing action to the class with careful planning (Farmer 2021). By going through the stages in this CAR cycle, the teacher as a researcher will reflect on how effectively s/he overcomes the class problems through his/her actions and understand what action to be applied further to find the most appropriate solution to class problems (Skyhar 2021).

DOI 10.1201/9781003347798-26

Several previous researchers have researched CAR regarding its effect on student learning outcomes (Kong et al. 2021). However, research that explains teachers about CAR skills such as the ability of teachers to solve real learning problems, the quality of the learning process and learning outcomes, and the quality of teachers in solving problems and the quality of learning from a teacher improvement program are rarely explored. For example, there is research about the provision of materials and training on the importance of CAR conducted by teachers (Julia et al. 2020). By conducting training, it can improve teachers' ability to solve learning problems in the classroom and increase knowledge and skills regarding selecting appropriate learning strategies. CAR training activities not only train CAR skills but also prepare CAR articles. According to Noermnzah (2020), CAR was conducted to increase the role of teachers as researchers and contribute their research results in national and international publications. CAR does not only provide action to the class with variables in the form of learning models but can also use types of teaching materials, learning media, learning methods, and types of assessments to assess instruments. Research by Barnawi et al. (2019) shows that training can facilitate and assist teachers in carrying out CAR in a more well-planned way. It includes the provision of material and assistance in implementing CAR from planning, implementation, and evaluation.

The importance of teachers in carrying out CAR is unfortunately not following the conditions of teacher understanding and teacher experience in the field. Based on Barnawi et al. (2019), junior high school teachers still do not understand and have not implemented CAR. The junior high school teacher admitted that his class also experienced obstacles such as decreased student motivation and uneven student learning outcomes above the minimum completeness criteria. Therefore, activities were carried out to provide material and assistance in CAR for junior high school teachers to overcome this issue. This research follows the process of designing workshop activities to improve professionalism and teaching skills. Numerous studies have been conducted on the implementation of teacher skills workshops. However, it has not discussed the impact of improving these skills, especially in CAR. Therefore, this study aimed to describe the effect of the workshop implementation on improving the understanding and skills of teachers in implementing CAR. Our research questions are as follows:

1. How is the implementation of CAR training workshops for junior high school teachers?
2. What are the forms of improving the skills of implementing CAR by teachers due to the workshop?

2 LITERATURE REVIEW

One of the competencies that teachers must master is pedagogic competence. Pedagogical competence is the ability of teachers to understand students and plan, teach, and learn. With teachers having these competencies, the learning process will be efficient and fun (Novianti & Nurlaelawati 2019). According to Law of the Republic of Indonesia Number 14 of 2005 concerning teachers and lecturers, pedagogic competence is defined as the ability to manage student learning. It means teachers need to have philosophical beliefs about teaching that adapt their background knowledge, experience, personality situation, and learning environment as learning objectives. There are two related corrective actions in this case: a gradual review of the teaching program and continuous improvement for further learning (Mills et al. 2021). In realizing this, teachers need to have good pedagogic skills and professionalism. Even with the times that make changes in education, teachers need to continue to update their skills and professionalism (Chang et al. 2020).

CAR is a process of teachers reviewing and evaluating learning. CAR is the type of research conducted by the teacher as a classroom researcher to determine the consequences of the actions applied to the class (Meesuk et al. 2020). Kurt Lewin first introduced CAR in 1946, which was further developed by other experts such as Stephen Kemmis, Robin Mc Taggart, John Elliot, and Dave Ebbutt. Thus, the concept of CAR is growing under the development of science and technology. CAR can mean that research is carried out for reflection by the perpetrators of the action to increase the stability of the action. With the teacher doing CAR, the teacher can evaluate the learning process in his class from planning implementation to evaluation. It is what teachers need to have to support their pedagogical competence (Susilowati 2018).

In CAR, treatment was given to the class. After being given the action, the teacher measures the target or research objective achievement (Chen et al. 2021). The main goal is to solve the learning problem. If the problem is still not resolved, the research will continue again. It is called a cycle of CAR (Karyamulya 2020). By conducting this cycle, it provides relevant literature research. The teacher can determine whether their action helps in overcoming students' problems. The teacher can then decide what actions need to be taken for further learning. Therefore, teachers need to have competence in carrying out CAR. CAR conducted by teachers can become material for making research articles that can then be published in journals or proceedings, both nationally and internationally (Mahardika et al. 2019). The importance of conducting CAR for teachers is unfortunately not following the conditions in the field (Setiawati & Sunarti 2021). Research by Laudonia et al. (2017) states that teachers still lack an understanding of CAR, very few students have carried out CAR. It is the same as the fact found through interviews in one city in East Java, Indonesia, that teachers still do not completely understand and implement CAR (Mahardika et al. 2019).

Efforts that can be made to improve the understanding and skills of teachers in carrying out CAR are by participating in training or workshops (Barnawi et al. 2019). Workshops are usually held in MGMP forums and school forums (Rery et al. 2019). Workshop activities containing the delivery of material and assigning participants and mentoring can provide meaningful knowledge learning for teachers. The teacher understands the knowledge of wishful thinking and has experience in applying that knowledge (Barnawi et al. 2019). CAR training has been conducted by Barnawi et al. (2019) and contains a description of the training process from planning to evaluation. However, it still does not explain the results of increasing teacher competence in implementing CAR in detail and clearly. Teacher competence in carrying out CAR can increase because the training provides scheduled assistance by an expert to answer and guide all teacher misunderstandings and difficulties (Meesuk et al. 2020). Based on the previous research, there are problems with teachers' understanding and skills in implementing CAR. In addition to getting assistance and materials related to CAR, teachers are also assisted in creating articles related to CAR. Therefore, to increase teachers' professionalism in doing CAR, the output of this workshop is CAR articles.

3 METHOD

This study is based on evidence gathered from community service in classroom action training for junior high school teachers. It was conducted in a junior high school in East Java, Indonesia. The research participants were 65 junior high school teachers (27 males, 38 females; 25-55 years old; 7-15 years of teaching experience) who teach different subjects and agreed to participate in this research. Participants were obtained by conducting an initial survey of teachers using Google Forms which were 'distributed randomly in one area for both public and private school teachers. This ethnographic study uses interview sessions and the distribution of questionnaires to junior high school teachers (Naidoo 2015; Lestiyanawati 2020). The data collection instruments were observation sheets of training activities, participant response questionnaires post-training, and observations on teachers' CAR implementation. The data analysis technique was carried out with thematic analysis and the first step is preparation. In this stage, preparations are made in carrying out research, like observations, interviews with junior high school teachers, literacy studies, and preparing all instruments required. In the second stage, implementation activities are carried out based on the training activity plan, collecting training participant response data during training and after training, and collecting valuable data from observers on class action research of trainees. In the last stage, an evaluation is carried out for analyzing the data that has been collected to answer the objectives of the research. Activity evaluation is also carried out.

4 FINDING AND DISCUSSION

This study describes the effect of the implementation of workshops on improving teachers' skills in carrying out CAR. The explanation in this study is based on a series of community service activities carried out at a junior high school in a city in East Java, Indonesia. Therefore, the initial findings on teacher CAR skills, workshop activities to improve teacher CAR skills, and how the real evidence of improving teacher skills in implementing CAR after the workshop will be explained.

4.1 *The implementation of car training*

Based on the data of this research, most junior high school teachers still did not understand and had not carried out CAR activities. There are many obstacles in the classroom, especially the implementation of online learning as an effect of the COVID-19 pandemic, such as low student motivation and decreased student learning outcomes. This research also found that the concern of the teachers is the lack of understanding in carrying out CAR. It is the same as the previous research, which states that teachers at the early, elementary to upper secondary levels still need to improve their understanding and implementation of CAR (Fitriah & Ita 2021; Laksana et al. 2021; Sulaiman 2020). Not all teachers can conduct CAR, so this competency needs improvement.

In this study, the teachers were junior high school teachers who teach different subjects. The training activities contained two main activities: the provision of materials and assistance. The material provision was carried out by explaining the material regarding teacher professionalism, theories, and procedures for implementing CAR. Mentoring activities were carried out by dividing the training participants into several groups. One supervisor from the training committee fosters each group. In groups, participants were guided on planning the implementation of CAR, evaluating the results of implementing CAR, and making scientific articles on CAR. This training stage is following training activities carried out in previous studies that contain material theory and participants acting in carrying out the application of the theory or material described (Declaro-ruedas & Ruedas 2020). Participants who apply theory can provide experience and the theory more firmly embedded in participants' minds.

4.2 *The forms of improving skill*

Based on observations, junior high school teachers are enthusiastic about participating in the workshop activities. It could be said that judging by the number of participants who asked questions, within 60 minutes of discussion, 5 participants asked the initial question, and 14 other participants gave their responses. This enthusiasm is seen in the whole process where the teacher carefully pays attention to the presenters in explaining the material about CAR. The successful delivery of messages between teachers and students or between presenters and audiences is the goal of discussion and presentation activities. This message can be transferred verbally or non-verbally (Syah 2016). After the explanation, the question-and-answer discussion session also resulted in many questions from the training participants. An example of a question from participants was their confusion about how to carry out CAR during online learning. Meanwhile, the results of the mentoring activities from the implementation of CAR by training participants can be seen in Figure 1.

Based on Figure 1, there is an increase in the teacher's ability to understand the concept of CAR, make action research plans, implement, and follow up on the teacher's class. Training activities can increase the trainees' understanding and skills according to the training topic (Barnawi et al. 2019). This proven improvement (Figure 1) answers the research question about the improving forms of teachers. The participants are also enthusiastic about holding more training in other topics such as developing HOTS questions, innovative learning media, and others.

Training activities have been proven to overcome the problems of teachers who do not understand and have good competence in CAR. Schools and teacher associations (MGMP) and higher education institutions must continuously carry out this kind of activity to manifest the Tri Dharma of Higher

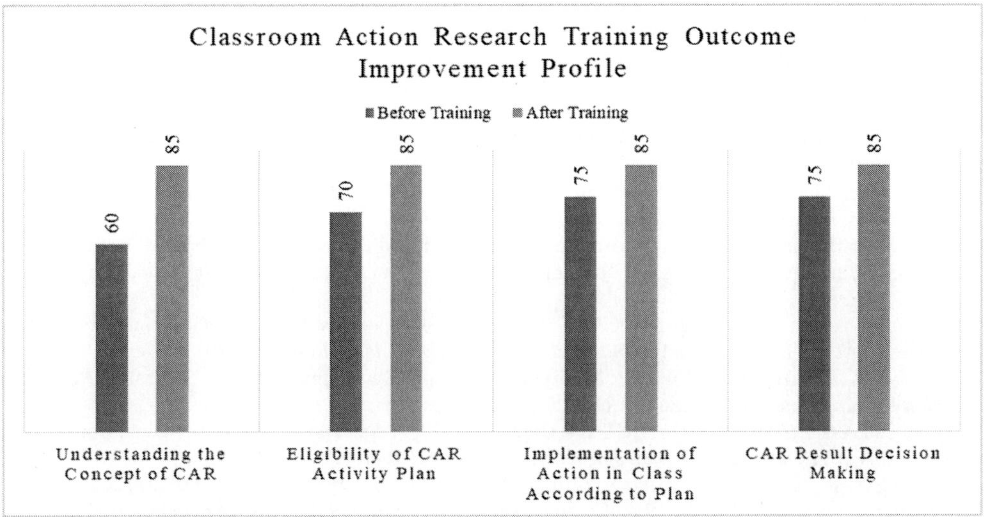

Figure 1. Comparison of teacher understanding and skills measurement in classroom action research between before training and after training.

Education (education and teaching, research, and community service). Training activities can provide more knowledge, strengthen brotherhood, and expand relations between educators and the community. This is following research conducted by Young (2018), that the training activities he carries out can minimize the occurrence of misconceptions about the material that the teacher will teach to students. In addition, it can be used as a forum or facility for sharing information so that the audience who initially did not receive this knowledge will receive new information (Young 2018).

5 CONCLUSION AND RECOMMENDATION

This study aimed to describe the effect of workshop activities on teacher skills in implementing CAR. Research shows that the skills of teachers in carrying out classroom action assessments are still lacking. However, after the workshop, the teachers' classroom action assessment skills improved understanding, planning, implementation, and teachers' follow-up. The teacher's ability and skill in implementing CAR were improved. It was shown from the score improvement before and after participating in the program, which is inseparable from the monitoring activities at the workshop. The pedagogical implications of teachers are also seen in the use of information and communication technology for the benefit of learning activities and facilitating the development of students' potential to actualize their various potentials. For further research, it is hoped that there will be more workshops for teachers with other topics to improve teacher professionalism.

REFERENCES

Barnawi, Junaedi, & Rido, R. (2019). Improve Teachers' Ability in Compiling Classroom Action Research Through Workshop Activities. *Action Research Journal Indonesia (ARJI)*, *1*(2), 1–12.

Chang, S. H., Ward, P., & Goodway, J. D. (2020). The effect of a content knowledge teacher professional workshop on enacted pedagogical content knowledge and student learning in a throwing unit. *Physical Education and Sport Pedagogy*, *0*(0), 1–16.

Chen, H., Li, M., Ni, X., Zheng, Q., & Li, L. (2021). Studies in Educational Evaluation Teacher Effectiveness and Teacher Growth from Student Ratings: An Action Research of School-based Teacher Evaluation. *Studies in Educational Evaluation*, *70*(4), 1–16.

Colognesi, S., Van Nieuwenhoven, C., & Beausaert, S. (2020). Supporting Newly-qualified Teachers' Professional Development and Perseverance in Secondary Education: On The Role of Informal Learning. *European Journal of Teacher Education*, *43*(2), 258–276.

Declaro-ruedas, M. Y. A., & Ruedas, E. G. (2020). Public School Teachers' Attitude towards Action Research in Magsaysay, Occidental Mindoro. *Asian Journal of Education and Social Studies*, *7*(1), 11–16.

Farmer, S. (2021). Developing a Teacher Professional Learning Workshop on the General Theory of Relativity. *Journal of Physics: Conference Series*, *1929*(1), 1-9.

Fitriah, L., & Ita. (2021). Workshop Penelitian Tindakan Kelas (PTK) bagi Guru Madrasah Aliyah SMIP 1946 Banjarmasin. Bubungan Tinggi: *Jurnal Pengabdian Masyarakat*, *3*(3), 300–310.

Joung, Y., & Anderman, E. M. (2020). Profiles of Perfectionism and Their Relations to Educational Outcomes in College Students: The Moderating Role of Achievement Goals. *Learning and Individual Differences*, *77*(4), 22-28.

Julia, J., Supriyadi, T., & Iswara, P. D. (2020). Training the Non-Specialist Music Teacher?: Insights from an Indonesian Action Research Study. *Universal Journal of Educational Research*, *8*(2), 547–558.

Karyamulya, K. (2020). Improve Teacher Ability in Curriculum Development through Workshop Activities. *Action Research Journal Indonesia (ARJI)*, *2*(3), 138–146.

Kong, Y., Seo, Y.S & Zhai, L. (2021). ICT and Digital Reading Achievement: A Cross-national Comparison using PISA 2018 Data. *International Journal of Education Research*, *111*(1), 16-23.

Laksana, D. N. L., Lawe, Y. U., & Ita, E. (2021). Pendampingan Guru TK dan SD Dalam Melaksanakan Penelitian Tindakan Di Masa Pandemi Covid-19. *Jurnal Abdimas Ilmiah Citra Bakti*, *2*(1), 24–32.

Laudonia, I., Mamlok-naaman, R., Abels, S., & Eilks, I. (2017). Action Research in Science Education – An Analytical Review of The Literature. *Educational Action Research*, *0792*(3), 1–16.

Lestiyanawati, R. (2020). The Strategies and Problems Faced by Indonesian Teachers in Conducting e-learning during COVID-19 Outbreak. *CLLiENT (Culture, Literature, Linguistics, and English Teaching)*, *2*(1), 71–82.

Mahardika, A. I., Arifuddin, M., M, A. S., & Sari, D. P. (2019). Study on Effectiveness of Classroom Action Research Workshop to Improve Teacher Research Understanding in Balangan District, South Kalimantan, Indonesia. *International Journal of Innovative Science and Research Technology*, *4*(9), 440–442.

Meesuk, P., Sramoon, B., & Wongrugsa, A. (2020). Classroom Action Research-based Instruction: The Sustainable Teacher Professional Development Strategy. *Journal of Teacher Education for Sustainability*, *22*(1), 98–110.

Mills, M., Mockler, N., Stacey, M., & Taylor, B. (2021). Teachers' Orientations to Educational Research and Data in England and Australia: Implications for Teacher Professionalism. *Teaching Education*, *32*(1), 77–98.

Naidoo, L. 2015. Ethnography: An Introduction to Definition and Method. *An Ethnography of Global Landscapes and Coridors Journal, 3(2)*, 1-8.

Niemi, R. (2018). Five Approaches to Pedagogical Action Research. *Educational Action Research*, *00*(00), 1–16.

Noermanzah. (2020). Workshop on Classroom Action Research and Preparation of Scientific Articles on Research Results in Rupit District North Musi Rawas Regency. *Journal of Community Service and Research*, *5*(1), 1–9.

Novianti, N., & Nurlaelawati, I. (2019). Pedagogical Competence Development of University Teachers With Non-Education Background: the Case of a Large University of Education in Indonesia. *International Journal of Education*, *11*(2), 169–177.

Peker, B.G., & Erdemir, N. (2021). Does Compensation Strategy Instruction Work? An Action Research?Study. *ELT Journal, 75*(3), 351-361.

Rery, R. U., Copriady, J., & Albeta, S. (2019). Role of Forum on Professional Development (MGMP) in Increasing Chemistry Knowledge for Chemistry Teachers. *International Journal of Science and Applied Technology Role*, *4*(1), 2–6.

Syah, S.(2016). Classroom Action Research as Professional Development of Teachers in Indonesia. *Junral Tarbawi, 13*(1), 1–16.

Setiawati, E., & Sunarti. (2021). Training Class Action Research, School Action Research, and Writing of Scientific Articles for Teachers and Principals of Basic School in Banyuurip Purworejo. *Journal of Physics: Conference Series*, *1823*(1), 1–8.

Sinurat, L., Rahman, A., Mantondang, Z., & Sari, F. (2021). Inquiries-Based Learning Workshop One of Teacher Development to Improve the Quality of Learning in Santa Lusia Sei Rattan Private Junior High School. *Advances in Social Science, Education and Humanities Research*, *591*, 126–131.

Skyhar, C. (2021). Teacher-Directed Collaborative Action Research as a Mediating Tool for Professional Learning in Rural Contexts. *Australian and International Journal of Rural Education*, *31*(1), 12–29.

Sulaiman, A. A. (2020). Profil Kemampuan dan Kreativitas Guru IPA dalam Menyusun Karya Tulis Ilmiah Pada Pembelajaran Kurikulum 2013. *Journal of Biology Education* *3*(1), 1–7.

Susilowati, D. (2018). Penelitian Tindakan Kelas (PTK) Solusi Alternatif Problematika Pembelajaran. *Edunomika, 2(1),* 36–46.

Young, M. (2018). Indonesia's Active, Crwative, Effective and Joyful Learning: From a University Teacher Training Program to High School Classrooms. *Indonesia Research Journal in Education (IRJE), 2*(1), 7–31.

In-service English teachers' reflection in an online professional development program in Indonesia

Mei Lusiana Sitorus, Ilza Mayuni & Ifan Iskandar
Universitas Negeri Jakarta, Jakarta, Indonesia

Yinghuei Chen
Asia University, Taichung, Taiwan

ABSTRACT: This mixed-method case study aimed to explore the implementation of the online reflective journal and the levels of reflection in a teacher professional development program. The participants were 31 in-service Junior High School English teachers from 16 provinces in Indonesia. The data were collected through observations and the completion of reflective journals by the teachers. The data analyses include content and frequency analyses. The results showed five identified types of reflections, namely reflection-on-practice, reflection-for-action, non-reflection, reflection-in-action, and reflection-with-action. Additionally, reflection levels of 1, 2, and 3 were identified in the 248 written responses measured using the Framework of Four Levels of Reflection for Teacher Education by Harland and Wondra. A slight increase was identified in the average levels of reflection within the two-cycles program. However, further analysis showed no significant correlation. Recommendations include the provisions of feedback, collaborative reflection and explicit instructions for reflective practice.

Keywords: Levels of reflection, in-service teachers, reflective practice, teacher development

1 INTRODUCTION

Recent research shows that the inclusion of reflective practice in teacher education impacts learning positively. It is widely acknowledged that a well-designed reflective practice results in improved teaching practice and students' progress. Darling-Hammond, Hyler, & Gardner (2017) stated that incorporating reflective practice is one of the critical components in teacher continuous professional development as it provides teachers with opportunities to digest and discuss the newly obtained knowledge. Furthermore, for reflective practice to successfully contribute to improved teaching practice, it should allow different types of reflections and a certain level of reflections to be generated within the professional development program (Ghaye 2010; Larrivee 2008; Schon & DeSanctis 1986).

To generate specific types and levels of reflection, reflective thinking needs to be intentionally exercised by incorporating reflective practice tools such as journaling. Studies on reflective journaling have shown that students across disciplines generally perceive the use of reflective journal positively and the implementation of reflective journaling as learning activity have improved their grades and metacognitive ability (Ahmed 2019; Cengiz & Karataş 2015; Hussein 2018; Kheirzadeh & Sistani 2018). Additionally, it is essential for the reflective journal to be introduced, structured and given feedback to yield effective results. (Cheng 2017; Veine et al. 2019, Weber-Mayrer et al. 2018). Those studies emphasized the importance of systematic reflective journaling in order to yield effective results. In other words, an unstructured reflective journaling will result in superficial types and levels of reflection.

 DOI 10.1201/9781003347798-27

Despite its practical benefits, recent research in teacher education tends to focus more on pre-service teachers than in-service teachers. In the Indonesian context, a qualitative study on Indonesian in-service teachers' perception showed that teachers' journals improved their professional skills (Sari et al. 2021). Meanwhile, in the context of the continuous professional development (CPD) program provided by the Indonesian government, there is a lack of systematic reflective practice (Cirocki & Widodo 2019). The teacher development program in this study was both continuous and involved a structured reflective practice as a key component of this government-initiated program. Three research questions were proposed to guide this study:

1) How was the online reflective journal in the program implemented?
2) What types of reflections were generated by the reflective journal?
3) What were the levels of reflection shown by teachers through the reflective journal?

2 LITERATURE REVIEW

The challenge in implementing reflective thinking in learning and professional development is the lack of awareness that thinking reflectively is a skill that needs to be taught (Atkinson & Lim 2013) and prompted (Brownhill 2021). Therefore, reflective thinking needs to be intentionally exercised in any development programs. The incorporation of reflective practice may utilize reflection tools such as journaling, peer discussions, videotaping and coaching (Ruffinelli et al. 2021) . The activities designed to purposefully implement the reflection is referred to as reflective practice.

In deciding the type of reflective practice and the tools used, there should be an awareness of the different types of reflection to be generated within the professional development. There are four kinds of reflections that are widely recognised (Ghaye 2010; Schon & DeSanctis 1986):

1) *Reflection-in-action*: This refers to the process of reflection within a specific situation, such as a workplace or classroom as well as during the action or on-the-spot assessment of the situation.
2) *Reflection-on-practice* refers to the process of reflecting on past experience. In this case, the focuses are on the timing and the specific practice.
3) *Reflection-for-action* is also called anticipatory reflection. The key difference between this type of reflection compared to the first two is that it results in improvement plans.
4) *Reflection-with-action* refers to reflecting on one's ability and skills while making certain decisions and then executing that decision. It can also refer to reflecting *with* another person or a group to gain better results in achieving a certain task.

Exposure to all four types of reflection and the habit of critical reflection needs to be reinforced to generate improvement for the individuals within the professional development program.

In Indonesian context previous studies have identified the implementation and benefits of reflective practice including reflective journal (Cirocki & Widodo 2019; Sari et al. 2021) These studies have investigated qualitatively the levels of reflection in both continuous or experimental teacher professional development. On the other hand, the need for more systematic and guided reflective practice is still called for (Lubis 2018). This is consistent with previous studies which highlight collaboration, as an important component to deepen reflection (Akiba et al. 2019; Darling-Hammond et al. 2017; Lubis 2018; Ong et al. 2021). This means an effective reflective practice requires methodical approach as well as external influence to generate progress in teaching practice.

Additionally, efforts to measure levels of reflection quantitatively have been made. Reflective practice is generally perceived to be developing in stages though not always linear (Larrivee 2008). With its complex nature, experts have developed and improved mechanisms to classify and code reflective thinking practice. Harland & Wondra (2011) developed the Framework of Four Levels of Reflection for Teacher Education to assess the written reflections in the forms of final papers and blogs of pre-service teachers. The Four Levels from the lowest to the highest level are referred to as non-reflective level, understanding level, reflection level and critical reflection, respectively. Due to its simplistic but comprehensive descriptors, this framework can serve as a scoring rubric to measure the levels of reflection shown within written responses. Hence, this study utilized Harland & Wondra's (2011) framework in converting the collected written reflection into numerical data.

3 METHOD

The professional development program in this study referred to an online literacy coaching program for 31 in-service junior high schools (SMP) English teachers from 16 different province in Indonesia. The program was conducted online using a synchronous meeting platform from August to November 2020 and was divided into two training cycles. The data was collected from the reflective journals completed by the in-service teachers at the end of each training session in both cycles.

This study employed a mixed-method case study design. Mixed-method case studies seek to comprehend the components that make up a case and how they work utilizing interpretation of quantitative and qualitative data (Fraenkel et al. 2012). Thus, it is suitable for the context of the study in which the reflective practice as the *case* is examined in depth. To answer the first research question, the general implementation of the online reflective journal was described based on the description of the program guide, the results of observation and the content of the online reflective journal. Discrepancies between the proposed program and its implementation were identified.

Meanwhile, to answer the second research question, a content analysis was conducted on the prompt questions to see the kinds of reflection they generated, namely *reflection-in-action, reflection-on-practice, reflection-for-action* and *reflection-with-action* (Ghaye 2010; Schon & DeSanctis 1986). Lastly, the researcher used the Framework of Four Levels of Reflection for Teacher Education by Harland & Wondra (2011) to rate the overall responses made by each teacher participating in the program. This framework allows conversion of the responses serving as qualitative data into numerical data. The results were further analysed with frequency analysis in two facets, as an overall and a cyclical result. Additionally, a correlation analysis was conducted to see if there was any correlation between the results of the two cycles.

4 RESULT AND DISCUSSION

This section presents the answers to the three research questions with the identification of prompt questions, the types of reflections and the levels of reflection shown in online reflective journals. Results of the first and third research questions are broken down per cycle following the format of the professional development program while the second research question is presented as an overall result of both cycles.

4.1 *The implementation of the online reflective journal*

The online literacy coaching program consisted of two compulsory cycles in which the teachers were asked to complete a reflective journal. The reflective journal consisted of a list of open-ended questions related to the session the teachers had attended. It was filled out and collected through online forms. In cycle 1, the prompt questions were labelled as six different categories, namely description, feeling, evaluation, analysis, text and conclusion. This set of questions were used for seven corresponding sessions in cycle one. However, the numbers of responses varied each meeting due to teachers' absence. There was no specific time restriction for the completion period but the teachers were expected to complete the reflective journal before attending the following session.

In cycle 2, there were five sessions requiring the completion of a reflective journal. There were eighteen different questions included during cycle 2. However, not all of them were required for every session. There were two sets of questions used. The first set was used for the first four meetings of cycle 2; meanwhile, the other set was only used once in the final session. There was also a discrepancy between the number and the categorisation of questions within the two sets of questions. Inherently, the sessions in both cycle 1 and 2 consisted of somewhat similar activities with some modification presented. The three sets of questions had gauged different types of reflections.

The implementation of reflective journals in this study is consistent with previous studies, which showed that the use of reflective journaling has fostered reflective thinking in teacher training (Darling-Hammond et al. 2017; Sari et al. 2021). However, in contrast with previous study, the reflective journal in this study involved three different sets of compulsory prompt questions instead of optional guided questions to gauge students' individual reflection. Finally, despite emphasizing the importance of repetition to foster habit and intensify the reflection like previous studies (Atkinson & Lim 2013; Veine et al. 2019), the use of three sets of questions instead of one and the cyclical design of the teaching training made the utilization of reflective journals in this study distinctive from those of previous studies.

4.2 *Types of reflections generated by the reflective journal*

Of the three sets of prompt questions listed above, some questions were recycled. When those were reduced, 17 out of the 29 questions were inherently different from each other. Based on the kinds of reflection they generated, the prompt questions used can be divided into four types of reflections, namely *reflection-in-action, reflection-on-practice, reflection-for-action,* and *reflection-with-action* (Ghaye 2010; Schon & DeSanctis 1986). However, there were two questions identified as non-reflective prompts. The percentages of the questions based on the kinds of reflections they generated are illustrated in Figure 1 below.

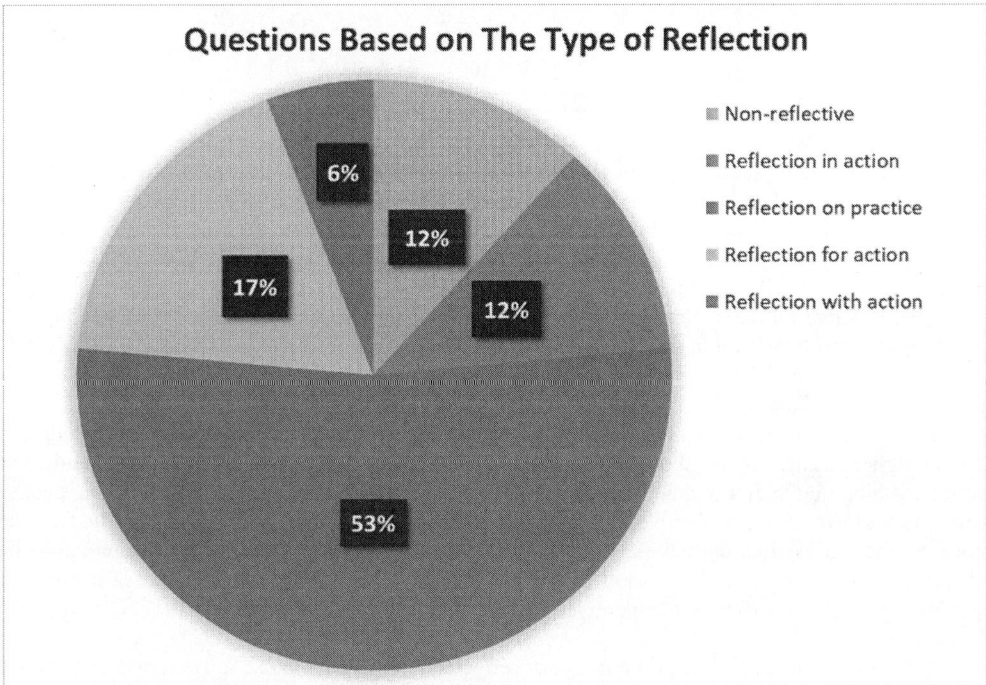

Figure 1. The type of questions based on the type of reflections generated.

The result showed that the majority of the questions (53%) were aimed at the reflection on practice followed by reflection for action (17%), reflection in action and non-reflective questions (12%, respectively) and reflection with action (6%). Additionally, there were two questions in the reflective journal that could be considered as a part of need analysis instead of reflection. However, this study did not aim to identify which type of reflection was better attained or supported the achievement of the program outcomes. Nevertheless, it is generally believed that all type of reflection is necessary since the main goal of reflection practice in teacher education is to form a deeper understanding which allows for critical view on repetitive experiences of a specialised practice and make new of sense of the experiences (Brookfield 2017; Larrivee 2008; Schon & DeSanctis 1986). Therefore, as long as the teachers achieved a deeper understanding of their different types of experience and used it to improve their practice, it could be regarded as a successful reflection.

4.3 *The levels of reflection shown in the reflective journals*

During the coaching program, 248 submissions of reflective journals by the teachers had been collected by the program developer. There were 146 responses made in Cycle 1 and 103 responses in Cycle 2. This data then analysed using the descriptors provided in the Framework of Four Levels of Reflection for Teacher Education (Harland & Wondra 2011). The conversion resulted in the leveling for each of the 248 responses. Once it was scored, the frequency of the responses was analysed using descriptive analysis in two facets: as a whole and per cycle. The summary of the data is depicted in the table below.

Table 1. Overall levels of reflection.

		Frequency	Percent	Valid Percent	Cumulative Percent
Level	1.00	190	76.6	76.6	76.6
	2.00	56	22.6	22.6	99.2
	3.00	2	.8	.8	100.0
	Total	248	100.0	100.0	

The result of the frequency statistical analysis showed most of the responses made was of level 1 reflection, which is also known as non-reflective level as teachers merely stating what they learned or what happened factually (Harland & Wondra 2011). The reflective journals computed were of 248 responses with 190 (76.7%) of them were of Level 1 reflection, 56 (22.6%) were of Level 2 reflection, and 2 (0.8%) were of Level 3 reflection. Further dissecting the results into each cycle, Cycle 1 reflective journals showed 85.5% and 14.5% of the teachers in the program produced a level 1 and level 2 reflection, respectively. Meanwhile, Cycle 2 consisted of 64.1% of Level 1, 14.1% of Level 2 and 0.8% of Level 3. It should be highlighted that level 3 Reflection produced only in Cycle 2 of the program.

A similar case could be found in Harland & Wondra's (2011) studies on reflective blogging, but with a higher dominating level of reflection at level 2. Both the reflective journal in this study and reflective blogging in Harland & Wondra's (2011) involved successive practice to generate deeper reflection. However, the reflective blogging did not include prompt questions, but it did include peer interaction as students were encouraged to comment on each other's blogs. This could be an influencing factor as collaboration in reflection contributes to deeper reflection as it involves getting feedback and different perspectives (Atkinson & Lim 2013; Brookfield 2017; Cheng 2017; Veine et al. 2019).

Meanwhile, a closer look into the levels of reflection in cycles 1 and 2 showed that there was a slight increase in the average score. To investigate this further, a descriptive statistical analysis was conducted. The results showed that the average level of reflection in cycle 1 with total responses of 145 was 1.15; meanwhile, in cycle 2 with N=103 was 1.38. In other words, there was a 0.23 gap.

However, to see if this gap was a result of a possible positive correlation, a correlation analysis was conducted (Fraenkel et al. 2012). Due to the irregularities in attendance and to maintain data validity (Fraenkel et al. 2012), the data analysed was limited to only those who had submitted more than half responses for each cycle. This resulted in 19 participants who submitted more than 50% reflective journals for each cycle. The result showed that there was no significant correlation

Table 2. Descriptive statistics of the levels of reflection.

	N	Min	Max	Mean	Std. Deviation
Levels of Reflection Cycle 1	145	1.00	2.00	1.1448	.35315
Levels of Reflection Cycle 2	103	1.00	3.00	1.3786	.52611
Both Cycles Overall Levels of Reflection	248	1.00	3.00	1.2419	.44759

between the level of reflection in cycle 1 and 2 with sig. (2-tailed) 0.201>0.05. On one hand, this result can serve as evidence of the theoretical claim of reflective practice perceived to be developing in stages though not always linear (Larrivee 2008). This means that it is possible for teachers' reflection to experience a slight decline. On the other hand, the use of compulsory prompt questions instead of optional guided questions as well as the lack of collaboration might have limited the scope and depth of the responses. This supposition allows room for future studies.

5 CONCLUSION AND RECOMMENDATION

The purpose of this study is to explore the incorporation of reflective journaling to promote teacher education. Two out of three sets of prompt questions used in the reflective journal were used repetitively to gauge deeper reflection. The prompt questions were also designed to generate four different types of reflection: reflection-on-practice, reflection-for-action, reflection-in-action and reflection-with-action. However, some questions have been identified as non-reflective questions. Furthermore, the study showed that Level 1 reflection was the most frequent response made. To improve the implementation of the reflective journal the incorporation of feedback and collaboration within the reflective journal and explicit teaching of the implementation of reflective thinking are recommended. Feedback and collaboration in reflective practice allow for outside perspectives to enrich the teachers' current knowledge as well as deepening it. Meanwhile, the teaching of reflective thinking is necessary to result in meaningful reflection which can be done through the introduction of the reflective journal, the type of reflection expected, and the use of rubrics as a guidance to generate a deeper level of reflection. Further research on reflective practice involving the recommendation made in this study will give valuable inputs for improved teacher education programs.

ACKNOWLEDGEMENT

The authors would like to thank the EFL teachers participating in the study as well as the online literacy coaching program organizers at the Universitas Negeri Jakarta in collaboration with the Asia University (Taiwan).

REFERENCES

Ahmed, A. M. (2019). Students' reflective journaling: An impactful strategy that informs instructional practices in an EFL writing university context in Qatar. *Reflective Practice*, *20*(4), 483–500.

Akiba, M., Murata, A., Howard, C. C., & Wilkinson, B. (2019). Lesson study design features for supporting collaborative teacher learning. *Teaching and Teacher Education*, 77.

Atkinson, D., & Lim, S. L. (2013). Improving assessment processes in higher education: Student and teacher perceptions of the effectiveness of a rubric embedded in a LMS. *Australasian Journal of Educational Technology*, *29*(5), 651–666.

Brookfield, S. D. (2017). *Becoming a critically reflective teacher (2nd ed.)*. Jossey-Bass.

Brownhill, S. (2021). Asking key questions of self-reflection. *Reflective Practice*, *23*(1), 57–67.

Cengiz, C., & Karataş, F. Ö. (2015). Examining the effects of reflective journals on pre-service science teacher' general chemistry laboratory achievement*Australian Journal of Teacher Education*, *40*(10).
Cheng, G. (2017). The impact of online automated feedback on students' reflective journal writing in an EFL course. *Internet and Higher Education*, *34*.
Cirocki, A., & Widodo, H. P. (2019). Reflective practice in English language teaching in Indonesia: Shared practices from two teacher educators. *Iranian Journal of Language Teaching Research*, *7*(3), 15–35.
Darling-Hammond, L., Hyler, M. E., & Gardner, M. (2017). *Effective teacher professional development*. Learning Policy Institute.
Fraenkel, J., Wallen, N., & Hyun, H. (2012). *How to design and evaluate research in education* (8th ed.). McGraw-Hill.
Ghaye, T. (2010). *Teaching and learning through reflective practice: A practical guide for positive action* (2nd ed.). Routledge.
Harland, D. J., & Wondra, J. D. (2011). Preservice teachers' reflection on clinical experiences: A comparison of blog and final paper assignments. *Journal of Digital Learning in Teacher Education*, *27*(4), 128–133.
Hussein, H. (2018). Examining the effects of reflective journals on students' growth mindset: A case study of tertiary level EFL students in the United Arab Emirates. *IAFOR Journal of Education*, *6*(2).
Kheirzadeh, S., & Sistani, N. (2018). The effect of reflective teaching on Iranian EFL students achievement: The case of teaching experience and level of education. *Australian Journal of Teacher Education*, *43*(2), 143–156.
Larrivee, B. (2008). Development of a tool to assess teachers' level of reflective practice. *Reflective Practice*, *9*(3), 341–360.
Lubis, A. H. (2018). Reflective teaching toward EFL teachers' professional autonomy: Revisiting its development in Indonesia. *International Journal of Education*, *11*(1), 35.
Ong, W. A., Swanto, S., Alsaqqaf, A., & Ong, J. W. (2021). Promoting reflective practice via the use of 5-step copora reflective model: A case study of east Malaysian ESL pre-service teachers. *Teflin Journal*, *32*(1), 72–96.
Ruffinelli, A., Álvarez Valdés, C., & Salas Aguayo, M. (2021). Strategies to promote generative reflection in practicum tutorials in teacher training: The representations of tutors and practicum students. *Reflective Practice, 23*(1), 30–43.
Sari, Y. R., Drajati, N. A., So, H.-J., & Sumardi, S. (2021). Enhancing EFL teachers' Technological Pedagogical Content Knowledge (TPACK) competence through reflective practice. TEFLIN Journal - A Publication on *the Teaching and Learning of English*, *32*(1), 117.
Schon, D. A., & DeSanctis, V. (1986). The reflective practitioner: How professionals think in action. *The Journal of Continuing Higher Education*, *34*(3).
Veine, S., Anderson, M. K., Andersen, N. H., Espenes, T. C., Søyland, T. B., Wallin, P., & Reams, J. (2019). Reflection as a core student learning activity in higher education - Insights from nearly two decades of academic development. *International Journal for Academic Development*, *25*(2), 147–161.
Weber-Mayrer, M. M., Piasta, S. B., Ottley, J. R., Justice, L. M., & O'Connell, A. A. (2018). Early childhood literacy coaching: An examination of coaching intensity and changes in educators' literacy knowledge and practice. *Teaching and Teacher Education*, *76*, 14–24.

Teacher Education and Teacher Professional Development in the COVID-19 Turn – Drajati & Putra (Eds)
© 2023 The Author(s), ISBN: 978-1-032-38970-7

Teacher outcomes from a training program to build Hots-based language assessment literacy: EFL teachers' voices

Sumardi Sumardi
Universitas Sebelas Maret, Surakarta, Indonesia

Junjun Muhamad Ramdani
The University of New South Wales, Kensington, Australia

ABSTRACT: The present study aims to explore EFL teachers' outcomes from a training program to build HOTS-based language assessment literacy with reference to EFL teachers' voices. This study is a qualitative study using ethnographic action research design. Forty-three EFL teachers were voluntary to be the participants of the study. The data garnered from questionnaires and interviews revealed that EFL teachers viewed training programs had helped them build their familiarity with and awareness of HOTS-based language assessment and how to enact it in EFL classrooms. They believed that HOTS-based language assessment has increased learners' potentiality to foster their critical and logical thinking skills. However, their comprehension of HOTS-based language assessment was not necessarily in line with their skills to develop this type of assessment in practice. The implication is that a training program needs to be built through effective communication between trainers and trainees as well as through mentoring in sufficient time. Besides, it is supposed to be well-designed by taking into account the needs of teachers to promote their pedagogical competence.

Keywords: EFL teachers; HOTS-based language assessment literacy; need analysis; teacher outcomes, training program

1 INTRODUCTION

Assessment is an inseparable part of the overall instructional practices, including in English as a foreign language (EFL) classrooms. Through an assessment, teachers have an opportunity to ameliorate their language instructional practices (Giraldo 2018) and at the same time language learners are likely to be motivated to learn better (Mellati & Khademi 2018). However, EFL teachers are generally stuck with traditional assessments in the form of multiple-choice, matching-type, true-false, and short answer tests to measure language learners' learning outcomes. These formats of assessment measure mostly the ability of EFL learners to simply think and comprehend the contents being tested and also measure other low-level cognitive abilities. It means that many EFL teachers have not yet possessed a good HOTS-based language assessment literacy. Although many teachers elucidate that they have known and understood the concept of HOTS-based assessment (Retnowati et al 2018), but their knowledge did not delineate their actual assessment practices (Sisman & Buyukkarci 2019; Tsagari & Vogt 2017; Vogt & Tsagari 2014). To build teachers' language assessment literacy, they need to engage in a contemporary professional development program (Cirocki & Farrell 2019; Fan et al 2020; Gotch & McLean 2019; Haug & Mork 2021) which ought to be designed systematically in response to their real-life needs (Tsagari 2017; Vogt & Tsagari 2014). That is why need analysis must be firstly undertaken before the training program is commenced (Firoozi et al. 2019; Vogt & Tsagari 2014). Besides, teachers' active engagement in a program can be a critical part of embodied actions and it is actually what is meant by teachers'

DOI 10.1201/9781003347798-28

voices. Teacher's voice is a process to empower teachers through active engagement with those who have the power to express their views, intent, beliefs, motivation, and motives related to their training program experiences as trainer co-collaborator (Widodo 2015). Based on this definition, the trainer and the trainee can jointly prescribe the contents of the training so that it fits the teachers' needs.

Research on teachers' language assessment literacy has been the consents of many practitioners (Shams & Iqbal 2019; Singh & Shaari 2019; Yoke et al. 2015) and it was concluded that there was still a need for a training program that ought to be carried out constantly for EFL teachers so that they could always update their knowledge and skills in terms of assessment in EFL classrooms. Research by Jannati (2015) reports that teaching experience does not mean that EFL teachers make some significant differences in their assessment practices. These findings reveal that the assessment literacy of EFL teachers still needs to be improved extensively (Tsagari 2017; Vogt & Tsagari 2014). In addition, as a result of the educational system in the world that is constantly changing, EFL teachers are expected to make a radical shift from simply relying on traditional assessments to alternative assessments (Stiggins 2006) and HOTS-based assessment is one of the alternative assessments which is currently becoming mainstream in various educational landscapes. Thus, a training program becomes a starting point for improving teachers' HOTS-based language assessment literacy. However, many training programs do not significantly enhance EFL teachers' assessment literacy due to the mismatch of the training contents and the needs of teachers. Many studies recommend that training programs should be implemented based on a need analysis (Hasselgreen et al. 2004; Vogt & Tsagari 2014) so that the contents of the training delineate the EFL teachers' real-life assessment activities (Fulcher 2012).

Although research on language assessment literacy has been well-documented in various contexts and geographies, little empirical study has looked at how HOTS-based language assessment literacy is built in the totality of training for EFL teachers. Research on this issue is important as this assessment approach has been a relatively new mainstream due to the requisitions of the existing curriculum. In response to the importance of HOTS-based assessment in EFL classrooms, this study aims to explore EFL teacher outcomes from a training program to build HOTS-based language assessment literacy. This study might provide new insight to policymakers or training program providers by designing a training program that is effective to build HOTS-based language assessment literacy for EFL teachers. To this end, the pursued research questions are formulated as follows:

1) What are EFL teachers' voices of a training program on HOTS-based language assessment?
2) To what extent are EFL teachers' skills to develop HOTS-based language assessment subsequent to their participation in a training program?

2 LITERATURE REVIEW

Assessment is an important component of instruction as teachers might improve the quality of their instructions and learners might be encouraged to learn better (Kerry 2002). In other words, assessment has a pivotal role in the whole instructional practice. A well-designed assessment might provide valuable information on the learners' ways of learning and their learning outcomes. The results of the assessment might ultimately provide information to stakeholders (teachers, schools, parents) on what learners have learned, how well they have learned, and in what areas learners still have to learn. In addition, the assessment aims to collect relevant information on the performance of learners and to identify the extent to which the effectiveness of the instructional methods used by teachers to facilitate learners to learn better.

Given the importance of assessment in language instructions, the consequences are that language teachers must always enhance their knowledge and skills regarding how to design and carry out an assessment as well as how to analyze the results of an assessment. It is important as the language instructions could not be separated from the assessment itself. Coombe et al (2012) estimate that about 30 to 50 percent of instructional practices are closely related to assessment. However, studies empirically found that teachers do not yet have adequate skills to carry out assessment practices (Fulcher 2010; Lam 2014; Retnowati et al. 2018) and it means that language assessment literacy by many teachers is still relatively inadequate. Teachers' language assessment literacy has been a pivotal issue in the area of language testing and assessment (Hasselgreen et al. 2004; Vogt & Tsagari 2014). Language assessment literacy refers to teachers' knowledge, skills, and abilities to design, develop, and evaluate the test and how to administer the test based on the agreed codes of practices, including how to identify the impact of the test on society, institutions, and individuals (Fulcher 2012; Gotch & McLean 2019). Although language assessment literacy is important in language instructions (Coombe et al. 2012; Turner 2012), teachers' language assessment literacy has been solemnly in questioned (Vogt & Tsagari 2014).

On the other hand, many language teachers are merely proficient in designing and developing traditional assessments (Tsagari & Vogt 2017; Vogt & Tsagari 2014). Most language teachers tend to implement lower-order thinking skills (LOTS) rather than HOTS due to a lack of skills regarding HOTS-based assessment (Shafeei et al 2017). Furthermore, Shafeei et al (2017) suggest that language teachers call for incorporating HOTS-based assessment in EFL classrooms to produce learners' critical thinking and prepare them for the increasingly fierce competition in the 21st-century (2017). HOTS refers to mind expansion in which learners have to analyze or interpret the answers or to manipulate the information as the correct answers are not always gathered from daily routines or knowledge (Onosko & Newman 1994; Shafeei et al. 2017). Furthermore, Brookhart's definition of HOTS falls into three categories, transferring information, critical thinking, and problem-solving (2010). It means that HOTS requires learners to be information transferors, critical thinkers, and problem solvers and HOTS-based assessment could be directed to achieve this end.

HOTS-based assessment and language assessment literacy are two crucial issues in this study and those two terminologies are incorporated to form a new perspective, namely HOTS-based language assessment literacy. For the present study, HOTS-based language assessment literacy is defined as EFL teachers' ability to design, develop, and evaluate the language test and how to administer the language test, as well as the ability to score the learners' performance based on critical thinking, problem-solving and transferring information. Empirically, the literacy level of the teacher's HOTS-based language assessment still needs to be built extensively (Hasselgreen et al. 2004; Tsagari & Vogt 2017; Vogt & Tsagari 2014; Shafeei et al. 2017) and to build HOTS-based language assessment literacy, EFL teachers call for a training program (Aziz 2015; Gotch & McLean 2019; Fan et al. 2020). However, the extent to which a training program might have an empirical impact on enhancing EFL teachers' HOTS-based language assessment still needs to be investigated to ensure that the training program is effective in achieving the prescribed goals.

3 METHOD

Forty-three EFL teachers were voluntary to be the participants of the current research. All were EFL teachers who had participated in a one month training program organized by a training provider that was merely intended to build HOTS-based language assessment literacy. Although they were selected purposely, the EFL teachers had to meet two crucial conditions. First, they had already participated in a training program on HOTS-based language assessment and how to apply it in EFL classrooms. Second, the participants had to serve as EFL teachers in secondary schools in which English is not students' first language. The participants had a variety of English teaching experiences ranging from five to thirty years. Similar criteria were also applied in the way of recruiting informants for the interviews.

This current study was qualitative in nature and was epistemologically carried out on the basis of interpretive and emancipatory traditions. Grounded in a tradition of interpretive, any data garnered from research participants were interpreted to multi-layered interpretations (Widodo 2015). While in emancipatory tradition, EFL teachers as the research participants are viewed as co-collaborator of the researcher and they were actively embroiled in both research and processes of pedagogical interventions (Widodo 2016). This paradigm fits greatly with the current training program to build HOTS-based language assessment literacy in which the trainer and EFL teachers were equally recognized as both parties played an active role in assigning the training contents through need analysis. In the current study, ethnography and action research were brought together to particularize naturalistic phenomena that took place in a training landscape. Before the training program was commenced, the research participants were informed that the existing training program was part of an investigative enterprise and a pedagogical innovation. Empirical data were collected through questionnaires and interviews. The qualitative data were then coded using thematic analysis and subject to interpretive analysis. All the selected data were coded using Braun and Clarke's (2006) thematic analysis. The thematic analysis was carried out through six stages, i.e. familiarizing with the data; creating initial codes; constructing themes; examining the generated themes; defining the themes; and writing a report. No absolute interpretations of the qualitative data as data themselves were open, dynamic, and fluid (Wodak 1999). Furthermore, the data garnered from the questionnaire were presented in percentages delineating a number of EFL teachers who were in need of particular content as part of the training contents as a whole.

4 FINDING AND DISCUSSION

I presented two pivotal findings from my analysis: (1) EFL teachers' voices of a training program on HOTS-based language assessment: the nature of HOTS assessment, stages to develop HOTS language assessment, table of specification of HOTS language assessment, indicators of HOTS items, and learners' cognitive levels; and (2) EFL teachers' skills to develop HOTS-based assessment. These themes shed light on the HOTS-based language assessment literacy of EFL teachers who are involved in a training program. Each finding addresses how EFL teachers are supposed to plan, enact, and evaluate their assessment to measure learners' critical thinking in the EFL classroom. The knowledge and skills of teachers in planning, enacting, and evaluating their HOTS-based assessments in EFL classrooms represented the level of HOTS-based language assessment literacy they possessed.

4.1 *EFL teachers' voices of a training program on HOTS-based language assessment*

Before the training commenced, both the trainer and EFL teachers negotiated what training contents or materials ought to be addressed, but those contents were not supplied in previous training programs. In doing so, the trainer who also acted as a researcher firstly distributed a questionnaire to which EFL teachers had to respond online and it was intended to identify what contents EFL teachers need to enhance their HOTS-based language assessment literacy. The results showed that most EFL teachers were in need of training contents in five areas, i.e. the nature of HOTS language assessment, stages to develop HOTS language assessment, table of specification of HOTS assessment, indicators of HOTS items, and learners' cognitive levels. An overview of the training contents required by EFL teachers regarding the assessment of HOTS is presented in Table 1 below.

Grounded in the data, it was known that the majority of EFL teachers were in need of these contents as part of the training contents as a whole. They stated that these contents were important as these led them to understand the procedures of HOTS-based language assessment in-depth and they could eventually develop this type of assessment well. Identifying training contents of EFL teachers in the different areas are urgent (Hasselgreen et al. 2004; Vogt & Tsagari 2014; Tsagari & Voght 2017), but those studies focused on language assessment in general (e.g. classroom-focused language assessment, purpose of testing, and contents of concepts of language assessment). However, some areas are considered more urgent than others (e.g. some aspects of HOTS language assessment). In the current study, such tangible areas are needed by EFL teachers as those areas have become mainstream in educational assessment in many countries.

EFL teachers' voice in a training program is another important aspect as teachers' voices lead to engagement in a training program because they are entirely aware of their potential and capacity to

Table 1. Overview of training contents required by EFL teachers.

No.	Training contents	Not necessary	Necessary, but basic level of training only	Necessary, even advanced level of training
1	The nature of HOTS language assessment	4.5	38.6	56.8
2	Stages to develop HOTS language assessment	4.5	36.4	59.1
3	Developing table of specifications of HOTS language assessment	2.2	36.4	61.4
4	Developing indicators of HOTS test items	2.2	36.4	61.4
5	Identifying learners' cognitive levels	2.2	38.6	59.1

*Numbers are in percentages.

engage in the training. Teachers' voices are aligned with their rights to get what they really need from a training program. In the training program practices, teachers are supposed to be afforded some opportunities to engage in making some decisions on what training contents and how to determine the objectives to be achieved from the training. For the current study, the majority of EFL teachers explicitly stated that if they were allowed to determine the contents of the training, they would be more actively engaged during the training. They would be more enthusiastic in participating in all training activities. That is why accommodating teachers' voices is crucial because teachers are social agents who better understand their own training needs (Widodo 2016; 2017). Self-determination and a sense of ownership of the training program could be built effectively by accommodating teachers' voices. By doing so, teachers feel in need of learning from a training program and they eventually put their great efforts into their training activities. Therefore, a liberating mode of training can build democratic participation and emancipation (Taylor & Robinson 2009) which at the end of the training is effectively able to achieve the expected goals. Anchored on this idea, teachers as trainees need to be placed as co-collaborators of the trainers and are given a trust to set the objectives and contents of the training jointly with the trainers.

4.2 *EFL teachers' skills to develop HOTS-based language assessment*

The ability of EFL teachers to develop tests, interpret test results, use interpretation to improve the quality of instructions, and evaluate the quality of the test itself is an important indicator of good HOTS-based language assessment literacy. In addition, with good HOTS-based language assessment literacy, they will be able to carry out accurate assessment practices to measure the learners' ability to transfer information, think critically, and cope with any problems that learners may encounter in their real-life (Brookhart 2010; Mansouri, Molana, & Nazari 2021). The improvement of EFL teachers' knowledge and skills to develop HOTS-based assessments is a pivotal indicator of the success of the training program. For the current research, EFL teachers' comprehension of HOTS-based language assessment is generally quite good. Teachers' knowledge related to the nature of HOTS-based language assessments and how to develop it is getting better, but teachers' skills to develop HOTS-based language test items still need to be improved extensively. That is why Coniam (2009) claims that tests the teachers had previously produced had not provided them with usable, accurate information about their students' abilities. Based on the results of the actual practices of developing test items, it could be known that many test items developed by EFL teachers were still limited to measuring learners' knowledge and understanding. In other words, their knowledge of the nature of the HOTS-based language assessment improved better, but their skills to develop HOTS-based language test items were still unsatisfactory. Thus, it can be understood that their HOTS-based language assessment literacy is still not entirely good. Effective communication between trainer and trainee (Fan et al. 2020) and sufficient time are needed to strengthen their skills and experience to develop and implement HOTS-based language assessments so that in the end their HOTS-based language assessment literacy will also enhance extensively.

Given the previous description, it is concluded that a well-designed training program has factually built teachers' knowledge and awareness related to the concept of HOTS-based language assessment. However, further action needs to be taken so that the knowledge can be put into practice in EFL classrooms. Teachers' better knowledge of HOTS-based language assessment is not necessarily able to change their practices of assessment in the EFL classroom context. They still cannot escape the 'shadows' of traditional assessments which only measure students' rot memory and understanding of the contents being tested. It signifies that to change the practices of assessment carried out by teachers, it is still necessary to provide systematic assistance by knowledgeable others in sufficient time.

5 CONCLUSION AND RECOMMENDATION

EFL teachers who engaged intensively in a training program on HOTS-based language assessment go beyond the basic level of knowledge and comprehension. After attending the training, they know and comprehend well what HOTS-based language assessment is; how to develop a table of specifications and indicators of HOTS test items; and how to identify learners' cognitive levels. They can make judgments and demonstrate a high level of insight on what HOTS-based language assessments are supposed to develop and implement in language classrooms. However, in actual language assessment practices, there are still many teachers who have not been able to develop test items that truly delineate the HOTS-based language assessment to measure learners' learning outcomes. In other words, the enhancement of EFL teachers' knowledge and comprehension of HOTS-based language assessment is not necessarily in line with their skills to develop this type of assessment. It shows that the EFL teachers' HOTS-based language assessment literacy is not yet entirely good. However, this research provides much for similar studies and other training program providers to elevate EFL teachers' literacy of HOTS-based language assessment. I agree with Fan, Frost, & Liu (2020) and recommend that training on HOTS-based language assessment needs to be built through more effective communication between trainers and trainees as well as through mentoring in sufficient time. Besides, the training program needs to be well-designed by taking into account the needs of teachers to promote their pedagogical competencies. Lastly, it is recommended that teacher training providers need to accommodate the needs of teachers in a bottom-up rather than top-down manner so that training can effectively enhance the pedagogical competencies of teachers.

REFERENCES

Aziz, M.M.A. (2015). A Training Program for Developing Saudi English Language Prospective Teachers' Successful Intelligence and Higher Order Thinking Reading Skills in the Light of the Successful Intelligence Theory. *Occasional Papers*, 59, 145–196.

Braun, V., & Clarke, V. (2006). Using thematic analysis in psychology. *Qualitative Research in Psychology*, *3*, 77–101.

Brookhart, S.M. (2010). *How to assess higher-order thinking skills in your classroom*. Alexandria, VA: ASCD.

Cirocki, A., & Farrell, T.S.C. (2019). Professional development of secondary school EFL teachers: Voices from Indonesia. *System*, 85, 1–14.

Coniam, D. (2009). Investigating the quality of teacher-produced tests for EFL students and the effects of training in test development principles and practices on improving test quality. *System*, 37(2), 226–242.

Coombe, C., S. Troudi & Al-Hamly, M. (2012). Foreign and Second Language Teacher Assessment Literacy: Issues, Challenges, and Recommendations. In C. Coombe, P. Davidson, B. O'Sullivan & S. Stoynoff (Eds.), *The Cambridge Guide to Second Language Assessment* (pp. 20-29). Cambridge: Cambridge University Press.

Fan, J., Frost, K., & Liu, B. (2020). Teachers' involvement in high-stakes language assessment reforms: The case of Test for English Majors (TEM) in China. *Studies in Educational Evaluation,* 66, 1–10.

Fulcher, G. (2012). Assessment literacy for the language classroom. *Language Assessment Quarterly*, 9(2), 113–132.

Giraldo, F. (2019). Language assessment practices and beliefs: Implications for language assessment literacy. *HOW*, 26(1), 35–61.

Gotch, C.M., & McLean, C. (2019). Teacher outcomes from a statewide initiative to build assessment literacy. *Studies in Educational Evaluation*, 62, 30–36.

Hasselgreen, A., Carlsen, C., & Helness, H. (2004). *European Survey of Language Testing and Assessment Needs*. Retrieved from: www.ealta.eu.org/resources.htm

Haug, B.S. & Mork, S.M. (2021). Taking 21st-century skills from vision to classroom: what teachers highlight as supportive professional development in the light of new demands from educational reforms. *Teaching and Teacher Education*, 100, 1–12.

Jannati, S. (2015). ELT Teachers' Language Assessment Literacy: Perceptions and Practices. Educational Research Association. *The International Journal of Research in Teacher Education*, 6(2), 26–37.

Kerry, T. 2002. *Explaining and Questioning.* Cheltenham, UK: Nelson Thornes.

Mansouri, B., Molana, K., & Nazari, M. (2021). The interconnection between second language teachers' language assessment literacy and professional agency: The mediating role of institutional policies. *System*, 103, 1–11.

Mellati, M., & Khademi, M. (2018). Exploring Teachers' Assessment Literacy: Impact on Learners' Writing Achievements and Implications for Teacher Development. *Australian Journal of Teacher Education*, 43(6), 1–19.

Onosko, J. J. & Newmann, F. M. 1994. Creating More Thoughtful Learning Environments. J. N.Mangieri & C. C. Blocks (Eds.). *Creating Powerful Thinking in Teachers and Leaners: Diverse Perspectives.* Fort Worth: Harcourt Brace College Publishers. 27–49.

Retnowati, H., Djidu, H., Kartianom, Apino, E., & Anazifa, R.D. (2018). Teachers' knowledge about Higher-order thinking skills and their learning strategy. *Problems of Education in the 21st Century,* 77(2), 215–230.

Singh, R.K.V., & Shaari, A.H. (2019). The analysis of Higher-Order Thinking skills in English reading comprehension tests in Malaysia. *Malaysian Journal of Society and Space*, 15(1), 12–26.

Shafeei, K.N., Hassan, H., Ismail, F., & Aziz, A.A. (2017). Incorporating Higher Order Thinking Skill (HOTS) Questions in ESL Classroom Contexts. *LSP International Journal*, 4(1), 101–116.

Shams, J.A., & Iqbal, M.Z. (2019). Development of Classroom Assessment Literacy Training Program for University Teachers in Punjab. *Bulletin of Education and Research*, 41 (2), 41–52.

Sisman, E.P., & Buyukkarci, K. (2019). A Review of Foreign Language Teachers' Assessment Literacy. *Sakarya University Journal Of Education*, 9 (3), 628–650.

Stiggins, R.J. (2001). *Student-Involved Classroom Assessment (3rd edition.).* Upper Saddle River, NJ: Prentice-Hall.

Taylor, C., & Robinson, C. (2009). Student voice: Theorising power and participation. *Pedagogy, Culture & Society*, *17*, 161–175.

Tsagari, D., & Vogt, K. (2017). Assessment literacy of foreign language teachers around Europe: research, challenges and future prospects. *Language Testing and Assessment*, 6 (1), 41–60.

Turner, C. (2012). Classroom assessment. In G. Fulcher & F. Davidson (Eds.), *The Routledge handbook of language testing* (pp. 65–78). London, UK: Routledge.

Vogt, K., & Tsagari, D. (2014). Assessment literacy of foreign language teachers: findings of a European study. *Language Assessment Quarterly,* 11, 37–402.

Widodo, H. P. (2015). Designing and implementing task-based vocational English materials: Text, language, task, and context in Indonesia. In M. Thomas & H. Reinders (Eds.), *Contemporary task-based language teaching in Asia* (pp. 291–312). London: Bloomsbury Press.

Widodo, H. P. (2016). Language policy in practice: Reframing the English language curriculum in the Indonesian secondary education sector. In R. Kirkpatrick (Ed.), *English language education policy in Asia* (pp. 127–151). New York: Springer.

Widodo, H. P. (2017). Framing vocational English materials from a social semiotic perspective: The design and use of accounting English materials. In B. Tomlinson (Ed.), *SLA research and materials development for language learning* (pp. 232–249). New York: Routledge.

Wodak, R. (1999). Critical discourse analysis at the end of the 20th century. *Research on Language and Social Interaction*, *32*, 185–193.

Yoke, S.K., Hasan, N.H., Jangga, R., & Kamal, S.N.M. (2015). Innovating with HOTS for the ESL reading class. *English Language Teaching*. 8(8), 10–17.

Teacher Education and Teacher Professional Development in the COVID-19 Turn – Drajati & Putra (Eds)

Impact of Covid-19: A need analysis of digital literacy for elementary school teachers

Sri Marmuah, Jenny I.S. Poerwanti & Suharno Suharno
Universitas Sebelas Maret, Surakarta, Indonesia

Diana Sinziana Duca
Stefan cel Mare University of Suceava, Suceava, Romania

ABSTRACT: This study aims to analyze the digital literacy needs of elementary school teachers. The research methodology used in this research is qualitative research. Data collection techniques were questionnaires, interviews, observations, and documentation. The respondents are 47 elementary school teachers from ten elementary schools. The data analysis used the (Miles et al. 2014) technique, namely data reduction, data display, and concluding. The results ICT-based activities and materials used for teaching were in the Rare Category about 62.52%. Then, the teachers' skills which include operating specific digital technology and interpreting data, information, and digital content related to fields, belong to the rare category of about 65.24%. Based on this analysis, a solution is provided by conducting digital literacy training for elementary school teachers. This research implies that such digital literacy for elementary school teachers can be a good reference and guideline for the other schools.

Keywords: Digital Literacy, Need Analysis, School Management, Covid-19

1 INTRODUCTION

Coronavirus disease (Covid-19) has caused many changes in various aspects, including education. Distance learning policies were eventually taken as a solution to reduce the spread of this virus (Hill et al. 2021). This policy certainly changes the learning system at various levels of education, especially in elementary schools. Distant learning requires digital skills for teachers, students, and parents. Digital literacy ability is the ability to search, read and view data collections from the internet to be processed as needed (Nair et al. 2012). According to Koh et al. (2015) and LaForce et al. (2016), A teacher needs analytical, communication, and digital skills in 21st-century education by learning the internet. Moreover, it can not necessarily be clarified. It can lead to errors in the concept of implantation that lead to misunderstandings, especially in elementary school-age who have long-term memory.

Digital literacy was echoed even before the current pandemic hit. However, before the pandemic, teachers did not use the internet as a learning tool. Based on research conducted by Sujendra Diputra et al. (2020), the teacher's internet use was dominated by access to social media such as Facebook, chat, Instagram, and YouTube. Teacher knowledge is still limited to Microsoft PowerPoint (PPT) regarding technology-based media. Some even said they never used it because there was no need for face-to-face learning. After the pandemic, circumstances forced teachers to adjust to using specific platforms that support the learning process, such as web meetings and other media that can only be used digitally (Hasbi 2020). This lies behind the need for digital literacy for teachers during the pandemic in distance learning for elementary school children who have not been able to

 DOI 10.1201/9781003347798-29

utilize digital technology optimally due to limitations in terms of ability compared to high school and university level children.

Teachers are required to be professional and continue to improve the quality of teaching. Therefore, training is needed to increase the potential of existing literacy as a form of informal learning. This is in line with Walker et al. (2015) and Martínez & Fernández (2015). Several studies have been conducted, namely Martínez & Fernández (2015) and McGuinness & Fulton (2019). The research is relevant for teacher training guidelines and assessing coaching in digital literacy. Research results show that digital literacy training positively affects digital literacy skills.

This research is fundamental to creating teachers who think critically are responsive to educational changes and provide satisfactory results for all stakeholders. This research contributes to recommending and correcting teachers' potential shortcomings regarding digital literacy. The main problem, in this case, is that teachers do not have adequate digital literacy that can be proven from teachers, especially at the elementary school level, who do not have techniques to find information sources on the internet and tend to use information as a reference to develop learning resources where information is not necessarily valid. This study aims to identify teacher needs related to digital literacy competencies and provide solutions to these problems. The question that guides this research is how elementary school teachers can use digital literacy.

2 LITERATURE REVIEW

The rapid development of technology has changed the way of life and learning. The concept of digital literacy includes knowledge that comes from various sources, both from the internet and about computers as learning and communication tools (Fraillon et al. 2014; Liang et al. 2021). Digital literacy skills support inclusive teachers to take advantage of various aspects, such as health and social awareness (Nedungadi & Raman 2018). A digital literacy campaign is designed for teachers to improve the teachers' learning teaching quality. The teacher's needs are also parts of digital literacy. Digital literacy is the ability to discover, evaluate, create, and communicate information through information technology and communication that needs technical and cognitive competence. Today, teachers must create, collaborate, and share digital content responsibly. Therefore, principals, learning resources, and teachers need to understand the importance of digital literacy in teacher learning (McGuinness, C., & Fulton 2019). Hobbs (2010) explains that digital literacy is started with analysis and evaluation ability.

Several international studies have shown that the importance of teacher attitudes towards the potential of digital literacy for teaching is a relevant factor in its application (Ainley 2018; Drossel 2017; Hatlevik 2017; Petko et al. 2017). It also reinforces the importance of competence in using digital literacy (Hatlevik 2018), so it always needs encouragement to improve potential deficiencies in teachers' digital literacy abilities. The analysis covers analyzing the news to identify the writer, the purpose, opinion, and evaluate the contents' quality and credibility. Teachers with low-level of digital literacy cannot do these tasks. Then, the information obtained risks low credibility (Heo & Toomey 2020). Qualified facilities must support the improvement of digital literacy to support the teacher's competence. Previous studies found that infrastructure affects student success because facilities are facilities for the teaching and learning process and can stimulate students' interest in learning (Baridoma & Jacob 2020).

Previous research conducted by Diputra et al. (2020) only focused on the use of ICT for learning resources and the preparation of learning devices while also only using one animation video creation application which was limited to animated video design features so that its function was less extensive when used for daily learning media. Activity in general at elementary school-age children get bored quickly using less varied learning media. In contrast, in this study, digital literacy, in addition to covering aspects of using the internet as a learning resource, also includes other elements such as the use of ICT in teacher management to manage administration and the use of Canva, Microsoft PowerPoint (PPT), and Kinemaster applications which of course have broader features and functions than the applications used in previous studies.

3 METHOD

This research used a qualitative approach with a descriptive type. This research used a case study research design. This research was conducted in ten public elementary schools in Indonesia. The study respondents were 47 teachers consisting of 30 women and 17 men. The data were collected through questionnaires, direct interviews, observation, and documentation with respondents. The instruments using questionnaires with Likert scale were made on Google Form. After the data is collected, the information is displayed in proportions. Then analyzed using an inductive model; this model is intended to clarify the data reduction process to create meaning in Likert scale data from the raw data that has been collected. The findings were obtained to interpret the basic information; the researchers used interviews to follow up on the findings. Interviews were conducted to find out the problems that occurred, namely deepening the results of the lowest questionnaire and deepening the causes of the weakest results. The interview used is a structured interview. In the end, the researcher also added observation and documentation as a vital source in data triangulation.

In this study, the researcher used Miles et al.'s (2014) theory to analyze the data, analyze the data. Firstly, data reduction (selecting and focusing, and transforming the data obtained from a bracket. In this study, the data were in questionnaires, interviews, observation, and documentation with respondents. Then, a summary was made more accessible to analyze the data and proceed to the next step. Secondly, data display (displays organized, compressed data from a collection of information so it can be concluded later, In this step, the researcher shows or displays data according to the data that has been prepared. Obtained and presented sequentially. Finally, researchers did conclusion drawing (after completing data reduction and displaying the data obtained, the researcher concludes the findings in the field and based on the research problem).

4 FINDING AND DISCUSSION

Digital literacy for teachers has become a widely studied topic in recent years; new educational paradigms are increasingly being put forward by research and education experts in 21st-century learning. The need for teaching skills, especially in digital literacy, was to overcome the current challenges with the rapid development of technology (Carneiro et al. 2012). Also, Hobbs (2010) explained that digital literacy begins with analyzing and evaluating. It analyzed messages in various forms by identifying the author, purpose, and point of view and assessing the content quality and credibility. Teachers with poor digital literacy could not do this task, with the risk that the information obtained had low credibility.

Based on the results of data collection that have been analyzed, it was found that there were problems related to the low literacy rate of teachers' digital literacy. It was proven by the results of interviews about how to teach online learning by teachers during the pandemic.

> "In online learning, we only use WhatsApp, the teacher sending assignments, students collecting assignments at the specified time" *(A. Doni, personal interview, February 20, 2021).*

> "Online learning is limited to collecting assignments due to the condition of some teachers who have difficulty using the internet and operating laptops and gadgets properly" *(Hasan, personal interview, February 22, 2021).*

> "During the COVID-19 pandemic, teachers gave assignments, and students collected assignments by coming to school once a week. In addition to the teacher factor, there are factors from students and parents" *(Saiful, personal interview, February 25, 2021).*

> "Teachers who have low IT need help from teachers who are more proficient in IT; it has become a tendency to depend on other teachers regarding the use and operation of IT" *(K. Muchlis, personal interview, February 25,2021).*

Researchers conducted interviews with school principals in improving teachers' digital literacy so that the quality of schools could be better.

> "The way to improve the quality of schools is to provide support for teachers to take part in digital literacy training. The problem is that the teacher training has not been implemented properly. These factors are age and technological stuttering" *(S. Inayah, personal interview, March 2, 2021).*

Based on interviews with teachers in the Simo area, Boyolali, there were problems related to online learning. First, internet access was slow. Second, the learning quota provided by the government could not be used by teachers, so teachers had to pay for the internet quota. Third, school infrastructure has not been supported; some schools have installed Wi-Fi. Fourth, some teachers have not been able to operate a laptop. Fifth, some students only had androids or gadgets; one android was shared for all family members. This finding contrasted with previous research that mentioned that digital literacy is only identified with the smooth use of technology. It reflected that the quality of technology must be in line with digital literacy (List & Klee 2020).

From the results of observations, it was found that teachers had difficulties in IT-based learning. Teachers only used WhatsApp groups and communication via telephone between teachers and parents to send material by providing photos related to the material being taught, giving assignments related to the material, and collecting duties by students only by offering pictures of student work within the allotted time. Even though digital literacy is not limited to just that, Eshet (2004) explained that literacy could make students analyze and be critical of any source. Then this ability is used to meet the task concept, consistent with the goal literacy category for the desired purpose (Bawden 2008; Eshet 2004; Ng 2012).

The researchers distributed a questionnaire to find out the results of the percentage of teachers' digital literacy. The indicators were used to experience in using ICT in the teaching-learning process, accessing information from the internet in the teaching-learning process, supports for the teachers to use ICT, ICT-based materials and activities for the teaching-learning process, problems in the use of ICT in the teaching-learning process, teacher's skills, ICT in teacher management. This questionnaire was given to 47 primary school teachers based on their experiences related to IT and teachers' digital literacy.

After the data were taken, the results of the research for each indicator are displayed in Table 1 below.

Table 1. The hand of teachers' digital literacy.

No.	Indicator	Sub-indicator
1.	Experience in using ICT in the teaching-learning process	1.1 Directly operating digital devices 1.2 Operating software in digital devices
2.	Accessing information from the internet in the teaching-learning process	2.1 Browsing, searching data and filtering data for information and digital content 2.2 Evaluating data, information, and digital content 2.3 Managing data, information, and digital content
3.	Supports for the teachers to use ICT	3.1 School supports 3.2 Peer (other teachers) supports
4.	ICT-based materials and activities for the teaching-learning process	4.1 Interaction through digital technology 4.2 Sharing through digital technology 4.3 Involved in citizenship through digital technology to support the teaching-learning process 4.4 Collaborating through digital technology 4.5 *Netiquette* (the etiquettes in using internet) 4.6 Managing digital identities
5.	Problems in the use of ICT in the teaching-learning process	5.1 Internal problems from the teachers 5.2 External problems from the teachers
6.	Teacher's skills	6.1 Specifically operating digital technology related to a specific field 6.2 Interpreting data, information, and digital content related to specific information
7.	ICT in teacher management	7.1 The use of ICT by the teachers to manage student administration 7.2 The use of ICT by the teachers to manage the teaching-learning

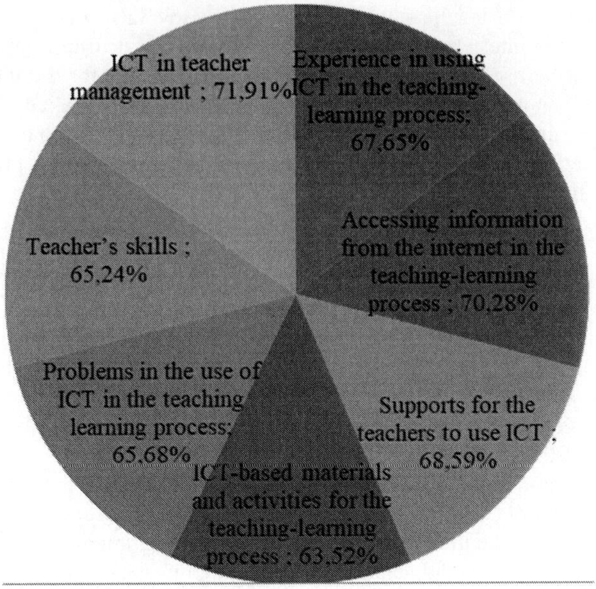

Figure 1. Analysis of skills teacher's digital literacy.

Based on Figure 1, the researchers intend to highlight the low result of the teachers' digital literacy needs. Below are the explanations:

1. First, ICT-based material and activities used by the teachers in the teaching-learning process belong to the Low category of 62.52 %. Teachers had difficulty in IT-based learning. Based on the results of observations during education, it was found that teachers only use WhatsApp groups and communication via telephone between teachers and parents to send material by providing photos related to the material being taught, then giving assignments of material and collecting duties by students just by offering pictures of the results. Students work in the allotted time.
2. This learning process occurred during the pandemic, causing boredom for some students because learning only took place in one direction. In addition, teachers still use conventional methods by using learning resources in teacher and student books during the learning process. The questionnaire results showed 63.52%, so the researchers interviewed several teachers at random to determine the cause of the low use of IT in teaching process activities on the fourth indicator.

> "Internet access at my school is not optimal and cannot be used." Therefore, the teacher only sends material via WhatsApp for students to study" (*M. Abdul, personal interview, March 7, 2021*).
>
> "Some students are getting tired of sending assignments via WhatsApp groups. I still have problems operating using video conferencing such as google meet and zoom meeting. I still need assistance from teachers who are experts in IT" *(Susi, personal interview, March 8, 2021)*

Digital literacy included the ability to search, share, and use technology and the digital capacity to create and develop critical information (Siddiq et al. 2017). Teachers also needed the proper application to be used in the teaching and learning process and understand the material contained in the information obtained through digital. Based on the research result, the elementary school teachers' ability of digital literacy in Simo District, Boyolali Regency, e.g. to manage and operate the software, was relatively low. Many teachers were confused about online learning so they created a teaching problem. Online learning or distance learning during the pandemic required the teachers to master the use of digital media both in accessing materials, finding materials, sharing materials, and creating information. The teachers were expected to encourage the students to be actively involved and contribute to the teaching-learning process during this pandemic.

Digital literacy has a vital role in achieving learning goals. Digital literacy can expand a teacher's digital knowledge, choose teaching or reference resources, and filter information. Several articles have been reviewed, e.g. Bozionelos, (2004); Fraillon et al., (2020). They explained nine categories in digital literacy as below: the availability of websites and internet connections, the ability to use different platforms, having the ability to protect privacy when using social media, using the identity correctly, having the ability to publish education content in applications, managing and identifying content, having the ability to create new content from digital media, having the ability to search, access, filter, and choose the information correctly, and having the ability to share learning ideas or personal scientific works.

School facilities that were supported to develop digital literacy would accommodate teaching and learning to find digital information and learn new technologies as part of digital literacy. In line with this finding, Blau et al. (2020) found that the availability of facilities and infrastructure was proven to improve the quality of schools. An explanation supported by digital media such as audio-visual was easier to be delivered by the teachers. It brought some benefits as below: teaching-learning process is easier to understand, creating interactive learning, saving time and energy, improving the quality of learning outcomes, creating students' positive behavior towards the materials, and overcoming the students' problems.

65.24% (Rarely category) of the teachers could operate digital technology skills specifically and interpret data, information, and digital content related to a specific field. This research indicated that the digital literacy level in elementary teachers in Simo District, Boyolali Regency was relatively low. Based on observations, some senior teachers were not yet proficient in operating laptops or the internet in finding information related to learning. The age factor influenced senior teachers to ask for help from teachers who master IT. The teacher's ability to use IT had less impact on online learning. They still found it challenging to operate digital devices and evaluate the content obtained from the internet, social media, or other digital media. This research showed that they rarely use digital devices such as smartphones, tablets, personal computers, or others. The teachers used smartphones mainly to communicate and access the internet's contents. However, they had difficulty using or typing in Microsoft Word for many purposes, such as making teaching instruments, printing documents, or copying data from another device.

In spreading or collecting the information through digital technology, the teachers had to evaluate every content wisely. The elementary school teachers in Simo District, Boyolali Regency, actively guided the students to filter the internet and not get false information. The teachers also explained the side effects of accessing inappropriate content or understanding information without checking the truth. According to (Szymkowiak et al. 2021), when the students used information technology carelessly, it would negatively impact them, especially in character building. The information that could be accessed easily gave a chance for irresponsible parties to spread misleading content, provoke hatred, and share harmful content. It was a different opinion from Ivanov et al., (2019) that technology affected industrial conditions because technology-facilitated industrial processes. Therefore, these things could be prevented through the training and assignment on operating digital technology as digital literacy media.

Another thing was found that some schools had not made policies requiring technology in learning. This impacted teachers, i.e. they only focussed on WhatsApp groups. Teachers had not been motivated to learn technology in education. Furthermore, researchers conducted interviews with teachers regarding the causes of the low questionnaire results related to teachers' ability in digital technology and not providing interesting content during learning, affecting the low level of digital literacy. "The school has not provided training related to digital literacy, or the use of technology in learning during the pandemic" (Teacher 2 and Teacher 6). "School and stakeholder support does not yet exist, so teachers have not been able to create interactive learning media and have not used online learning platforms" (Teacher 4 and Teacher 1).

The age factor of the teacher affected the technological ability, so it affected the teacher's creativity in creating content on online learning media. This caused low motivation to learn about digital literacy. Creativity was essential in digital literacy; this was also supported by the principal's policy (Glazzard & Stokoe 2015). School policies providing digital literacy training could increase teacher creativity in making interactive learning media in online learning. The teachers' digital literacy technical skills could be seen from the way they surf the internet, using some features provided in digital applications, the ability to optimize the services provided on the internet to support critical thinking, encouraging creativity in the teaching-learning process, and creating innovation, which they have not optimized yet.

The finding was slightly different from some previous findings, where teachers were required to have the ability to operate various digital literacy features (Fauth et al. 2019). Teachers had to learn it themselves by being self-taught or watching social media about computer operations. The principal's policy required that teachers master IT, and using technology in online learning during a pandemic could improve teacher digital literacy and affect the school's goal, vision, and mission. In addition, teachers are more motivated to learn.

Based on the questionnaire and interview result, the researchers then held digital literacy training to improve the teachers' ability to operate digital technology in the teaching-learning process. Digital technology would help the teachers to deliver the material efficiently and effectively. The materials arranged can be distributed to the students through applications such as e-mail, WhatsApp, Messenger, Telegram, or even virtual meetings like Zoom Meeting, Google Meet, etc. The student can obtain the materials more accessible and faster, anytime and anywhere, as long as they have internet access.

The teachers could also deliver the material quicker and easier so that they can save time and cost in conducting a study from home during the Covid-19 pandemic. In addition, teachers could share materials between them easily while for the needs of analysis or evaluation of objective, accountable, and fast learning, teachers use Microsoft Excel. Besides, the teachers could share the material among them easily. Meanwhile, teachers used Microsoft Excel for the needs of objective, accountable, and fast learning analysis or evaluation. It is in line with Lai & Bower's (2019) finding that digital technology facilitates flexibility and can emphasize the influence of motivation and optimism in learning.

Based on the observation, questionnaire, and interview results above, these findings further explained the use of technology globally to facilitate teachers' learning and learning process who constantly strive to improve digital literacy skills to achieve the desired learning goals. This was in line with (List 2019; Tsatsou 2018) that teachers must be aware of and express interest in the great need for continuous training and self-development to promote technology-driven independent digital literacy to later be utilized in understanding community needs. The paragraph related to the state's role on a global scale to maintain a national digital transformation ecosystem in building an innovative digital economy. This research was relevant with Hanna's (2018) study. This research had the potential to open new insights that the use of literacy digitization can predict literacy skills based on competence digitally globally.

Based on the problems explained, the researchers were provided a solution related to the low ability in teachers' digital literacy by conducting digital literacy training for the elementary school teachers. This training was held on 23rd June 2021, and it was followed by the elementary teachers in Simo District, Boyolali Regency. This training will help teachers implement ICT and interpret data, information, and digital content in their teaching activities. This training allows the teachers to create teaching materials using PowerPoint Presentation, Canva application, and videos in the Kinemaster application.

5 CONCLUSION AND RECOMMENDATION

This study aims to identify teacher needs regarding digital literacy competencies and provide solutions to the problems. The results showed that teachers' ICT-based materials and activities were included in the rare category around 62.52%. The teacher's skills, covering operating digital technology to interpret specific digital data, information, and content related to certain fields, were included in the rare category. This research has implications. Knowing the potential shortage of teachers related to several skill areas regarding digital literacy, all stakeholders must create retraining for teachers to improve existing weaknesses. This research is only limited to the regency area, so researchers did not yet know the potential flaws in big cities in Indonesia. It is hoped that further research will be able to follow up to expand the scope of the sample so that the data will be interesting and in-depth. This research can create education management quality involving digital literacy in elementary schools.

REFERENCES

Ainley, J. (2018). Students and their computer literacy: Evidence and curriculum implications. *Second handbook of information technology in primary and secondary education*, (pp. 69–88). Cham, Switzerland: Springer International Publishing AG.

Assidiqia, Muhamad Hasbi, W. S. (2020). Utilization of Digital Platforms during the Covid-19 Pandemic. *Proceedings of the UNNES Postgraduate National Seminar*, 298–303.

Baridoma, M. P., & Jacob, O.-O. N. (2020). Perceived Influence of School Facilities on Educational Attainment of Rural Farmers' Children in Abua/Odual Local Government Area of Rivers State. *International Journal of Contemporary Academic Research*, *1*(2), 125–138.

Bawden, D. (2008). Origins and concepts of digital literacy. *Digital literacies: Concepts, policies and practices*, 30(2008), 17–32.

Blau, I., Shamir-Inbal, T., & Avdiel, O. (2020). How does the pedagogical design of a technology-enhanced collaborative academic course promote digital literacies, self-regulation, and perceived learning of students? *The Internet and Higher Education*, *45*, 100722.

Bozionelos, N. (2004). Socio-economic background and computer use: the role of computer anxiety and computer experience in their relationship. *International Journal of Human-Computer Studies*, *61*(5), 725–746.

Carneiro, R., Toscano, J., &, & Díaz, T. (2012). *Los desafíos de las TIC para el cambio educativo*. OEI & Fundación Santillana.

Chan, N. N., Walker, C., & Gleaves, A. (2015). An exploration of students' lived experiences of using smartphones in diverse learning contexts using a hermeneutic phenomenological approach. *Computers & Education*, 82, 96-106.

Drossel, K., & Eickelmann, B. (2017). Teachers' participation in professional development concerning the implementation of new technologies in class: a latent class analysis of teachers and the relationship with the use of computers, ICT self-efficacy and emphasis on teaching ICT skills. *Large-scale Assessments in Education*, 5(1), 1-13.

Eshet, Y. (2004). Digital Literacy: A Conceptual Framework for Survival Skills in the DigChan, N. N., Walker, C., & Gleaves, A. (2015). An exploration of students' lived experiences of using smartphones in diverse learning contexts using a hermeneutic phenomenological approach. Computers & Education (Italic), 82, 96-106.ital era. *Journal of Educational Multimedia and Hypermedia*, *13*(1), 93–106.

Fauth, B., Decristan, J., Decker, A. T., Buettner, G., Hardy, I., & Klieme, E., & Kunter, M. (2019). The effects of teacher competence on student outcomes in elementary science education: The mediating role of teaching quality. *Teaching and Teacher Education*, *86*, 102882.

Fraillon, J., Ainley, J., Schulz, W., Friedman, T., & Duckworth, D. (2020). *Preparing for Life in a Digital World*. Springer International Publishing.

Fraillon, J., Ainley, J., Schulz, W., Friedman, T., &, & Gebhardt, E. (2014). *Preparing for Life in a Digital Age*. Springer International Publishing.

Glazzard, J., & Stokoe, A. (2015). *Digital Literacy for Primary Teachers*. Critical Publishing.

Hanna, N. (2018). A role for the state in the digital age. *Journal of Innovation and Entrepreneurship*, 7(1), 1-16.

Hatlevik, I. K. R., & Hatlevik, O. E. (2018). Examining the relationship between teachers' ICT self-efficacy for educational purposes, collegial collaboration, lack of facilitation and the use of ICT in teaching practice, *Frontiers in Psychology*, 9 (2018), p. 935

Hatlevik, O. E. (2017). Examining the relationship between teachers' self-efficacy, their digital competence, strategies to evaluate information, and use of ICT at school. *Scandinavian Journal of Educational Research*, 61(5), 555-567.

Heo, M., & Toomey, N. (2020). Learning with multimedia: The effects of gender, type of multimedia learning resources, and spatial ability, *Computers & Education*, 146, 103747.
Hill, G., Mason, J., & Dunn, A. (2021). Contract cheating: an increasing challenge for the global academic community arising from COVID-19. *Research and Practice in Technology Enhanced Learning*, *16*(1), 24.
Hobbs, R. (2010). Digital and Media Literacy: A Plan of Action. *A White Paper on the Digital and Media Literacy Recommendations of the Knight Commission on the Information Needs of Communities in a Democracy*. Aspen Institute. 1 Dupont Circle NW Suite 700, Washington, DC 20036.
Ivanov, D., Dolgui, A., & Sokolov, B. (2019). The impact of digital technology and Industry 4.0 on the ripple effect and supply chain risk analytics. *International Journal of Production Research*, *57*(3), 829–846.
Koh, J. H. L., Chai, C. S., Benjamin, W., &, & Hong, H.-Y. (2015). Technological Pedagogical Content Knowledge (TPACK) and Design Thinking: A Framework to Support ICT Lesson Design for 21st Century Learning. *The Asia-Pacific Education Researcher*, *24*(3), 535–543.
LaForce, M., Noble, E., King, H., Century, J., Blackwell, C., Holt, S., Ibrahim, A., &, & Loo, S. (2016). The eight essential elements of inclusive STEM high schools. *International Journal of STEM Education*, *3*(1), 1-11.
Lai, J. W. M., & Bower, M. (2019). How is the use of technology in education evaluated? A systematic review. *Computers & Education*, *133*, 27–42.
Liang, Q., de la Torre, J., &, & Law, N. (2021). Do background characteristics matter in Children's mastery of digital literacy? A cognitive diagnosis model analysis. *Computers in Human Behavior*, *122*, 106850.
List, A., Brante, E. W., & Klee, H. L. (2020). A framework of pre-service teachers' conceptions about digital literacy: Comparing the United States and Sweden. *Computers & Education*, *148*, 103788.
List, A. (2019). Defining digital literacy development: An examination of pre-service teachers' beliefs. *Computers & Education*, *138*, 146–158.
Martínez, R. A., & Fernández, R. L. (2015). Digital literacy for teachers in Cuban tourism and hotel management schools. Some experiences on its development. *RUSC Universities and Knowledge Society Journal*, *12*(3), 3–15.
Martínez, R. A., & Fernández, R. L. (2015). Digital literacy for teachers in Cuban Tourism and Hotel Management Schools. Some experiences on its development. *International Journal of Educational Technology in Higher Education*, 12(3), 3-16
McGuinness, C., & Fulton, C. (2019). Digital Literacy in Higher Education: A Case Study of Student Engagement with E-Tutorials Using Blended Learning. *Journal of Information Technology Education: Innovations in Practice*, *18*, 001–028.
Miles, M. B., Huberman, A. M., & Saldana, J. (2014). *Qualitative Data Analysis?: A Methods Sourcebook*. SAGE Publications, Inc.
Nair, G. K. S., Rahim, R. A., Setia, R., Adam, A. F. binti M., Husin, N., & Sabapathy, E., …Seman, N. A. (2012). Terengganu schools extent of computer literacy and internet usage. *Asian Social Science*, *8*(8), 74.
Nedungadi, P. P., Menon, R., Gutjahr, G., Erickson, L., & Raman, R. (2018). Towards an inclusive digital literacy framework for digital IndiaEducation and Training. *Education and Training*, *60*(6), 516–528.
Ng, W. (2012). Can we teach digital natives digital literacy? *Computers & Education*, *59*(3), 1065–1078.
Petko, D., Cantieni, A., &, & Prasse, D. (2017). Perceived Quality of Educational Technology Matters. *Journal of Educational Computing Research*, *54*(5).
Sujendra Diputra, K., Ketut Desia Tristiantari, N., & Nyoman Laba Jayanta, I. (2020). Digital Literacy Movement For Primary School Teachers. *Journal of Character Education Society*, *3*(1), 118–128.
Szymkowiak, A., Melović, B., Dabć, M., Jeganathan, K., & Kundi, G. S. (2021). Information technology and Gen Z: The role of teachers, the internet, and technology in the education of young people. *Technology in Society*, *65*, 101565.
Tsatsou, P. (2018). Literacy and training in digital research: Researchers' views in five social science and humanities disciplines. *New Media & Society*, *20*(3), 1240–1259.

Pre-Service Teacher Training and Education

The readiness of Indonesian prospective vocational teachers towards TPACK for 21st-century learning

Anis Rahmawati, Nunuk Suryani, Muhammad Akhyar & Sukarmin
Universitas Sebelas Maret, Surakarta, Indonesia

ABSTRACT: This research aimed to portray the prospective vocational teacher's readiness toward TPACK for 21st-Century Learning (CL). A descriptive survey methodology was applied with 215 prospective vocational teachers as respondents came from four vocational teacher training in Indonesia. The instrument used was a Likert-scale self-assessment questionnaire and lesson plan rubric. The results from the descriptive analysis of self-assessment and lesson plan score showed that the prospective vocational teacher showed readiness for TPACK 21st CL at the intermediate level. The lesson plans prepared by prospective teachers during their apprenticeship programs showed less technology integration. Therefore, enriching the vocational prospective teacher's experience of learning that integrates technology suitable for the pedagogical approach and vocational content during their studies in the vocational teacher training education through the TPACK framework is suggested.

Keywords: Apprenticeship programs, Prospective vocational teacher, Lesson plan, Self-assessment, TPACK 21st CL

1 INTRODUCTION

The Teacher Training and Education Institutions face the challenge to equip their graduates with all the professional teachers' competencies and build teachers' professional identity. Teachers' professional identity includes teachers' perspectives of the teacher's role and how they interpret that role and act in the teaching profession (Lai & Jin 2021). The vocational teacher candidates who will teach prospective professional workers also need to have the competence to educate their students so that they can contribute to 21st-century work. The integration of technology into the learning process is part of the competencies required by the work of the 21st-century world. Teachers and prospective teachers need specialized professional knowledge to integrate technology, pedagogy, and content, which is well known as TPACK (technological pedagogical content knowledge) (Mishra & Koehler 2006), into instructional design. TPACK is a framework related to teacher knowledge; however, the efforts to establish TPACK need to start from the period of the teacher education program. Nelson (2017) stated how technology integration is applied in the learning process in classrooms during teacher education programs has a significant impact on the teacher's confidence in integrating technology into learning later.

Vocational (prospective) teachers need to have TPACK competency to be able to plan and implement effective teaching and learning that supports the mastery of 21st-century skills under their respective vocational skills (Sally et al. 2011; Teo et al. 2021). Learning vocational content that integrates appropriate technology can be a path for achieving vocational teaching goals and improving teaching efficiency (Dai & Fan 2012). Therefore, mastery of TPACK among vocational (prospective) teachers needs to get attention. Chua (2012) conducted research to assess the level of professional knowledge competency based on the TPACK model among TVET instructors in Malaysia and concluded that they were at a satisfactory level. To map the TPACK for Vocational High School prospective teachers during their education, Nurhadi et al. (2019) developed a model

DOI 10.1201/9781003347798-30

for the setting of vocational prospective teachers' educational curriculum integrating elements of TPACK.

Studies on TPACK readiness for vocational teacher candidates in Indonesia have not been widely carried out. Vocational learning has a distinctive character according to their respective fields of work, thus influencing the appropriate form of learning approach and learning technology. This study was conducted in a vocational context especially in the technological vocational field to measure the readiness of prospective vocational teachers' TPACK. Knowledge related to TPACK readiness of prospective teachers becomes essential information for educational institutions to take corrective steps to increase the level of TPACK readiness of their leavers. Through a survey study, this research was going on to answer the question: how was the prospective vocational teacher's readiness toward TPACK for 21st CL? This study drew upon the TPACK assessment from the work of Valtonen et al. (2017) which is grounded on the 21st-century skill from the perspective of TPACK.

2 LITERATURE REVIEW

Mishra & Koehler (2006) stated that TPACK as a framework for the use of technology in teaching and learning consists of three fundamental knowledge, namely, knowledge of content (CK), pedagogy (PK), and technology (TK). The integration of the three fundamental elements then forms the three intermediate knowledge that is knowledge of Pedagogy Content (PCK), Technology Content (TCK), and Technology Pedagogy (TPK). The inclusion of the three features creates the main aspects of TPACK. TPACK covers the understanding of parts of content that are easy or difficult to understand and how to use technology to help students get a better understanding. Several factors influence the level of TPACK's mastery; one factor came from a personal (prospective) teacher, including age and teaching experience. Generally, TPACK mastery achievement points based on self-evaluation of prospective teachers will be lower than active teachers who have had long teaching experience (Dong et al. 2015). Teacher mentors who often integrate technology in learning will make prospective teachers have a higher tendency to use technology in learning during the apprenticeship program (Nelson 2017). Koh, et al. (2014) reported that technology integration in learning is also influenced by student characteristics, the learning environment both within the scope of class and institutions, the situation of learning activities, and teacher perceptions.

This study focuses on vocational knowledge in content knowledge as one of the TPACK components. Vocational education must accommodate the combination of disciplinary expertise and situated learning (Gamble 2006). Disciplinary knowledge is related to particular scientific theories. In contrast, situational awareness is closely related to assignments in specific fields of work. Therefore Wheelahan (2015) defines vocational knowledge as applied disciplinary knowledge, in line with the purpose of the vocational curriculum, which is to direct students to practical activities as a basis for integrating and synthesizing theoretical expertise in specific fields of work. Vocational competence depends on all the attributes needed to carry out effective action, namely working with the use of propositional and theoretical knowledge in the context of the workplace and local ethical judgments (Winch 2010). Then the vocational knowledge in this study is described as theoretical knowledge as the basis for carrying out work practices especially in the technical vocational field.

Guidance from the Partnership for 21st-Century Skills (2008) mentions that the world of work has shifted from the dominance of the manufacturing sector to the service sector which is strongly influenced by information, knowledge, and innovation. The 21st-century work requires workers to have the ability in terms of skills and knowledge according to specific fields of work and must be able to use ICT as a tool in areas related to 21st-century skills, including creative and critical thinking, collaborating, communication, and problem-solving (Valtonen et al. 2017). Therefore, all vocational teacher candidates should be competent to develop learning designs for 21st CL. The 21st CL provides opportunities for students to strengthen their competencies in the fields of technology, problem-solving, collaboration, and knowledge to play a role in the work of the 21st-century world (Koh et al. 2015).

Several studies have been conducted in the context of TPACK in vocational education. Nurhadi et al. (2019) developed the Model of MPDCA (Map, Plan, Do, Check, and Action) cycle for preparing TPACK among prospective Vocational High School teachers in the era of Industrial Revolution 4.0. The TPACK integration is divided into five curriculum categories, namely educational technology, vocational content, learning methods, integrated courses, and teaching internships. Further Nurhadi et al. (2019) stated that the implementation of this model requires collaboration between the leadership of teacher training institutions, administration, lecturers, prospective teachers, and schools. Arifin et al. (2020) put forward a new concept in technology integration in vocational learning in the form of the TAWOCK model which combines work, content, andragogy, and technology. They claim that work knowledge and andragogy knowledge are theoretically more in line with the character of vocational education but still need further empirical evidence. Therefore, this study still relies on the TPACK model.

3 METHOD

The research methodology applied in this study was a descriptive survey. The instrument used was a self-assessment questionnaire (Rahmawati et al. 2020) and a lesson plan rubric that was adapted from Harris et al. (2010). The self-assessment questionnaire used a 6-point Likert-type scale to measure the seven elements of the TPACK framework in the vocational field, namely: pedagogical knowledge (PK; 10 items), content knowledge (CK; 6 items), technology knowledge (TK; 5 items), vocational pedagogical knowledge (PCK; 12 items), technological pedagogical knowledge (TPK; 6 items), technical vocational knowledge (TCK; 5 items) and technological pedagogical vocational knowledge (TPACK; 7 items). The assessment of the lesson plan was conducted using a rubric that portraits technology integration based on four criteria, that are: 1) Curriculum goals & technologies, 2) Instructional strategies & technologies, 3) Technology selection(s), and 4) "Fit" (Content, pedagogy, and technology together). Lesson plan documents came from prospective teachers' work results during their teaching apprenticeship program. Participants in this study were 215 prospective teachers consisting of 46% female (n=99) and 54% (n=116) male which were determined randomly from four vocational teacher educations in Indonesia that carry out a Vocational Building Engineering education program. Participants differentiated based on their teaching experience, that was who have participated in a teaching apprenticeship (35%, n=76) and who have not (65%, n=139).

Researchers had given direct instruction to the respondent before they were answering the questionnaire to avoid misunderstanding regarding the question. Data from the self-assessment questionnaire was then analyzed statistically descriptively in terms of percentage (%), mode, mean (M), and standard deviation (SD). The content analysis was used to analyze data from the open-ended questionnaire by founding the central theme of each question. Data resulted from lesson plan assessment explained using inter-rater reliability with intraclass correlations in a second percent agreement procedure. The Cronbach alpha level, the intraclass relationship on a single judge, and the average judgments were used to determine the reliability of data. Then the reliable data was analyzed statistically descriptively in terms of mode.

4 FINDING AND DISCUSSION

This section of this paper describes the results of measuring the TPACK ability of prospective teachers which is carried out through self-assessment and assessment of lesson plan documents. The measurement results through self-assessment are distinguished between prospective teachers who have participated in teaching apprenticeship and those who have not, followed by a description of the differences between the two. The results of the assessment of the lesson plan document include the forms of technology used by prospective teachers during their teaching apprenticeship program.

4.1 *Measurement of TPACK through a self-assessment questionnaire*

4.1.1 *Among prospective pre-apprenticeship teacher*

The results of the TPACK survey on prospective pre-apprenticeship teachers showed that for the fundamental element, TK reaches the highest point with a mean of 3.33 (SD= 1.10), followed by PK (M=3.13, SD= 1.14) and CK (M=3.01, SD=1.03). While for the intermediate elements, TCK as the best performance (M=3.41, SD= 1.10) followed by TPK (M= 3.30, SD= 1.05) and PCK (M= 3.12, SD= 1.05). The TPACK itself reached a mean of 3.30 (SD=1.00).

In all the components of the TPACK, most prospective teachers who had not yet participated in the teaching apprenticeship program considered that they still needed additional knowledge related to these components. This result was indicated by the highest average percentage in majority variables is the third (need a little additional knowledge). The fundamental element with the highest mastery was TK, while the lowest mastery was CK. The previous study by Khan and Pattern (2018) confirmed this condition, where prospective teachers have high mean points in TK. The prospective teachers today were those who were born in an era where information technology was developing rapidly. Their closeness to technology has been established from an early age. This generation stands out with information technology. Therefore, they felt confident with technological knowledge. While most of them only noticed vocational expertise when they were starting their higher education in teacher training.

The mastery of the fundamental elements affects the ability of intermediate items, where PCK was found as the lowest point. The low of this intermediate element affected the lower TPACK overall. This result aligned with previous studies performed by Valtonen et al. (2017) which concluded that the roles of intermediate elements built by CK and PK were more reliable than the other middle features. The intersection of understanding of CK and PK was mostly influenced by the ability to comprehend the context. The prospective pre-apprentice teachers with little experience and knowledge of the vocational field, met the difficulties in finding learning strategies that are most appropriate with the vocational content.

4.1.2 *Among prospective post-apprenticeship teacher*

The results of the TPACK survey on prospective post-apprenticeship teachers showed that for the fundamental element, TK still reaches the highest point with a mean of 3.77 (SD= 1.08), followed by PK (M=3.61, SD= 1.05) and CK (M=3.17, SD=0.91). For the intermediate elements, TPK in the first place with mean 3.61 (SD= 0.97) followed by PCK (M= 3.47, SD= 0.93) and TCK (M=3.47, SD= 1.06). While the TPACK reach means 3.45 (SD=0.86).

Vocational content (CK) was the dominant component in which prospective teachers still needed additional knowledge. While components that include technological capabilities (TK, TPK, TCK, TPACK) were almost equal between those who already have the knowledge and those who still need additional knowledge. However, those who claimed to have knowledge related to TPACK were still dominated (overall average 78%) at the level of merely having knowledge (point 4). They were not yet at the level of mastering it well (point 5), let alone having excellent knowledge (point 6).

4.1.3 *The differences between the two groups of prospective teachers*

Data from the two groups of prospective teachers showed an increase in TPACK points between pre- and post-apprenticeship prospective teachers in all TPACK elements. PK shows the most significant difference. Before undergoing an apprenticeship program at school, prospective teachers had already taken a micro-teaching course. For one semester, they practiced preparing lesson plans and carrying out teaching practicum. Then for three months following the apprenticeship program, they would prepare lesson plans and practice them in real classes at Vocational Secondary School. Microteaching courses along with an educational apprenticeship certainly provide valuable pedagogical experience to shape better pedagogical knowledge skills among prospective teachers. This result supported Guthrie et al. (2009) statement that the development of (prospective) teachers' knowledge is influenced by their on-job training experience. Chua (2012) reported the same condition that the knowledge of teachers varied significantly across teaching experience.

In both groups of students, among the three essential elements of the TPACK namely PK, TK, and CK, CK in the vocational field was the ability with an enormous difference between the number who already "have knowledge" and those who "still need additional knowledge." Even the perception of CK can be said to be no difference between the two groups of students. For the fundamental element of PK, there was an increase from those who had not yet had an apprenticeship, although the number was not significant (a 3% increase in point $\geq$4). Likewise, the PCK element experienced a rise in the same amount. The elements related to technology show an almost equal number of prospective teachers who claim they still need additional knowledge with those who already have knowledge related to technology.

4.2 *Assessment of lesson plan documents prepared by the prospective teacher*

This study assessed several lesson plan documents compiled by prospective teachers. Fifteen lesson plans from three Teacher Trainer Education were analyzed using rubrics from Harris et al. (2010). Two researchers each lesson plan document. Inter-rater reliability using intraclass correlations was used to analyze the lesson plan assessment report from two researchers. The Cronbach's alpha for inter-rater reliably in four criteria were 0.784, 0.870, 0,621, and 0.737, responsively. The intraclass correlation showed that the reliability of a single judge level in four criteria was 0.645, 0.769, 0.450, and 0.583, and the average reliability for the two researchers was 0.784, 0.870, 0.621, and 0.737, responsively. The results of the lesson plans assessment showed that the lesson plans prepared by prospective teachers had not shown the integration of technology in accordance with the curriculum objectives and learning strategies. The value of the statistic mode in all components is at point one (1) which meant that: technology that was used in lesson plans was not in accordance with one or more curriculum objectives; the use of technology did not support learning strategies; the technology chosen was not following curriculum objectives and learning strategies; and materials, learning strategies, and technology together were not suitable for its application in the lesson plan.

The low technology integration in the lesson plan prepared by prospective teachers was not in line with the results of the TPACK self-assessment measurement, which showed that almost half of them said they "have knowledge" of TPACK (46%, n = 35). The results indicated that they thought that they knew about the TPACK elements but still had difficulties on how to design such a lesson. The same condition was found by Dong et al. (2015) in their study, which met a situation where the teachers wish to conduct ICT integrated lessons supported by technology, but they are not sure that they can design it. Therefore Dong et al. (2015) suggested the development of teachers' design capacity to solve the problem. Likewise, Isiyaku et al. (2018) stated that it was necessary to train the integration of technology in the curriculum for (prospective) teachers to help them overcome the difficulties of using ICT in learning. How prospective teachers design and implement the learning that integrated technology was greatly influenced by their mentor teachers who supervised them during an apprentice program. Mentor teachers who had excellent TPACK skills would stimulate prospective teachers to use technology more as they saw their mentor teachers more frequently (Nelson 2017).

4.2.1 *The forms of technology used in learning*

Based on surveys related to students' experiences in integrating technology into learning, information was obtained that the form of technology they used was almost the same as that of instructors. This condition was in line with the opinion from Voithofer et al. (2019), which stated that the possibility of teachers adopting technology in their learning increased when they saw technology as a part of their lives and saw the value of technology to enhance learning. The type of technology most widely used by the prospective teacher was presentation slides with PowerPoint. AutoCAD was the graphical application that occupied the next highest number, followed by other computer applications in general that are not explicitly mentioned. The most used technology was video; both videos were obtained directly from the internet or videos made by the teacher. Most video content is in the form of tutorials and practical work performance and field conditions of building construction. The use of unique applications mentioned by respondents other than AutoCAD was

Sketchup as an image application, SAP as a structural analysis calculation application, and Revit as a building modeling application. The mentioned social media used in learning are the web in general and Facebook.

The learning game technology that has been used by instructors did not appear in the learning planned and carried out by prospective teachers during the apprenticeship programs, likewise with various e-learning portals, including Edmodo, and social media such as Facebook. Applications that were rarely used by instructors but are used by prospective teachers are Sketchup applications. Huizenga et al. (2017) reported that learning media using interactive games support learning and enhance students' motivation because of their attractive characteristics for school-age students. Further, Huizenga et al. (2017) stated creating a game that matched the subject material was not easy for just anyone, especially those unfamiliar with programming languages. Whereas finding online open-access games that were suitable for particular vocational content was also not straightforward. That condition might make game applications not used by prospective teacher students when implementing an education apprenticeship program. The distance learning portal with the Learning Management System (LMS) did not apply to prospective teachers because of their unfamiliarity with the management of free and closed source LMS. The LMS that was used by instructors during the teacher training program was mostly open-source LMS as reported by Dağ (2019), which was managed by the IT team of higher education institutions. Not all Secondary Vocational Schools have provided this kind of open-source LMS.

This study showed that the level of mastery of TPACK for prospective vocational teachers was at an intermediate level from the results of self-assessment measurements, both for prospective teachers who have not participated in teaching apprenticeship or have already. The integration of technology in teaching vocational content was rarely found in the lesson plan documents prepared by prospective teachers when participating in teaching internships. The difference between the results of the self-assessment and lesson plan measurements indicated that prospective teachers have knowledge related to the integration of technology in learning but still do not know how to apply it in actual learning.

5 CONCLUSION AND RECOMMENDATION

This study determines the readiness of TPACK 21 CL among Indonesian prospective vocational teachers and is carried out through self-assessment of TPACK abilities and assess the lesson plan documents prepared by prospective vocational teachers. 215 respondents involved in this study came from four Indonesian vocational teacher educations with a specialization in building engineering. The TPACK self-assessment measurement resulted in the prospective vocational teacher's readiness for TPACK 21st CL at the intermediate level with the most mastered was TK and the lowest mastery was CK. The lesson plans prepared by prospective teachers during their teaching internship showed less integration of technology. Based on the results of this research, vocational teacher training education institutions need to force their faculty members to give the vocational prospective teacher's more experience in integrating technology which is suitable with pedagogical approach and vocational content through the learning process. Therefore, further research is needed to develop instructional designs that integrate technology effectively into vocational content learning.

REFERENCES

Arifin, Z., Nurtanto, M., Priatna, A., Kholifah, N., & Fawaid, M. (2020). Technology Andragogy Work Content Knowledge Model as a new framework in vocational education: revised Technology Pedagogy Content Knowledge Model. *TEM Journal*, *9*(2), 786–791.

Chua, H. (2012). Factors influencing the Technological Pedagogical Content Knowledge (TPACK) among TVET instructors in Malaysian TVET Institution. *Procedia - Social and Behavioral Sciences*, *69* (Iceepsy), 1539–1547.

Dağ, F. (2019). Prepare pre-service teachers to teach computer programming skills at K-12 level: experiences in a course. *Journal of Computers in Education, March*, 1–37.
Dai, W., & Fan, L. (2012). Discussion about the pros and cons and recommendations for multimedia teaching in local vocational schools. *Physics Procedia*, *33*, 1144–1148.
Dong, Y., Chai, C. S., Sang, G., Koh, J. H. L., & Tsai, C. (2015). Exploring the profiles and interplays of pre-service and in-service teachers' Technological Pedagogical Content Knowledge (TPACK) in China. *Educational Technology & Society*, *18*(1), 158–169.
Gamble, J. (2006). Theory and practice in the vocational curriculum. In M. Young, & J. Gamble (Eds.), *Knowledge, curriculum and qualifications for South African further education.* Cape Town: HSRC Press.
Guthrie, H., Harris, R., Simons, M., & Karmel, T. (2009). Teaching for Technical and Vocational Education and Training (TVET). In S. L.J. & D. A.G. (Eds.), *International Handbook of Research on Teachers and Teaching,* 21, 851–863.
Harris, J., Grandgenett, N., & Hofer, M. (2010). Testing a TPACK-based technology integration assessment rubric. In C. D. Maddux, D. Gibson, & B. Dodge (Eds.), *Research highlights in technology and teacher education,* 1–8.
Harris, J. B., & Hofer, M. J. (2011). Technological Pedagogical Content Knowledge (TPACK) in action: A descriptive study of secondary teachers' curriculum-based, technology-related instructional planning. *Journal of Research on Technology in Education*, *43*(3), 211–229.
Huizenga, J. C. T. ten Dam, G. T. M., Voogt, J. M., & Admiraal, W. F. (2017). Teacher perceptions of the value of game-based learning in secondary education. *Computers & Education*, *110*(July 2017), 105–115.
Isiyaku, D. D., Ayub, M. A. F., & AbdulKadir, S. (2018). Antecedents to teachers' perceptions of the usefulness of ICTs for business education classroom instructions in Nigerian tertiary institutions. *Asia Pacific Education Review*, *19*(3), 337–352.
Khan, Z., & Pattern, R. (2018). Measuring the Technological Pedagogical Content Knowledge (TPACK) of pre-service teachers in relation to their gender and stream. *American International Journal of Research in Humanities, Arts and Social Science*, *22*(1), 50–55.
Koh, J. H. L., Chai, C. S., & Tay, L. Y. (2014). TPACK-in-Action: Unpacking the contextual influences of teachers' construction of Technological Pedagogical Content Knowledge (TPACK). *Computers & Education*, *78*, 20–29.
Koh, J. H. L., Chai, C. S., Benjamin, W., & Huang-Yao Hong. (2015). Technological Pedagogical Content Knowledge (TPACK) and Design Thinking: A framework to support ICT lesson design for 21st-century learning. *The Asia-Pacific Education Researcher*, *24*(3), 535–543.
Lai, C., & Jin, T. (2021). Teacher professional identity and the nature of technology integration. *Computers and Education*, *175*(April), 104314.
Mishra, P., & Koehler, M. J. (2006). Technological Pedagogical Content Knowledge: A framework for teacher knowledge. *Teacher College Record*, *108*(6), 1017–1054.
Nelson, M. (2017). The role of a mentor teacher's TPACK in prospective preservice teachers' intentions to integrate technology. *Journal of Technology and Teacher Education*, *25*(4), 449–473.
Nurhadi, D., Purwaningsih, E., Masjkur, K., & Nyan-Myau, L. (2019). Using TPACK to map teaching and learning skills for vocational high school teacher candidates in Indonesia. *Advances in Social Science, Education and Humanities Research*, *299*(38–40), 38–40.
Partnership For 21st Century Skills. (2008). *21st Century Skills, Education & Competitiveness*. Partnership for 21st Century Skills. www.21st century.org
Rahmawati, A., Suryani, N., Akhyar, M., & Sukarmin. (2020). Developing self-assessment instrument for measuring pre-service teachers' technological pedagogical vocational knowledge. *Humanities & Social Sciences Reviews*, *8*(3), 676–685.
Sally, F., Carole, O., & Sarah, C. (2011). *Effective teaching and learning in vocational education.* www.lsnlearning.org.uk
Teo, T., Unwin, S., Scherer, R., & Gardiner, V. (2021). Initial teacher training for twenty-first-century skills in the Fourth Industrial Revolution (IR 4.0): A scoping review. In *Computers and Education*, *170*, 1–21.
Valtonen, T., Sointu, E., Kukkonen, J., Kontkanen, S., Lambert, M. C., & Mäkitalo-siegl, K. (2017). TPACK updated to measure pre-service teachers' twenty-first-century skills. *Australian Journal of Educational Technology*, *33*(3), 15–31.
Voithofer, R., Nelson, M. J., Han, G., & Caines, A. (2019). Factors that influence TPACK adoption by teacher educators in the US. *Educational Technology Research and Development.*
Wheelahan, L. (2015). Not just skills: what a focus on knowledge means for vocational education. *Journal of Curriculum Studies*, *47*(6), 750–762.
Winch, C. (2010). *Dimensions of Expertise: A conceptual exploration of vocational knowledge*. Continuum.

Teacher Education and Teacher Professional Development in the COVID-19 Turn – Drajati & Putra (Eds)

The contribution of language teacher identity formation to English teacher professional development viewed from community of practice

Dewi Sri Wahyuni, Joko Nurkamto, Munawir Yusuf & Dewi Rochsantiningsih
Universitas Sebelas Maret, Surakarta, Indonesia

ABSTRACT: This study aims at describing the contribution of language teacher identity formation to English teacher professional development viewed from membership of community practice. The participant involved is Ms. DR who teaches English at a state senior high school in a suburb area in Central Java. In this study, the researchers apply narrative inquiry as it legitimates to investigate Ms. DR language teacher identity formation based on her continuous experiences and performances in the community (Barkhuizen et al. 2014). The data were collected using narrative written and focus discussion techniques and analyzed using interactive models of analysis (Miles et al. 2014). The result showed that Ms. DR's formation of language teacher identity changed gradually as she interacted and negotiated the meanings within her community (Wenger 1998). It contributed to her professional development. Since language teacher identity provokes positive attitudes toward teacher professional development, an English language teacher hopefully does it narratively, reflectively, and regularly in the community of practice.

Keywords: community of practice, language teacher identity, teacher professional development, narrative inquiry

1 INTRODUCTION

Teacher identity (TI) is viewed in various ways due to many affecting factors. For some, TI is an approach toward either personal or community practice development resulting from interaction between individual and social through negotiating some meanings (Canagarajah 2017; Tsui 2007; Teng 2019). TI is also seen as a belief that is central to teacher education and teacher learning (Barcelos 2017) that provides a highly individual sense characterizing who we are as an English teacher (Benson 2017). Varghese (2005, 2017) brings the concept of TI as subject of the study widely in English teaching and learning as an emerging subject of interest in research on language, teacher education and teacher development. TI provides some awareness of position in the standardized professional teacher provided by government and community as well (Farrell 2017; Leibowitz 2017; Richards 2017). Teaching English as foreign language needs an identity on the level of awareness, so that teachers can reflect on their usefulness (Farrell 2017). The way TI contributes to professional development can be seen as nexus membership entwined social and professional membership in community of practice, especially English teacher community (Canagarajah 2017; Richards 2017; Wenger 1998).

The issue of language teacher identity (LTI) becomes an interesting subject for studies in non-native English's speaking countries since English is both of medium instruction and subject of teaching (Nunan 2017). Teachers for EFL find more difficulties in giving instructions and knowledge using English while English is not exposed widely in daily communication. This phenomenon underlines the importance of investigating language teacher identity. More studies focus on the formation than relate the formation to further issues. This study wants to fill this gap by relating identity formation to teacher professional development since personal identity engages to membership of community practice due to recognition of professional identity (Pennington & Richards 2016). This

DOI 10.1201/9781003347798-31

study is questioning how LTI formation of a state senior high school English teacher in Central Java contributes to teacher professional development viewed from membership of community practice?

2 LITERATURE REVIEW

Language teacher identity (LTI) focuses on knowing who I am as a teacher of language but also what kind of language teacher I am (Benson 2017; Richards 2017). The forming sense of self-image and self-awareness is a requirement of being and becoming a language teacher (Barcelos 2017). It is not only shaped by changes in social, teachers positioning in the community of practice and how this position leads to reaction (Leibowitz 2017), but also formed by historical experiences that are negotiated to take action in the present days (Norton 2013; Oda 2017). As multifaceted, layered and dynamic acknowledgement, LTI provides emotional aspects, interaction between personal and social, experiences and professional community of language teachers (Barcelos 2017; Richards 2017; Teng 2019). The dynamic concept of LTI emerges its functions in the area of teacher teaching and learning. LTI gives sign that the professional concept always changes since LTI contributes to the dynamism and vitality to the profession as teacher of language (Canagarajah 2017). Directly and indirectly, LTI provides information reflective and participative in the teaching process that will lead to self-understanding or self-awareness (Leibowitz 2017; Pennington & Richards 2016) which contributes to the teacher professional development (TPD). This TPD is done voluntarily and consciously when teachers have self-awareness of the position they belong to this present time. Reflection of LTI becomes the basis for taking TPD activities (Pennington & Richards 2016).

Investigating LTI on its formation and contribution is challenging to educator researchers. Some researchers relate the issue of LTI formation with the community of practice concept proposed by Wenger (1998). It relates how one negotiates meaning in various contexts of community of practices. Tsui (2007) explores how the identity of language teachers is formed using social theory of identity formation proposed by Wenger (1998), particularly the concept of community of practice. Tsui investigates that the identity formation of becoming an EFL teacher in a non-English-speaking country is a dual process of identification and negotiation of meanings. Tsui finds that the participant of research is struggling in maintaining his professional identity as a language teacher and identifies that there are complex relations in practice and processes of identity formation. In general, Tsui describes how an EFL teacher forming accepted and recognized his or her identity involves many aspects ranging from personal to part of membership in a community in which the ability of negotiating and consolidating meanings determine also the position as a member in community of practice. Meanwhile, Li (2020) notes that identity formation as EFL teacher is more complex in the non-native English' speaking countries. In her study, Li recognizes that personal histories and social contexts shaped beliefs and emotions about learning and teaching English; at last, the beliefs and emotions contribute to the teacher's identity formation. Extracting the theory proposed by Barcelos of dynamic construction in beliefs, identity, and emotions, Li unpacks the complexity of identity formation from two Chinese English non-state teachers. These two studies highlight that teacher identity formation is a complex process faced by EFL teachers as non-native speakers. The process captures the intertwined of personal (history, beliefs, identity, and emotions), professional (knowledge on English language teaching) and social interactions (within community of practice).

Since most researchers observe LTI formation as the main issue, this research tries to investigate how LTI formation contributes to professional development as EFL teacher. Wenger (1998) notes that teacher faces negotiable situations by which she or he can learn meanings that can change the identity. Teacher learning is different from student learning since there are no definitive rules requiring teachers to learn. Teacher learning is a matter of willingness to develop knowledge and skills in performing classroom teaching (Wenger, 1998). The process of teachers' learning can be observed in detail by making use of the Community of Practice (CoP) concept. In CoP, the process of learning takes place on informal and social situated interaction (Huiyin 2019) which is collective, relational, and social process. The process of acquiring identity is gained through regular interactions by involving and contributing in the practice (Chen & Cheng 2014; Huiyin 2019) and also the artifacts. It is proved to be effective in increasing teaching and learning process for higher education (Viskovic 2006) when it is applied in teacher professional development (Wenger 1998).

3 METHOD

This study took place at the state senior high school of a well-known district in Central Java by the academic year 2019/2020 which has more than one English teacher. The method for sampling in this study was purposive sampling since it works with small samples and intends to have in-depth context (Miles et al. 2014; Yin 2011). The researchers set the boundaries and frames to define participants of this study by giving some criteria, such as: experienced for more than five years in teaching English, joining actively in an English teachers' community, certified as professional teacher by the government, recognized as professional teacher by students, colleagues, institution, and community of practice. There was one teacher who could meet these criteria, i.e. Ms. DR. This 42 years old teacher kindly shared her experiences in developing her professional career by narrative written document and depth focus discussion on her belief and emotion in professional membership of community

The researchers used narrative inquiry to get the aim of this study; that was gaining depth knowledge on how the language teacher identity formation contributes to the teacher professional development viewed from the membership in the community of practice. Narrative inquiry is commonly used in researching teaching identity in various contexts (Barkhuizen 2014; Barcelos 2017; Pennington & Richards 2016; Tsui 2007; Varghese 2005). This method points at handling in-depth interviews, diaries, visual narratives and; as techniques for data collecting (Barcelos 2017). In this method, the participant was asked to write her experience in forming identity as a professional teacher narratively including personal and professional aspects and the process of negotiation in the social context of community (Barkhuizen 2014; Barcelos 2017). In the depth-interview technique, the participant is interviewed on her emotion and belief in the context professional. To get more detailed data and trustworthiness, the researchers used focus group discussion with the colleagues, the chief of her institution, the members of her community of practice, and also with students in her classes. In analyzing the data collected, the researchers adopted an interactive model of analysis as proposed by Milles, et al. (2014) in three paths of analysis, namely condensation, displaying, and concluding/verification. The data were analyzed on the criteria of personal engagement within the community, negotiating professional meaning as members of the practice community, and social interaction as part of community membership.

4 FINDING AND DISCUSSION

Answering the questions on how LTI formation contributes to the professional development of English teachers in a state senior high school of a suburb area, this research runs on the Community of Practice theory proposed by Wenger (1998). How did the participant (Ms. DR) acquired her professional LTI through simultaneous interaction within members of the community and how she contributed to the practice (Chen & Cheng 2014; Huiyin 2019). The findings of this study informed that Ms. DR formed her LTI on the personal identification in community, professional negotiation meaning in community and active interaction with gradual membership in community of English teachers in which these processes gave contribution to personal and communal professional development.

4.1 *The process of identification as peripheral participant*

Ms. DR had been engaged to have her identity as teacher, as she acknowledged herself to be pre-service teacher although she realized that she did not have not enough competence to teach. She believed that studying in the college would increase her competencies in teaching English. Barcelos

(2017) notes that the process of forming LTI involves beliefs of tacit knowledge, experiences of personal, social, and professional, and emotion on facing both beliefs and experiences. As Ms. DR moved from pre-service teacher into in-service teacher, she had to join the English teachers' community formally. She came to the process of identification whether she identified herself or by other members of the community. This process involved three modes; engagement, imagination and alignment (Tsui 2007; Wenger 1998). In the early moments of joining the community, Ms. DR found that it was very difficult to be engaged. It was hard to determine what should be or not should be within the professional community as Wenger notes that the process of engagement is investing ourselves in what we do within the community (1998) and gained through the practice. Tsui states that the process of engagement determines the membership of a community (2007).

The position as new participant provoked her intense emotion and affected her teaching practices and interaction with students and colleagues (Barcelos 2015; Li 2020). Her belief that teaching remarks as a process of self-enjoyment did not match with the community practice' belief when she learned that teaching remarks also in political issue. It arouses emotion of insecurity on cognitive dissonance of ideal teacher belief and reality (Li 2020). Ms. DR found that the process of formatting English teacher identity was struggling in the early years since she was recognized as new comer. With the process of engagement as part of community, imagination of being good teacher for students, and alignment with senior members in good communication and interaction, she proved that her identity as good English teacher was confirmed by other members (Borg 2017).

The first step of the forming LTI in community was the identification process for peripheral member. Peripheral participant tried to be identified personally as part of community in which there was a recognition gap between former and new members. To limit the gap and be engaged in community, the peripheral member needed alignment with senior members. The process of alignment could be developed through good communication and interaction. The conflict of emotion came to surface because there was a gap of ideal personal beliefs and reality of communal beliefs. The peripheral member needed to reimagine her beliefs through negotiation process in her professional activities.

4.2 *Negotiating meaning in professional and social interaction as active participant*

The second step of forming LTI was negotiation meaning that determined the extent of contribution and shaped the invested meaning within community (Wenger 1998). The changing status of Ms. DR, from substitute teacher to state teacher, contributed to the shift of membership in community, from peripheral into active participant who had an active role (Huiyin 2019). She negotiated her imagination on ideal teaching English with the community's practical belief. This reimagination process minimized emotion conflict as she perceived herself for her new identity of being and becoming language teacher (Feng 2019). She was doing negotiation for maintaining her meanings in community since the community was not developed in one solid and fix context; but it is dynamic and changeable context based on social interaction (Canagarajah 2017; Richards 2017). The ability to negotiate meaning in community brought Ms. DR far from the sense of marginal participant (Tsui 2007). The context of situation while teaching in class became her concern to negotiate the role that should be taken in certain time. When the students needed more explanation and attention, she put herself to be manager instead of facilitator in learning process. This role would be dynamic based on her negotiation of the situation (Burns & Richards 2005).

As younger active member, she was pointed to present some professional workshops, events, and so on either by her institution or community of practice. She accepted these orders as her responsibility to develop her LTI, institution, and community as well (Pennington & Richards 2016). When she attended some professional activities, she negotiated herself to be learner who acknowledged new knowledge. Her meaning as learner changed when she was asked to share knowledge. She meant herself as informer in the institution and community. Negotiation in meaning determined the extent to which one was able to contribute and shape the meanings in which one was invested as the basic process in identity formation (Tsui 2007; Wenger 1998).

This second step of forming LTI focused on reimagination of belief by negotiating meaning between personal ideal belief and practical of community belief in active membership. The ability of negotiation meaning in various social contexts led to the easiness of role shifting in professional context and community interaction. The changing role contributed to self-professional development as Mr. DR became active in seeking new knowledge and shared it in the community of practice for communal professional development.

4.3 *Becoming teacher agent in core membership*

Ms. DR's LTI brought awareness and self-reflections of usefulness in her community (Farell 2017; Richards 2017) and shifted her participant from active into core of membership in accordance with her professional development growth (Huiyin 2019, Tsui 2007; Wenger 1998). Her focus moved from individual professional development to community professional development. This shifting of an autonomy teacher whose focus was mostly on the personal' development became individual whose acting focused on the environment or community development was the process of becoming teacher agent (Kalaja et al. 2015). Handling and managing professional activities in community, promoting other members of community to develop their professional identities, sharing new knowledge in teaching English to the community, becoming inspiring and motivating teacher for her students were her roles as core member in community of practice (Wenger 1998).

The changing membership from active into core member in community was the temporary highest step of LTI formation since LTI was the reflection of dynamic process involving personal (including beliefs, emotions, and experience), professional (knowledge in teaching and English), and social aspect (interaction within community negotiated through the contexts in the query of searching the meaning). As LTI was confirmed by the government, institution, community, and students; there was a changing role in the professional context. Identity as an autonomous teacher shifted into the identity of teacher agent that led to community professional development.

5 CONCLUSION AND RECOMMENDATION

Focusing on how LTI formation contributes to English teacher professional development, this study remarked that LTI formation serves as the base for self-understanding and self-awareness of the teacher to develop her professionalism within the community. Since LTI involves interaction in the community, it becomes a dynamic and unstable process. This study revealed that the participant moved from peripheral member to the highest rank of membership in the community. She showed the process of personal identification by engaging herself in the community of English teachers, reimagining belief by negotiating personal belief and community belief, and aligning her social interaction actively in the community. These processes of identification were gradually experienced from peripheral member to core member. LTI made a contribution to professional development by shifting the role not only from learner to informer but also from an autonomous teacher to a teacher agent.

This study implied that it was important for teachers to know their own identity which could be done narratively, reflectively, and regularly. Teachers would realize the position they belong to in the community; as confirmed participant or marginal participant. The ability to involve and negotiate meaning in community became an important aspect to develop personal professionalism. The further research could be run on the strategies of forming English teacher agents, how it would affect the community and the psychological challenges faced by English teacher agents in the community.

REFERENCES

Barcelos, A. M. F. (2015). Unveiling the relationship between language learning beliefs, emotions, and identities. *Studies in second language learning and teaching.*

Barcelos, A. M. F. (2017). Identities as emotioning and beliefs. In Barkhuizen, Gary (editor). *Reflections on language teacher identity research, pp: 145–150*. New York: Routledge.
Barkhuizen, G., Benson, P., & Chik, A. (2014). *Narrative inquiry in language teaching and learning research*. New York: Routledge.
Barkhuizen, G. (2017). *Reflections on language teacher identity research*. New York: Routledge.
Beauchamp, C., & Thomas, L. (2009). Understanding teacher identity: an overview of issues in the literature and implications for teacher education. *Cambridge Journal of Education*, 39:2. 175–189.
Benson, P. (2017). Teacher autonomy and teacher agency. In Barkhuizen, G (editor). *Reflections on language teacher identity research,* pp: 18–23. New York: Routledge.
Canagarajah, S. (2017). Multilingual identity in teaching multilingual writing. In Barkhuizen, G (editor). *Reflections on language teacher identity research, pp: 67–73*. New York: Routledge.
Castañeda, J. F. A. 2011. *Teacher Identity Construction: exploring the nature of becoming a primary school language teacher*. Dissertation of the PhD degree, Faculty of Humanities and Social Sciences School of Education, Communication and Language Sciences, University of New Castle.
Chen, C. W., & Cheng, Y. (2014). Learning from team teaching and beyond: a case study on EFL teachers' professional development. *Pan-Pacific Association of Applied Linguistics, 18*(1), 33–47. emotions, and identities. *Studies in second language learning and teaching*.
Coldwell, M. (2017). Exploring the influence of professional development on teacher careers: A path model approach. *Teaching and Teacher Education, 61*, 189–198.
Creswell, J. W. (2014). *Educational research: planning, conducting, and evaluating quantitative and evaluating quantitative and qualitative research, fourth edition*. Boston: Pearson Education Inc.
Farrell, T. (2017). "Who I am is how I teach": reflecting on language professional role identity. In Barkhuizen, G (editor). *Reflections on language teacher identity research, pp: 183–188*. New York: Routledge.
Huiyin, L. (2019). From Periphery to the Core: Promoting College English Teachers' Professional Development from the Perspective of Community of Practice. *English Language Teaching*, Vol. 12(9), 74–8.
Kalaja, P., Barcelos, A. M. F., Aro, M., & Ruohotie-Lyhty, M. (2015). *Beliefs, agency and identity in foreign language learning and teaching*. Springer.
Leibowitz, B. (2017). Language teaching identity in troubled times. In. Barkhuizen, G (editor). *Reflections on language teacher identity research, pp: 74–79*. New York: Routledge.
Li, W. (2020). Unpacking the complexities of teacher identity: narratives of two Chinese teachers of English in China. *Language Teaching Research*: 1–19.
Miles, M. B., Huberman, M. A., & Saldana, J. (2014). *Qualitative data analysis: a method sourcebook, (3rd ed)*. Sage Publisher.
Norton, B. (2017). Learner investment and language teacher identity. In Barkhuizen, G (editor). *Reflections on language teacher identity research, pp: 80–86*. New York: Routledge
Oda, M. (2017). Reflecting on my flight path. In. Barkhuizen, G (editor). *Reflections on language teacher identity research, pp: 222–227*. New York: Routledge.
Pennington, M. C. & Richards, J. C. (2016). Teacher identity in language teaching: integrating personal, contextual, and professional factors. *RELC Journal*, 1–19.
Richards, J. C. (2017). Teacher identity in second language teacher education. In Barkhuizen, (editor). *Reflections on language teacher identity research, pp: 139–144*. New York: Routledge.
Teng, F. (2019). *Autonomy. agency, and identity in teaching and learning English as a foreign language*. Hongkong: Springer.
Tsui, A. B. M. (2007). Complexities of identity formation: a narrative inquiry of an EFL Teacher. *TESOL Quarterly*, (41), 4.
Varghese, M., Morgan, B., Johnston, B., & Johnson, K. A. (2005). Theorizing language teacher identity: three perspectives and beyond. *Journal of Language, Identity & Education*, 4:1, 21–44.
Varghese, M. (2017). Language teacher educator identity and language teacher identity: towards a social justice perspective. In Barkhuizen, G (editor). *Reflections on language teacher identity research, pp: 43–48*. New York: Routledge.
Wenger, E. (1998). *Communities ofpractice: learning, meaning, and identity*. New York: Cambridge University Press.
Yin, R. K. (2011). *Qualitative research from start to finish*. New York: Guilford Press.

Teacher Education and Teacher Professional Development in the COVID-19 Turn – Drajati & Putra (Eds)
© 2023 The Author(s), ISBN: 978-1-032-38970-7

A student teacher's agency construction during the COVID-19 pandemic

Intan Pradita, Salma Maulani & Afrianto Akhmad Mukhamirudin
Islamic University of Indonesia, Yogyakarta, Indonesia

ABSTRACT: This study aims to investigate a student teacher's experiences in constructing her teacher agency. The data were collected through diaries for five weeks of teaching practicum in schools and four weeks of micro-teaching practicum. This study employed a qualitative narrative research design conducted in a high-qualified senior high school in Yogyakarta. The data were analyzed using Appraisal theory for the lexicogrammar analysis and then linked the coding into the theory of teacher agency. Findings showed that flexible and informal interactions are needed to construct teacher agency instead of a rigid curriculum. The implications for higher education stakeholders in shaping teacher agency include accommodating student teachers to be involved in self-assessment, informal dialogue with the students at schools, and informal peer discussion.

Keywords: Learning engagement, teacher agency, teacher agency construction

1 INTRODUCTION

During the COVID-19 pandemic, student-teachers were forced to have online learning and teaching. This experience could be challenging since they should manage themselves as learners during the microteaching stage and apply the knowledge to their teaching practices at schools. However, the method of bringing theory into practice does not directly determine the promotion of teacher agency. "There should be sufficient collective development as well as a firm consideration from their environment" (Biesta et al. 2015). Thus, it is necessary to research the process of student-teacher agency construction under the perception of the student-teachers. This research considers that the learning process could extensively affect how to construct the agency. By referring to Griffiths' (2015) concept, a successful learning process would bring a significant achievement, so this research valued successful learning through the student-teachers engagement in their learning process. Thus, further discussion in this study examined teacher agency construction under the lens of student-teachers' *iterational* dimension, especially through their learning engagement.

Moreover, the narration of growing teacher agency through the learning experience was still rarely discussed in the Indonesian context, especially by using qualitative research. Current empirical findings on previous studies tend to capture quantitative results (Mardiah & Yulhendri 2020) and (Triastuti 2020). Instead of narrating the development of teacher agency, they focus on measuring the pedagogical competencies of student teachers. Whereas, focusing too much on student teachers' pedagogical competence tends to make student teachers consider what they do as just a job rather than a meaningful profession (Leijen et al. 2019). Thus, it is necessary to consider not only the pedagogical competencies of the student teachers, but also to explore the extent of teacher agency construction of the students' teachers.

Although the number of English language teacher departments in Indonesia is high, research on teacher agency construction is limited (Hamid & Nguyen 2016; Yusra et al. 2021). Meanwhile, to perceive teachers as a meaningful profession requires the process of teacher agency construction. Thus, this research focused on reflective practice as means to construct teacher agency rather than on the quantitative measurement of pedagogical competence. The findings of this research are

 DOI 10.1201/9781003347798-32

expected to shed light on the teacher agency construction as it is a qualitative indicator of pre-service teachers' pedagogical competence. The research question underlying this research is: "how is construction of teacher agency from micro teaching and teaching internship experiences?"

2 LITERATURE REVIEW

2.1 *Teachers' agency construction*

The term teacher agency is broadly defined as the way teachers' understanding and intention are enacted and shaped critically in particular situations whether physical, emotional, social, pedagogical, or professional (Pappa et al. 2019). Even in more constraining circumstances, teacher agency becomes apparent and occlusive (Damşa et al. 2021). Additionally, there are some contributing factors to construct teacher agency. Biesta et al. (2015) formulated three dimensions that contributed to the construction of the teacher's agency. Those are; 1) *iterational* dimension which deals with the past like teachers' life histories and professional histories, 2) practical-evaluative dimension which deals with present such as cultural (ideas, values, beliefs, discourses, languages), social structures (relationship, roles, power, trust) and material (resources, physical environment), lastly 3) projective dimension which deals with the future whether in the short term or long. The previous study found that by succeeding in teacher agency construction, teachers would consider what they do as a meaningful profession rather than just a job (Tao & Gao 2017; Leijen et al. 2019). Additionally, teacher agency construction can be succeeded through dialogic inquiry which makes student-teachers aware of their capacity for agency (Wallen & Tormey 2019). When they are able to fully practice agency, they could strengthen their commitment and increase their contribution. Imants and Van der Wal (2019) even explained and revealed how teacher agency can be a parameter to identify professional development.

2.2 *Learning engagement to build the agency*

Many scholars have approved the essentials of learning engagement. Consequently, many researches have been conducted over years. Martin and Bolliger (2018) reveal the benefits of engagements such as increasing student satisfaction, enhancing students' motivation to learn, reducing student isolation and improving student performance. The benefits, in the case of initial teacher education, could raise a possibility to foster the construction of teacher agency. Learning engagement, according to Fredericks et al. (2004), are divided into three types; behavioral, emotional, and cognitive engagement. For example, when student-teachers perform cognitive engagement, they tend to do problem-solving better. To this extent, practical-evaluative dimensions were shaped. It was because they got a meaning-making process of practical and evaluative dimensions when the student teachers used their knowledge and understanding about teaching English. The same case was found in utilizing technology in language teaching. By demonstrating this ability, it could foster student-teacher agency, especially in projective dimensions when the student teachers need to adjust the current needs and the possible solutions (Bond et al. 2020).

The further stage of learning engagement is learning agency (Rajala et al. 2016). Mingmei (2019) explains that when completing a task, learners' agency relies on the process of dynamic negotiation between the learner and the social and cultural background in order to build a personal sense of agency. Mercer (2012) even further explained that nurturing the sense of agency will benefit students to help them become independent and active learners. For example, students who actively engaged to get learning experiences were able to contribute in real situations efficiently and perform valuable work (Triastuti 2020). Although it was found that the student-teacher needs to have their personal sense of agency, the previous studies have not elaborated the process of constructing personal sense of teacher agency.

3 METHOD

The setting of this study was in Indonesia and the participant in this study was Hayu (pseudonym). She was 20 years old, and she had 3 years of learning experience. She had two experiences of teaching English for a private course with two students. Moreover, she also experienced a fully online Micro Teaching practicum with her classmate in the university. The researchers got her consent due to the fact that she was one of our Micro Teaching laboratory members. Even though she had a little experience of teaching English fully online and teaching a whole class simultaneously in a month. However, after the practicum was done, Hayu should teach English fully online in a senior high school during the pandemic. When the time came, Hayu was assigned by her department to teach English for four weeks in a high-qualified senior high school in Yogyakarta. The data were generated during these four weeks.

This study is a narrative inquiry (Clandinin 2006). The data collection was conducted through accessing the participants' teaching diary in Google Sites. There were four weeks of teaching diary starting from July, 1st 2020 until August, 5th 2020; and two weeks of micro-teaching learning reflection from March, 24th until April, 7th 2020. There were 30.047 words in the data in this study. Each clause of the diary was transferred into a data analysis table. The data were then analyzed by 1) reading and rereading the content of Hayu's teaching diary to get a sense of the data, 2) doing axial-coding upon the diary by referring to language appraisals theory, 3) interpreting the coding to investigate the sequence of the feelings and perceived emotions that Hayu experienced in micro-teaching and teaching practicum, and 4) sequencing the events that begin with the character, setting, and time frame. These themes were generated by combining several codes and re-narrating the codes into a story. The re-narration process is done by meaning-making of the data. Finally, we did a focused-group discussion to confirm our data analysis to Hayu to gain data trustworthiness.

4 FINDINGS AND DISCUSSION

This section firstly presents Hayu's narratives and her reflective teaching agency. The interpretation of the narratives was elaborated into three emerging themes: cognitive engagement as meaningful experience, self-evaluation and the shared responsibility, and the sequence of teacher agency construction.

4.1 *Hayu's story: "I want to be a teacher"*

Through the narratives, Hayu perceived herself to be well known as a timid student teacher by her friends. She had a good linguistic competence as proven by her B1 CEFR level. However, teaching English by infusing her linguistic competence was not enough for Hayu. Once she experienced micro-teaching class, she became insecure because apparently teaching was not a simple job. There were many things to prepare just for ninety minutes of teaching practice. Ms. Carla, her micro-teaching lecturer, provided a learning circle with three other student teachers. At the beginning, Hayu did not really understand the function of this circle unless it was going to be annoying. Hayu murmured that group work should have never been invented. To her surprise, when she dealt with every single instruction by Ms. Carla, Hayu found that although it seemed hard, she tried to follow each stage and task. Hayu enjoyed the feedback-revision sessions until she found herself better in teaching performance. In the school, Hayu tried to implement all the knowledge that she has gotten from her lecturers. Whenever she found a difficult situation, she preferred telling this case to her learning circle. After twice teaching and a marathon of discussion with the supervisor and her peers, Hayu found a moment when her students noded a lot, interacted with her, and enjoyed her class. Her students' reactions and engagement became Hayu's main priority. She finally had a teaching model. Hayu has decided to be a teacher. Although it might not be easy in the future, she will always want to be a good teacher.

4.2 *Cognitive engagement as a meaningful learning experience*

Although the pandemic forced Hayu to have face-to-face supervision with her lecturer, Hayu was considered an engaging and attentive learner. This represents the type of behavioral engagement (Frederick et al. 2004). Moreover, she did the whole process of learning engagement by investing her time to keep learning until midnight, empowering her understanding, and practicing teaching. It shows that Hayu made progress from behavioral engagement to emotional engagement due to her willingness to keep learning until midnight and she also made progress to cognitive engagement since she was able to implement her understanding to practice. However, as Biesta et al. (2015) previously emphasized, self-progression should be accommodated not only by the student-teacher herself, but also by a supportive environment. Robinson & Hullinger (2008) found learning engagement could be managed by providing meaningful learning through multimodal networks or multichannel communication and by creating learning communities for their students. To this extent, Hayu also mentioned that she got some feedback for three teaching cycles. Not only the lecturer encouraged her, but also her peers in microteaching class gave feedback. Thus, Hayu's micro-teaching lecturer was considered to be a part of a supportive environment due to the provision. Hayu perceived the feedback as helpful to develop her teaching skills. It could be identified through the coding that Hayu tended to be satisfied with her self-progression when she could revise. Through the feedback and her learning engagement, Hayu grew to be a student-teacher who did preventive actions to anticipate any possible problems.

4.3 *Self-evaluation and the shared responsibilities*

Hayu continued to manage herself as an engaging learner during the school-based teaching practicum. She kept being attentive when she got some feedback from her supervisor teacher at school. Hayu maintained her self-evaluation habit that she has built since she was in microteaching practicum. In this case, Hayu's teacher agency started to emerge, especially in the practical-evaluative dimension (Biesta et al 2015). Through the data codification process, it was found that Hayu's narration tends to represent her appreciation in each process. Martin & White (2005) explained that appreciation discourse tends to highlight the evaluations of semiotic and natural phenomena. The majority of her evaluative lexis was about how she perceived her teaching practice as worthy. She had been in the lowest moment when she found that her first teaching practicum at the school was not satisfying. Hayu felt disappointed when her instructional design was not scored less than her peers during school-based teaching practice. She then devoted herself to keep practicing and sometimes she asked her peers how to develop an effective instructional design. Hayu could invest her time by doing the revision in a quiet area in the school. Hayu chose not to give up. To this extent, Hayu's teacher agency has been constructed. The willingness and abilities to act upon and transform educational activities and circumstances (Rajala et al. 2016) were narrated by Hayu in her diary. Thus, the students at school could be the determiner of successful teaching practice. More attention should be provided to developing informal peer dialogue that tends to elevate their problem-solving skills.

4.4 *Teacher agency construction*

Biesta et al. (2015) emphasized that teacher agency was not something that can be owned like property, or to be achieved like capacity or competence. Teacher agency dealt with something that the teacher as the actor did. In Hayu's story, her most important consideration was her students' reactions and engagement. The students in schools, who used to be less considered, apparently have contributed to shaping the teacher agency. Hayu described that after having a timely feedback and revision session from her supervisors, she felt glad because her students gave a positive response to her latest instructional design. In this respect, when a student-teacher puts a strong consideration on the school students, s/he has already had a call as a teacher. The call was that Hayu did not perceive her internship duty as simply a way to get a score and to pass the internship program. She did meaning-making by taking every step of teaching preparation as a serious matter in order to make her students engaged with the learning process. To this extent, Hayu's teacher agency was not about how well she taught her students or how effectively she implemented teaching strategies. It was about the process that she felt and she dealt with by being a student-teacher. Hayu had significant development. Hayu deconstructed her own thought from '*being a teacher should not be difficult*' into '*I will keep trying to be a good teacher*'. Hayu grew to be a teacher who did not easily give up. She even closed her diary with some suggestions for student-teachers who had online teaching practice. The figure of this process was displayed below:

Figure 1. The construction of teacher agency.

As derived from Figure 1, there is one remarkable contributing factor to shape teacher agency. It is the school students' responses and attitudes during their teaching internship program. To some extent, Hayu felt insecure at the beginning when she had to deal with this factor. However, empowering the other contributing factors enabled her to switch her insecurity into a teacher agency. The contributing factors were the lecturer's role modeling, a recall of previous teaching experiences, and peer learning. Hayu was practicing practical evaluation by referring to her past experiences in micro-teaching class. She also kept evaluating her previous teaching performance. This process was continuing until she came into the longing of motivation for not giving up. When she narrated that she wanted to be a teacher, the projective dimension was constructed. Hayu did not perceive her teaching internship as just a task accomplishment. Being a teacher became her projective dimension in the future.

5 CONCLUSION AND RECOMMENDATION

This research aims to narrate the construction of a student-teacher teacher agency during the pandemic. The construction process does not fully rely on student-teachers' learning engagement in microteaching class. There are more informal issues that need to be considered such as peer dialogue, role modeling, and concerning students' needs. As an insight, higher education stakeholders should reconsider the role of school students in shaping teacher agency by; 1) providing more time to discuss with the student teachers as well as to listen to their teaching reflection, 2) accommodating the student teachers to keep evaluating and, if necessary, to revise the lesson plan before every teaching practice, and 3) suggesting the student teachers to have informal interactions with their peers to share thoughts and feelings during teaching practicum. Further studies could provide extensive research on the role of students at school to develop the teacher agency of student teachers.

REFERENCES

Biesta, G, Priestley, M., & Robinson, S. (2015). The role of beliefs in teacher agency. *Teachers and Teaching, 21*(6). 624–640.

Bond, M., Buntins, K., Bedenlier, S., Zawacki-Richter, O., & Kerres, M. (2020). Mapping research in student engagement and educational technology in higher education: a systematic evidence map. *International Journal of Educational Technology in Higher Education, 17*(2), 1–30.

Clandinin, D. J. (2006). Narrative inquiry: A methodology for studying lived experience. *Research studies in music education, 27*(1), 44–54.

Damşa, C., Langford, M., Uehara, D., & Scherer, R. (2021). Teachers' agency and online education in times of crisis. *Computers in Human Behavior.* 121, 1–16.

Fredricks, J. A., Blumenfeld, P. C., & Paris,A. H. (2004). School engagement: Potential of the concept, state of the evidence. *Review on Educational Research, 74*(1), 59–109.

Griffiths, C. (2015). What have we learnt from 'good language learners'?. *Elt Journal, 69*(4), 425–433.

Hamid, M. O., & Nguyen, H. T. M. (2016). Globalization, English Language Policy, and Teacher Agency: Focus on Asia. *International Education Journal: Comparative Perspectives, 15*(1), 26–43.

Imants, J. & Van der Wal, M.M. (2019). A model of teacher agency in professional development and school reform. *Journal of Curriculum Studies. 52*(1), 1–14.

Leijen, A., Pedaste, M., & Lepp, L. 2019. Teacher Agency Following the Ecological Model: How It Is Achieved and How It Could be Strengthened by different types of reflection. *British Journal of Educational Studies. 68*(3), 295–310.

Mardiah & Yulhendri. (2020). Pengaruh IPK, Micro Teaching, dan Praktik Pengalaman Lapangan Kompetensi Pedagogik Mahasiswa Calon Guru Jurusan Pendidikan UNP. *Jurnal Ecogen. 3*(1). 165–175.

Martin, F. & Bolliger, D.U. (2018). Engagement matters: Student perceptions on the importance of engagement strategies in the online learning environment. *Online Learning, 22*(1), 205–222.

Martin, J. R., & White, P. R. (2005). *The language of evaluation* (Vol. 2). Palgrave Macmillan.

Mercer, S. (2012). The Complexity of Learner Agency. *Journal of Applied Language Studies, 6*(2), 41–59.

Mingmei, D. (2019, May). An Empirical Study of Learners Agency and Identity in Project-based Learning. In *2019 5th International Conference on Humanities and Social Science Research (ICHSSR 2019)* (pp. 655–659). Atlantis Press.

Pappa, S., Moate, J., Ruohotie-Lyhty, M., & Eteläpelto, A. (2019). Teacher agency within the Finnish CLIL context: Tensions and resources. *International Journal of Bilingual Education and Bilingualism, 22*(5), 593–613.

Rajala, A., Martin, J., & Kumpulainen, K. (2016). Agency and learning: Researching agency in educational interactions. *Learning, Culture and Social Interaction, 10*, 1–3.

Robinson, C. C., & Hullinger, H. (2008). New benchmarks in higher education: Student engagement in online learning. *Journal of Education for Business, 84*(2), 101–109.

Triastuti, A. 2020. Assessing English Pre-Service Teachers' Knowledge Base of Teaching: Linking Knowledge and Self-Portrayal. *TEFLIN Journal.* 31 (1). 108–131.

Tao, J., & Gao, X. (2017). Teacher agency and identity commitment in curricular reform. *Teaching and Teacher Education, 63*, 346–355.

Wallen, M., & Tormey, R. (2019). Developing teacher agency through dialogue. *Teaching and Teacher Education, 82*, 129–139.

Yusra, K., Lestari, Y. B., & Hamid, M. O. (2021). Teacher agency and the implementation of CEFR-like policies for English for tourism and hospitality: insights from local vocational high schools in Indonesia. *Current Issues in Language Planning*, 1–21.

Teacher Education and Teacher Professional Development in the COVID-19 Turn – Drajati & Putra (Eds)

The student teachers' emotional experiences in online learning at an Indonesian university during the COVID-19 pandemic

Banun Havifah Cahyo Khosiyono
Universitas Sarjanawiyata Tamansiswa, Yogyakarta, Indonesia

Nico Irawan
Rajamangala University of Technology Krungthep, Bangkok, Thailand

ABSTRACT: This interview study aims to explore the emotional experiences of student teachers in an Indonesian university during the COVID-19 pandemic. Data in this study were garnered from the results of interviews with 11 Indonesian student teachers' experiences of the virtual learning process. The interview data were thematically coded with Hargreaves's emotional geography framework. Findings showed that the COVID-19 pandemic affected the emotional experiences of Indonesian student teachers in terms of the virtual learning process, psychological mental state, and student teachers' interaction. This study also contributes to the theory that teachers or lectures may change and evolve as a cognitive process of acquiring information and skills, and that the social character of teaching and the basic relevance of emotion in the online learning process are both recognized in this study.

Keywords: EFL student teachers, emotional geography, virtual learning during COVID-19 pandemic

1 INTRODUCTION

Customary systems were thrown into disarray, resulting in unexpected regulatory adjustments during COVID-19. (Bryson et al 2020). Face-to-face interaction is used to implement several academic programs. Many university academic teaching personnel were taken off guard when institutions switched to online learning after a few weeks of face-to-face sessions during the recent COVID-19 shutdown. Despite the spectacular expansion of online learning, many academics prefer conventional face-to-face methods of student involvement and find online participation intimidating (Robinson & Hullinger 2008). Azorin (2020) demonstrates that the COVID-19 pandemic results in a temporary loss of classroom learning and the potential for human resource loss if it lasts for a long time. Online learning for student teachers comes with its own set of difficulties. During online learning, student teachers may have challenges with discipline, mental and physical health, motivation, and feelings of isolation (Raaper & Brown 2020). The COVID-19 pandemic, like any other disaster, has elicited a range of emotional problems and reactions. Faculty and student instructors may be overwhelmed by the stress and panic caused by the coronavirus pandemic. The closeness of student teachers to locations where COVID-19 outbreaks are common might lead to unanswered concerns, worry, and concern for their health and well-being (Sundarasen et al. 2020). Remote learning was quickly established in reaction to the crisis, and many student teachers were exposed to online classrooms for the first time.

Previous studies on the geography of emotions have been conducted. Liu (2016), for example, investigates the emotional geography of English as an ESL teacher with an immigrant background in England's professional lecturer community. The findings of Liu's study demonstrate that lecturers feel a range of good and negative emotions while teaching English as a second language. As a result, the teacher must be able to devise emotional geography management solutions. Marvell

 DOI 10.1201/9781003347798-33

and Simm (2018) did another study on emotional geography, this time looking at the emotional geography of students performing fieldwork practice. The study's findings indicated that students thought they were frequently dismissed at work. This condition has a significant impact on students' emotional geography; for example, they are always fearful, nervous, and insecure. Sidhu and Ishikawa (2020) did a recent study on the emotional geography of international students in East Asia. Sidhu and Ishikawa discovered that education, political criticism, and compassion influenced students' emotions. Albin-Clark (2020) describes how a peer-mentoring initiative helped an English ECE teacher alter her practice by including parents. With Hargreaves' (2001, 2005) emotional geographies approach, he agrees that reflexive processes are entwined inside emotional responses and reactions, even within the confines of a single case study. From the views and psychological experiences of graduate students, Kee (2021) investigates the influence of COVID-19. Five major topics emerged as a result of the investigation. Accepting virtual learning and teaching; dealing with disappointment; losing power and control; feeling nervous and afraid; using coping mechanisms; and finding relief.

Based on the prior studies already reported various empirical problems portrayed through Hargreaves's emotional geographies theory, however, there was a little discussion with graduate students in the EFL context and impact of COVID-19 pandemic in online learning process. Therefore, this study aims to explore the emotional experiences of student teachers' online learning in EFL context during the COVID-19 pandemic. The study specifically addresses two questions:

1. What kind of geographic emotional experiences do students experience during online learning?
2. How do students overcome geographic emotional problems that arise during online learning?

2 LITERATURE REVIEW

Non-representational theory explores occurrences, events, moods, feelings, flow (Thrift 2008; Vannini 2015), and rhythms (Latham & McCormack 2009) in everyday life through emotional landscapes. Through "a geography of what occurs," non-representational theory seeks meaning in a landscape by stressing processes, movements, flows, and emotions (Thrift 2008, p. 2). This method focuses on the creation, lived experiences, and growth of a place, and as a result, places are "not merely seen, but also heard, smelled, and felt," generating an emotional reaction (Lorimer 2005, p. 86). Emotional geography is a branch of geographical science that investigates emotion in the social domain (Davidson et al. 2005). Hargreaves' (2000, 2001a, 2001b, 2005) study on emotional geographies theorizes that emotions function as a closing and distancing mechanism between the social groups with whom instructors interact. Liu (2016), Albin-Clark (2020), Supriyatin et al. (2020), Susanto et al. (2020), and Taloko et al. (2020) are examples of previous research from 2015 to 2021 that have previously utilized Andy Hargreaves' notion of emotional geography (2020). Hargreaves' emotional geography framework was utilized to evaluate their study data thematically in their investigations.

The present study adopts Andy Hargreaves's concept of emotional geography to examine the student teachers' emotions on their experiences in online learning during the COVID-19 pandemic (2000a, 2001a, 2001b, 2005). This theory consists of five mutually emotional dimensions related, namely professional geography, physical/personal geography, moral geography, socio-cultural geography, and political geography.

3 METHOD

This research was conducted in one of Indonesian universities. The participants of this research were 11 graduate students, from the English education student teachers program. They consisted of three females and eight males from different islands (e.g., Java, Sumatra, Sulawesi, and Kalimantan) in Indonesia. Their age is around 23–35 years old with TOEFL score around 500-550 as requirements

Table 1. Hargreaves's (2000, 2001a, 2001b, 2005) *The Emotional Geographies Theory.*

Emotional Geographies	Conceptual Description
Professional Geographies Distance	In this dimension, professional norms influence relationship/social interaction between lecturers, students, and other campus communities. Each campus has professional norms that must be obeyed/followed by the campus community, including international students.
Physical Geographies Distance	This geographic dimension affects social interactions limited by factors, such as emotional closeness occurs in the relationship between students, supervisors, and other campus communities. Distance shows that this emotional connection is strongly influenced by the intensity of the relationship made by students and lecturers supervisors.
Moral Geographies Distance	The discourse of proximity or social inequality is influenced by moral values and norms possessed by students, lecturers, and other campus communities in achieving their visions and goals which may differ from one another.
Sociocultural Geographies Distance	This dimension creates a discourse of proximity or social inequality due to differences in race, culture, gender, and disability, including differences in how they feel and express emotions that can create distance between students, lecturers, and other campus communities. If all these things are not understood socially and culturally, the campus community will treat students internationally based on existing stereotypes or norms of the socio-cultural campus community and the local community.
Political Geographies Distance	In this dimension, hierarchical power or power causes closeness or social inequality among the campus community. For example, the campus authority has the power in controlling the campus community, including the enactment of online learning policies for lecturers and students.

to join the graduate program. Data were taken for two weeks of May 2021, and I positioned myself as the insider of this research, a lecturer. Therefore, I played a role as an interviewer to collect the data, and I also took part in analyzing and reporting the data of this research.

This research used a qualitative interview method by Roller & Lavrakas (2015). They suggest 3 types of interview (structured, semi-structured, and unstructured). This study used semi-structured data collection. There are 9 steps on conducting semi-structured: define the research questions, identify the appropriate interviewees, select the appropriate modes, select an appropriate platform for interview, determine how data will be recorded, develop interview guide, conduct pilot interview, obtain consent from interviewees, and use good interviewing techniques. Data of the study were collected through English spoken interviews with 11 student teachers' via WhatApps using voice notes. All student teachers agreed to participate in the study voluntarily. The participants were invited to participate in semi-structured interviews. These interviews lasted for 20-30 minutes. Ethical approval and written consent were obtained from participants for the recording of interviews and the use of anonymized quotes in reports and publications. Before the researchers conducted this research, researchers asked participants' consent to fill out the form willingness to be involved in this research. The theoretical framework used to analyze the data in this study was adopted from the theory of the geography of emotions formulated by Hargreaves (2000, 2001a, 2001b, 2005). Thematic analysis is performed on data that has been transcribed and translated from Indonesian to English by recognizing patterns that recur repeatedly in the material (Widodo 2014). The process of coding particular words/sentences from various categories or themes that correspond to geographic theory emotions leads to the discovery of this pattern. The process of discovering applicable topics is easily identified when this code is given, such as the line "*I feel like a passive learner*" on the professional geography theme.

4 FINDING AND DISCUSSION

4.1 *Professional geography: I felt uncomfortable at first with online learning using Zoom, but now all changes*

Face-to-face classroom interactions were considered more comfortable for student-teacher interaction purposes because learning activities went smoothly because all elements on campus (e.g. library facilitation, laboratory, etc.) were quite supportive. The learning situation changed completely when the pandemic was going on. Considering the high rising case of the COVID-19 situation in Indonesia, student teachers are more comfortable in their homes. Despite having no prior exposure to online education, these student teachers seemed to adjust easily to the COVID-19 pandemic's unexpected move to the online learning process.

> *I feel comfortable face-to-face. It makes us interact directly with our lectures. However, on Covid-19, everything changed. I'm also nervous and self-conscious about voicing my thoughts for the first time throughout the online learning process. I felt uncomfortable at first with online learning using Zoom, but now allchange. I believe it's because different communication techniques are impacted by the participants' and lecturers' cultures... But now I'm relaxed and at ease. We will continue to employ Zoom in online learning in the future. (Participant 2, WhatsApp Interview, 3 May 2021)*

The empirical data shows that offline classes are more rewarding since there is a direct connection between lecturers and students, as opposed to online learning without this direct engagement. Along with the difficulties students had during the deployment of online learning during COVID-19 period. Many students prefer to turn off the camera during classes and discussions, according to Moorhouse (2020). This empirical data also demonstrates participants' discontent and discomfort with the COVID-19 pandemic's circumstances. In this context, Baba (2020) asserts that a person's ability to maintain emotional stability in the face of a pandemic is the skill to control emotions during a pandemic. This study suggests that the students can control their emotions while learning online to show their professional development.

4.2 *Physical geography: I feel distance from my teacher and my friends*

Emotionally, the gap between teacher and student teachers in the online classroom became wider because they reported that they have felt physically distant from their teacher during the online learning process.

> *The distance between lecturers and students is very influential, both of them will get their satisfaction when they can meet in person. Because direct discussion will have a positive effect on students and lecturers. When I study online, I feel different and unsatisfying. ...I also feel frustrated and disappointed in the online learning process.... Sometimes, unstable connection, there is no sound, or I have automatically left the meeting. (Participant 1, WhatsApp Interview, 3 May 2021)*

The proximity and/or distance caused by time and space is referred to as physical geography. Collegiality is especially significant for Indonesian student teachers because it allows them to discuss disparities in their knowledge of online teaching and learning. In this situation, virtual learning struggles to duplicate the physical interaction of in-person classroom learning. Students were upset and unsatisfied due to a lack of in-person experiences and restricted peer relationships. Students' conversations, group work, and participatory projects were interrupted by online and remote learning styles (Kee 2021). As a result, to prepare for this, teachers must think of new ways to engage students in online learning, such as collaborative learning groups (Walti & Blaschke 2009) or synchronous and asynchronous online learning environments (Fabriz, et al. 2021). This study complemented that students' technical problems should be solved and supported by the government to create mutual intelligibility and equal distance among students with the teacher.

4.3 *Moral geography: Student teachers are struggling to see moral values in online learning*

Student teachers are struggling to see the lecturer's moral value given during the online learning process and provided implicitly on the material. It indicated students' negative attitudes toward online learning, so it made them see fewer moral values or negative emotions.

> *I am afraid if it is from a negative point of view, that perhaps online learning is focusing on knowledge and sometimes the moral or attitude do not exist ... During the online learning process, I feel only learning knowledge and limited time. (Participant 5, WhatsApp Interview, 4 May 2021)*

Negative emotion can also arise when there is a significant moral divide between teachers and students, such as when teachers believe their goals are being jeopardized or have been lost (Hargreaves 2001b). In professional practice, moral geography refers to the proximity and/or distance established by various goals and perceptions of achievement. Because they come from a distinct social and educational structure, Indonesian students and instructors have a different set of moral norms (e.g. respect, not to lie, tolerate cultural, religious differences, etc.). Moral education also contributes to the moral life of classrooms, students' moral lives, the character of society (Fenstermacher et al. 2009), and as "notions of good and bad, better and worse," can be shown in a teacher's choice of curricular subject and where they stand while dealing with students (Hansen 2001). Empirical evidence shows that researchers and teacher educators, particularly teacher trainer centers, continue to support the moral goal of teaching and education (Sanger & Osguthorpe 2011). Moral geography of student teachers in online learning can be achieved if there is collaboration and tolerance between lectures and student teachers.

4.4 *Sociocultural geography: I feel tolerance around various friends from different cultures with*

Coming from various places in Indonesia (e.g. Java, Sumatra, Sulawesi, and Kalimantan), students faced various norms in the student-student interactions. Where their tolerances were trained by facing various cultures and norms, student teachers indicated online learning process required tolerance to build friendships with their friends with different cultures and local languages. Therefore, mutual understanding of English language learning around students is achieved by exposing them into native cultures around students.

> *In my class, there are male and female students. We come from various regions in Indonesia which have various cultures, races, ethnicity, and language. We are from Java, Sumatra, Sulawesi, and Kalimantan. So different language accents to require extra effort in reaching mutual understanding during the learning process. Especially, we use English in the online learning process.(Participant 10, WhatsApp Interview, 6 May 2021)*

Differences in gender, color, ethnicity, language, and culture produce closeness and/or distance in sociocultural geography. Student teachers come from various regions including Java, Sumatra, Sulawesi, and Kalimantan and with diffrent Indonesian background knowledge significantly from the rest of their peers in the community. . As a result, they speak indigenous languages as their mother tongue (e.g. Javanese, Sundanese, etc.). For example, students speak a variety of dialects, but while chatting with their peers, they modify their accents to Indonesian. Little is known, however, regarding how cultural variations influence students' learning behavior patterns (Tlili et al. 2021). In online learning, taking into account and understanding a student's culture may result in a more effective learning design. A study in online learning culture, although they only looked at specific learning actions/behaviors (e.g., accessing a course, time reading a given learning material, etc.), such as according to Shang et al. (2020), cannot adequately capture the students' cognitive engagement and learning behavior traits. Students' sociocultural geography on the online classroom can be created by developing tolerance cultures in communicating using the national language to promote their cultures, so each student built intercultural knowledge to respect similarities dan differences among students.

4.5 *Political geography: I feel there are differences in the attitude of lecturers to differences in student teachers' social status*

The pandemic has indeed affected all lines of life, including in terms of status social. The impact of the pandemic on this side was also felt by my participants. The impact was expressed by participant 11.

> *I feel so far it has been very influential. This can be seen from the fast and slow pace of the teacher is responding to students or students in the interests of online interactions. If those who have this interest have a high status socially or intellectually, they will get a quicker response, but if the opposite is true, then it is delaying or giving slow responses for various reasons, perhaps it can be said that they are due to other activities so they cannot respond. But, thankfully, where I study at this time, the lecturers give quick responses even though they are very short and dense. (Participant 11, WhatsApp Interview, 6 May 2021)*

The proximity and/or distance caused by differing understandings of power is referred to as political geography (Hargreaves 2001b; Liu 2016). The sequence of engagement and practice in professional societies is determined by a power structure. The multiple challenges of being an Indonesian background student teacher and a new member of the community. They need the identification of one's own and others' obligations and statuses, as well as the formation of a nuanced understanding of the 'feeling norms'. Self and others may experience unpleasant feelings such as guilt, humiliation, and embarrassment as a result of misunderstanding or breaching these principles. Teachers are frequently picked from the top quartiles of accomplishment in their school systems when they have a great reputation because the academic quality of persons entering the sector appears to be a critical feature (Hargreaves 2009). This outcome was predicted, and it backs with prior studies that showed a high number of respondents had an unfavorable view of teacher education programs (Sang 2018). In terms of politics, inequalities in social position impact student-teacher relations. However, teachers acquire a distinct attitude toward students, one that does not separate individuals based on their socioeconomic level.

5 CONCLUSION AND RECOMMENDATION

The purpose of this study is to explore the emotional reactions of Indonesian student teachers' online teaching and learning process during the COVID-19 pandemic. The COVID-19 pandemic had an impact on the emotional experiences of Indonesian student teachers in terms of virtual learning, psychological mental state, and student-teacher interaction where lecturers or teachers must consider the emotions of student teachers during the online learning process. Moreover, because this online learning is being implemented swiftly during the COVID-19 pandemic, student teachers' fear must be alleviated in a variety of methods (eg. asynchronous, synchronous, and hybrid online learning) so that they may participate actively and successfully in online learning. This study also contributes to the theory that teachers or lecturers may change and evolve as a cognitive process of acquiring information and skills and that the social character of teaching and the basic relevance of emotion in the online learning process are both recognized in this study.

REFERENCES

Albin-Clark, J. (2020). '*I felt uncomfortable because I know what it can be*': The emotional geographies and implicit activisms of reflexive practices for early childhood teachers. *Early Childhood*, 21(1), 20–32.

Azorin, C. (2020). Beyond COVID-19 supernova: Is another education coming? *Journal of Professional Capital and Community*, 1(1).

Bryson, J. R., Andres, L., & Davies, A. (2020). COVID-19, virtual church services and a new temporary geography of home. *Tijdschrift Voor Economische En Sociale Geografie, 111(3), 360–372.*

Baba, M. M. (2020). Navigating COVID-19 with emotional intelligence. *International Journal of Social Psychiatry, 66(8), 1–11.*

Cambridge Assessment English. (2020). *Qualifications for higher education: Cambridge English qualifications for admissions purposes*. Cambridge: University of Cambridge.
Davidson, J., Bondi, L. & Smith, M. (2005). *Emotional geographies*. Aldershot: Ashgate.
Duraku, H. Z., & Hoxa, L. (2020). The impact of COVID-19 on education and on the well-being teachers, parents, and students: Challenges related to remote (online) learning and opportunities for advancing the quality of education. *Journal of Education,* 2(3), 1–10.
Fabriz, S. et al. (2021). Impact of synchronous and asynchronous settings of online teaching and learning in higher education on students' learning experience during COVID-19. *Frontiers in Psychology, 12*, 1–16.
Fenstermacher, G.D., Osguthorpe, R.D., & Sanger, M.N., (2009). Teaching morally and teaching morality. *Teacher education quarterly, 36 (3), 7–19.*
Hansen, D.T. (2001). Teaching as a moral activity. In V. Richardson (Ed.),*Handbook of research on teaching*, 4th Ed. (pp. 826-857).Washington, D.C.: AERA.
Hargreaves, A. (2000a). Mixed emotions: Teachers' perception of their interaction with students. *Teaching and Teacher Education*, 16(8), 811–826.
Hargreaves, A. (2000b, April). *The emotional geographies of teaching.* Paper presented at a symposium at the American Educational Research Association annual conference, New Orleans, A 21–28.
Hargreaves, A. (2001a). Emotional geographies of teaching. *Teachers College Record,* 103(6), 1056–1080.
Hargreaves, A. (2001b). The emotional geographies of teachers' relations with colleagues. *International Journal of Educational Research*, 35(5), 503–527.
Hargreaves, A. (2005). Educational change takes ages: Life, career, and generational factors in teachers' emotional responses to educational change. *Teaching and Teacher Education*, 21(8), 967–983.
Hargreaves, L. (2009). The status and prestige of teachers and teaching. In *international handbook of research on teachers and teaching*, L.J. Saha, A.G. Dworkin (eds.), 217–229. New York: Springer.
Kee, C.E. (2021). The impact of COVID-19: Graduate students' emotional and psychological experiences. *Journal of Human Behavior in the Social Environment, 31 (1), 476–488.*
Latham, A., & McCormack, D. P. (2009). Thinking with images in non-representational cities: Vignettes from Berlin. *Area*, 41(3), 252–263.
Liu, Y. (2016). The emotional geographies of teaching. *Teacher Development*, 20(4), 482–497.
Lorimer, H. (2005). Cultural geography: The busyness of being "more-than-representational". *Progress in Human Geography*, 29(1), 83–94.
Marvell, A., & Simm, D. (2018). Emotional geographies experienced during international fieldwork: An evaluation of teaching and learning strategies for reflective assessment. *Journal of Geography in Higher Education*, 42 (4), 515–530.
Moorhouse, B. L. (2020). Adaptations to a face-to-face initial teacher education course 'forced' online due to the COVID-19 pandemic. *Journal of Education for Teaching*, 46 (4), 1–3.
Raaper, R., & Brown, C. (2020). The Covid-19 pandemic and the dissolution of the university campus: Implications for student support practice. *Journal of Professional Capital and Community,* 5(3), 343–349.
Robinson, C.C. & Hullinger, H. (2008).New benchmarks in higher education: Student engagement in online learning. *Journal of Education for Business*, 84 (2), 101–108.
Sang, H.C. (2018). Teacher educators' and student teachers' attitude towards teacher education programmes in universities: A case of selected private and public universities in Kenya. *International Journal of Scientific and Research Publications,* 8 (5), 186–196.
Roller, M. R., & Lavrakas, P. L. (2015). *Applied qualitative research design: A total quality framework approach.* New York: Guilford.
Sanger, M. & Osguthorpe, R., 2005. Making sense of approaches to moral education. *Journal of moral education, 34(1), 57–71. Blended learning. education in a smart learning environment.*
Shang, J., et al. (2020). A sequential analysis on the online learning behaviors of Chinese adult learners: Take the KGC learning platform as an example. *International Conference on Blended Learning ICBL 2020: Blended Learning. Education in a Smart Learning Environment*, 61–76.
Sidhu, R., & Ishikawa, M. (2020). Destined for Asia: Hospitality and emotions in international student mobilities. *Compare: A Journal of Comparative and International Education*, 52 (3), 399–417.
Stengel, B. S., & Weems, L. (2010). Questioning Safe Space: An Introduction. *Studies in Philosophy and Education, 29 (6), 505–507.*
Sundarasen et al. (2020). Psychological impact of COVID-19 and lockdown among university students in Malaysia: Implications and policy recommendations. *International Journal of Environmental Research and Public Health, 17 (17), 6206.*

Suprihatin (Kehok), Istiqomah, L., Meilani, R. I., & Khoiriyah. (2020). Exploring the emotions of single international students in Hong Kong facing the COVID-19 pandemic. *Journal of International Students, 10(3), 91–107.*
Susanto, G., Suparmi, & Rahayu, E. Y. (2020). The emotional geography of international students in online bahasa indonesia learning during the COVID-19 pandemic. *Journal of International Students, 10(3), 161–179.*
Taloko, J. L., Putra, M. S., & Hartanto, Y. (2020). Emotional geographies experienced by an indonesian doctoral student pursuing her PhD in New Zealand during the COVID-19 pandemic. *Journal of International Students, 10(3), 126–141.*
Thrift, N. (2008). *Non-representational theory: Space/politics/affect.* Routledge.
Tlili et al. (2021). Impact of cultural diversity on students' learning behavioral patterns in open and online courses: a lag sequential analysis approach. *Interactive Learning Environments*, 29(1), 1–20.
Vannini, P. (Ed.). (2015). *Non-representational methodologies: Re-envisioning research.* Routledge.
Walti, C. & Brindley, J.E. (2009). Creating effective collaborative learning group in an online environment. *International review of research in open and distance learning, 10* (3), 1–18.
Widodo, H. P. (2014). Methodological considerations in interview data transcription. *International Journal of Innovation in English Language Teaching and Research*, 3(1), 101–109.

Author index

For Product Safety Concerns and Information please contact our
EU representative GPSR@taylorandfrancis.com Taylor & Francis
Verlag GmbH, Kaufingerstraße 24, 80331 München, Germany

Public Financial Management, State Capacity, and Public Services in India

INSTITUTIONS AND DEVELOPMENT IN SOUTH ASIA SERIES

SERIES EDITOR
Rahul Mukherji

Institutions are a central facet of development in South Asia. They can promote competition, spur democracy, and engender citizen well-being. They can equally be the harbingers of monopolies, authoritarianism, and social inequities. Institutions embody norms that influence the behaviour and lives of citizens in South Asia. How do democratic institutions succeed and fail in South Asia? This question is central to the exploration of the mechanisms of South Asian development.

Citizen well-being and economic growth within a sustainable framework are equally constitutive of the process of South Asian development. Economic growth often requires regulated market systems that promote internal and global competition. How can regulated competitive capitalism be promoted in South Asia? Can regulations structure markets to create a sustainable future? Can the State in South Asia ensure a dignified life for the average South Asian citizen?

In seeking answers to these questions, this series will interrogate the interplay of global and local forces within and outside the State that create and perpetuate norms that spur and deter South Asian development.

Rahul Mukherji is Professor and Head, Department of Political Science, South Asia Institute, Heidelberg University, Germany.

Public Financial Management, State Capacity, and Public Services in India

SANTHOSH MATHEW
BHUMI PUROHIT
DEVESH SHARMA

OXFORD
UNIVERSITY PRESS

OXFORD
UNIVERSITY PRESS

Great Clarendon Street, Oxford, OX2 6DP,
United Kingdom

Oxford University Press is a department of the University of Oxford.
It furthers the University's objective of excellence in research, scholarship,
and education by publishing worldwide. Oxford is a registered trade mark of
Oxford University Press in the UK and in certain other countries.

Published in the United States of America by Oxford University Press
198 Madison Avenue, New York, NY 10016, United States of America.

British Library Cataloguing in Publication Data
Data available

Library of Congress Control Number: 2025942614

ISBN 9780198973188

DOI: 10.1093/9780198973195.001.0001

Printed on demand

The manufacturer's authorised representative in the EU for product safety is
Oxford University Press España S.A. of Parque Empresarial San Fernando de Henares,
Avenida de Castilla, 2 – 28830 Madrid (www.oup.es/en or product.safety@oup.com).
OUP España S.A. also acts as importer into Spain of products made by the manufacturer.

Dedicated to our families, for all you've given us and above all, joy.

Santhosh Mathew: To my wife, Sulu

Bhumi Purohit: To my family, for their unconditional support

Devesh Sharma:
To Anita Sharma and Narendra Sharma, it's all been possible because of your sacrifices, patience, and love

Disclaimer

All views are the authors' own and do not represent the views of their respective employers, present or past.

Preface

Some ideas take time. Others, a lifetime.

As early as 1989, I began experiencing the underbelly of India's public financial management. It was the year I took office as the Deputy Development Commissioner of the undivided border district of Saharsa in the Indian state of Bihar. Rather than creatively solving the structural issues that were hindering growth and development of my district, I was spending an inordinate amount of time and effort replying to queries from the state capital to get funds that were already earmarked for my district. A large amount of this effort involved gathering information from officers in different parts of the district so that we could assuage concerns in a timely manner, which would enable us to secure and use the funds by the end of the financial year. You may be surprised to hear that the questions that were tying me up in knots were very simple ones: How much money had we received already? How much of it had we spent? How much physical progress had we made?

Most of us who have spent long years in government running districts and departments have stories to tell about what we had to do to keep the ship running. I can recount a midnight call from the Principal Secretary to the Chief Minister of Bihar in 1993 when, as the District Magistrate of Palamau, he asked me how much money from the bank account of the District Rural Development Agency I could take out before 2 a.m. by issuing a cheque and depositing it into the state government account. The situation warranted immediate action in the middle of the night to ensure that when the RBI offices opened in the morning, they would not order the stoppage of all Government of Bihar payments because the state had crossed its ways and means advance threshold. The local State Bank of India manager had to fax the details of my deposit to the Central Accounts Section of the Reserve Bank of India in Nagpur because it is the Nagpur office that gives ways and means advances to state governments. The District Police chief was asked to find the bank manager and the head accountant of the DRDA, and bring them to the official residence of the District Magistrate so that the transaction could be executed.

While technological changes no longer warrant such jugglery, managing public finance in India, with poor observability and burdensome administration, continues to be an enormous task. Think of the fodder scam that first came to light in 1996. It involved the fraudulent withdrawal of over Rs. 9 billion from the government treasury, even in excess of what was provided for in the approved budget. Think of the Srijan Mahila Sahyog Samiti scam in Bhagalpur detected in 2017 that involved misuse of over Rs. 7 billion of government funds using forgery. Think of the admission by Government of India in the Economic Survey presented to the Indian Parliament in February 2016 that '. . . the estimated stock of unspent balances in government accounts is at least Rs 1 lakh [1 trillion] crore and leads to an annual cost of Rs 8,500 crore [85 billion]'. Commenting on the death of children in Gorakhpur due to shortage of oxygen caused by delay in payments, Bhumi Purohit and I had written in *Mint* on 25 August 2017 that 'India's public finance architecture is the real culprit in Gorakhpur and Bhagalpur'.

By 2004, and after 19 years in government, it was clear that it made no sense for me to carry on if I could not have at least a framework or a plausible theory on why governments underperform and what could be done about it. With the help of many well-wishers, I embarked on a five-year period of academic scholarship during which time I completed a D.Phil in Development Studies. My thesis, 'State Incapacity by Design', explained how low state capacity is often a political choice. Armed with fresh insights and new energy, I came back to the government in 2009. In the many postings that followed—in the Health Department of Bihar as Executive Director of the State Health Society, in the Finance Department of Bihar as Expenditure Secretary, in the Rural Development and Welfare Departments of Bihar as Principal Secretary, in the Rural Development in Government of India as Joint Secretary, and in the National Council for Teacher Education in Government of India as Chairperson—I worked on fixing transversal issues that were impeding state execution capacity. These efforts helped move my thinking from 'State Incapacity by Design' to 'State Capacity by Design', or a set of concrete reforms in finance, people, and processes in government that can make a difference in the lives of ordinary citizens.

An inflexion point in this journey was my work in the Rural Development Department both in the state government and in the

Government of India. While working on 'State Incapacity by Design', I found that the poorer states of India got fewer resources from Centrally Sponsored Schemes, and subsequently also used less than richer states. My efforts at designing and testing a new fund and information flow mechanism for MGNREGS (Mahatma Gandhi National Rural Employment Guarantee Scheme) with the help of Abhijit Banerjee, Esther Duflo, Rohini Pande, and Clément Imbert were driven by the hypothesis that if we could smoothen the flow of funds by delayering and smoothen the flow of information through digitization that reduced information asymmetry, we could improve India's social protection regime by getting poor states of India to use more of MGNREGS funds. This is how my path crossed with my co-authors, with Devesh Sharma in 2010 and Bhumi Purohit in 2016. We have been thinking together ever since. This book is the result of our partnership. While I have been fortunate to continue working at the Gates Foundation with Devesh, our views here are our own and do not represent views of our employers then and now.

The year 2025 marks the 15th since we began work on fund and information flow reform in MGNREGS and 36 years since it became a bee in my bonnet. Though my goal in reforming fund and information flows in MGNREGS was to show that we will be able to improve the uptake of MGNREGS, especially in the poorer states of India, results from the randomized control trial showed that our efforts resulted in a reduction in programme expenditure. This was without a reduction in payouts to workers. We also found that our efforts led to delays in payments. Why were the payments delayed? Answers abound: the lack of interoperability between two softwares—CPSMS and NREGASoft, piled up payment instructions with wet signatures at a bank branch in Patna, my reluctance to ask *mukhiyas* and panchayat secretaries across more than 8,400 panchayats in Bihar to use digital signatures, and the angst of Deputy Development Commissioners being deprived of using their power to distribute funds. We could not even pay for a fund management system and worked with the only bank that was willing to offer the services for free. The idea that India's system of PFM needed reform wasn't invalid, but the conditions at the time didn't allow for this myriad of issues to be fixed. We had to work within the constraints.

These issues, ideas, and constraints stayed with me. In the Ministry of Rural Development, we were able to support my friend and batchmate R. Subrahmanyam in designing and rolling out a new, country-wide fund management system for MGNREGS. This time, we were able to make what we thought were improvements to the system that we had trialled in Bihar. Another opportunity came about in late 2015, when the Prime Minister's office tasked a constitution of multiple committees of secretaries to look at different opportunities for reforms. As part of these reform efforts, I presented the idea of 'Just in Time' funding to the coordinator of one of the committees, R.N. Choubey, which had the task of looking at 'Innovative Budgeting and Effective Implementation'. I also managed to get an audience with Jayant Sinha, the then Minister of State for Finance, to discuss the same topic. On 15 January 2016, and on behalf of the committee, Mr Choubey presented the ideas on 'Just in Time' to the Prime Minister of India, following which a decision was taken to 'create extensive IT platforms across center, states and RBI for "just-in-time" fund releases' that would make Utilization Certificates, an artefact of the public finance system, redundant. NITI Aayog was tasked to create the technology to achieve this goal. Upon learning of this decision, then Secretary of Ministry of Rural Development in the Government of India, S.M. Vijayanand, wrote to the Principal Secretary to the Prime Minister on 5 April 2016 to draw attention to my work on advancing PFM reform. In the meantime, two pieces I had written on the subject were published. One was on 'Modernizing Government Payment Systems to Deinstitutionalize Corruption' in the 11 April 2016 edition of *Outlook* magazine, and the other was a paper co-written with Dwijo Goswami, another of my long-term collaborators, on 'Doing More with Less: Developing JAM+ to Reform Public Finance Management' in the 23 April 2016 edition of *Economic and Political Weekly*.

In the wake of this renewed interest and the decision to implement just-in-time fund releases and make Utilization Certificates redundant, I sought and got three opportunities. The first was a set of discussions with the then Joint Secretary to the Prime Minister, T.V. Somanathan, on ways in which the decision of the Prime Minister could be effectively implemented. The second was a presentation to the then Vice Chair of NITI Aayog, Dr Arvind Panagariya, on 20 June 2016, and the third was to the then Expenditure Secretary, Ashok Lavasa, on 22 June 2016. I cannot claim that what followed was because of what I did because the wheels of government are too complex for any attribution. However, I can state that three breakthroughs did happen.

The first breakthrough was on 1 June 2016, when T.V. Somanathan wrote to the CEO of NITI Aayog suggesting that individual ministries should follow up on the development of IT systems that facilitate real-time fund release without the submission of Utilization Certificates. The second breakthrough was on 24 June 2016 by the Department of Expenditure conveying the decision of government to implement just-in-time fund releases by expanding the coverage of the Public Financial Management System (PFMS) for Central Sector schemes and Central Assistance to State Plans schemes. The letter explained how integrating PFMS with State Treasuries and use of the PFMS-EAT module can contribute to enhanced utilization of funds. The third breakthrough was on 1 July 2016, when the minutes of the meeting chaired by Dr Panagariya were released and stated that, 'The Vice Chairman concluded the meeting by saying that both PMO and NITI Aayog are on board with the proposal for just in time fund management and to get it started, a successful pilot as suggested above be done under the lead of DoE'.

Much has happened in India's public finance since then. What began as a software called CPSMS became a much more highly reformed PFMS. The Department of Expenditure issued new guidelines on Single Nodal Account and SNA-SPARSH. On annual Civil Accounts Day on 1 March 2025, the Finance Minister and others on the dais spoke about PFMS more than anything else, and discussed just-in-time funding, the cause I have been advocating for since 2010. Just-in-time funding marks a significant transition in how PFM is managed in India. The recognition that digital technologies can help the country better manage public finance is now much more common than it was in 2010. Many state governments are paying much more attention and investing in building IT systems that reduce friction and smoothen information flows. I am hopeful that the ideas we put forth in this book will give wind to the sails of all of these efforts by so many.

To those who might find the ideas in this book remotely interesting, I do not claim that they will be realized anytime soon. But with the journey I explain above, the gnarly issues make themselves all the more a reason why we need to continue working on them—someone has to.

Santhosh Mathew
2 March 2025

Acknowledgments

This work is truly a collective effort over many years. Mohan Gopalakrishnan, J.K. Mohapatra, Anup Mukerji, Bhanu Pratap Sharma, and T.V. Somanathan gave us the earliest opportunities to develop and test our ideas that form the foundations of this book. With her moral courage and thought leadership, Kalpana Kochhar gave us the fortitude to publish the ideas in this book. These are debts that can't be paid and we hold them dear to our hearts.

Sunil Bhandari, A.M. Corper, Jayant Dasgupta, Abhishek Jain, Anu Joshi, Lalitesh Katragadda, Kerim, Stuti Khemani, Shankar Maruwada, Mick Moore, Nandan Nilekani, Sanjay Purohit, Manish Srivastava, and Pramod Varma gave us the inspiration to continue to work on our ideas over the years. Their work and sage guidance forms the core of how we have come to think on many things we write about. We stand on the shoulders of giants.

There have been countless thinkers and collaborators who have helped bring the ideas in this book to fruition. We are especially thankful to Vikram Srinivas, who contributed a section of the appendix which details reforms in Kerala. Reuben Abraham, Ronald Abraham, Anindita Adhikari, Junaid Kamal Ahmad, Yamini Aiyar, Matt Andrews, Abhijit Banerjee, Arvind Kumar Chaudhary, Stephen R. Davenport, Iqbal Dhaliwal, Esther Duflo, Vanita Falcao, Siddharth E. George, Venkat Goli, Dwijo Goswami, Anil Gupta, Dan Honig, Clement Imbert, Shruti Jagirdar, Manoj Jain, Lauren Kahn, Miral Kalyani, Srinivas Katikithala, Philipp Krause, Mekhala Krishnamurthy, Cathal Long, C.V. Madhukar, Sharan Mamidipudi, Maria May, Rajiv Mehrishi, Tanvi Ravel Mehta, Emily Middleton, Kriti Mittal, Charity Troyer Moore, Mick Moore, Karthik Muralidharan, Gulzar Natarajan, Daniel Ortega Nieto, Vishnu Padmanabhan, Rohini Pande, Varad Pande, Sailendra Pattanayak, Lant Pritchett, Daniel Radcliffe, Nirmala Ravishankar, Natalia Rigol, Lorena Rivero del Paso, Manish Sabharwal, Rajeev Sadanandan, Ajay Shah, Satish Ranjan Sinha, Vidhu Sood, James Stewart, Arvind Subramanian,

R. Subrahmanyam, Rakshita Swamy, Mitul Thapliyal, Susan Thomas, Viraj Tyagi, Gerardo Una, Alok Verma, S.M. Vijayanand, Srikanth Viswanathan, and Jeffrey Witsoe have given us invaluable thought partnership to make these ideas stronger over many years. The critiques, assurances, and revelations they have shared have helped us make the seed of our ideas into a growing plant. We're additionally indebted to the many bureaucrats across Bihar, Gujarat, Karnataka, and Rajasthan who provided invaluable time to help us understand the policy landscape in India.

Through their own work and friendship, Divya Agarwal, Yamini Atmavilas, Arielle Bernhardt, Anil Bharadwaj, Susna De, Balamurugan D, Ashish Dhawan, Avantika Dhingra, Gargee Ghosh, Victoria Griffith, Sakshi Gudwani, Vishal Gujadhur, Ronald Heifetz, Sanjay Jain, Anmol Kamra, Elizabeth Kellison, Praveen Khanghta, Dave Kim, Suneeta Krishnan, Saleh El Machnouk, Michelle Maziar, Ritambhara Mehta, Hari Menon, Navya Mishra, Himanshu Nagpal, Karan Nagpal, Keyzom Ngodup Massally, John Osborne, Nidhi Parekh, Alok Rajan, Surabhi Rajaram, Anshu Singh, Kanwaljit Singh, Surendra Tripathi, Shashi Verma, Rodger Voorhies, Priya Vora, and Archna Vyas have cemented our beliefs and given us courage to continue to work on these ideas.

Many of these ideas were refined thanks to inputs from the working group on South Asian politics at University of California, Berkeley and we're grateful for the space they provided to develop our thoughts. We are particularly indebted to Anustubh Agnihotri, Jennifer Bussell, Pradeep Chhibber, Thad Dunning, Pranav Gupta, Tanu Kumar, and Irfan Nooruddin for their guidance.

Unbeknownst to them, for we have never met, we have learnt much from works of David Graeber, Albert O. Hirschman, Douglass North, Jennifer Pahlka, Venkatesh G. Rao, James C. Scott, and Dave Snowden. We have not cited most of their celebrated work in this book, for they didn't write much about public finance. Nonetheless, the depth of their work influences how we think about public institutions and state capacity.

Our families and friends have everything to do with this work, except for the errors and incorrect assumptions and ideas that will be discovered by wiser eyes over the course of time. Mathew, Nida, John, and Leontine, thank you for being our strength. Chinmay, Keshav, Trapti, and Vimal, thank you for all your selfless love and support. Bhalchandra, Chetna,

Kanak, Priyank, Naksh, Niki, Ramila, and Vinay, thank you for your continual and never-ending support over the years.

We will undoubtedly disagree with some of the ideas we've put forth in this book in the near future as new learnings emerge and some of these ideas are tested empirically. Indeed, in the course of ten years over which we wrote this book, we've questioned and re-questioned our thinking to come to the ideas we present here. We are happy to change our minds in the face of compelling evidence, for it means that public financial management or public service delivery reforms have been empirically tested and become more evidence-based. We look forward to continuing the conversation, and hope that this book brings many more people to the table to discuss public finance in India.

To all the people across the world who have put us and many others through abysses of uncertainty, waiting, and meaningless form-filling, we express our gratitude for the reminders of why we do what we do.

Contents

1
Introduction

In August 2017, tragedy unfolded at the Baba Raghav Das Medical College Hospital, a public institution serving communities around Gorakhpur, Uttar Pradesh in India. Over a two-week period, over 60 children lost their lives, sparking national outrage and bringing scrutiny to the state of care in the state. Doctors, bureaucrats, and politicians initially attributed the deaths to encephalitis and various other illnesses in the resource-deprived areas from where residents came to Gorakhpur for treatment (Rashid 2017). In reality, the government hospital faced a critical shortage of oxygen for patients. For over six months, the hospital owed over Rs. 6.3 million (USD 77,778)[1] to Pushpa Private Limited, the oxygen tank supplier. This financial delinquency stemmed from the state government's failure to make the necessary funds available to the hospital.

Dr Rajeev Mishra, the principal of the college, elaborated on such hindrances as a continual problem. 'There is always delay in the payment when you deal with the government. That is a fact', Mishra told a reporter (Rao 2017). This resignation to inevitable delays in government transactions extends far beyond this specific case, resonating with businesses or citizens who have interacted with public service delivery in India. The incident at Baba Raghav Das Medical College Hospital exposed deep-rooted issues in India's public health system, including inadequate funding, poor management, and a lack of accountability. Yet these issues permeate across many public service sectors in the country.

Government inefficiency in public funding is a symptom of a pervasive and structural issue. Although the public financial management (PFM) system in India has vastly improved over the years, more needs

[1] At the average rate of USD 1 = Rs. 81, the approximate average conversion rate between 2021 and 2025. We use this conversion rate in the remainder of the book unless otherwise listed.

Public Financial Management, State Capacity, and Public Services in India. Santhosh Mathew, Bhumi Purohit, and Devesh Sharma, Oxford University Press.
 DOI: 10.1093/9780198973195.003.0001

to be done. Its current architecture leads not only to delays in payments but also to persistent failures in the public service delivery system in producing outputs and outcomes. Public financial management, or PFM, refers to the rules, systems, and processes that govern revenues, allocations of funds, expenditure, accounting, and auditing (Potter and Diamond 1999). In this book, we focus on public financial management after revenue is collected.

Take the Gorakhpur case as an example. Dr Rajeev Mishra, attempting to address the hospital's fund shortage, sent three letters to the state medical education department in July 2017, seeking the release of allocated funds. The process followed a labyrinth of bureaucratic steps. After hearing back from the medical education department on 7 August, the college sent a payment request to the treasury, containing line-item lists of its proposed payments. The next day, the treasury released a token for funds to the hospital, and the hospital sent the token to the bank two days thereafter. The token allowed the bank to transfer funds to Pushpa Private Limited, but because the state does not use a fully integrated, real-time payments system from the treasury to the end recipient, funds only reached Pushpa Private Limited on 10 August, nearly a month after Dr Mishra sought the funds (Rao 2017). Unfortunately, these delays proved fatal for many children who needed the oxygen supplies.

This convoluted and antiquated process that the government undertakes to fulfil its payment obligations—whether to government employees, government pensioners, government vendors, or citizens who rely on public services—consistently creates public service delivery failures. The issue is rooted in the government's low expenditure capacity, an oft understudied aspect in both public service delivery and state capacity literatures. Expenditure capacity refers to a state's ability to create a coherent budget, manage its expenditure, scrutinize expenditure against outcomes and outputs, and create expenditure systems that are transparent to politicians, administrators, citizens, and government contractors. Holistically, expenditure capacity largely depends on the architecture of public financial management after revenue generation.

Though existing literature focuses on revenue capacity as a form of state capacity, we argue that expenditure capacity is also a core component. While there is disagreement on a uniform definition, we borrow from Lindvall and Teorell (2016, 7), who define state capacity as 'the state's power to achieve intended outcomes', including how states use their resources to showcase their power. We posit that a state's ability to use its resources to achieve outcomes—particularly policy outcomes—depends on the effectiveness of its expenditure systems. In other words, how well a state manages and deploys its financial resources directly impacts its capacity to deliver on its promises and meet societal needs. We acknowledge, as scholars have rightly pointed out, that viewing state capacity through any single function of the state can be limiting. However, while our focus on expenditure capacity may seem narrow, it provides critical insights into a state's operational effectiveness.

At the core of expenditure capacity issues—and our book—lie two architectural problems within India's PFM system. First, the country relies on several unsynchronized systems across most of the government to collate information on fiscal and programmatic data for managing its public funds. These systems are not always interconnected, resulting in fiscal and programmatic data that do not link to each other; at times, even fiscal data from one part of government does not connect to another part of government. The consequences of such fragmentation are far-reaching. For instance, unconnected systems can mean that for certain programmes, the state government may be unaware of the funds held by one of its districts, or the specific programmatic functions for which those funds are being utilized. This lack of transparency and linkages severely hampers effective resource allocation and programme management.

The second architectural flaw involves inadequate access to critical fiscal and programmatic data for key government officials. Programme managers and other stakeholders have to constantly ask for information from those above or below them in the administrative hierarchy to make timely decisions.

These two core problems of *synchroneity* and *observability* of fiscal and programmatic data in the PFM architecture make it challenging for programme administrators to efficiently run public services. As a result, citizens receive unreliable and inconsistent public services, eroding their trust in government's ability to deliver on its promises.

In this book, we highlight how these two architectural issues—poor synchroneity and poor observability—contribute to a litany of problems in public service delivery. Between funds being allocated by a government and being received by the end recipient (individual or institution), they either leak through the system or are delayed somewhere in the pipeline. At the same time, fund and programme managers alike struggle to trace and utilize funds within the system, resulting in money being left unused and parked in various accounts. Lack of clearly linked data further restrains fund managers from verifying whether beneficiaries have satisfied the conditions to receive payments, making it onerous for them to make informed funding decisions. These problems exacerbate burdens on already resource-constrained administrators, who spend an inordinate amount of time on paperwork for obtaining funds rather than on the implementation of policies. Similarly, auditors also find it time-consuming and, at times, impossible to verify data, and there is little transparency regarding how financial inputs translate into policy-level outputs and outcomes. As we describe in this book, governments in India and those who rely on them for funding have faced such issues for decades, and much has been done to improve systems. Yet in many parts of the Union[2] and in state governments, more can be done and needs to be done.

Despite the far-reaching consequences of PFM, a holistic understanding of how expenditure functions operate in India is scattered amongst a few specialists, even though PFM is governed by a vast number of actors across a variety of institutions such as the Union government, states, district, and sub-district entities. Other than researchers who study expenditure and revenue management, these specialists are often administrators within governments or experts in specialized divisions within multilateral institutions such as the World Bank or the IMF. Lacking a rounded understanding, public policy analysts and social scientists have largely ignored the role of PFM in public service delivery, often pointing to other causes for India's weak state capacity.

While some scholars (Aiyar and Raghunandan 2015; Bhatia 2018; Jena 2013; Mohanty and Bhanumurthy 2021; Rao 2022; Singh 2017) and practitioners (Ahluwalia et al. 2019; Banerjee et al. 2020; CGA 2012; Nilekani and Shah 2015; Watal 2014) have made significant contributions in highlighting the different facets of PFM issues in the

[2] In this book, we refer to the 'central' or national government as the Union as it is colloquially known in India. This choice reflects the fact that 'central' government often connotes centralized planning and control, which is unreflective of the federal and decentralized nature of Indian governance.

country, we provide a comprehensive view on how public expenditure in India functions. At the heart of our analysis is unpacking the ways in which India's public expenditure architecture exacerbates inefficient public service delivery. We first detail how PFM is designed to fund expenditure, and why. We then contend that the architecture of PFM can be redesigned and implemented to modernize the flow of funds and information in government. When fiscal and programmatic data sync up and are transparently available, PFM can enable rather than hinder public service delivery.

Currently, India's PFM system employs a fragmented and unsynchronized 'push-and-wait' architecture. As we explain in Chapters 3 and 4, this architecture requires actors at various points in the chain to submit requests for three key elements of policy implementation—funds, information, and decisions—and then wait for those requests to be fulfilled. This fragmented nature of transfer of funds, information processing, and decision-making generates significant burdens on administrators, who face heavy co-ordination costs and uncertainty while implementing policies.

The process begins when legislators consult programme managers and approve departmental budget allocations. Programme implementers then send or 'push' requests for budget allocations—and sometimes, reallocations—and wait. Before implementers can get the money, they need approval from a range of actors including finance departments, line departments, and various levels of government (Union, state, district). These levels of government push procedural guidelines further down the implementation chain and wait for policies and processes to be followed before the disbursal of funds. When various levels of government finally push funds down to implementers, they continually ask for and wait for expenditure reports to be passed back up, along with any missing information. Programme implementers push information from public programming (such as outputs, contractual obligations, invoices received for delivery of goods and services, outcomes, or uptakes) to accountants and treasuries, and wait for payments, accounting, and reporting. Auditors, citizens, legislature, and civil society actors push requests for information and wait for this data to promote accountability. Throughout each step of this push-and-wait chain, no single actor in this chain has visibility on what has happened, or is happening, in other fragments of the ecosystem. As we write this book, this process has started to improve in some parts of government. For certain types of schemes known as Centrally Sponsored Schemes, the government has notified departments to introduce a system called Treasury Single Account (TSA) for autonomous bodies, and SNA-SPARSH for other parts of government, which allows implementers to 'pull' or draw money directly from the Union government so long as that request is under an allocation ceiling.

To illustrate how convoluted the process can be, take the example of ordering a pizza for delivery. The push-and-wait architecture is akin to ordering your pizza, and then having no knowledge of whether the restaurant has accepted your order, whether the pizza is being made, whether it has been picked up, and whether the delivery associate is heading in the right direction and can find your delivery location. In such a world, you would have no information until the pizza gets to your door—if it does at all. This is the limit of this metaphor, for people usually do not begin to hoard perishable pizzas. People do however begin to stock up on funds and undertake accounting jugglery because funds hardly ever reach implementers in time to be used effectively.

In this book, we break down the nature of the push-and-wait architecture, at the centre of which are unsynchronized and unobservable PFM systems. We then propose architectural shifts that can help provide funds as well as information about where, why, and when the funds are released per the needs of stakeholders. We call this an *on-demand public finance architecture* (henceforth, referred to as on-demand architecture), which proposes a way to redesign PFM so that the focus is on public service delivery and not just checking boxes to meet accounting and auditing requirements.

By supplementing PFM reforms with information systems to ensure all data used for public service inputs and outputs are linked, accurate, and audited, the reforms we suggest can provide a comprehensive, data-driven way to govern. We use existing literature, our experience with and within government, and primary data from government officials and citizens to provide a framework for PFM reforms with two key criteria in mind: How can PFM be structured to ensure good public service delivery? Can PFM be structured in a way that allows stakeholders to hold public service delivery providers accountable?

1.1 Book Organization

The book is organized into nine chapters. In the first part of the book, we define the problems with public service delivery as they relate to PFM. To this end, Chapter 2 examines the traditional explanations for low state capacity in public service delivery, highlighting evidence from India. We focus specifically on two sets of challenges that explain poor public service delivery: political economy failures and institutional failures. While existing explanations partly describe why public services challenges arise, they tend to neglect an additional reason—poor fund flow systems in government. We tie this literature with state capacity literature, which has partially focused on fiscal and administrative capacity. We contribute to this literature by addressing a gap: the role of expenditure capacity in enabling public service delivery.

In Chapter 3, we first outline trends in global PFM reforms before providing a history of public finance reforms in India. In Chapter 4, we combine knowledge from primary interviews, our personal experiences, and existing literature to explain precisely how the PFM system works across the cycle of budget allocation, budget execution, and budget monitoring and evaluation. We argue that though contemporary reforms such as SNA-SPARSH have made improvements, more needs to be done to address the fundamental issues of PFM architecture. We end Chapter 4 by analysing the two key architectural issues in PFM: lack of synchroneity within fiscal data and between fiscal data and programmatic data, and poor observability of these data.

Chapter 5 analyses challenges that synchroneity and observability issues in PFM create for public service delivery. One, the system creates artificial scarcity, where some agencies do not receive government funds even though others are sitting on excess money that goes unused, called float. Two, poor synchroneity and observability disincentivize decision makers to approve payments out of fear of being held accountable if incorrect payments are made. Three, the burdensome fund flow system creates significant administrative burden for government officials, who must go to great lengths to obtain or report payments. Four, a convoluted system creates constant delays and uncertainty in funds reaching recipients. Five, the expenditure process exacerbates problems of corruption at multiple levels of governance. Six, many parts of the PFM system rely on data that is unreliable, untimely, and difficult to use for connecting financial inputs to programmatic outputs. Seven, the system produces data that is insufficient for auditors, making their basic tasks cumbersome and undermining their ability to hold government officials accountable.

Upon reviewing these issues in Chapter 5, the second part of the book provides ideas for changes in architecture of India's PFM system. In discussing these ideas, we keep an eye towards designing an on-demand architecture, which provides stakeholders with funds and information on programmatic inputs and outputs as they take place.

Chapter 6 puts forth three key PFM architectural changes to make this happen: (a) co-ordinating fiscal and programmatic data to allow for two-way visibility into both parts; (b) coupling events such that programmatic events trigger automatic payments as soon as proof for payment or fund transfer requirements are uploaded onto a system; and (c) making unit-level fiscal and programmatic data available on demand to authorized stakeholders. We further offer suggestions for changes to the way data is produced and managed in the system to enable these architectural changes, including a particular emphasis on data privacy and security to protect citizens' private information held in the public domain.

With such an on-demand architecture, we propose three possible outcomes of a new PFM system in Chapter 7. One, an on-demand architecture can promote just-in-time funding, which allows transfers to happen between a funding agency and the end recipient as needed, without delays or advances. We detail how just-in-time funding is additive to existing versions of this idea, such as SNA-SPARSH or Treasury Single Account. Two, architectural changes can enable smart payments that automatically trigger fund releases as soon as the prescribed conditions to receive those funds are met. Three, architectural changes can enable departments to create user-level platforms such as programme-specific IT platforms, auditing platforms, and public transparency platforms, which enable access to and analysis of fiscal and programmatic data.

After explaining the details of these ideas, we explain how they can be enacted in Chapter 8. We focus on three key ideas—using incentives to get people on board; agile development; and asynchronous onboarding. Agile development focuses on building systems using adaptive planning, fast delivery, and rapid response to needs. Rather than creating monolithic platforms and forcing them upon users, agile development emphasizes constant testing and revising of ideas to make them more apt for the end user. Asynchronous onboarding, on the other hand, refers to the idea that stakeholders can adopt and integrate the overall PFM ecosystem in their own time. The chapter explains how, using these ideas, stakeholders can improve India's PFM—either by using existing institutional design or by creating a new institution for enacting architectural changes. This approach to change focuses on decentralized implementation because outlays, outputs, and outcomes will vary across time, geographies, programmes, and political contexts.

We conclude in Chapter 9 by identifying the risks and opportunities linked to changes in architecture of PFM and highlight potential areas for research. Specifically, we advocate that while the architecture of PFM does not need to be changed in one go, future changes in PFM systems should keep the overall architecture in mind. Only with a design philosophy that builds unified systems instead of centralized ones can gradual changes—whether at the Union or state level—coalesce into a cohesive PFM system that works to enable, rather than detract from, public service delivery.

To be sure, this book is not about proposing new fiscal policies or programmatic policies. We focus our work on how PFM is connected to public service delivery. Therefore, we do not speak to the traditional and important concerns of accountants and PFM experts such as upstream processes of budgeting (e.g. medium-term expenditure frameworks, participatory budgeting, and linkages with planning). Also outside the scope of this book are ideas such as fiscal responsibility legislation, fiscal rules, medium-term budget frameworks, fiscal councils, fiscal risk management techniques, performance budgeting, and accrual reporting and accounting, which Cangiano et al. (2013) have termed as the most important innovations in public financial management in the first two decades of the twenty-first century. As Cangiano et al. (2013) note, ensuring that a state has good public financial management is not a substitute for these types of reforms or questions about the optimal level of public debt, proper size of government, and fair distribution of resources. Instead, the role of good PFM is situated within the larger objective of efficient and effective fiscal policy and management in government.

With these scope conditions, our book speaks to underlying architecture and system properties of PFM related to public services when the budgeting is complete, funds are ready to be disbursed, and expenditures have been incurred. This process includes expenditure management, auditing, monitoring of funds, and transparency for stakeholders and beneficiaries.

1.2 Situating PFM within a Broader Set of Reforms

The ideas in the book are situated within a broader set of changes for public services in India over the past two decades. Some of the biggest citizen-facing reforms include the Right to Information Act and the Right to Public Services legislation. The Right to Information Act has allowed citizens to demand and receive information from government and government-funded entities, inspiring a host of activists now working towards improving transparency. Moreover, twenty states have enacted the Right to Public Services legislation, which legally empowers citizens to receive time-bound public services. The Right to Public Services allows citizens opportunities to appeal any denial of services and penalizes officers for failing to provide the same.

Beyond these reforms, other government actions have improved public financial processes. The Directorate General of Supplies and Disposal has created the Government e-Marketplace (GeM), an e-procurement portal for suppliers of goods and services to government offices that reduces information asymmetries and corruption. The government is expanding the 'Jan Dhan, Aadhaar, and Mobile' (JAM) reforms to improve financial inclusion and transparency through bank accounts, biometric identity services, and mobile banking. In a reform most relevant to PFM, the Controller General of Accounts (CGA) has set up the Public Financial Management System (PFMS) to track Union government funds used for financing Central Sector[3] (CS) and Centrally Sponsored Schemes[4] (CSS). New systems such as TSA and SNA-SPARSH also allow implementation agencies to directly draw from the national-level Consolidated Fund of India and/or the Consolidated Funds of the State,

[3] Central Sector Schemes are those financed and implemented by the Union government.
[4] Centrally Sponsored Schemes are those financed by both the Union and state governments but implemented by the state governments.

provided that such withdrawals are below budgetary allocations. These reforms, and many others, have improved transparency for citizens and built the government's execution capacity, though they are not without their critics. Yet, more can be done by reimagining PFM as a pathway to fuel public services and sustain governance at a deeper level. Current efforts work within the confines of the push-and-wait architecture of PFM, and therefore have limited impacts on service delivery.

India's PFM system and the problems it creates are not unique. Many countries and/or programmes have faced the same dilemmas, but reforms focused on high transparency to beneficiaries and policymakers, combined with easy access to funds through remote banking services, have led to dramatic outcomes.[5] In some countries, for example Mexico and Brazil, funds were consolidated into a system of bank accounts and made traceable from source to spending. Based on lessons from such programmes, our ideas concentrate on developing an observable, interoperable, and evolvable PFM ecosystem that consists of stakeholders both within and outside governments at Union, state, and local level.

Alongside global lessons, our ideas also build on capabilities and learnings from India's digital transformation. A reformed, on-demand architecture of PFM would allow various agencies across Union, state, or local government levels to exchange funds and information with each other in close to real time, and with instant access to authorized data when needed through the use of digital platforms.[6] According to the architecture we discuss—which works within India's federal structure while keeping individual data protection in mind—each government programme or department will have its own platform that sends and receives data through secure repositories.

Given India's federal structure where states have autonomy over a set of revenue decisions, we advocate for a decentralized model of data sharing. We discuss automated fund management systems that use data to process and analyse payment requests and fulfil payments when prescribed conditions are met. Such a model has multiple benefits. For one, it need not require a centralized data repository; the data is captured according to

[5] We further discuss reforms in other countries in Chapter 3.

[6] In theory, this is also possible without the use of information technology, but entails high time, co-ordination, and compliance costs; we discuss this in Chapter 6.

priorly agreed-upon standards by stakeholders that allow the data to be used and reused in a variety of ways. In doing so, it prevents data consolidation which may lead to data misuse. Two, this model allows states, departments, and programmes to maintain autonomy over the design and use of data platforms so long as the data they share conforms to basic, agreed-upon standards. Through such a model (in addition to others we discuss in the book), our goal is that an on-demand architecture connects information silos that currently clog the fund flow pipeline and paves a way for policymakers to maximize transparency and accountability of public expenditure, ensuring it fulfils its intended purpose of running India's public services.

A big concern such vast data systems raise is that of data protection. Data protection refers to protecting data from unauthorized access, use, disclosure and modification. We think of data protection as being in service of data privacy, which concerns individuals' ability to control how their data is used or shared. Any technological data ecosystem that can yield databases of individual-level information should be created with data protection at the forefront, and with legal protections in place. While this book doesn't detail out ways in which data protection should be ensured—a task we leave to experts—we explore existing solutions in Chapter 6. We argue that information about payments can indeed be made available to authorized administrators without providing information about individuals' personally identifiable information. In fact, we argue that an on-demand architecture can only work with data protections engrained into the system, otherwise public finance can be used to hollow out public services rather than augment them, as has been done historically (Levitsky and Ziblatt 2019; Suryanarayan and White 2021).

While it may seem expansive, an on-demand architecture is aimed at ensuring that public services are funded on time, securely, and with more certainty. An on-demand architecture seeks to ensure that once programme recipients meet requirements to be paid, they do not wait on programme officers to process paperwork and make discretionary decisions on how and when to process the funds. Likewise, once funds are authorized and disbursed, the architecture ensures that administrators with legitimate access to budget allocation or expenditure data no longer need to wait for someone's discretionary approval to access that information.

These reforms do not intend to remove discretion on the many other necessary factors of a bureaucrat's job, such as determining how to produce evidence that payment requirements are met, having the ability to flag fraudulent payments, or intervening when someone eligible to receive benefits has been excluded from a list of recipients. In fact, in a place like India, which presents bureaucrats with a diversity of challenges in policy implementation, knowing when to use discretion to ensure that citizens obtain their rights is a desirable function of state officials (Lipsky 2010; Mangla 2022). Instead, our proposal for on-demand architecture seeks to expand optionality, as we note in Chapter 7, and makes unintended use of discretion observable to those holding bureaucrats accountable.

An on-demand architecture, though seemingly complex, is not new in its vision. In their 2011 report, the Technology Advisory Group for Unique Projects (TAGUP) led by Nandan Nilekani recommended the creation of the Expenditure Information Network (EIN), a special-purpose vehicle aiming to transform the hierarchical fund and expenditure information transfer model to a hub-and-spoke one (TAGUP 2011). In other words, the EIN would consist of an information 'hub' integrated with various levels of governance, such as monitoring agencies, implementation agencies, and district governments. These nodes, or spokes, would integrate with the hub, allowing for real-time data transfer, budget adjustments, and fund transfers. While we do not discuss an on-demand architecture model with a centralized hub, our proposal echoes the need for a network of nodes that can share data with each other across various parts of government.

An on-demand architecture has also been pushed by the Union government. In January 2016, the Committee of Secretaries, which oversees the Union government's business transactions and allocations, recommended a system akin to the EIN to the then Prime Minister, Narendra Modi. This followed a presentation by Santhosh Mathew, as Joint Secretary, Ministry of Rural Development, to R.N. Choubey, co-ordinator of the Group of Secretaries on 'innovative budgeting and effective implementation' constituted by Prime Minister Narendra Modi. This group proposed that the government ought to 'develop a system of "just in time" budgetary releases through extensive IT platforms covering the

Union government, States and RBI' (Press Information Bureau 2016a). On a suggestion from the Prime Minister's Office, Santhosh Mathew, as Joint Secretary, Ministry of Rural Development, also prepared an implementation plan for this proposal in May 2016, on the basis of which the Prime Minister's Office wrote a letter to NITI Aayog to create an action plan for implementing just-in-time budgetary releases. In August of the same year, following a presentation to NITI Aayog and officials of the Ministry of Finance, the Ministry of Finance requested all central ministries to create an action plan for moving all government spending to just-in-time releases. The Finance Ministry issued further orders in July 2023 requiring all Centrally Sponsored Schemes to move to just-in-time fund releases. These ideas have found much momentum in government. Notably, even after people initially involved in bringing about just-in-time funding have gone on to take other jobs and roles, the idea has continued to be implemented, with the Ministry of Rural Development and the Office of the Controller General of Accounts being front-runners in making these reforms possible.

Alongside, many states have also been modernizing their PFM systems. Most notably, Kerala, Karnataka, Odisha, Assam, and Rajasthan are states that have digitized parts of their PFM system, with the development of an ambitious plan in Bihar as well. For states to adapt their existing systems to an on-demand architecture based on their own idiosyncratic and contextual needs, however, they will need nationwide data standards governing the exchange of funds, information, and decisions. This would allow various stakeholders to integrate their own PFM systems with those of others at a time of their choosing. Such an architecture will allow for systems that can be built for each stakeholder's needs while being able to speak to each other through machine-readable, standardized protocols and data standards.

Given that different states and departments have varying levels of digitization in their PFM processes, ideas like an on-demand architecture need an approach that is amenable to multi-year, phased, and nuanced implementation, rather than being implemented at a single point in time. With administrative leadership changing every two years, it is unlikely that government officials have administrative incentives to lead long-term changes. Similarly, implementing changes at a national or state level may take longer than a political party's term in government, yielding limited returns before the next election. Even if a party were to revise PFM's architecture, short-term pains for administrators coping with new technology may bring parties bad publicity. It is no surprise, then, that similar changes across India have not materialized beyond the ideation phase so far.

In addressing these concerns, however, we need not look further than the Mahatma Gandhi National Rural Employment Guarantee Scheme (MGNREGS), which has gradually automated many of the systems that make payments to beneficiaries. The rural works scheme has seen corruption decrease by 17% (Banerjee et al. 2020); transparency on the scheme's expenditure has likewise increased. MGNREGS has consolidated its national expenditure and monitoring data into a national database, spurring both improved oversight by policymakers and a plethora of research by academics.[7]

Eventually, all of India's public services and government payments will need to harness the power of an on-demand architecture for PFM, and there is no better time than now to actualize these reforms. Both government and private-sector efforts are already in place for expanding financial inclusion and transparency to citizens, particularly for Direct Benefit Transfer (DBT) schemes. An on-demand architecture offers a way to do just that, and can be built while keeping individual-level data protection and data privacy in mind. In fact, enacting these ideas is necessary to ensure budget allocations have their intended outcomes, not just for DBT schemes but for all government payments. Ultimately, if India's growth is to become inclusive and sustainable and if dire poverty is to be eliminated, India will need to dramatically improve the quality of its public health, education, and social protection systems. This cannot happen without extensive changes to create an on-demand architecture in PFM systems.

[7] Though the scheme requires architectural shifts and system properties that we describe later—leading to fewer delays in payments and easier withdrawal mechanisms for beneficiaries—more direct fund transfers have vastly improved both the quality of implementation and transparency.

1.3 Conclusion

Ideas to improve PFM systems are not new. They have been studied by public finance researchers and prominently facilitated by entities such as the World Bank since the 1980s. However, these studies and projects have largely been relegated as work for specialists. We seek to demystify this field for researchers and policymakers. In doing so, we elucidate how antiquated architectures for PFM affect public service delivery. Consequently, the first part of the book is particularly geared towards explaining this theory. While policymakers will find this section of use, we write this section keeping existing scholarship and research in mind and situate our theory within this work.

The second half of the book is primarily tailored for policymakers. In this section, we step aside as objective researchers and take a normative stance on how improved PFM has the potential to reduce the public service delivery problems that we outline in the first half of the book. Where applicable, we highlight the work of scholars who have shown that reforms are not a panacea and, in certain cases, even have adverse effects. While we provide evidence where it is available, it is important to note that much research to date on national or even state-level reforms is undertaken through case studies, which may be biased depending on the goals of the organization writing them. We acknowledge that the architectural shifts and system properties we argue for have not been applied comprehensively to public financial management in India. However, we claim they may have merit and cite others in academia and government advocating for similar ideas.

While this is largely a descriptive book based on primary fieldwork and secondary research, we hope that it provides a grounded framework for linking India's PFM to public service delivery. We also hope it inspires a host of empirical work, both qualitative and quantitative, on the link between PFM and public services, as well as on the policy components of any changes that materialize. The ideas put forth in this book are done so with a strong belief in research and monitoring of policy implementation and outcomes. We aim to be a part of such research and invite others to join.

2

The Existing Narratives of India's Public Service Delivery Challenges

Despite having an impressive administrative service, a functional democracy, and a budget of over Rs. 47.6 trillion (USD 588.37 billion) in FY 2024–25, India consistently lags in the quality of delivery of its public services. According to the 2022 Human Development Index (HDI), India ranks 134th out of 191 countries in human development outcomes. While poverty rates have significantly come down to 14.96% of the population in 2019–21 (NITI Aayog 2023) and 11.28% in 2022–23 (NITI Aayog 2024), some dimensions of human development across India still remain low. For instance, 28% of women and 16% of men are illiterate, and approximately a third of all children under the age of five are underweight or stunted (International Institute for Population Sciences and ICF 2021). Notably, these figures reflect significant improvements over time. Even still, the poor actively participate in voting, have access to over 100 nationally sponsored schemes for social protection, health, education, and other benefits, and are constantly targeted by transfers of payments and subsidies. Yet the spending is not translating into commensurate improvements in human development outcomes. It raises the question: Where does the money for welfare programming go?

Public service delivery, or the extent to which services and activities of the government meet citizens' expectations, has received much attention in the last two decades. The dominant narratives of the failure of public service delivery frequently centre on policy implementation problems. They highlight deep-seated issues: a politicized bureaucracy, an overburdened administration, corruption, underwhelming monitoring and accountability structures, and poor decentralization. While these issues could be contributing to public service delivery failures, a larger

Public Financial Management, State Capacity, and Public Services in India. Santhosh Mathew, Bhumi Purohit, and Devesh Sharma, Oxford University Press.
 DOI: 10.1093/9780198973195.003.0002

structural strain pervades the ecosystem: the systematic nature of how funds, payments, and information flow, which slows or hinders policy implementation for even the most effective of officials. In this chapter, we first review the core explanations for poor public service delivery. We then argue that the public financial management (PFM) structure is an often overlooked yet key component of state capacity that influences public service delivery.

2.1 Political Economy Challenges

Conventional narratives of public service delivery failures have several theoretical strands, but a common thread revolves around the incentives of stakeholders who occupy the political structure. These political narratives focus on the implications of political interests and incentives on public service delivery.[1]

One theory for poor public service delivery centres on electoral systems.[2] In a system such as India's where politicians only need a plurality—more than any other candidate—they have to establish credibility with specific groups rather than the majority to successfully get elected (Chhibber and Nooruddin 2004). Rather than providing public services, politicians can provide targeted transfers to groups that may help them obtain the most electoral clout (Keefer and Khemani 2004). Alternatively, they can use other methods to obtain votes that negate the need to provide public services: engaging in a quid pro quo for votes known as clientelism (Hicken and Nathan 2020; Stokes et al. 2013), buying votes directly through cash (Cruz et al. 2020), or giving patronage through government jobs (Chandra 2007).

Such measures, when systemic and engrained, debilitate the ability of voters to seek credible change through voting. Indeed, recent research across a series of experiments in various countries finds that informing voters about politicians who perpetuate such problems does little to persuade them to vote against them—partially because such information

[1] See Grossman and Slough (2022) for a review of the accountability mechanism between voters and citizens.

[2] See Healy and Malhotra (2013) for a review of research on a tangential area, retrospective voting.

confirms their prior expectations about the poor quality of politicians and partially because they have little faith that alternative candidates will be any better (Dunning et al. 2019). In rare cases, information does seem to matter. In an experiment in the Philippines, Cruz et al. (2024) provide voters with policy preferences of mayoral candidates while experimentally manipulating whether they receive information on (a) what candidates propose to do if elected, and (b) whether incumbent candidates followed through on prior policy promises. They find that respondents are more likely to vote for candidates who mirror their own policy preferences, as well as for candidates who fulfilled prior policy promises. Though scholars find such results in controlled settings, voters are often inundated with non-policy-related communication atop policy information from parties, making it difficult to assess what ultimately makes the biggest difference.

Bureaucrats who are hired on the basis of merit rather than patronage are also influenced by the political economy of the state, despite being some of the main implementers of policies. In models of bureaucratic accountability, politicians are typically in charge of monitoring bureaucrats for policy implementation, and exerting sanctions where necessary. Yet personal relationships between politicians and bureaucrats can render such monitoring ineffective. This is especially true if the bureaucrats can counter sanctions with their own bargaining (Agnihotri 2022), or if politicians do not have clout with the party in power (Raffler 2022). Furthermore, even when bureaucrats are monitored by upper-level politicians, they can still enact discretion with lower-level politicians who do not have the power to sanction them. Significantly, this is more likely to happen with women and marginalized-caste politicians (Purohit 2024).

Furthermore, the number of politicians monitoring bureaucrats can also influence the quality of public service delivery. This is most effectively shown by Gulzar and Pasquale (2017), who find that when bureaucrats are supervised by a single politician (as opposed to multiple ones), politicians can incur the electoral benefits of motivating strong public service delivery through the bureaucracy. This is not always the case in India, where electoral and bureaucratic boundaries only overlap in some cases. If multiple politicians can claim credit for work, then the incentive for a single politician to put in the effort for expanding such quality of public service delivery is reduced. Similarly, political oversight of bureaucrats may be entirely ineffective if, in systems with few institutional constraints, politicians use their power to extract rents from bureaucrats (Brierley 2020).

Overall, accountability mechanisms between voters and politicians, as well as those between politicians and bureaucrats, ought to theoretically improve public service delivery outcomes. However, power exerted through political networks between these groups can weaken such accountability mechanisms and render them ineffective.

2.2 Institutional Challenges

Political failures in public service delivery are exacerbated by institutional failures. Frontline service providers face pressures to undertake a large amount of work which—along with institutional limitations such as inadequate resources—increase corruption and contribute to the failure of public service delivery. Government agencies charged with implementing public services are often overburdened, poorly staffed, and poorly resourced to meet policy objectives (Dasgupta and Kapur 2020). Contrary to popular perception, India has one of the lowest ratios of government employees to its population at 146 employees per 10,000 residents (Sukhtankar and Vaishnav 2015, 203). Comparatively, China (536), Brazil (approximately 600), Mexico (approximately 450), and South Africa (approximately 400) all fare significantly better. Subsequently, a District Collector, the officer overseeing one of the lower administrative units, is on average in charge of 2 million residents (Sukhtankar and Vaishnav 2015, 203); the Block Development Officer in a sub-district is in charge of 150,000 citizens, on average (Dasgupta and Kapur 2020).

The skewed ratio aside, the salary structure for public office is also asymmetrical. When compared to the private sector, entry-level and top-level wages are high for low-skill jobs, with limited rewards or punishments based on performance (Muralidharan and Sundararaman 2013). For instance, according to a 2015 study, an entry-level driver can earn an average of Rs. 25,000 (USD 390 in 2015) per month working with the government, compared to an average of Rs. 13,000 (USD 203 in 2015) per month in the private sector. With 25 years of experience, the salary can go up to an average of Rs. 59,000 (USD 920 in 2015) per month in the public sector compared to Rs. 51,000 (USD 795 in 2015) per month in the private sector. While highly skilled public employees earn about the same or more in entry-level jobs compared to their private sector counterparts, comparative salaries go down with the experience accrued (Singh et al. 2015). An entry-level doctor with a master's degree earns an average of Rs. 95,000 (USD 1,481 in 2015) per month with the government and Rs. 105,000 (USD 1,637 in 2015) per month in the private sector. With 15 years of experience, however, monthly government salaries only go up to an average of Rs. 160,000 (USD 2,494 in 2015) compared to Rs. 370,000 (USD 5,768 in 2015) in the private sector (Singh et al. 2015). While public jobs demand working with a larger group of people and more bureaucratic burdens, their salary structures do not attract long-term retention for high-skill jobs.

The institutional structure, in turn, creates incentives for corruption when beneficiaries are immediately interacting with the state, which Sukhtankar and Vaishnav (2015) categorize into three kinds: facilitative, collusive, and extractive. In facilitative corruption, officials charge fees or bribes to expedite services they already have the responsibility to provide, for instance the provision of food grains through the public distribution system (PDS). In terms of its coverage, India's PDS is the world's largest social assistance programme and covers over 800 million people (Das et al. 2024). Like many welfare schemes in India, the PDS has evolved over the years in terms of its size, structure, and efficacy. For example, the High-Level Committee on 'Re-Orienting the Role and Re-Structuring of Food Corporation of India', chaired by the former Food Minister Shanta Kumar, estimated that, based on the 2011–12 Consumption Expenditure Survey, the food grain distribution through the PDS suffered from leakages amounting to about 46% at an all-India level. While the 2022–23 estimates showed improvement, suggesting that approximately 28% of grains failed to reach beneficiaries (Das et al. 2024), such leakages can ultimately have an impact on developmental and health outcomes for citizens.

The second type, collusive corruption, involves the breaking of rules to benefit those who are paying the bribe. This type of corruption may involve creating procurement contracts for a particular agency or allowing individuals to bypass fines or fees. In a study of the tendering process for road contracts in Uttar Pradesh, for instance, scholars find that most firms were disqualified to apply based on 'technical' grounds (Lewis-Faupel et al. 2016). In 95% of the examined cases, the government only evaluated the bid of a single firm to procure the contract, leading to almost no competition based on previous experience or success in public service delivery performance.

The third kind, extractive collusion, involves the administrator extracting funds from the government or a private party. Embezzlement, harassment, or—as seen in the previous passage—absenteeism from work fall into this category. In India, for instance, Niehaus and Sukhtankar (2013) undertook a study of the Mahatma Gandhi National Rural Employment Guarantee Scheme (MGNREGS) in districts of Odisha and Andhra Pradesh and found that approximately 70–80% of the labour budget is embezzled before it reaches beneficiaries. Even when the government increases the wage budget, Niehaus and Sukhtankar find that none of the increases are passed onto beneficiaries. These institutional failures to curb embezzlement are compounded by the fact that in an overburdened administrative system, frontline workers may opt to spend their time on activities that maximize the extraction of rents (Gibbons and Murphy 1992; Wade 1982). Similarly, they may opt to take advantage of political networks and shirk responsibility altogether while extracting salaries. Ultimately, these problems lead to a huge gap between scheme budgeting and actual public service delivery outcomes. With low institutional capacity to monitor and overcome such types of facilitative, collusive, and extractive corruption, the quality of public service delivery can suffer.

Another factor that can ameliorate accountability between the government and its citizens to improve public service delivery is decentralization. Decentralization involves the transfer of decision-making power of centrally produced and provided goods to subnational levels, usually to local governments.[3] In principle, such a transfer of power ought to increase the level of accountability—both between central and local governments as well as between local governments and beneficiaries of public services, who can hold local administrators and politicians

[3] See Gadenne and Singhal (2014) and Mookherjee (2015) for reviews of the literature on decentralization in developing economies.

accountable. The political economy literature reviewed earlier highlights the problems with the latter theory.

Decentralization can still improve accountability between higher levels of government and local government in meeting public service delivery goals provided that it is also accompanied by an increased ability of local governments to better allocate financial resources. In India, though the state generously allows local governments to plan for public services, it has yet to implement fiscal decentralization. As noted in Raghunandan (2017, 493), 'local tax collections are next to nothing, very little money filters down to the local government via intergovernmental fiscal transfers, and LGs are shackled with strict administrative, financial, and audit controls'. According to a United Nations study, fiscal decentralization must include transparency of allocations, predictability of when fund transfers to local governments will occur, and local government autonomy in the decision to utilize funds (Svensson and Rosenbaum 1996). An extensive study in Karnataka by Accountability Initiative, however, finds that these three defining factors fail to hold true (Raghunandan et al. 2016).

In FY 2014–15, Accountability Initiative found that local governments failed to receive Rs. 162.4 million (USD 2.59 million in 2014–15, on average) of a total state budget outlay of Rs. 15 billion (USD 239.62 million in 2014–15, on average) that was meant for devolution. Additionally, a long-term analysis showed that while state budget outlays have gradually increased over the last two decades, devolution to local governments has decreased. In FY 1991–92, for instance, local governments received 35% of the outlay, but this number went down to 16% in FY 2014–15. Despite a reduction in fiscal devolution, the share of devolved funds for non-plan funds—spending on routine operations and maintenance of existing infrastructure such as salaries, pensions, or interest payments—has increased. This means that though local governments have an increasing say in areas like the allocation of salaries, they have little control over planning and expenditure for local programmes or central programmes (Singh 2017).

These factors may be reflected by, or complicated by, the low levels of procedural knowledge that many local politicians seem to have. In a study of small-town politicians in Rajasthan, Auerbach et al. (2024) find that respondents correctly answered questions on municipal spending 16% of the time; on taxation, 60% of the time. Such low levels of knowledge are concerning. However, even if local governments are offered more control and knowledge, they currently lack the ability to clearly track their allocations, releases, and spending that connects to the state or the Union. When funds do get disbursed, they are often given in the last quarter of the financial year, making it difficult for local governments to effectively plan for efficient utilization.

The outlined issues of decentralization are vital to understanding why it may not produce more effective public service delivery. At the same time, however, they also illuminate a larger problem in the implementation ecosystem that is entirely ignored in the literature: the transfer mechanisms, accounting, and auditing of funds and information about those funds between the Union and state governments on one hand, and the implementation agencies on the other. Unless funds reach frontline workers and implementers in a transparent, efficient, and leakage-proof way, decentralization of public service delivery planning or public service delivery in general will remain a fruitless exercise. Not only will inefficient fund transfers continue to create decentralization challenges, they will also, alongside uncertainty around actual use of funds, exacerbate the corruption and monitoring mechanisms that are central to public service delivery failure.

Political economy and institutional narratives go a long way in explaining why public service expenditures fail to meet the expected outcomes. While these explanations—a politicized bureaucracy, an overburdened administration, rampant corruption, and poor decentralization—all contribute to mismanagement of programme implementation, they miss a core component of programmatic failures: the public financial management architecture that plays a role in weakening state capacity.

Our goal is not to argue that the existing political economy and institutional factors do not explain poor service delivery. They in fact do so, in part. Instead, we posit that even if these issues were to be resolved, a feeble public financial management architecture would still yield poor expenditure capacity in a state and therefore contribute to weak public service delivery. A poor capacity of the state to expend its funds is ultimately a problem with state capacity, yet a study of the state capacity literature indicates that scholars have largely focused on revenue rather than expenditure capacity. A key element of note is whether a state can use the generated revenue to expend money in a meaningful, transparent way.

2.3 Expenditure Management as Integral to State Fiscal Capacity

Political scientists, sociologists, and economists have long discussed the importance of state capacity in development. Referring to a state's ability to co-ordinate resources and effectively produce social and economic outcomes, state capacity has been defined and empirically tested through varying interpretations.[4] While older literature on state capacity has focused on macro-historical questions of how a state reaches across its territory (Mann 1984; Skocpol 1985; Tilly 1992), recent literature has connected it to a variety of development outcomes (Besley and Persson 2009; Hendrix 2010; Muralidharan et al. 2016; Soifer 2008). Regardless, scholars define state capacity as the ability, potential, or capability of state institutions to function, typically in four areas: state autonomy, integrity, bureaucratic refinement, and resources (Skocpol 1985). Relevant to our work is the state's capacity through an autonomous, skilled bureaucracy or administrative capacity and a state's ability to gather resources, what scholars refer to as fiscal capacity.

Administrative capacity has been described both broadly and narrowly. For instance, Fukuyama (2004, 7) traces it in general as 'the ability of states to plan and execute policies and to enforce laws cleanly and transparently'. In contrast, Bersch et al. (2017) describe administrative capacity as the existence of a professional bureaucracy that can implement policy without external influences. As Williams (2021) rightly notes, however, in this stream of literature, bureaucratic 'capacity' becomes a catch-all term that doesn't tackle the problems that arise from the multiple agents and multiple principals that make up these institutions in reality, who are pulled in varying directions. Still, the importance placed

[4] Mann (1984, 189) refers to state capacity as '"infrastructural power," or "the capacity of the state to actually penetrate civil society, and to implement logistically political decisions throughout the realm."' Soifer (2008, 232) contends that state capacity has been thought of in three different ways: as the capabilities of the central state, the state's territorial reach, and the effects of the state on society.

on bureaucratic competency in building administrative capacity has led to a Weberian focus on measurements, well described by Hanson and Sigman (2013). These measurements include outcomes for public goods and services such as literacy rates; the Government Effectiveness ratings; the International Country Risk Guide's Bureaucratic Quality rating; and an Administrative Efficiency measure (Hanson and Sigman 2013).

While measures of policy outcomes can be confounded by a variety of factors such as policy priorities, the environment, and social norms, Weberian measurements on the other hand have a narrow scope, with a focus on bureaucrats. What these measures lose is a wide-ranging focus on the institutions that govern bureaucrats and the institutional capacities that may enable bureaucrats to plan and implement policies. For instance, do bureaucrats have accurate data to effectively plan policies? Does the country have a robust accountability mechanism to reward bureaucrats for employing honest techniques in deploying public goods and services and, likewise, penalize them for engaging in corruption? These accountability and transparency mechanisms, especially as they relate to understanding PFM, are one of the key focus areas of this book.

An alternate definition of state capacity as fiscal capacity is equally insufficient. Fiscal capacity has been primarily defined as a state's ability to produce wealth through GDP (Levi 1989) and extract it through taxation (Besley and Persson 2009; Shen and Williamson 1997). In a manner, the notion of fiscal capacity extends beyond revenue generation because it encompasses states' ability to 'reach their populations, collect and manage information, possess trustworthy agents to manage the revenue, and ensure popular compliance with tax policy' (Hanson and Sigman 2013, 5). These roles encompass a broader set of functions that are correlated with the potential of a state to create rules and organize society to better optimize its extractive capacity. While wide in its scope, the notion of fiscal capacity as it has been defined misses out on half the aspect of what Skocpol (1985) described as key to a state realizing its goals, the state's ability to deploy its resources. As Skocpol (1985, 17) argued: 'A state's means of raising and deploying financial resources tells us more than any other single factor about its existing (and immediately potential) capacities to strengthen state organizations, to employ personnel, to co-opt political support, to subsidize economic enterprises, and to fund social programmes.'

Skocpol's claim—the ability of a state to use its resources, or expenditure capacity—is central to our argument that PFM institutions and the key accountability and transparency functions related to PFM are as central to fiscal capacity as revenue generation. We systematically dissect the Indian case to show how its expenditure capacity can improve the state's ability to plan and fund its programmes, direct its bureaucracy to productively plan and implement policies, and create transparent systems that maximize the ability of citizens to benefit from social programming. If a state can expand its ability to raise revenue but cannot effectively manage, monitor, and measure the impact of its expenditure, its ability to accrue wealth through resources is only a partial and misleading measure of state capacity.

India serves as a primary example for this argument. In June 2017, the country enacted the Goods and Services Tax (GST), a singular indirect tax that replaced multiple taxes at the state and central level. The tax was enacted to decrease tax complexity and improve the government's ability to bring informal businesses into the formal economy. To an extent, it has had success in achieving these goals (Ministry of Finance 2023). While the GST may have achieved its objectives and increased tax capacity—a key measure of state capacity—it does not necessarily indicate that the government has increased its ability to allocate this revenue back into public services. This limitation, we argue, is primarily a result of the way finances are managed within many governments, which creates challenges in effectively utilizing and monitoring public funds. The current system creates a host of inefficiencies which fuel corruption, prevent transparency, weaken the bureaucracy, harm accountability, and lead to excess wealth accumulation by the government. We further expand on these challenges in Chapter 5.

Overall, we put forward the argument that fiscal capacity cannot merely be thought of as the state's capability of collecting revenue. Instead, it ought to encapsulate the state's expenditure capacity, or its ability to use, account for, and transparently show financial outlays and outputs, which is equally important to holistically understanding fiscal state capacity. Separate literature on public financial management rather than state capacity has urged policymakers to consider this key role of the state: that of managing public expenditure. In the next section, we connect ideas from public service delivery, state capacity, and public financial management to substantiate that public financial management reform is essential to improving public service delivery.

2.4 Our Wedge in the Conversation: Public Financial Management for Public Service Delivery

What is public financial management or PFM? Public financial management systems relate to how governments manage public resources—both revenue and expenditure—and the results that these financial flows produce (Andrews et al. 2014, 2). They are a combination of underlying laws, regulations, practices, people, and tools that underlie how public finances are overseen in government across ministries, departments of finance, and other parts of the government. Such PFM systems are often described across the cycle of budget formation and allocation, budget execution, and budget monitoring and evaluation. This is a more procedural view of PFM systems which focuses on what PFM systems do. As per this perspective, strong PFM is the end goal. This is the trap in which many PFM reforms fall in assuming that a PFM system that has strong controls, by itself, will lead to good public service delivery; thus, the evidence on whether this hope materializes is inconclusive (Miller et al. 2021; NYU Wagner and ODI 2021).

Another view of PFM systems considers them as a means to an end for the effective functioning of public services and governance. The focus here is determining if a PFM system is effective or not based on whether it achieves effective public service delivery and meets the goals of governance. The remainder of this section describes the objectives of this interpretation of the PFM system.

2.4.1 Objectives of PFM

Traditionally, experts envisaged PFM as being narrowly concerned with how governments manage budgets across the PFM cycle at the level of the Union government (Cangiano et al. 2013). Over time, PFM has become increasingly relevant to not only fiscal policy-making but also public service delivery. Cangiano et al. (2013, 2) note that the main goals of PFM include 'maintaining a sustainable fiscal position, the effective allocation of resources, and the efficient delivery of public goods and services'. The Joint PEFA/World Bank/IMF Public Expenditure Working Group has posited that '[e]ffective PFM systems are crucial to countries making progress in reducing poverty' (Commonwealth Secretariat 2005, 27, Appendix B).

To understand PFM as means to good public service delivery, PFM systems must focus on function—what they do, instead of what they look like. In other words, PFM systems need to have not only procedures and tools that are desirable, but those that enable public service delivery and good governance. How do current PFM experts envision good functionality?[5] Andrews et al. (2014, 6) suggest that functional PFM systems ought to promote four things. One, PFM systems should allow governments to make prudent fiscal decisions, such as ensuring that spending decisions are affordable and do not threaten the state's solvency. Two, they should ensure that budgets are credible by aligning actual revenue and collection with budget plans. Three, these systems should enable a reliable and efficient flow of funds to spending agencies. This includes creating systems that allow for timely and effective procurement, contract management, and payments. Four, PFM systems ought to have institutionalized mechanisms for reporting and oversight by administrators, citizens, politicians, and auditors.

We agree with Andrews et al. (2014) that these four tasks are crucial, though implementation of these mechanisms is insufficient where the goal is to ensure strong public service delivery. In practice, these goals are often addressed in silos with the aim of individually improving PFM systems by themselves. Such improvements are not wrong, as the incentive of PFM practitioners is indeed in having efficient systems. However, if PFM is to become a means to improve public service delivery, these four tasks must be adjoined with an integrative view of how PFM enables or hinders good policy implementation.

[5] In taking this view, Andrews et al. (2014, 7) contend that 'a direct measure of PFM functionality should reflect on the direct impacts of the PFM system (asking whether it fosters prudent fiscal decisions, credible budgets, reliable and efficient resource flows and transactions, and institutionalized accountability) more than whether the country has macroeconomic stability or allocative efficiency (which are only influenced partly by PFM systems)'. Please see an introductory blog by one of the co-authors here: https://www.internationalbudget.org/2014/08/what-is-pfm/.

We are not alone in calling for connecting PFM to public services. Health scholars have assessed whether PFM changes can lead to differing health outcomes. Some work, reviewed by Welham et al. (2017), indicates that corruption is consistently and negatively linked to health delivery. Their review also suggests that both budgetary allocations and some cash-budgeting practices are linked to poor health delivery outcomes. They further note that improvements in transparency, accountability, and decentralization have a positive correlation with progress in these outcomes. However, both Welham et al. (2017) and Goryakin et al. (2020) emphasize the data and empirical challenges associated with connecting PFM to public service delivery. Given the complexity of PFM, and the siloed nature of reforms that can only transform a small aspect of the cycle, it is difficult to causally link changes in PFM to changes in health services.

In this book, we do not claim to make causal linkages between PFM and public services because, as others have noted, PFM is complex and multifaceted. There are insufficient studies and data to isolate all aspects of PFM to examine its causal impact. Additionally, PFM and state capacity are often correlated, and the direction of causality is difficult to establish—especially with the dearth of data on PFM itself.

Rather, we take a process view to illustrate how public services have been limited by processes in the expenditure, auditing, and monitoring of public funds. This process-oriented view that connects PFM to improvements in public service delivery has been undertaken by others as well. For instance, Ravishankar et al. (2023) argue that timely, credible, and granular information about flow of funds is crucial for improving health services. In putting forth this claim, they connect the cross-linkages within parts of PFM and health systems. We make similar links, which are expanded upon in Chapter 4 and accompanied by detailed examples, that show precisely why PFM is correlated with poor instances of public services.

Taking a step further, we offer two core ways in which good PFM can enable good public service delivery. One, the PFM architecture ought to be designed such that key public finance functions (budgeting, expenditure management, monitoring, and accounting) are synchronized to share data with each other and with other parts of government around programmatic events that have a financial implication (key events or processes that occur within public programmes and policies, such as a vaccination delivered, or a subsidy received by the end recipient). Such architectural synchronization (expanded upon in Chapter 4) can ensure that financial outlays and programmatic outputs transparently align with policy expectations and can be monitored and audited seamlessly. Two, the architecture should support an underlying plumbing that seamlessly links fund allocations and expenditure to public service delivery experiences of end recipients, without funding delays, leakages, and other current manifestations of a poor architecture that we highlight in Chapter 5. PFM should stop being the sand in the gears of public administration.

In meeting its first goal of a synchronized architecture, an obvious role for PFM systems is to be unified in their varying functions and to better integrate with the functions they do support—namely, public services. In doing so, PFM systems can take burdensome 'management' out of 'public financial management'. To this end, an ideal system ought to connect financial outlays in the budget to its near to on-time expenditure at the micro level; accurate expenditure data to data that allows easy budgetary monitoring and accounting; and financial data to data for programmatic events for which the budgets are allocated in the first place. As we show in the next chapter, attempted reforms, both in India and globally, often do not improve the synchroneity and data accuracy challenges of these varying functions, leading to immense difficulty in understanding where money goes, how it is spent, and what its impact is, if any. Data in PFM systems therefore ought to enable three functions: (a) upstream planning and budgeting for the future; (b) near real-time feedback on government's financial operations which enables administrators to course correct when needed; and (c) implementation of downstream accountability mechanisms such as government audits conducted by government and social audits undertaken by citizens.

The second goal of a PFM system is to have seamless plumbing between budget allocation and public service delivery. These functions include on-time predictable budgetary releases to implementing agencies, optimal cash management, smooth contract management with reliable payments to employees, citizens, and vendors, credible and timely financial reporting, transparent data for stakeholders, and efficient audits. These roles, which are vital to not only effective public financial management but also public service delivery, should function in the background with minimal interference—and be supported by the underlying architectural design. To extend a metaphor, good public financial management is to public administration what a healthy heart is to the human body—essential, autonomous, mostly unnoticed, and with minimal needs for fixes.[6] We argue that two reforms—an integrated architecture and an unclogged plumbing pipeline—can help PFM systems go a long way in producing efficient public service delivery.

Existing literature has envisioned public finance in this broader way.[7] Scholars such as Miller et al. (2021) and Andrews (2013) have connected PFM to service delivery. In the Indian context, however, this discussion has not been very robust, save for a few voices (Aiyar and Raghunandan 2015 and Jain and Gurazada 2023). Experts who work on public financing of specific sectors such as health have also meditated on similar goals for PFM systems. For instance, speaking of aligning PFM and health financing, Cashin et al. (2017, 3) point out that 'a strong PFM system can ensure higher and more predictable budget allocations, reduced fragmentation in revenue streams and funding flows, timely budget execution, and better financial accountability and transparency'. In short, these goals show that a PFM system's function is to enable effective budget allocation, execution, and monitoring with an eye towards public service delivery at its core.

2.5 Conclusion

Various theories have emerged to explain the poor quality of public service delivery. Scholars have argued that political interference in bureaucracies and public policies prevent adequate accountability for good service delivery. They have further made the case for institutional weaknesses—poorly structured incentives for frontline workers and career bureaucrats, along with structural incentives for corruption within

[6] We owe this metaphor to Charity Troyer Moore.

[7] More recently, both the Public Expenditure and Financial Accountability programme (see https://www.pefa.org/resources/strengthening-links-between-pfm-and-service-delivery-sectors) and ODI have made connections between public finance and public service delivery (see https://odi.org/en/publications/public-finance-and-service-delivery-whats-new-whats-missing-whats-next/).

the bureaucracy, leading to weak public service delivery. While some have argued that decentralization increases accountability at the local level in solving some public service delivery challenges, others disagree. Moreover, another stream of literature looks at the need to improve state capacity, but it focuses on the ability of a state to collect revenues from its citizens rather than a state's ability to effectively spend such revenue.

While we agree that many of these explanations shape why citizens experience poor delivery of public services, we add that they do not sufficiently explain why financial outlays fail to produce public goods outcomes. We add to these streams of literature by arguing that the failure of government's expenditure to translate to outputs is also rooted in a slowly reforming PFM architecture. The current architecture needs further improvement in integrating PFM functions across the board, as well as with functions that are crucial for public service delivery. With changes, the internal PFM plumbing can ensure that allocated resources are actualized as intended. In ameliorating these architectural and plumbing issues, we posit, public service delivery can be drastically improved.

3

History of Public Financial Management Reforms

PFM, both globally and in India, has not always been geared towards public service delivery, but PFM reforms in the last few decades have inadvertently affected the way people experience public services. This chapter discusses the historical areas of focus for PFM reforms—globally and in India—and where these reforms are headed in the near future. Lessons from global reforms underline the need to account for context in how PFM is modified. Countries have differing needs, and the impact of a reform in one part of the world may not necessarily translate to another.

India has to some extent taken this lesson in its stride. There have been numerous changes to PFM's functioning in the country, including efforts to digitize processes to improve efficiency. Other, more citizen-centric developments in the system—such as Direct Benefit Transfer (DBT)—have improved the efficiency with which funds reach end recipients. Overall, India's PFM reforms, while making progress in areas such as cash management through SNA-SPARSH, cash transfers through DBT, and visibility through bank integration, continue to have two constraints that, if addressed, can help improve quality of delivery of public services. One, they are often designed by accountants without the big picture of challenges across various parts of government in mind. Second, they do not fundamentally change the underlying challenges with the way programmatic and fiscal data is captured, recorded, and used to help fully connect public finances to public service delivery. We highlight these challenges in this chapter by first summarizing global reforms.

Public Financial Management, State Capacity, and Public Services in India. Santhosh Mathew, Bhumi Purohit, and Devesh Sharma, Oxford University Press.
 DOI: 10.1093/9780198973195.003.0003

3.1 Global Reforms

PFM has been practised for as long as there has been a public purse. The current traction in PFM reforms—marked by heightened engagement amongst analysts, the public, civil society organizations, and international donors—dates roughly to 1980s, emerging in OECD countries. These reforms were a response to increases in government expenditure (Wencai 2016) and sought to make public finance more efficient. They largely fell under the umbrella term 'new financial management', consisting of five main types of changes: (a) new financial reporting systems such as accrual accounting; (b) market-oriented systems for pricing and public service provision; (c) performance measurement techniques; (d) devolved or decentralized budgets; and (e) changes to internal and external audits to look for 'value-for-money' (Guthrie et al. 1999). As Guthrie et al. (1999, 212) argue in their multinational comparative study, however, these reforms were aimed to keep up with the latest techniques, with a 'limited awareness of the capacity of such systems to change public services without any corresponding public debate over both their reliability and operational merits'.

Donor support for PFM reforms, especially in African, Asian, and Latin American countries, has risen even more rapidly since the 1990s and has largely focused on making investments in information technology systems. For instance, since 1984, the World Bank has funded Financial Management Information Systems (FMIS[1]) in 84 countries and mapped the development of 194 FMIS solutions across 198 economies worldwide, with the total cost of these projects being roughly USD 6 billion.[2] While these information technology initiatives have been impactful, they have rarely achieved their full potential. They have also taken much longer to be completed than expected; the average completion time for FMIS systems globally has been about eight years, twice their original forecast (Dener et al. 2011).

FMIS reforms have generated much analysis and introspection within the field of public finance. The main lessons from these reforms include due

[1] FMIS and IFMIS are two abbreviations used globally. The word 'integrated' usually is brought in to refer to a broader set of integrations in the public finance function (such as payroll, debt management, accounting). We use the term FMIS as interchangeable with IFMIS.

[2] See more at https://www.worldbank.org/en/topic/governance/brief/financial-management-information-systems-fmis.

consideration of the context and avoiding overspecified or oversimplified solutions; expanding the focus beyond technocrats in central departments such as departments of finance at Union or state levels (Andrews 2013); recognizing the complexity and political nature of PFM reforms (CABRI 2013; Dener et al. 2011); and avoiding big-bang approaches that solely rely on building large monolithic IT systems (Uña et al. 2019). The large IT systems, particularly when heavily customized, also lead to vendor lock-in, and are expensive to change when adapting to evolving needs of governments. Furthermore, many of these reforms are borrowed from OECD countries because they promote good governance, but they do not fit the needs of non-OECD countries (Andrews 2010).

In this book, we take these lessons seriously and propose reforms specific to the Indian context. Our suggestions are rooted in what we believe to be good for better expenditure management in India, as opposed to what is usually considered good by external agents. In doing so, we do not advocate for a singular approach or movement, but find common ground across many lessons. For instance, there has been a large push to make finances 'open'. An increased push for transparency has led to developments such as the Open Budget Index (OBI), Open Government Partnership (OGP), Open Government Data Platform in India, and data.gov in the United States. We value such movements for transparency, and the other shifts in PFM such as global reporting benchmarks like the Public Expenditure and Financial Accountability (PEFA) programme. We aim to integrate learnings from such practices to the Indian context and offer suggestions for how governments in India can improve their PFM systems that align with these movements.

Global PFM reforms to date have been created by, and for, public finance experts. We value the inherent need for such expertise, and for PFM to speak to these needs. Where we diverge from the existing direction in global reforms is in promoting the view that PFM should be built to complement both financial efficiency and public service delivery. We therefore offer a lens of how PFM is directly tied to public service delivery, and how such reforms can improve delivery.[3]

[3] The book does not provide guidance on fiscal policy or programmatic policy. Instead, the foundational shifts we propose are focused on the architectural level of how financial data is recorded, linked, shared, and used. As we expand later, the shifts we propose in the data architecture can foster a series of reforms in how money, information about the money, and decisions about that money move through government, promoting greater visibility to stakeholders.

While our starting point is the set of challenges that we have witnessed across India, these are aligned with suggestions by researchers with global experiences who have faced similar issues and recommended solutions that are aligned with those we propose.[4] Some versions and parts of what we suggest in this book have already been implemented in other countries. The Philippine government has established the Convergent Hub, a coreless digital architecture made up of a collection of APIs (application programme interfaces) that sits atop existing FMIS technology to enable PFM functions such as budgeting, revenue management, accounts, disbursement, and reporting, across different levels of government (Department of Budget and Management 2023). Uganda's UGhub enables seamless data sharing across government systems in a secure, efficient, and sustainable manner. It is aimed at improving service delivery to citizens as well as reporting, planning, and decision-making across government and private entities, including financial institutions. Countries such as Benin, Brazil, Cabo Verde, Cambodia, China, Costa Rica, Eswatini, Georgia, Guinea, Guinea-Bissau, Honduras, Kenya, Liberia, Maldives, Mongolia, Rwanda, Senegal, South Africa, Togo, Uruguay, and Uzbekistan are undertaking business process automation, data exchange, and digital payments that are changing PFM systems. While these changes have improved parts of these countries' PFM systems, we propose a holistic redesign for the Indian case.

This redesign emerges from the context specificity required for PFM. Digitizing existing systems necessitates a fit-for-purpose transformation process that addresses constraints specific to each policy area,

[4] For instance, Paso et al. (2023) recognize challenges associated with legacy IT systems that do not support PFM reforms and the lack of process design to leverage digital tools. Their suggestions for design of PFM systems include the use of principles such as automation of controls; data exchange, flexibility, and scalability; and the once-only principle for data capture, amongst others. Similarly, Long et al. (2023), recognizing the current nature of PFM's siloed architecture, note that, 'to be more flexible and responsive to users' needs (policymakers, civil servants, and citizens), PFM should embrace the new paradigm for digital government'. To achieve this goal, they recommend the usage of tools such as registers, common platforms, agile systems, and different yet connected layers of data, platforms, and services. Furthermore, Richard Pope's *Platformland: An Anatomy of Next-Generation Public Services* (2024), which has been called a nuanced homage to O'Reilly's seminal work on 'Government as a Platform', offers ideas on reinventing public services, including eliminating administrative burden, building and using common components, and imagining data as infrastructure. While the ideas in this book use a different frame that speaks specifically to India's public financial management, our ideas connect with the advice that authors in the field of public financial management and digital technologies offer.

geography, and local incentives. We propose a redesigned PFM in India in Chapters 6–8. In the subsequent subsections, we first highlight how India has historically thought about PFM.

3.2 A Centralized British Legacy in India

Notwithstanding Kautilya's *Arthashastra*—a treatise on economics and finance—from the ancient times, India's PFM system as it stands now is a result of efforts put in place by the British Crown during colonization in the nineteenth century. Since then, independent India has enacted various improvements to PFM, its core functions, and its capabilities to meet the needs of an ever-evolving nation. Yet even these changes find their roots in a colonial legacy that emphasized centralized control over funds.

In the nineteenth century, the British Crown established finance systems to exercise strict control over budget execution. In their system, budgeting began when a spending department sent proposed expenditure to the Finance Department for scrutiny and approval. After approval, the Finance Department requested the legislature to pass the budget. Once the budget was approved by the legislature, spending departments then waited for the Finance Department to examine expenditure proposals yet again. This allowed the Finance Department to scrutinize the spending department before allowing them to spend the money, leading to a long chain of files and letters exchanged between the two entities. This system allowed the British government to control and monitor budget execution in its colonies starting as far back as 1858 in India (Handa 1990). Even today, in many parts of governments across India, this system remains more or less the same, including the long exchanges of files and letters.

The administrative culture that drives PFM in India is still not divorced from the logic of centralized institutional structures established during British colonization. The British Crown had three main goals when it established PFM in 1858 (Handa 1990). One, it sought control over the 'loose' financial administration of the East India Company. In fact, the saying went that anyone who could get a job with the Company could return to England sufficiently rich to live the rest of his life like a lord. By vesting financial powers with a few people at the top levels of the administration, the Crown sought to tighten its reins on the Company. Two, with the control of administrative expenditure, the colonial administration sought control over policies in the Indian territory and retained a surplus. Having a process that required excessive scrutiny to execute and subsequently manage budgets gave the Crown the steering wheel. Three, the centralization of PFM allowed the foreign rulers to oversee the local Indian officials, who were perceived to be untrustworthy. The system allowed the British administration to retain power away from any local officials.

This was not unique to India. In fact, the British Crown used centralization to wrest control elsewhere, even during adverse circumstances. A letter dated 11 August 1812 from the Duke of Wellington to the British Foreign Office in London points to the extent of scrutiny that may have been levied by London-based officials. In it, the Duke regrets being unable to account for one shilling and ninepence (Natsios 2010, 1–2):[5]

> Whilst marching from Portugal to a position which commands the approach to Madrid and the French forces, my officers have been diligently complying with your requests which have been sent by His Majesty's ship from London to Lisbon and thence by dispatch to our headquarters.
>
> We have enumerated our saddles, bridles, tents and tent poles, and all manner of sundry items for which His Majesty's Government holds me accountable. I have dispatched reports on the character, wit and spleen of every officer. Each item and every farthing has been accounted for with two regrettable exceptions for which I beg your indulgence.
>
> Unfortunately the sum of one shilling and ninepence remains unaccounted for in one infantry battalion's petty cash and there has been a hideous confusion as to the number of jars of raspberry jam issued to one cavalry regiment during a sandstorm in western Spain. This reprehensible carelessness may be related to the pressure of circumstance, since we are at war with France, a fact which may come as a bit of a surprise to you gentlemen in Whitehall.

[5] Attributed to the Duke of Wellington, during the Peninsular Campaign, in a message to the British Foreign Office in London, 11 August 1812, as cited in Natsios (2010).

There was a moment where policies could have shifted away from strict centralization. The Montagu–Chelmsford reforms,[6] enacted in 1919, had the potential to lead to a more decentralized India in line with the demands of provinces like Bombay and Bengal that wanted more autonomy in fiscal decisions. The reforms increased the powers given to central and provincial legislative councils but ensured that the ultimate responsibility remained in London. Despite legislative expansion, the fiscally constrained and inter-war Government of India chose the reforms to be in service of centralization, and increased central financial control (Charlesworth 1985, 547).

This centralist point of view continues to date with Centrally Sponsored Schemes, for which the state and local governments act more as representatives of the Union government rather than autonomous decision makers promoting the needs of their local communities. Some observers have also objected to the use of the phrase 'Central government', for it alludes to centralized planning and control—preferring the use of the term 'Union government' instead.[7] This highly centralized system fits well with a tradition of structuring human relationships in hierarchical terms, a tendency encouraged by the caste system and by the tradition of deference towards authority (Taub 1969, 161 as quoted by Dwivedi et al. 1989, 256).

British rulers may have gravitated towards such a centralized system with tight controls when the Crown was mostly interested in collecting revenue and maintaining law and order. Even when economic development was undertaken, it was largely to serve British trade interests. Hence, most departments were organized to ensure compliance to rules without much concern for friction and delays in fund flows or public service delivery. Consequently, a large part of the centralized nature of India's public administration systems today does not speak to the needs of its current welfare state, even with calls from within the administration to decentralize (Agarwal and Somanathan 2005, 12, 16).

[6] Also known as Mont–Ford reforms or Government of India Act, 1919.

[7] As Kelkar and Shah (2019, 183) note, 'We wince every time the expression "central government" is used in the popular discourse as this is suggestive of central planning and an exaggerated conception of the role of New Delhi. The phrase "Union government" is an accurate description of the constitutional scheme, and a more modest phrase.'

The sticky, centralized nature of India's PFM architecture under the British Raj—and more broadly, India's public administration—did not evolve to serve the needs of a welfare state that needed social and economic development. For instance, even the modern-day budget year from April to March is a colonial import from 1866, and continues to be unsuitable for development work (Committee on Change in Financial Year 1985; Press Information Bureau 2016b); by the time the sanctions reach spending agencies, monsoon hits most parts of the country and work comes to a halt (Handa 1985, 648).

Despite parallels to British-enacted PFM systems, various parts of modern PFM in India have undergone new designs and reforms since its independence, which we highlight next.

3.3 Modern Day Reforms in India

Significant reforms in treasuries, budget management, and direct benefit transfer have led to key changes to PFM in modern-day India. Though these changes have digitized existing processes and led to efficiencies in many parts of PFM, particularly in cash transfers, they can do more for public service delivery. These reforms in PFM are shaped by well-intentioned accounting and finance professionals, who design PFM systems primarily for, and which are almost exclusively implemented by, other accounting and finance professionals. While accounting experts are undoubtedly essential to PFM reforms, relying solely on them overlooks the needs of the very stakeholders who use public finances for public services. Regardless, these reforms showcase the drive in the government to address the challenges of India's PFM architecture.

3.3.1 Digitized Treasury and Payment Systems

In 2009, the Government of India launched the treasury computerization project as a part of the National e-Governance Plan (NeGP). Treasury computerization, aimed at state and district treasuries, was expected to bring about a series of reforms: make budgeting processes more efficient, improve cash flow management, promote real-time reconciliation of accounts, strengthen Management Information Systems (MIS), improve accuracy and timeliness in accounts preparation, bring about transparency and efficiency in public delivery systems, and provide better financial management in states and Union Territories (Monga 2010). Over the years, many states have gone beyond treasury computerization to introduce Integrated Financial Management Information Systems (IFMIS) that incorporate many functions in PFM beyond treasury management, such as budget management, accounting, public debt management, and preparation of financial statements (Uña and Pimenta 2016).

Additionally, the launch and growth of the Public Financial Management System (PFMS), a financial management system run by the Controller General of Accounts, has been phenomenal. Set up in 2009 as the Central Plan Scheme Monitoring System (CPSMS) to track the use of funds under the Government of India's Plan Schemes and undertake real-time reporting of expenditure, the scope of PFMS has since widely expanded. For agencies across the country, it is the default payment gateway for all Direct Benefit Transfer schemes; it is envisaged that it will facilitate the compilation and preparation of accounts of the Union government, accounting for receipt of non-tax revenue. It will integrate with state treasuries for exchange of data on budgets, allocation, and expenditure against the central transfer of funds for Centrally Sponsored Schemes of the Government of India (CGA 2012).

PFMS has developed wide-ranging capabilities and services for public finance at the vanguard of digital technologies for public financial management. It can communicate with over 360 banks across the country to verify names of an account before money is transferred, obtain real-time bank statements of government accounts, and send payment instructions to these banks for operating government accounts. PFMS also has modules which computerize payments, receipts, and accounting in the Pay and Accounts Offices. Moreover, it can enable various departments in the Government of India to collect non-tax revenue (MicroSave Consulting 2022; Sitharaman 2020; Somanathan 2023).

These changes have led to a number of benefits for the Government of India. The Finance Minister at the time, Nirmala Sitharaman, in her speech on the 49th Civil Accounts Day, noted that PFMS' use of Treasury Single Account for autonomous bodies has saved the Government of India Rs. 150 billion, amounting to approximately USD

2.33 billion at the time, in interest costs since 2017–18 due to reduced float (Sitharaman 2025). Furthermore, Sitharaman noted that PFMS' use of Single Nodal Agency, used for disbursements to state governments, has saved the Government of India Rs. 110 billion since 2021–22, or approximately USD 1.48 billion at the time. These savings come with an additional benefit; the use of just-in-time fund release through use of PFMS helps ensure that the government does not borrow more than it needs.

These benefits, among others, have made PFMS a foundational step towards improving public fund management in India. Yet PFMS can go even further in streamlining solutions for concerns faced by public service providers. For one, PFMS is a Union government software—not an independent protocol or a platform—which limits its adoption by state governments. State or local implementation managers cannot adapt and tailor PFMS to their specific needs. Furthermore, the system has limitations in how far it can go to help managers understand public service delivery. For instance, PFMS can be used to digitally trigger a payment through the banking system. However, due to the absence of standardized linkages with programmatic data, the system cannot determine whether a payment accurately reflects the amount of work completed, or even how much work it pertains to—an issue observed first hand by one of the authors, Santhosh Mathew, during his tenure in government, that still persists today. Any programme manager seeking to link programmatic data with PFMS must co-ordinate with many actors to manually extract and match various datasets. Additionally, even though PFMS is used across the country for Centrally Sponsored Schemes, its data is not linked with policy output and outcome data, nor with state and local government data systems (with the exception of state treasuries). This renders PFMS inadequate at scale in helping generate information for effective programme management, or for external scrutiny with tools such as public expenditure reviews and public expenditure tracking surveys that can meaningfully inform policy.[8]

[8] For more information on public expenditure tracking surveys (PETS), see Gurkan et al. (2009).

3.3.2 Budget Management

The Indian government has also made changes in budget management. The first organized and centralized form of budgeting consisted of line-item budgeting, which classified expenditure and the purpose of the expenditure (e.g. the goods or services being procured such as supplies, personnel, or equipment) (Dutta 1975). Following the recommendations of the Administrative Reforms Commission and in line with global trends, the government introduced performance budgeting in 1969. Performance budgeting shifted the focus from 'inputs received' (i.e. supplies, equipment, and personnel) to 'work to be done'—in other words, outputs being produced using these inputs. It also brought to the fore the usage of measurement tools and techniques such as unit costing and performance standards. However, performance budgeting failed to connect the budgets to outcomes—rather than just outputs—of government programmes. To remedy this, the government went a step further and introduced Outcome Budgets in 2005, and an Output–Outcome Monitoring Framework in 2017–18 (Dutta 1975; Gayithri 2015; Ministry of Finance 2006; Pandey 2005; Satpathy 2018).

Despite the benefits of outcome-based budgeting, many issues hinder its effective use, not just in India but elsewhere. In its report, the High Level Expert Committee on Efficient Management of Public Expenditure chaired by Dr C. Rangarajan identified various challenges to implementing outcome-based budgeting, including: 'segmentation of expenditure between Plan and Non-Plan,[9] absence of a hard budget constraint, lack of incentives to Ministries to reallocate resources, non-availability of information on costing of services, problems in budget and accounting classification, inadequate information systems on transfers of resources to States and absence of robust financial management information systems' (Planning Commission 2011, xiii). Some of these challenges have been addressed; there is no longer a distinction between

[9] Non-plan expenditure was usually what the government spent on the committed expenditure, such as salaries, subsidies, loans, and interest, considered core to the functioning of government. Plan expenditure pertained to the money set aside for proposed expenditure on various ministry or department projects. This distinction remained while the Planning Commission was active. The Commission was disbanded in 2015 and many of the duties were shifted to the newly formed NITI Aayog (for non-financial roles) and the Ministry of Finance (for financial roles). The distinction was removed in FY 2017–18.

plan and non-plan expenditure, for instance, and all central transfers to implementing agencies are now routed through state treasuries. Still, the other issues remain.

By design, outcome-based budgeting necessitates a consolidated and accurate data infrastructure to regularly identify the key outputs and outcomes, collate data on them, and connect them to expenditures (Asian Development Bank 2017). And if outcome-based budgeting is to truly represent the relationship between outlays, outputs, and outcomes that will vary across time, geographies, programmes, and political contexts, this data infrastructure needs to support effective policy-making and implementation. Today, reliable and high-quality data constrain India's ability to use outcome-based budget management (Bhunia 2020).

3.3.3 Direct Benefit Transfer and Aadhaar

In recent times, Direct Benefit Transfer, or DBT, is an important reform that India has successfully implemented. DBT removes intermediaries in the fund flow pipeline such that money is sent directly from the initial level of government which owes the funds to the employee or beneficiary. DBT is enabled by two core additional functions: fund recipients need to have bank accounts and verifiable identification, which ensures that funds are sent correctly and withdrawn by the correct individual. To ensure that low-income individuals also have access to DBT, the government has opened low-savings bank accounts for them through a scheme known as the Pradhan Mantri Jan Dhan Yojana. These advances have reduced leakages in the fund flow between the government and end recipients (Muralidharan et al. 2016).

DBT has undoubtedly eased the way recipients, on average, interact with the government to receive funds. While it has resolved issues that arise after government stakeholders authorize or approve a payment, reducing the number of steps that go into the decision to make a payment in the first place can make the process even better. Steps prior to making payments include determining whether eligibility requirements are met to warrant a payment, capturing data accurately to show such requirements are met to have the payments authorized by managers, and ensuring that the correct amount is paid. Effective public service delivery rests on smooth functioning of this chain of activities that precede DBT.[10] Additionally, the strides that India has made with DBT show that reforms in payments outside of cash transfers—such as vendors that supply goods and services to government or to citizens on behalf of government—are possible.

DBT has frequently been used with Aadhaar, which is a form of biometric authentication that is used to verify if the intended beneficiary of a cash transfer is authorized to obtain the funds. Aadhaar has become increasingly common across India, with over 99% of the adult population having the ID. While the Supreme Court of India has issued varying orders that do not suggest Aadhaar is necessary for obtaining government services, it remains a common form of identity verification for both receiving payments and enrolling in government programmes. The use of Aadhaar, much like use of digital identity across the world, is also widely debated, particularly due to concerns that requiring biometric authentication may both raise data privacy concerns and prevent certain types of citizens from obtaining public entitlements. As we detail in Chapter 6 on the use of data, especially personally identifiable information, these concerns are legitimate and require a healthy debate to address the underlying reasons behind issues of entitlement, data protection, and privacy.

3.3.4 Other Reforms

Other successes have focused on improving the timeliness of fund disbursal between funding and implementation agencies by reducing the number of layers that funds travel through before reaching the beneficiaries. These include the integration of state treasuries with PFMS for smoother reporting of central assistance to states; Single Nodal Account to improve visibility of the funds taken out of the consolidated fund of the Union and states and put into the bank accounts of implementing

[10] We do not cover the issues that citizens face enrolling in Aadhaar-linked DBT. Scholars such as Rithika Khera have documented the issues citizens face in obtaining Aadhaar itself (Khera 2019). Even when individuals link DBT with Aadhaar, there is limited transparency on what benefits they are eligible to receive, and how to meet those eligibility criteria. In recent years, Haqdarshak has created a tech-based platform to help citizens identify schemes and apply for them.

or intermediary agencies (societies); and just-in-time payment initiatives to expedite flow of funds and eliminate float for Centrally Sponsored Schemes.

India has also created the Government eMarketplace (GeM) to streamline public procurement, and the Tax Information Network (TIN) and Goods and Services Tax Network (GSTN) to streamline revenue collection. While we do not cover revenue administration within the ambit of this book, these systems are important examples of how government can use digital technologies to solve public finance problems (Shah 2006).

Irrespective of the architecture of solutions suggested, these reforms offer valuable insights into implementing PFM through digital technology. With the vast number of actors involved in expenditure management at various levels of government, each with different needs, we mainly diverge in the way we propose architecting PFM in a decentralized manner. Rather than creating a central system or a central data repository, we propose an architecture for expenditure management later in the book which allows stakeholders to build their own microsystems that speak to a larger digital infrastructure using common data standards and data exchange. This decentralized model allows each stakeholder to maintain a level of autonomy over how they use data systems, while adding a layer of data protection.

3.3.5 Unrealized Reforms

Though some parts of Union and state governments have enacted a growing number of reforms, there are many more recommendations yet to be implemented. These pending reforms—primarily focused on improving data management—hold significant potential to address public service delivery challenges. Although some have yet to materialize, the government's engagement with these proposals suggests a clear recognition of the need for change.

In 2019, the Prime Minister of India, speaking at the Conclave of the Accountants General and Deputy Accountants General convened by the Comptroller and Auditor General of India (CAG), called for making data usable for auditing:

> We aim to make Evidence Based Policy-Making an integral part of governance by 2022 . . . Right now the data that is being generated here is very large and it is stored with the various agencies and departments. The data is also collected by these agencies and departments for their use. But it is also true that often this data is not shared with each other. This huge data is not linked. Hence, in natural course there is an accountability gap. We have to bridge it.

Prime Minister Modi's speech signals that political leadership finds value in the large amounts of data that exist but are currently non-sharable, non-usable, and hence not optimal for improving public services.

Soon after in 2020, the CAG, by invoking Article 150 of the Indian Constitution, advised the President of India on the importance of reforming accounting practices in India in an advisory letter calling for better data systems (Thakur 2020). In an op-ed, Sabharwal and Phatak (2020) refer to the CAG's letter and note:

> The Union budget grew from Rs 197 crore in 1947 to Rs 30 lakh crore last year and total government expenditure may be higher than Rs 70 lakh crore. But the form and manner of keeping accounts have more or less remained unchanged since Independence; manual transactions and manual payments often lead to manually entered data at different stages in different databases on different systems. This makes data unreliable, violates the principle of 'single source of truth' and sabotages transparency and good governance. DATA (Data Accountability and Transparency Act) recognises that digitally empowered citizens require digital public utilities that not only provide e-services but make all government revenue and expenditure data electronic, machine-readable, granular, comprehensive, purpose linked, non-repudiable, reliable, accessible and searchable.

These calls for reforms,[11] along with many others, are a sign that there is initiative within government for improvements. While these have

[11] Suggestions for reforms, such as moving to a multidimensional chart of accounts (Sundaramurti 2012), need for linking outlays with outcomes (XIV Finance Commission 2014), and need for an overarching legal fiscal framework across Union and state governments (XV Finance Commission 2019), have also been given in the past.

not necessarily been implemented, they show that the government recognizes that there is a fundamental problem in the manner data are used within PFM and between PFM and public service delivery. If there are core architectural shifts in the way such data is integrated within PFM systems, there is potential to improve management of funds and information, public service delivery, and the link between them.

3.4 Conclusion

To date, reforms implemented globally have pointed to an important lesson—that context matters. Monolithic, one-size-fits-all solutions have therefore had muted impact, while advancing PFM nonetheless. The improvements made in PFM systems which we have covered in this chapter do not completely detail those across various parts of the world, nor do they cover the full breadth of changes in India's PFM. Yet they are representative of the energy and the motivation in the political and administrative system to make structural changes to PFM.

However, these reforms also lack scope—they are limited in their capacity to address critical binding constraints on public service delivery. In the next part of the book, we propose a series of improvements that restructure PFM such that it drives improvements in public service delivery. Before we explain these ideas, in the next chapter we provide an overview of India's PFM and its challenges, by exploring the integral connections between PFM and public service delivery failures.

4

Architectural Challenges in India's Public Financial Management

Despite India's growth across different metrics over the last three decades, major problems persist in public service delivery. One of the two key arguments in this book is that these problems are connected to India's poor public financial management (PFM), which has the opportunity to enable good public service delivery. For PFM to function as a support for public service delivery, its architecture must be internally coherent so that various parts of PFM speak to each other, as well as externally aligned with the functions of public service delivery. Prior to detailing our ideas, this chapter first steps back and explains the various components of the PFM architecture, which we call push-and-wait architecture, as it relates to public service delivery today. We then explain why this existing architecture—which forces officials across government to push requests for funds, information, and decisions and requires officials to wait for funds, information, and decisions—fails to enable efficient delivery and implementation of public goods and services.

We review PFM in India by combining secondary research with extensive interviews with functionaries at the central, state, and sub-state levels in Bihar, Delhi, and Rajasthan. We combine these data with personal experiences from one of us, Santhosh Mathew, who worked in the Indian Administrative Service at the sub-district, district, state, and Union levels of government for over 32 years. The process described below may vary from state to state given that states may have changed a select part of the process. Hence, much of this description is a stylized view to facilitate ease of understanding. To our knowledge, some states—Assam, Chhattisgarh, Himachal Pradesh, Karnataka, Kerala, Odisha, Rajasthan, and Uttarakhand, amongst others—have enacted reforms to

Public Financial Management, State Capacity, and Public Services in India. Santhosh Mathew, Bhumi Purohit, and Devesh Sharma, Oxford University Press.
 DOI: 10.1093/9780198973195.003.0004

improve their PFM systems. As of July 2023, with the initiation of SNA-SPARSH (which we describe later), the Union government has also begun implementing some of the ideas in line with this book. Outside of select states, the process holds true for most departments of most states, as well as the Union government. Notable exceptions are Mahatma Gandhi National Rural Employment Guarantee Scheme (MGNREGS), Ayushman Bharat Pradhan Mantri Jan Arogya Yojana (AB-PMJAY), Pradhan Mantri Awas Yojana (PMAY), Pradhan Mantri Gram Sadak Yojana (PMGSY); we describe MGNREGS in detail in Chapter 7.

Before detailing the problems with the push-and-wait PFM architecture, we first provide a background of the main components of PFM, beginning with the budget formation and allocation process.

4.1 Budget Formation and Allocation

The PFM cycle starts with budgeting, which in its big-picture form has been described as 'a political act, a plan of work, a prediction, a source of enlightenment, a means of obfuscation, a mechanism of control, an escape from restrictions, a means to action, a brake on progress, even a prayer that the powers that be will deal gently with the best aspirations of fallible men' (Wildavsky 1964, xxiii). For political actors, budgets achieve the objective to restrict, control, or obfuscate, depending on the type of budgeting technique used. Budgets vary, ranging from line-item budgeting to performance budgeting, programme budgeting, zero-based budgeting, or sunset legislation—and have had various degrees of success in countries around the world (Handa 1980, 648, 649). At its core, however, a budget is a contract between multiple principals and agents. The way these contracts are negotiated, monitored, and enforced is shaped by formal and informal institutions at various levels (Moe 1984, 765; Patashnik 1996; Williamson 1998, 26). In this section, we focus on the mechanisms through which budgets are formed and allocated in India.

The processes for budget allocation followed by the Union and state governments are duly defined in the Constitution. Beyond these government levels, processes at the local level are guided by distinct laws of different states. For instance, local bodies in Maharashtra and Gujarat follow the Bombay Provincial Municipal Corporation Act, 1949, which governs the incorporation and operations of municipal corporations such as general administration, revenue, and expenditure.

At the Union level, the Budget Division in the Department of Economic Affairs at the Ministry of Finance drives the budget process, which starts in August or September. The Division creates estimates of expenditure and revenue based on inputs from various ministries and their counterparts in states and field offices. These estimates include expected revenue from direct taxes, indirect taxes, non-tax revenue, and capital receipts such as divestments in public sector undertakings and repayment of loans to the government. The budget estimates additionally plan for the fiscal deficit by adjusting revenue and expenditure. They also account for additional borrowing and debt service obligations guided by the Fiscal Responsibility and Budget Management Act, 2003 (FRBM Act). The budget at the state level incorporates states' share of divisible taxes and grants received from the Union government.[1]

The Division creates budget documents,[2] which include plans for many social welfare schemes, a focus of this book. There are three types of government schemes: Centrally Sponsored Schemes (CSS) that share the funding costs between the Union and state governments; Central Sector Schemes (CS), entirely funded by the Union government; and State Sponsored Schemes (SSS), funded entirely by state governments. The Union budget incorporates funding needs for CSS and CS.

Of specific relevance to these schemes are the Demand for Grants and the Annual Financial Statement, which are part of the budget. The Demands for Grants are generally presented as requests for each ministry or department; they are scrutinized by various Standing Committees.[3]

[1] Since 2011–12, owing to growth in states' own revenue and increased central transfers in the share of taxes (untied) and central grants (tied) (Mishra and Singh 2017, 2), the aggregate expenditure of state governments has grown substantially as has the importance of state budgets. For instance, in 2019–20, the states were estimated to spend 64% more than the Union government, up from 46% in 2014–15 (Tiwari and Surya 2019).

[2] Key documents that constitute the Union budget are Annual Financial Statement, Demand for Grants, Finance Bill, statements mandated under FRBM Act viz. Macro-Economic Framework Statement and Medium-Term Fiscal Policy cum Fiscal Policy Strategy Statement, Expenditure Budget, Receipt Budget, Expenditure Profile, Budget at a Glance, Memorandum Explaining the Provisions in the Finance Bill, Output Outcome Monitoring Framework, and Key Features of Budget (Ministry of Finance 2020).

[3] An exception is the state of Haryana, which, in 2023–24, consolidated budgetary grants across departments into eight thematic areas to meet the Sustainable Development Goals (SDGs).

These Standing Committees oversee the work of ministries/departments and, upon reviewing the Demands for Grants, provide reports to the Lower House of Parliament or *Lok Sabha*. Thereafter, parliamentary discussions begin.

The Demand for Grants is incorporated in the Annual Financial Statement, which shows distinct figures for three consecutive years. One, the 'Budget Estimate', is the estimated sum of receipts and expenditure of the Union government for the upcoming financial year. Two, the 'Revised Estimate' for the financial year is the adjusted budget for the present financial year, reflecting new needs that have emerged since the last budget. Three, 'Actuals' are the final amounts received and spent under different budget heads and are thus usually one year old; for instance, the Union budget 2022–23 presented the actuals from FY 2021–22 (Ministry of Finance 2022).[4]

Upon finalization of discussions, Parliament approves the Union Budget. The budget process concludes with the passing of the Finance Bill[5] and the Appropriation Bill[6] to start the financial proposals for raising resources and drawing on them. The Appropriation Bill seeks the approval of the Parliament for withdrawing funds from the Consolidated Fund of India as prescribed under Article 114 of the Constitution to meet the expenditures stated in the Union budget's Demand for Grants. If the funds appropriated by Parliament are found insufficient during the year, it can approve subsequent Appropriation Bills for supplementary grants under the Constitution. Each state government, similarly, passes an Appropriation Bill for withdrawal from the Consolidated Fund of the State. As we describe later, the process exists because drawing from the Consolidated Fund requires approval of the legislature. The Finance Bill, which generally seeks the approval of the Parliament to raise resources, is usually used to enact changes in other laws (such as Income Tax Act, 1961

[4] However, the Controller General of Accounts now produces monthly aggregate expenditure and revenue reports, but these data are only available publicly as totals, and do not detail even department-level expenditures or revenues, let alone line items.

[5] The 'Finance Bill' is not the same as a 'Financial Bill' as defined in the Constitution. For more details on a variety of bills, see https://prsindia.org/theprsblog/money-bills-vs-other-bills?page=67&per-page=1.

[6] Both the Finance Bill and the Appropriation Bill are introduced as money bills as defined in the Constitution and, thus, do not need to be passed by the Rajya Sabha, which can only make recommendations on money bills.

and Customs Act, 1962) to change specific provisions that relate to revenue. After accepting or rejecting the recommendations from the Rajya Sabha and receiving the assent of the President, both these money bills become Acts.[7] We summarize this process in Figure 1.

State legislatures also go through a similar budgeting process, allocating their share of the funds for both CSS and SSS.[8] For these two types of schemes, states allocate Union and/or state funds to specific departments of the state government headed by a minister who is a member of the legislature. After the state legislature approves the withdrawal of funds, the departments in turn make allocations to their units in the districts, which are not a part of the legislative process but are involved in policy execution. District units further allocate funds to lower units such as those in blocks. Blocks may then disburse the funds to local units such as the panchayat (village-level government) or until the last implementation agency in charge of making payments to the end recipient is reached.

While the process of budgeting itself is streamlined, challenges abound. Primarily, budget formation today is prone to poor planning, which has downstream effects for policy implementation and leads to differences between budget estimates and actuals.

4.2 Budget Execution

In the previous section, we discussed the ways in which funds are allocated and approved. Next, we discuss the institutional arrangements for how budgets are disbursed for spending based on the source and destination of funds.

A variety of institutions manage the disbursal of resources raised by various levels of government. The institutions and processes that direct the flow of budgetary resources and their spending in India vary based on the following factors: (a) the level of government raising the resources; (b) the mechanisms under the Constitution; and (c) a legal arrangement

[7] For more information on the budget, refer to PRS India's guide at https://prsindia.org/files/budget/Union_Budget_Primer.pdf and https://www.business-standard.com/budget/article/union-budget-2018-19-of-india-what-it-is-and-why-it-is-so-important-117122800447_1.html.

[8] See, for instance, Madhya Pradesh's budget manual: https://web.archive.org/web/20191222112611/https://finance.mp.gov.in/manual.htm.

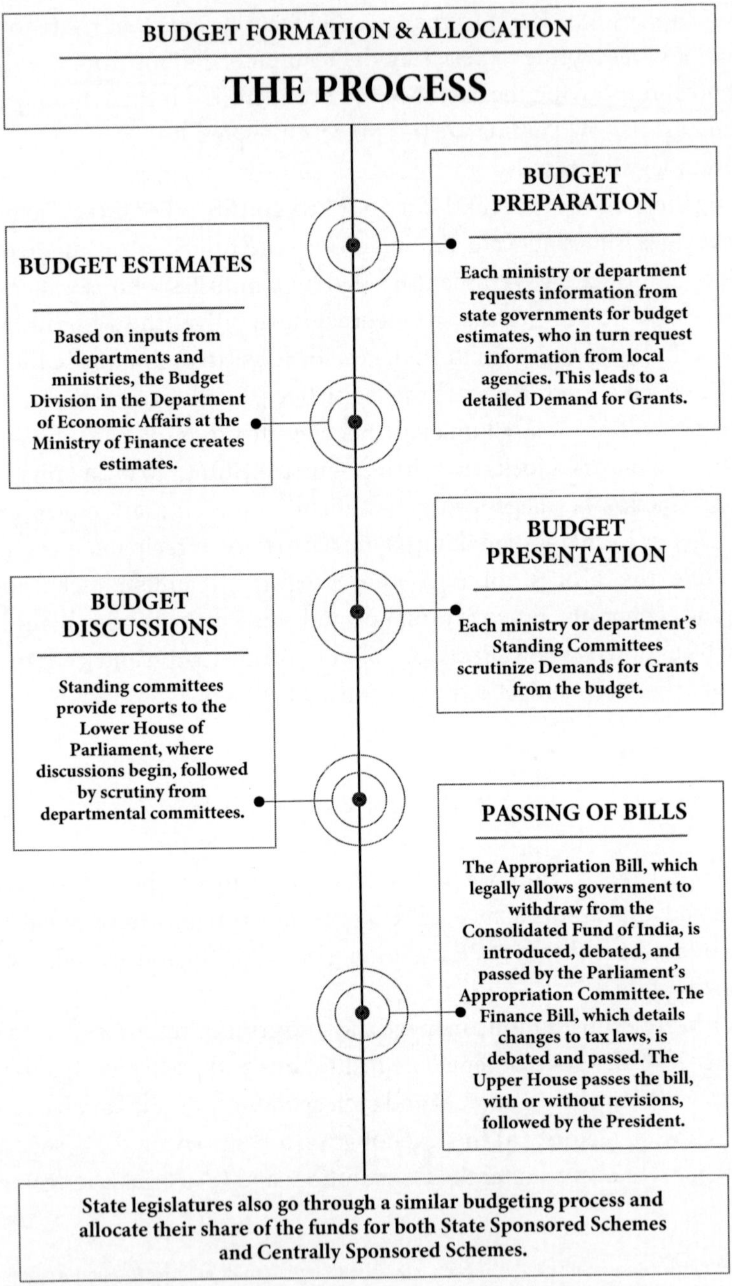

Figure 1 India's Budgeting Process. This chart outlines the budget creation process, from its formation to its approval.

or a loan agreement with an external aid agency. Together, these factors define where the budgets are ultimately executed and the power that the receiving ministries, departments, or agencies have on the further distribution and ultimate use of those resources.

Below, we outline the process of disbursal of resources from Union to state governments, and from state governments to local bodies, before detailing the two modes in which these resources are spent.

4.2.1 Intergovernmental Transfers

States and other government bodies receive funds from the Union government in one of two ways.[9] One, they receive general purpose transfers that are not reserved for any specific areas of expenditure. Two, they receive specific purpose transfers earmarked for specific areas of expenditure such as education or health. While specific purpose transfers ensure minimum standards of public services, general purpose transfers allow states to provide commensurate public services compared to their tax efforts (Rao 2019, 52).[10]

All general purpose transfers, also known as untied funds, come from the devolution of states' share of all taxes and duties that the Union government collects, such as Income Tax and Goods and Services Tax.[11] These transfers are based on recommendations of the Finance Commission, set up every five years.[12] The XIV Finance Commission

[9] More broadly, states in India depend on central transfers, own tax, and non-tax revenue and borrowings to raise resources. In 2019–20, the estimated proportion between the three was 35:44:21, respectively (Tiwari and Surya 2019).

[10] The Union budget does not include the devolution of taxes and duties as part of Union government expenditure. However, general purpose transfers received by states are included in the state budgets along with non-tax revenue, specific purpose transfers as described below, and borrowings. Specific purpose grants are a part of the Union government's annually presented expenditure budget.

[11] To a small extent, this could also be in the form of special arrangements such as GST Compensation Cess. See more at National Academy of Customs, Indirect Taxes & Narcotics (2018), Sharma (2020).

[12] The Finance Commission's mandate is defined in Article 280 of the Constitution: 'The Central Finance Commission is set up every 5 years by the Government of India under Article 280 of the Constitution to recommend (a) sharing of the net proceeds of taxes between the Union and the states and the allocation between the states of the respective shares of these proceeds and (b) the principles which should govern the grants in aid of the revenues of the state out of the Consolidated Fund of India' (Ahluwalia et al. 2019, 19).

(2014) recommended that the share of the states be 42%, while the XV Finance Commission (2019) recommended this share be 41% for all states (and 1% for the Union Territories of Jammu and Kashmir, and Ladakh).

Specific purpose transfers usually take three forms. One, they can consist of the Union's share for Centrally Sponsored Schemes (CSS) and funds for Central Sector Schemes (CS); as of 2022, there were 28 CSS and 45 CS (Rao 2022, 201). Two, the Finance Commission can recommend the specific purpose such as grants for 'revenue deficit, disaster relief, local bodies, sector-specific schemes and state-specific schemes' (Panda et al. 2019, 43). Three, they can comprise other loans and grants given under Article 282 and 293 of the Constitution, which were designed as residuary provisions to deal with unforeseen contingencies (Singh 2019).

The Union government generally releases funds to states and agencies for specific purpose transfers in instalments. All bodies receive the first instalment of funds at the beginning of the financial year. Once bodies certify that they have met particular conditions, such as having used a proportion of previous transfers, they are eligible to receive a second instalment. Some of the funds from the central government go through the state and trickle down to urban and rural local bodies or other government agencies depending on their purpose. However, these bodies[13] also rely on a mix of resources: general purpose, specific purpose, and performance-based grants from the Central Finance Commission[14] and State Finance Commissions[15]; specific purpose transfers channelled

[13] For rural local bodies, see more at tinyurl.com/y7qxwqdg.

[14] 'The 74th Amendment to the Constitution inserted a new subsection to Article 280 adding the responsibility to "recommend measures to augment the Consolidated Fund of a State to supplement the resources of its municipalities"' (Ahluwalia et al. 2019, 19).

[15] 'Article 243 (1) of the constitution, the State governments are required to appoint a State Finance Commission (SFC) every five years to determine: the distribution between the State government and the municipalities of the net proceeds of the taxes, duties, tolls, and fees that can be levied by the State which may be divided between them, and the allocation of such proceeds between the municipalities at all levels; the determination of the taxes, duties, tolls and fees which may be assigned to, or appropriated by the municipalities; the grants-in-aid to the municipalities from the consolidated fund of the state; the measures needed to improve the financial position of the municipalities; and any other matter referred to the SFC by the Governor in the interest of sound finance of municipalities'. After the 74th Amendment to the constitution, the terms of reference to the Central Finance Commissions (CFC) were amended under Article 280 (3) (cc) to include an additional clause to make recommendations on the 'measures needed to augment the Consolidated Fund of the State to supplement the resources of the municipalities in the State on the basis of the recommendations made by the Finance Commission of the State' (Rao and Bird 2010, 31).

from Union ministries and state governments; their own revenues; and market borrowing.

Since the approval of recommendations of the XIV Finance Commission, grants from the Union government are first transferred to the states, which route these resources to urban local bodies, rural local bodies, or other agencies such as State Health Societies.[16] Before, Union ministries also transferred funds directly to bodies such as the District Rural Development Agency and State Health Society.

4.2.2 Expenditure Management

Once funds are disbursed between government levels, government programmes use one of two modes of budget disbursal and budget management: Treasury mode[17] and Society mode[18].

The Treasury mode tasks treasuries with maintaining all accounts, exercising budget control where necessary, and making data available for audits. For obtaining funds, Drawing and Disbursing Officers (DDOs) are authorized to send payment requests to the treasury. When a DDO makes a payment request, treasury officers go through the paperwork to

[16] Some exceptions to the state treasury route include MGNREGS and MPLADS.

[17] 'In the Treasury mode, after the sanction of funds by the concerned administrative ministries/finance ministry of the Union Government, the RBI is intimated to transfer the funds to the State Government . . . As a confirmation of the fund transfer a clearance memo from the RBI is received by the State Government and the Accountant General of the State. The finance department of the State approves the budgetary allocation if required and conveys a sanction for withdrawal of funds. The concerned department/agency withdraws funds. The expenditure is routed through the treasury and is captured by the AG office through the vouchers received for the same. As accounts compilation in the States have been computerized by the State treasuries (States' treasury computerization projects) and AG offices (Voucher level computerization project, VLC), funds can be tracked till the State Government spends through State departments or transfers the fund to the Implementing Agencies (IAs) (mostly local bodies). The funds transferred to local bodies are captured at the time of release and booked as expenditure. The actual expenditure by the local bodies is not fed back into the treasury system' (Planning Commission 2011, 21).

[18] 'In case of the Society mode the funds are sanctioned by concerned administrative ministries and released by them. The funds are credited directly to the bank accounts of concerned agencies, i.e. DRDAs, Societies, NGOs (first recipients mostly functioning at the State level etc.) which function at the State level. These funds are subsequently released further by these first level recipients to their constituents at the district, block (taluk) or village level. The expenditure of funds is monitored by the concerned central administrative ministry/department by keeping a watch over the Utilization Certificates provided by the agencies. The audit of such bodies is conducted by chartered accountants' (Planning Commission 2011, 22).

verify whether the payment requests follow existing rules and processes. Specifically, they check if (a) the person requesting the payment has the authorization to make the payment, (b) the account from which the payment is requested is the correct one, and (c) there are sufficient allocated funds in the account to make the payment. If these three things indeed pass the checks, then the treasury authorizes the payment.

This process begins when DDOs at the implementation level request funds by producing a document, called a bill, to the state or district treasury. Through this bill, authorized officials can request funds in one of two ways. One, they can request a direct payment to a beneficiary or vendor—called a Detailed Contingent (DC) Bill. Two, they can request an advance withdrawal on behalf of an office at the state or sub-state level when exact, itemized payment amounts are not known or where cash disbursements are required—called the Abstract Contingent (AC) Bill. When AC bills are drawn, a DC bill must follow to explain the itemized expenditure within six months (Singh 2017). Based on an analysis by the Comptroller and Auditor General's office which audits all expenditures, the highest use of AC bills between 2011 and 2017 was for 'Capital Outlay on Other Rural Development Works' and 'Capital Outlay on Roads and Bridges'—activities that should ideally be planned and therefore use DC bills. Moreover, a majority of the analysed AC bills were not accompanied by a DC bill for six or more years after the submission of the AC bills (Singh 2017). We come back to look at this issue later in the book.

AC or DC bills go to the treasury under the signature of a DDO. In most schemes or departments, DDOs sit at various levels including the block, the district, or the state. In some instances, DDOs will sit at the gram panchayat level. These varying positions allow DDOs to engage with the level at which implementation work is done. In cases where expenditure is to be incurred at a level where there is no DDO—usually a gram panchayat—a request is sent to the authorized DDO above such an administrative level, such as the Block Development Officer, who will submit bills in aggregate to the treasury. Upon receiving the appropriate paperwork, the treasury instructs the bank managing the Consolidated Fund of the state to send funds to the end recipient if it is a DC bill or give the advance amount to the DDO for an AC bill.

The second, now less common, process for fund disbursal and management is the Society mode, still used by a few ministries, departments, and agencies to implement programmes for CS and CSS. Societies are autonomous and independent legal entities; they operate their own bank accounts outside of the Consolidated Fund of India or State Consolidated Fund. For CSS, the funds typically go from the state treasury to a state-level society's bank account. The state society then passes on the funds to district or block societies, which can send the designated amount of funds to other implementation agencies. Societies can use normal banking functions to withdraw the money without producing bills as described above in the Treasury mode. Therefore, they utilize varying fund transfer mechanisms such as direct transfers, checks, or cash withdrawals as needed and as expenditures come up. A stylized model for this process is summarized in Figure 2.

The Society mode of payments was developed in the mid-1980s to tackle a prominent issue for departments: though funds were made available to a state's Consolidated Fund from the Government of India, there were long delays in making them available to spending agencies—be it a society or a DDO.[19] This problem stemmed from the varying uses of the Consolidated Fund of the state. The state Consolidated Fund holds all the money of a state including bulk advances received from the Government of India for various schemes as well as the states' own revenues. At the same time, these funds, while parked with the state, are used to maintain the daily balance between revenues and expenditures, known as the 'ways and means', for the state government. Therefore, when agencies requested the state finance departments for funds, their release was delayed until the 'ways and means' position was comfortable enough for the transfer to be made (CAG 2015a, 157; Rath 2013, 62, 63; Raymus 2012, 21).

In order to overcome this hurdle, the Society mode was put in place for departments and implementing agencies to be able to receive funds when required. While it did reduce delays, it also decreased the accountability some societies had to face. Moreover, as noted above, all

[19] Special-purpose vehicles had been used earlier as well, such as District Rural Development Agencies that were established in 1980 (https://www.downtoearth.org.in/coverage/drda-faces-axe-38068), and before that, Small Farmers Development Agencies that were established in the 1970s (https://www.google.co.in/books/edition/Small_Farmers_Development_Agency_Bolangi/LkosAAAAMAAJ, page 36). However, most literature attributes the use of 'Society mode' for implementation of government programmes as emerging in the mid-1990s.

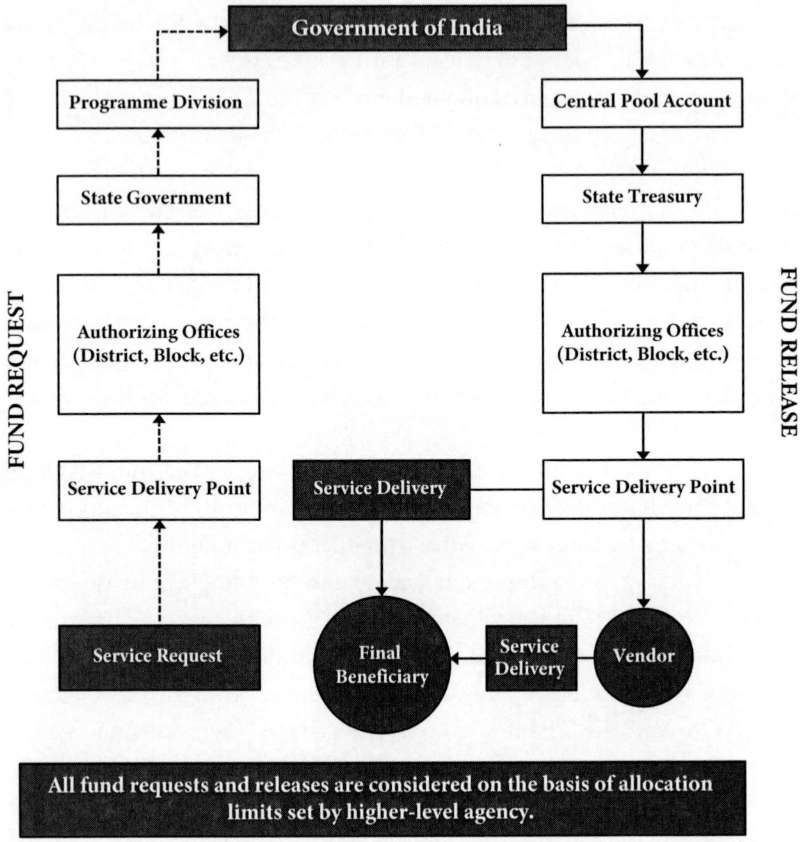

Figure 2 Society Mode Budget Execution Process. This figure outlines the fund disbursement process in Society mode. When services are requested, implementing agencies at the service delivery point request funds from authorizing agencies, who request funds from the state government. State governments request funds from programme divisions, who request them from the union government. Upon authorization, funds are released from the Central Consolidated Fund and/or Consolidated Fund of the State (depending on the type of scheme) and released through the State Treasury. The Treasury passes funds to authorizing offices, who then make payments at the point of service.

transfers to implementing agencies are now necessarily made through the Consolidated Fund of the state (and therefore the state treasury) before they reach the society. In this sense, a reason for the genesis of this mode—to get funds to implementing agencies quickly—is fading in significance since the Government of India has started implementing SNA-SPARSH.

4.3 Budget Monitoring and Evaluation

In the previous sections, we describe how funds are first allocated through budgeting, and then disbursed through Treasury or Society modes. Upon disbursement, funds need to be accounted for and audited. In this section, we describe these processes. Specifically, governments use vouchers to account for payments, Utilization Certificates (UCs) to report expenditure, and vouchers to audit expenditure.

4.3.1 Vouchers and Utilization Certificates

Once payments have been made, vouchers are the primary accounting method for these payments. A voucher is acceptable proof in government that a payment has been made on a particular day for a particular purpose. To illustrate, if an individual goes to a grocery store and obtains a receipt after making a payment, this receipt would be akin to a voucher. Except in government, the voucher would be stamped by the recipient and detail what the items in the receipt were used for. Ultimately, vouchers confirm that the end beneficiary, government official, or vendor has been paid.

Vouchers are used for reporting and accounting in both Treasury and Society modes, albeit with some differences. In the former, aggregated vouchers are used to report expenditures to government. Both the disbursing officer and their supervisor sign and authorize the vouchers (CGA 1982, 26) and send them to the district treasury for accounting. The district treasury sends the vouchers to the Accountant General's office for auditing and expenditure certification. The auditors check a sample of the payments claimed in a voucher, to confirm if they were ever made. They also check if the payment was for an authorized expenditure, if the expenditures were excessive beyond the market rate, and if the procurement was made by following due processes. If auditors discover any issues, they flag the corresponding expenditures and advise the Public Accounts Committee of the legislature to recover the funds after giving the concerned department or office an opportunity to explain itself. The auditors' role stops at recommending the recovery of funds to the legislature. The subsequent process is managed by the Public Accounts Committee in the Parliament or state legislature, and the rules of the Public Demand Recovery Act.

Compared to the Treasury mode, reporting in the Society mode varies. Agencies follow the accounting norms prescribed by their governing statute, such as the Societies Registration Act or the state-specific law governing urban local bodies. However, they report expenditure to government using Utilization Certificates rather than vouchers, which we describe next.

In both Treasury and Society modes, expenditures are reported[20] using Utilization Certificates (UCs). Simply put, a UC is a document that states how much a government entity has spent compared to the funding they received from another government body. The UC allows one government entity to obtain the next instalment of funds from another government entity once a certain threshold of expenditure has been met. At its core, the UC is an assurance by a government official that money has been used, but it is not backed by vouchers. Historically, officials submitted and accounted for vouchers manually. This created delays because vouchers had to be submitted, received, and reconciled before future funds could be issued—severely delaying fund flows. The UC solved this delay.

Typically, each implementation agency or society can produce a UC upon utilizing 60% of their funds. The UC lists the total expenditure used during that financial year and is authorized by an official with their signature. Sub-state administrative units such as a district or a block send their UC to the state when they are ready to receive their next instalment. Similarly, the state consolidates UCs from multiple sub-state units when the total amounts to 60% of the first instalment and submits the state UC to the Union government to receive their next instalment.

In theory, UCs must be backed by itemized vouchers because, ultimately, they state the fund amounts of aggregated vouchers. Whether or

[20] It is important to note that reporting is not accounting; a UC may meet the reporting requirements, particularly to be eligible for the next instalment, but it does not always meet the requirements for accounting and audit.

not this takes place depends on one of two government auditing bodies. In Treasury mode, the State Accountant General undertakes auditing in each state, and the Comptroller and Auditor General does so in the Union government. Because societies are legal entities, auditing in Society mode is different; a Chartered Accountant, usually from a list of auditors empanelled by the State Accountant General, takes on the task. While societies maintain vouchers internally, they do not submit these. Instead, their UCs have to be backed by audited accounts for final expenditure, even though in many cases these accounts verify the society's total expenditure rather than line-item expenditures.[21]

4.3.2 Accountant General and the Audit

Government auditors conduct three kinds of audits: compliance, financial attest, and performance audits. Compliance audits ensure that all activities and financial transactions follow the prescribed rules, regulations, and guidelines that are applicable (CAG 2016). In the UC and voucher systems explained above, a compliance audit can determine whether a UC is backed by corresponding vouchers that add up to the stated aggregate amount. Financial attest audits check if the financial transactions are legitimate or fraudulent. For instance, auditors verify whether the vouchers are authentic or not by scrutinizing line-item expenditure where possible. Lastly, performance audits determine whether there is 'value for money'. In other words, auditors examine expenditures against the targeted outputs and outcomes to determine if the money was used as intended. For instance, if 1 km of a road is allocated a particular amount with line-item expenditure for stones, concrete, machines, and water, the auditor examines the line-item costs for these materials, as well as their quality and utility (CAG 2016).

[21] These accounts are based on the chart of accounts. As Sundaramurti (2012, 8) explains: 'Expenditures are classified according to the function, programme, and their economic nature using a fifteen digit numerical code. Receipts are classified according to their nature and source. The classification system also applies to the State Governments, with the exception that they are given the flexibility to open their own heads below the third tier in the six tier hierarchy as per their respective needs.'

4.4 Architectural Design

Having outlined the budget allocation, execution, and monitoring and evaluation systems, we bring the focus back on the macro-level, push-and-wait PFM architecture of India. Specifically, we look at two issues: (a) the lack of observability for stakeholders and synchroneity within PFM's *fiscal events*, and (b) the lack of observability and synchroneity between data from fiscal events and the *programmatic events* they fund. By fiscal events, we refer to incidents or occurrences within the implementation of government programmes that involve any monetary or monetary-related exchanges, such as budget estimations, budget approvals, raising of an invoice, approval of a payment, fund transfers and payments, financial audits, etc. Programmatic events are the activities involved in the delivery of services, goods, or infrastructure that are funded by a government programme—such as vaccines delivered, institutional deliveries made, school toilets repaired, consulting services received, conditions being met for payments to vendors, employees, or beneficiaries, etc. By design, these programmatic events will differ with each policy, scheme, or programme. In public administration, occurrence of programmatic events and fiscal events are both routinely reported, referred to as the 'physical' and the 'financial' in administrative parlance.

Lastly, by observability and synchroneity, we refer to two things. One, observability indicates a range of search costs that individuals must incur to find information that they need to know as part of their jobs. When observability is perfect, individuals incur zero search cost to obtain information. As it decreases, individuals incur higher search costs. As a result of the asynchronous and unconnected architecture within PFM, India's public finance functions currently have a relatively low level of observability. It is a status quo that, in aggregate, renders everyone worse off.

Two, synchroneity refers to different data being linked to each other in two systems (e.g. data from a fiscal event is linked to data from a programmatic event), allowing both pieces of data to be concurrently observable to authorized stakeholders. In order for data to be synchronized, data updates must happen in near real-time such that all stakeholders have the same data as it becomes available.[22]

[22] Given these definitions, observability can exist without synchroneity, and vice versa. For instance, data can be visible to relevant stakeholders but out of date and therefore of little use to

The last part—whether data is captured and available to stakeholders in near real-time—is important because it ensures data credibility. Whether data is credible or not depends on the time between an event occurring and being reported. When the gap is longer, credibility is likely to be lower; this works similarly to recall bias for researchers. Other factors that affect data credibility include whether the data is reported by an observer of an event, or whether it is captured automatically as the event occurs; the former is more prone to over- or under-reporting than the latter. For example, many schemes have workers collect data manually, which is subsequently entered into a Management Information System or MIS. Others collect data through workers' interactions with electronic systems, such as the time of day. The time duration between an event happening and the data being captured will also affect its accuracy. The longer it takes to capture the data, the more its propensity to be less accurate.

4.4.1 Observability and Synchroneity within PFM Events

The functions of PFM involve a variety of stakeholders—legislators who allocate budgets, bureaucrats at various levels of government who execute programmes, and both accounting and auditing agencies who monitor and evaluate finances and programme implementation. In a streamlined system, any actor who needs access to information should be able to obtain it seamlessly and without delays, and the data should be credible such that all actors are looking at the same accurate data. Currently, as we explain below, this is not always the case.

PFM's push-and-wait architecture is highly fragmented both horizontally across ministries and programmes, as well as vertically across various levels of government, i.e. Union, state, district, block (sub-district), and village units (gram panchayats). Working with multiple systems imposes burdens on stakeholders, who must navigate these varying channels to successfully obtain information.

them. Similarly, data may be synced and updated in near real-time but not be made observable to the stakeholders who need it. As we detail later, both of these conditions are necessary for PFM to function appropriately.

Ease of access in obtaining funding-related data often depends on the sources of funding. Funds for a particular programme or to an agency may come from various budget heads with varying sources (e.g. Union, state government, aid agencies, corporate social responsibility initiatives, user fees, etc.). Consequently, for any public service initiative, the higher the number of funding sources, the greater the number of layers of intergovernmental transfers. With these multiple layers and sources at play, challenges borne by bureaucrats in exchanging information and securing money intensify. This is especially problematic when various institutions or government layers have disjointed information systems.

An example of a relatively simpler exchange of funds and information would be an urban development programme for developing transport infrastructure which is funded solely by a state government. Passing the funds to the implementing agency would likely involve one level of intergovernmental transfer (for instance, from the state treasury to urban local bodies or a parastatal body, such as a state bridge corporation used for construction). At the more elaborate end of the spectrum would be, say, a Centrally Sponsored Scheme for sanitation implemented by an urban or rural local body, where part of the funding comes from the Union government and part from a state government. Such a transfer would involve the Union government sending funds to the state government, which would then pass it on to the local body—perhaps by first routing it through an intermediary level like the district.

The number of layers in fund transfers—and the systems used to make these transfers—intensify in complexity when the state government combines a Centrally Sponsored Scheme with a state scheme to expand the scope of a policy. For instance, a state may pool funds to expand a scheme to not only households below the poverty line as mandated by the Union government, but all households. The complexity would increase further if local bodies were then expected to use their own untied funds (for instance, general-purpose transfers) to contribute to such a programme. In these cases, the Union government would use the Treasury mode to transfer funds to the state government using the Public Financial Management System (PFMS); the state government would then pass the funds to a state-level society using their own Financial Management Information System (FMIS); the state-level society would proceed to use its banking arrangements to transfer funds to the local bodies or other agencies implementing the programme; the local bodies would then have their own banking arrangements to make payments. If the programme is part of a Centrally Sponsored Scheme, the state-level agency and local bodies would also mandatorily use PFMS to send payment requests to their bank for any transfers or payments.[23] Reporting on expenditure and unspent balances for each of these specific funding sources would additionally differ.

As the process above shows, disjointed funding and accounting arrangements of the various institutions and levels of government lead to significant friction in the flow of funds, information, and decision-making. When data for transfers or expenditure is captured using various methods that are not designed to talk to each other—whether manually, through a centralized PFM software, or through a state-level PFM software— tracking unused funds at each government layer for a specific programme, as well as determining how much has been spent, becomes a complicated task. This is true for politicians, programme managers, fund managers, accountants, and auditors alike. The data for various fiscal events exist in disjointed systems, and they may not necessarily be linked to each other, nor updated at the same time. That means, for instance, that while local-level agencies may possess the latest data on the amount of residual money relative to the allocated budget for a particular scheme, state governments may not get this information until much later. Consequently, programme managers and other stakeholders must 'push' requests for information and wait for it to be collated, level by level.

This system means that for any stakeholder to obtain the data in the chain, they must send a request for information to their reportee, who then requests the person reporting to them, and so on. The information then gets sent back up the chain until it reaches the original requester. For instance, if someone in the state Finance Department wants to know the amount spent for a particular programme in a certain sub-district, then

[23] As stated in Rule 86 (3) for Central Schemes and Centrally Sponsored Schemes: 'All the payment, to the extent possible, shall be released "just-in-time" by the Ministries through PFMS' (Department of Expenditure 2017, 27). Rule 230 (7): '... The principles of "just in time release", should be applied for releases in respect of all payments to the extent possible. The following broad principles shall be adhered to: (a) Cash balance at a time should preferably not be more than 3 months of requirements, (b) Funds should be released as per actual requirements and that sanction may precede the release of funds, though its validity may be limited to that financial year' (Department of Expenditure 2017, 69).

they would first request this information from the sectoral or line department, who would then ask the District Collector, who would then pass on the request to the Block Development Officer (BDO) at the sub-district level. The BDO would obtain this information from the programme manager, and then send it back up, layer by layer, until it reaches the state Finance Department. Were the data from these various levels synchronized and connected, stakeholders who ought to have access to the data for their jobs would not need to go through such a convoluted, push-and-wait system to obtain information which should be readily accessible.

Overall, the lack of synchroneity and interconnectedness within PFM systems leads to observability issues for stakeholders. Beyond impacting state and sub-state-level financial officers who only have limited observability at their level, they also affect financial stakeholders across the board. For the sectoral Ministry and Department of Finance, lack of observability on cash position and upcoming expenditure commitments across government agencies—resulting from the same asynchronous PFM and unconnected structure described above—makes cash flow forecasts and management an almost impossible task. Cash management is necessary because even if an annual budget is balanced for revenue and expenditure, the timings and seasonality of revenues and expenditure may not match (Lienert 2009, 1). In other words, there is a lack of predictability in cash flow because revenues and expenditures come in and go out at different times. Predictable cash flows are important because those who manage debt for the government 'prefer to issue bonds with a stable and predictable pattern' (Williams 2010, 7). Lack of observability of reliable cash flows often pushes governments to borrow more money than needed, in anticipation of low cash flows.

Further, Ministries of Finance are unable to hold administrative departments and agencies accountable to their fiduciary duties when they lack observability. These include the duty to ensure that funds are used for their intended purposes, with due attention to the principles of economy, efficiency, effectiveness, transparency, and accountability.[24] When finance departments have clarity on exactly how the money is being spent, they can inherently allow departments to utilize funds flexibly because

[24] We owe this conceptualization to objectives of the World Bank's Integrated Fiduciary Systems Assessment (IFSA).

misuse can be detected easily. But in the face of a paucity of observability, opaqueness breeds mistrust and it manifests in budget inflexibility. Because finance departments are unsure about near real-time expenditures and only receive reports after a considerable time (often months), they put constraints on how funds can be spent. Budget inflexibility, often up to the line-item level, then prevents programme managers from responding to changes in needs over time, space, and context.

This issue further affects the administrative machinery that constitutes middle managers such as Block Development Officers, as well as frontline workers such as Accredited Social Health Activists (ASHAs), Auxiliary Nurse Midwives (ANMs), and Panchayat Secretaries. The administrative machinery has no visibility into upstream processes of budget formation, allocation, and distribution. To them, it is unclear for months at length whether a budget will be allocated for a programme, how much would be allocated, and for which activities. Once the budgets are allocated, they lack visibility into when funds will be released. Consequently, the adage of 'use it or lose it' holds true as any incentives for thoughtful planning and budgeting by the administrative machinery are significantly marred by this lack of observability and certainty.

Overall, then, the lack of interconnected or synchronized PFM systems—and the low observability within the PFM process—affect not just PFM stakeholders, who are unable to manage funds well, but also programme administrators, who need information on fund releases and budgeting.

4.4.2 Observability and Synchroneity between PFM and Programmatic Events

In the current system, stakeholders have to undergo a variety of arduous processes for seeking information on how a component of a budget connects to—or funds—specific parts of a programme, service, or good. There are, of course, exceptions to this where funds are transferred directly from the funding authority to the beneficiary. But in many programmes, fiscal events are not associated with programmatic events in near real-time. The information architecture underpinning the programme management and monitoring is fragmented; information on programme activities, outputs, budgets, and expenditure are all processed through non-concomitant and disconnected information systems. For instance, the programmatic management information systems of most government programmes in India remain, for the most part, disconnected from the Public Financial Management System (PFMS) at the Union level or the Integrated Financial Management System (IFMS) at the state level. Even where these systems do speak with each other, they show aggregated data instead of unit-level data, which may be insufficient for some stakeholders such as auditors or citizens demanding accountability from the government.

Finance managers do not always know how each budget line item supports programmatic outcomes unless they ask a variety of bureaucrats to send this data for their portion of the programme. Similarly, programme managers do not have full observability into how finances are spent, in near real-time, to guide implementation.

Of course, not every programmatic event has a corresponding fiscal event (for instance, a healthcare worker can be registered onto a system, but without any corresponding payment). Additionally, programmatic and fiscal linkages are not always possible for budget-making purposes (as many programmatic and fiscal events cannot be accurately predicted at the lowest level of granularity a year in advance), but two-way synchroneity and observability between fiscal and programmatic events are possible for budget execution and budget reporting or monitoring. We therefore focus on cases where programmatic and fiscal events can be linked.

Even though the current financial architecture does not adequately sync fiscal and programmatic events, the budgeting exercise does notionally connect expenditures to an activity or output within a programme. For instance, an appropriations bill might earmark a certain amount of money to provide wages in Bihar for a particular scheme. This money, at the state level, is allocated to districts for providing wages for that scheme. Districts allocate their funds to sub-district units or blocks, who allocate it to gram panchayats or villages. These villages can then use the funds allocated to them for providing wages for that scheme. Therefore, the fiscal event is linked from the budget allocation down to the programmatic event.

Accessing this information on how fiscal and programmatic events are linked poses some challenges in a push-and-wait architecture. The first major problem arises when someone above the gram panchayat or block needs to know (or, in our language, wants observability) on how allocations are being spent for a particular programmatic event. For instance, a bureaucrat in the state finance department wants information on the portion of the allocated funds for wages spent in a particular gram panchayat in Bihar. To obtain this data, the finance department bureaucrat will ask a bureaucrat in the department that houses that scheme at the state level. The state-level bureaucrat in charge of that scheme will then request this information from the district bureaucrat, who will in turn ask the block bureaucrat. The block bureaucrat will obtain the information for how many fiscal events supported the programmatic events within the gram panchayat—in other words, how much money was spent on wages in a particular gram panchayat for that scheme in a particular period. This information is then relayed back, layer by layer, up to the initial bureaucrat in the finance department who requested the original information. Without observability, which can stem from a lack of synchroneity and interconnectedness in a PFM system, obtaining data on how expenditure connects to programmatic events becomes an overwhelming and arduous task, and creates an environment where there is lack of trust about the accuracy of the data.

The second challenge in the architecture is that fiscal events (transfer of funds and recording and accounting of expenditure) do not necessarily get triggered once the underlying programmatic events have occurred, unless a series of stakeholders approve the fiscal event. While approvals themselves are not problematic, the set of criteria for approvals varies drastically. For instance, a community healthcare worker called an ASHA worker may bring an expecting mother to a delivery centre. To receive the payment due (the fiscal event) for this service (the programmatic event), the ASHA worker first reports the completion of the service, which her supervisor—ASHA facilitator—then verifies, and the block medical officer subsequently approves. However, one supervisor may deem that the verbal report from the ASHA worker is sufficient for approval, while another may deem that an approval requires a series of paperwork showing that the referred mother was taken in at the community healthcare centre. This can cause several issues—the supervisor can extract rent to approve a payment or delay it because superiors even higher up require additional proof that the ASHA worker did her job. Additionally, auditors may find it hard to track inconsistency within the same block due to the different criteria used by various supervisors to approve payments.

To solve such issues, in theory, a PFM system should undertake several functions. One, it should be able to produce expenditure reports that allows stakeholders to examine not only the aggregate expenditure for a programme, but also the unit-level expenditure for each programmatic output. Once the programmatic event is completed and enough proof to corroborate this event has been furnished, a PFM system should automatically trigger the fiscal event. When this fiscal event occurs, the system should capture data on when, how, and where money is spent, in addition to whom it is disbursed to and for what purpose. This data, if integrated into the architectural design, can not only ease payment transfers but also allow stakeholders to see individual items for data analysis or auditing, when these stakeholders have authorizations to see specific pieces of data they require. In essence, such linkages and data synchroneity, coupled with observability, can dispel the ambiguity and lack of insight that existing aggregate expenditure reports engender.

To expand on this idea, we revisit the example of a programmatic event—an ASHA worker who brings a pregnant mother for an institutional delivery. In an ideal system, the payment should be released to the ASHA worker in one of two ways. Once the ASHA worker completes the service and uploads the proof of completion to a digital system or to a non-digital one, the fiscal event should automatically occur. If the PFM architecture is digitally designed to verify the plausibility of data, the PFM system itself can automatically trigger a fiscal event directly to the ASHA worker as soon as the proof of work—which is set to be consistent for all ASHA workers in a geography—is uploaded. In the absence of a digital architecture, an ASHA facilitator can look at the data to determine if clearly defined conditions for payment are met and, subsequently, recommend that the payment be made. In fact, a facilitator need not verify every single claim; a random sample may suffice, as we describe later in the book. Before this payment is made, the PFM system should connect which ASHA worker brought which mother for an institutional delivery, and this information should be linked to the payment that the ASHA worker (and the mother) received for the event. The data should also selectively be available to the ASHA worker's direct supervisors and, as needed, to auditors to conduct checks. These pieces of information should be included in the system even if the ASHA worker receives an aggregate payment for multiple entitlements in a single transaction; for instance, for not only facilitating the delivery but also perhaps bringing a tuberculosis patient to a healthcare clinic. Unit-level records then allow for accurate, upwards aggregation of why funds are used. As we detail in section 6.2, all this data would further be protected as a requisite of the system.

Why is this important? The primary objectives of a different kind of PFM architecture are twofold. One, programmatic events, once completed, should automatically trigger fiscal events without a series of intermediaries making that determination via arbitrary conditions. Two, data should be captured at the granular level and inform the relationship between inputs (public expenditure), outputs (the work done with public expenditure), and outcomes (goals of public policy) for better iterative adaptation of policies and processes. Ultimately, this can lead to more predictive budgeting, better expenditure management, improved programmatic outputs and outcomes, and improved auditing or monitoring functions of programmes.

4.5 Conclusion

The current fund allocation, disbursal, and management system for India's push-and-wait PFM architecture was designed in an age where tools that enabled synchroneity and interconnectedness of data were unavailable. Funds therefore had to be pushed down from the Union government to implementation agencies. In order to function, the system today subsequently relies on both fund and knowledge transfer up and down a cumbersome number of layers, individuals, and approvals.

Ultimately, this push-and-wait architecture leads to two key issues. One, data from various functions of PFM—namely, budgeting, expenditure management, and monitoring and evaluations—are not synchronized with each other in near real-time or made observable to all who need this information, which requires stakeholders to rely on passing information through various layers to maintain their key functions. Two, fiscal events such as fund transfers or auditing are not linked to specific programmatic events such as the line-item task which the budget funds at the implementation level. This creates complexity not only in fund approvals and transfers, but also in budgeting for future public goods and either auditing or monitoring of funds.

In the next chapter, we describe in further detail how the two core architectural problems of public financial management identified above are at the heart of poor state capacity, and how they cause significant bottlenecks in the implementation of public schemes and programmes.

5

Consequences of Public Financial Management Architecture on Public Service Delivery

In the previous chapter, we described India's public financial management (PFM) system and how it relies on a series of administrative layers and intermediaries to approve, disburse, report, manage, and audit expenditures. We describe this system as having push-and-wait architecture, whereby each administrative layer pushes requests for funds, information, and decisions to another layer and waits for the results to travel back to them. The specifics of this architecture vary based on the type of scheme, but it broadly involves the transfer of funds from one layer of the government to another, and typically in advance. Afterwards, information on the disbursal of the funds is often uncertain and opaque. For instance, in the case of Centrally Sponsored Schemes, the Union government transfers funds to the states, which in turn send funds to the implementation agencies at the state, district, block, and gram panchayat levels on a need-by-need basis or in bulk as advances. Implementation agencies report expenditures back up the administrative layers to the state and the state then reports them to the Union government.

India's hierarchical and fragmented PFM push-and-wait architecture causes two main challenges: (a) the various functions of PFM such as budgeting, expenditure management, and monitoring do not share near real-time data with each other and are not interconnected, meaning stakeholders must request and pass data through various layers of government to obtain information; and (b) fiscal events are not linked, in near real-time, with specific programmatic events. In this chapter, we understand these two key architectural problems and how they lead to

Public Financial Management, State Capacity, and Public Services in India. Santhosh Mathew, Bhumi Purohit, and Devesh Sharma, Oxford University Press.
 DOI: 10.1093/9780198973195.003.0005

several public service delivery challenges: an artificial scarcity of funds due to float in an otherwise resourced system; a structure that incentivizes risk aversion in making payment decisions; administrative burden; uncertainty and delays in fund payments for recipients; misallocation or diversion of funds; poor auditing; and poor accountability mechanisms for civil society actors and citizens. In describing these issues that stem from the PFM architecture, we continue to maintain a goal-oriented view of PFM systems and explain how state capacity for service delivery is affected.

The challenges we highlight are by no means isolated. As we illustrate ahead, many of the symptoms are interconnected and arise from architectural problems. Given the complexity of PFM, it is virtually impossible to overcome the problem of endogeneity, even with descriptive explanations of why issues arise. For instance, delays can partly be caused by administrative burdens but, at the same time, are also rooted in architectural issues. Similarly, administrative burdens give rise to delays but also lead to larger programmatic implementation issues for bureaucrats. We therefore describe the key challenges PFM users face, in light of the knowledge that one problem can be interlinked with the existence of another.

The challenges we discuss in this chapter are not new. In *Arthashastra* (which translates to *The Science of Wealth*)—an ancient Sanskrit treatise thought to have been expanded between the second century BCE and the third century CE—Kautilya detailed 40 ways in which 'embezzlement' may be carried out (Choudhury 1982, 109):

> What has accrued first is realised afterwards, what is to accrue later is realised first, what is to be carried out is not carried out, what is not to be carried out is carried out, what is carried out is made out as not carried out, what is not carried out is made out as carried out, what is carried out a little is made out as much, what is carried out much is made out as little, one thing is carried out while another is made out . . ., what is carried out from one source is made out as from another, what is to be paid is not paid, what is not to be paid is paid, payment is not made in time, payment is made untimely, a little paid is made out as much, what is overpaid is made out as little, one thing is given while another is made out as given, what is paid to one is made out as paid to another, what is

> delivered . . . is made out as not delivered, what is not delivered is made out as delivered, forest produce for which the price has not been paid is delivered, that for which the price has been paid is not delivered, concentration . . . is made out as dispersal, or dispersal made out as concentration, an object of high value is changed for one of low value, or one of low value for one of high value, the price is raised, or reduced, the year is made discrepant as to months, or the month discrepant as to days, discrepancy as to source, discrepancy as to head (of income etc.), discrepancy as to workmen, discrepancy in performance, discrepancy in the sum-total, discrepancy in quality, discrepancy in price, discrepancy in weighing, discrepancy in measuring, . . . discrepancy as to . . . vessels.

While this excerpt focuses on varieties of embezzlement, wide-ranging issues of public finance have been a subject of scrutiny for a long time. As we have outlined above and detail in the remainder of the chapter, these observations ring true even today across many governments.

5.1 Float

Due to the lack of synchroneity and observability within PFM's push-and-wait architecture, funds are typically released in instalments for Central Schemes, Centrally Sponsored Schemes, and State Sponsored Schemes. The government will send its first instalment of an allocated sum of a budget to the administrative layer below it. Once the lower administrative level uses up a proportion of the funds, it can provide a document proving as much, before requesting the next instalment.[1] This back-and-forth of information, the lack of observability within one's own jurisdiction, and the lack of near to real-time data synchroneity of funds-related information across levels and systems has two major implications.

One, because governments do not have observability on fund use across levels, they send lump-sum instalments, which necessitates that, depending on their cash balance, governments borrow in advance of

[1] This problem does not occur when state governments insist that Drawing and Disbursing Officers use only DC bills for State Sponsored Schemes (e.g. advanced withdrawals are not allowed through AC bills).

their expected needs. Two, because lower-level governments are only required to use a proportion of the first instalment to request a second one, there is potential for money to sit parked in a bank account—called float—and/or be spent in a way other than designated. This float can create artificial scarcity at the micro-implementation level and fiscal deficit at the macro level. Due to the lack of synchroneity within PFM systems, governments do not always know if money remains unspent, and if unused money can be reallocated to other implementation agencies based on immediate needs.

Float occurs in four different instances. First, in Society mode, Union and state treasuries send money to state government agencies in lump-sum advances. Societies store this money in various bank accounts until an administrative layer below them, such as a district or block, qualifies for its next instalment. Once an implementation agency—say, the district-level society—is ready to receive funds, it leads to another type of float. This second type of float also occurs in the Society mode, when funds are sent from state-level agencies to implementation agencies in advance of expenditure. Third, in Treasury mode, the Union government sends an advance, lump-sum instalment for specific-purpose transfers to the consolidated fund of a state for Centrally Sponsored Schemes.[2] Fourth, also in Treasury mode, funds are sometimes taken out as a lump-sum advance without the itemized details, known as Abstract Contingent (AC) bills (see section 4.2), and parked outside the treasury in a bank account. This type of withdrawal is usually made for contingent

[2] Advance releases from the Union government to the state government also create a peculiar situation where float is used to earn interest. Given that the government has more expenditure than revenue, the Union government tends to finance this deficit by borrowing from the market and paying interest on these borrowings. States, which tend to invest money in the market when they have surplus cash holdings, could park some of these excess funds in treasury bills—one of the instruments the Union government uses to borrow money. This creates a loop where the Union government borrows money and pays interest on it, sends it to state governments, and the state government then earns interest by investing in Union treasury bills. Because of float, one level of government borrows while the other earns interest on it. The Expenditure Secretary T.V. Somanathan noted this issue in his letter to the states, dated 4 May 2020: 'This means that, while on the one hand, the Union pays interest on the borrowings [from the market], which are released to the state, on the other, for lack of utilization, states park these funds in Treasury Bills, earning interest' (Deka 2020; Gupta 2020). In 2016–17, for instance, state governments invested in Rs. 1 trillion worth of treasury bills of the Union (USD 28.36 billion using the average 2016–17 exchange rates) (Budget Division 2018, 32). Such investments can accrue significant revenue when interest rates fluctuate, drawing in money for governments to balance their ways and means and manage against other debts and liabilities (Saikat 2019).

expenditure, for which final classification of the payment and supporting vouchers are not available at the time of withdrawal of funds. However, AC bills are also prone to be misused as business-as-usual expenditures. Use of AC bills requires itemized payment information—known as the Detailed Contingent (DC) bill—to be furnished within timelines specified by each state's Treasury Code. However, in reality, a majority of these DC bills are never produced (Singh 2017), and money remains stagnant in accounts below the state level with limited accountability on how it is used.[3]

Float exists not only because funds are sent in lump-sum advances in the four ways described above, but also because implementing agencies do not use the disbursed funds completely in a timely manner. For example, according to the Standing Committee on Rural Development and Panchayati Raj, the Ministry allocated Rs. 1,844.61 billion to its Centrally Sponsored Schemes in 2024–25, but only spent around 50% of the funds (SCRDPR 2024, 17). Starting with the budget for 2025–26, the Ministry of Finance has also begun publishing Statement 4AA.[4] This provides details of sitting fund balances under Single Nodal Agency (SNA) accounts with state/eligible Union territory governments for select Centrally Sponsored Schemes. For instance, at the end of financial year 2024–25, the Samagra Shiksha, a scheme for education, had Rs. 111.5 billion (USD 1.2 billion) unspent funds and Human Resources for Health and Medical Education had Rs. 70 billion (USD 779 million). Such staggering sums can sit unused in accounts because the current system allows such float to be accrued rather than sending funds on a need-by-need basis in real-time. Because there is not near to real-time observability on how much money sits in various layers of government (unless requests for information are sent from top down), the lack of synchroneity and observability creates conditions for float to accrue. Were there synchroneity and interconnectedness between PFM systems across various levels—and observability on how money sits in various accounts and is needed to meet

[3] The CAG's audit reports on use of AC bills also provide details of this problem. See https://cag.gov.in/uploads/download_audit_report/2017/Chapter_3_Financial_Reporting_Report-No_2_of_2016_Punjab.pdf and https://cag.gov.in/uploads/download_audit_report/2013/Jharkhand_Report_1_SF_2013_Chap_3.pdf.

[4] See https://www.indiabudget.gov.in/doc/eb/stat4aa.pdf.

near-term tasks—money could be sent to implementing agencies as and when it is needed (we expand on this in Chapter 6).

The issue of unspent money, whether at the state or the agency level, can further create a problem of artificial scarcity. Take, for instance, a situation where three gram panchayat bodies, panchayat A, B, and C, are provided an annual allocation of Rs. 10 million each. Combined, they have an annual allocation of Rs. 30 million, so their common funding agency receives Rs. 7.5 million in its account—or 25%—for the first instalment. All three panchayats receive their initial allocation of Rs. 2.5 million each, leaving their common funding agency with nothing for the second allocation. The common funding agency is unable to ask for further instalments until it provides its own funding agency at a higher level with Utilization Certificates (UCs) amounting to 60% of the Rs. 7.5 million funds it received—Rs. 4.5 million. Similarly, the three panchayats must produce UCs amounting to Rs. 1.5 million each before being able to request additional funds.

In such a scenario, assume that panchayat A is more active in the first half of the year and utilizes 60% of its funds while the other two panchayats do not meet this requirement. Upon submitting its UC, panchayat A requests Rs. 2.5 million for its next instalment. However, the common funding agency has already disbursed all of its first instalment of Rs. 7.5 million to the three panchayats and cannot yet receive its second instalment—even with panchayat A's UC of Rs. 1.5 million, it is insufficient to total the funding body's UC requirement of Rs. 4.5 million. This problem occurs because panchayats B and C remain inactive at a time when panchayat A is incurring expenditures. Even though panchayats B and C have unused money lying in their accounts, their common funding agency cannot reallocate the funds to panchayat A.[5] The only way for the common funding agency to obtain its second instalment is when panchayats B and C each produce their UCs, perhaps in the second half of the year when they are more active. The common funding agency now reaches its 60% requirement for the UC after panchayats A,

[5] Technically, the common funding agency can pull back the money from panchayats B and C and give it to panchayat A. However, this is not practical given the complexity in accounting and the time it takes to execute the task. Therefore, funding agencies rarely do so.

B, and C have proved that they have utilized Rs. 1.5 million each, totalling Rs. 4.5 million. The agency can then request its second instalment of Rs. 7.5 million and disburse it.

The problem apparent in this example is that when panchayat A requests money in the first half of the year, their funding agency does not have any left to disburse even though the money is sitting higher up at another level, and lower down with panchayats B and C unused. The float with panchayats B and C therefore creates artificial scarcity wherein the common funding agency has no funds to make payments even though they lie unutilized with another body. While the scenario above uses implementing agencies as an example, artificial scarcity can similarly affect state departments, which can experience delays or lack of funding at a particular time even though funds sit unused in other departments.

Artificial scarcity, as seen in the example above, is a result of a system that relies on funding agencies in advance of payments and does so through UCs at each level of administration for disbursed funds, primarily because the system lacks observability on fund usage. UCs merely outline an agency's non-transparent claim about its latest total expenditure. They do not reveal past or current spending patterns, lack detailed and itemized accounts of their expenditures, and do not account for future needs. In the way fund instalments are allocated to implementation agencies, they fail to prioritize efficient usage or needs.

Given the vast number of central and state government schemes, the amount of float can be consequential. The 2015–16 Economic Survey of India estimated that, at the time, there was an estimated Rs. 1 trillion (USD 15.56 billion at that time) in float (Ministry of Finance 2016, 59, 60). Today, the way float operates has the potential to change; the Department of Expenditure in the Ministry of Finance, at the time of writing, has asked departments to implement 'SNA-SPARSH' for government departments and autonomous bodies to be able to use Treasury Single Account (TSA). These are akin to the global use of TSA, which is a cash management practice which ensures that funds are not parked across multiple government agencies. With the combination of SNA-SPARSH and TSA in the Government of India, implementing agencies can open accounts in the Reserve Bank of India. Instead of disbursing cash to implementing agencies, these systems give them digital financial ceilings, against which they can generate payment files. Through these systems, the government account is debited just in time as the need arises, so long as the total value of cumulative payment files is below the financial ceiling. SNA-SPARSH and TSA are only applicable for Centrally Sponsored and Central Sector Schemes, respectively.

As of July 2023, the Union government signalled its intent to transition funding systems to SNA-SPARSH and TSA. When fully implemented, which could take years, these systems can eliminate float. Rather than the money being held in bank accounts of state government implementation agencies, fund releases from Union government will be matched by states' share of funds for schemes on a daily basis, and in advance of payment. In order for float to be eliminated, though, these systems need to be implemented with speed.

Until changes are made, float will result in avoidable fiscal strain. To finance its expenditure not met by its revenue, the government must borrow money to meet its needs. Since actual expenditure estimates come up till a year after budgeting, the government borrows more than necessary to fund its budgeted expenditures. When this money sits in various bank accounts without being used, i.e. as float, it contributes to the fiscal deficit since the government is borrowing more than it needs to, based on real expenditure. The FY 2018–19 fiscal deficit was Rs. 6.45 trillion (USD 92.25 billion at the time) (ET Bureau 2019). If the estimated float of Rs. 1 trillion is eliminated from this figure, the fiscal deficit can see a significant one-time reduction of approximately 15%.

The impact of float is not limited to the public sphere. Fischer and Easterly (1990) and Mitra (2006) have both underscored that a fiscal deficit, even when sustainable, can reduce credit availability for the private sector, ultimately slowing economic growth. Similarly, they note an increase in interest rates for private-sector borrowers, which may adversely affect not only national businesses but also international investments.

Float is merely a result of the current financial architecture, which necessitates the government to plan for bulk funding in advance because PFM systems across various levels of government are not synchronized. If the system is redesigned to allow for this synchroneity and therefore create observability of real-time fund utilization, it can issue funding for payments based on needs.

5.2 Incentives for Deciding When to Pay Recipients

The lack of synchroneity and interconnectedness between PFM systems and how PFM connects to programmatic events not only has consequences for float but can also cause significant delays in funds reaching implementing agencies. Here, we highlight one of the root causes of delay: the incentive structure for fund managers, who—without synchroneity and predetermined criteria to connect fiscal events and programmatic events—face high risks if their approved funds end up being used fraudulently. Even though fund managers have to go through varying layers to obtain information on what is being implemented on the ground, by whom, on what unit-level basis, and for what purposes, they choose to navigate and enforce this system, but why? It is because they are ultimately held accountable if payments are found to be fraudulent or negligent. The consequence is that fund managers 'fatten the file' with multiple rounds of questions and answers as insurance to justify their approval for a payment, resulting in delays that affect public procurement and payments to beneficiaries.

In India, between 20 and 30% of GDP is estimated to be public procurement—defined as goods, services, and works purchased by various levels of government and public sector enterprises (Khan 2017; Ministry of Finance 2019)—most of which is used in the delivery of public goods and services.[6] Public procurement is undertaken through government contracting, which faces its own set of challenges.[7] Here, we highlight the key issue related to PFM: delayed payments to firms and enterprises that work with the government. Stuck payments are neither small in quantum, nor inconsequential in their adverse effects.

According to one report, private contractors were awaiting Rs. 1–3 trillion (USD 14.88–44.64 billion at the time) from the government in

[6] Of the global GDP, 12% is spent on public procurement (Bosio and Djankov 2020).

[7] The process starts with the first phase in the life of a government contract, called procurement, which consists of the preparation and publication of bids, the submission and evaluation of bids, and the awarding of a contract (Hazarika and Jena 2017, 17). Once the contract is awarded, the contractor executes it. The contracting Ministries, Departments, and Agencies (MDAs) monitor and review the contract execution, and release interim and final payments. Throughout this process, dispute resolution mechanisms are governed by the terms of the contract and the legal framework of many laws and rules. Challenges and inefficiencies throughout this contracting cycle often result in delayed payments for publicly procured goods, services, and works.

2016, largely because of delays or disputes (Mondal 2016). Contractors in industries as wide as chemical fertilisers, information technology, construction, and health services face these challenges (Agarwal 2020; Goliya 2016; HT Correspondent 2020). N.R. Narayana Murthy, the co-founder of one of India's largest firms, Infosys, noted this very fact in an interview: '[T]here is not a single project Infosys has done . . . where the company has not lost money, that's a reality, when dealing with the governments' (Press Trust of India 2015). He identified a host of issues that firms face: low procurement prices, delayed payments, uncertainty of requirements that change mid-stream, ill-defined requirements, unrealistic timelines, and, in some cases, corruption. On the other end, governments too have acknowledged the importance of resolving delays in payment (Ayub 2019; ET Now 2019; The Pioneer 2018). To this end, the government has set up the Samadhaan portal to track payment delays to micro, small, and medium enterprises (MSMEs) (Sharma 2020).[8]

Delayed payments act as a deterrent for private firms to engage with the government. The lack of competition in procurement processes may be associated with reduced quality of government work and, ultimately, reduced programme effectiveness (Mondal 2016; Prasad 2013). For firms, late payments also affect business viability and economic growth, and exacerbate non-performing assets for the banking sector (Goliya 2016; Ponappa 2019; Rajaraman 2019; Sen Gupta 2019).

The challenge of delayed payments is not only restricted to government procurement. Delays also exist in cash transfers to citizens and for payments to contractual workers. In cases where delays create uncertainty for citizens enrolled in welfare programmes, they can adversely impact programme goals of social protection, hampering citizen participation (Dhorajiwala 2018; Narayanan et al. 2017). When payments are held up for contractual workers, they can affect performance and motivation alike, especially if individuals are driven by monetary incentives or work in low-pay jobs (Kaur 2018a; 2018b).

As we have highlighted, the problem of delayed payments is not unique to certain kinds of 'releases'—payments for procurement contracts or

[8] According to the World Bank's Ease of Doing Business 2020 report (World Bank 2020), India is ranked 63rd amongst 186 countries in enforcing contracts. However, this ranking is based on de jure rules, while the de facto delays in getting paid are enormous and increase the difficulty of doing business with the government.

payments to contractual workers, for instance. If the problem was isolated, a systematic review for each type of release may reveal sources of delay. Rather, given the widespread occurrence across many kinds of releases, we review a common problem with all the aforementioned types of payments: a small number of individuals in the system make the 'decision to pay' a recipient in each process, and this decision is made in differing ways. The combined effect of information asymmetry, accountability, monopoly, and discretion of decision makers creates a system that is prone to delays. We highlight each of these issues below.

Information asymmetry: Individuals who make the decision to pay a recipient usually do so after ensuring that all the conditions to make the payment have been met. These include proof of delivery for goods or services, proof of eligibility for receiving social cash transfers, or a Utilization Certificate for intergovernmental transfers. Given the lack of synchroneity and interconnectedness between fiscal and programmatic events, two problems emerge. One, each officer may rely on differing types of information to make this decision; and two, officers may not know if the information they are given is accurate. The officers that make the decision to pay are rarely the ones to verify for themselves that conditions for payment have been met; instead, they rely on certifications from other officers below them in the administrative hierarchy. Based on information provided by others, senior officers decide if payment criteria have been met. This system creates information asymmetry, in which senior officers may not know if information provided by others is accurate before making a payment decision—especially because there is no near real-time data synchroneity and it is difficult to see the unit-level activity for which payments are requested. Combined with the high stakes of accountability that we describe next, the process creates delays because the incentive for the decision maker is to indulge in multiple rounds of questions and answers to demonstrate good intent.

Accountability: The incentives for senior decision-making officers are skewed towards seeking a high level of information. Ultimately, these individuals are accountable in cases of wrongful payments, which are investigated by auditors and agencies such as the Comptroller and Auditor General, Central Vigilance Commission, the Central Bureau of Investigation, state vigilance agencies (Somanathan and Natarajan 2022). Sanctions for wrongful payments include job demotions, dismissals, or even jail time. Given that officers who make the decision to pay do so based on certifications or data provided by others in the administrative hierarchy, they are assumed to be privy to any incorrect data or wrongful certification informing the decision to pay. This means that if lower-ranking officers provide incorrect data—whether intentionally due to factors such as corruption, or accidentally, due to errors—senior officers are ultimately held accountable, whether they participated in wrong-doing or not.

Former IAS officer Anil Swarup confirmed this accountability problem while discussing the aftermath of a CAG report on the audit of coal block allocations from 2004 to 2010. The report alleged that the government failed to allocate coal blocks for mining to private sector enterprises and firms through the typical process of competitive bidding, leading to lower revenue and lesser competition in the allocations process. Known as 'Coalgate', the allegations from the report were heavily publicized in the media, describing how individuals across the system failed to follow due process. The result was a series of investigations by the Parliament, the Supreme Court, and the Central Bureau of Investigations. The result of the report and the aftermath, Swarup argues, was disastrous for the 'bureaucratic "psyche"' and for 'honest civil servants'; 'officers refrained from committing themselves on the files. "They were playing it 'safe'" in the fear that they would be audited and found at fault' (Swarup 2019, 131). Here, Swarup refers to the tendency of bureaucrats to fatten a file with information as written evidence that they have undertaken due diligence before making any decisions for payments.

Monopoly and discretion: Because the decision-making power for payments lies in the hands of a few individuals, they exercise considerable monopoly in the system and, consequently, discretion. While discretion in itself may not be a bad thing and may even motivate bureaucrats to feel empowered (Honig 2024), in this case discretion often exists to an extreme, where the system incentivizes bureaucrats to be overly cautious with payment approvals even when it is warranted by the rules. Even when all the conditions to authorize a payment are clearly defined and have been met, the officer who is permitted to make payments has the full right to choose whether they release the funds. These functions stem from the General Financial Rules (GFR) (Department of Expenditure 2017, Rule 37) of the government itself, which state:

> An officer shall be held personally responsible for any loss sustained by the Government through fraud or negligence on his part. He will also be held personally responsible for any loss arising from fraud or negligence of any other officer to the extent to which it may be shown that he contributed to the loss by his own action or negligence.

This virtually limitless responsibility borne by the official means that they are blamed not only if work they themselves do is found to be negligent (note that we use the word negligent, not fraudulent), but also if the work of a supervisee is found to be negligent. This rule exists because officials deal with public funds and therefore they need to be accountable and have fiduciary responsibility. It thus makes sense to retain such a rule. But it should also be clear to those who authorize funds what exactly constitutes non-negligent behaviour for every type of payment in every programme or scheme. In other words, what is the basis under which those who administer schemes should make payments? Ultimately, it should be the responsibility of policymakers who design schemes to define these constraints clearly. If officials who authorize funds are unsure of these facts—which is possible in an unsynchronized and unconnected system within PFM and between PFM and programmatic events—they are bound to over-assure that payments are made under highly constrained conditions.

When combined, these factors—information asymmetry, accountability burdens, and monopoly or discretion—give rise to three possible responses that cause delays and uncertainty in payments. First, there is a large incentive for senior officers to have a rapacious appetite for additional information from implementers because it behoves them to have low trust in the accuracy of the initial payment-related data. Those seeking payments are therefore put in a position where they have to provide a variety of information, often over multiple iterations, to approval officers. These requests are usually done on the record and filed to show that all possible due diligence was indeed undertaken. The additional due diligence may not always assure the senior officers of the validity of the requested payment, but it allows future auditors to ensure that these officers sought the due avenues to verify information. Even though the accuracy of any additional information obtained is likely to be quite similar to the accuracy of the information originally provided, senior bureaucrats request this to hedge against the risk of investigations in the future. Moreover, many of these requests are for information not strictly required for payment approvals, so it is not always clear to stakeholders when payment decision makers will find themselves assured of the legitimacy of the payment request.

The second response to the three aforementioned factors is that senior officers take a calculated risk to request speed money to release payments. Those awaiting the funds expect government funding to be prone to significant delays, which is why they may choose to comply. Lastly, a third possible response is to not make the decision to pay at all or 'decision paralysis' (Rao 2022)—the belief that the best way to not make a mistake is to do nothing at all. It is possible for risk-averse senior officers to choose this third response since inaction and not achieving outcomes is less likely to be punishable than being a party to incorrect payments.

These responses are significant for programme managers in government and other parties who conduct business with the government. When the links between fiscal and programmatic events are unsynchronized and different stakeholders have access to different types of data, there is no way for parties who are requested to provide proof to find out whether the new demands for information and certification stem from payment makers' aversion to risk or if they signal the need for speedily acquiring funds. It creates uncertainty for programme managers and those working with government who have little information about whether the payment is likely to be released at all. It is indeed appropriate and necessary that every payment, whether to a vendor, programme beneficiary, or contractual worker, be established or certified as legitimate through appropriate means. But the decision to pay system creates a space for officers to add conditionalities that were not agreed upon when stakeholders agreed on a contract to work with the government or receive payments from the government.

5.3 Administrative Burdens

Public officials are heavily under-resourced relative to their expected duties (Dasgupta and Kapur 2020). The financial and accountability architecture only adds to these burdens. Because of the way accountability structures are set up in government, programme managers have to constantly meet demands of higher-ups for routine and emerging information needs of higher-ups in the administration (Agarwal and Somanathan 2005).[9] This is especially true when managers are not privy to granular information connecting fiscal and programmatic events. Along with fund managers, programme managers are stuck in a cycle of seeking more and more information to meet higher-up demands in conjunction with obtaining accurate lower-level data. The disjointed information architecture in government, especially between fiscal and programmatic events, creates what we call 'administrative burdens'. While scholars have used the term to mean the hurdles citizens face in navigating the bureaucracy (Herd and Moynihan 2019), here we deploy it to describe the burdens bureaucrats routinely undertake to meet the expectations of the administrative structure and fiscal architecture.

As noted in the previous subsection, if those who authorize payments are audited, and auditors find evidence of fund mismanagement, fund managers who approve payments risk sanctions. This leads fund managers to request information from their supervisees to create a paper trail demonstrating that they have undertaken due diligence before authorizing a payment. Given the lack of observability in the current layered system—combined with information asymmetry, accountability mechanisms, and monopoly in decision-making—those that approve payments are incentivized to ask for information repeatedly.

A study of any scheme's files for a vendor's request for an instalment would illuminate this fact. When a vendor is contracted for supplying goods or services, the criteria for payment are usually part of the contract. Let us take a hypothetical example in which three pieces of information—say A, B, C—are required to obtain payment for a contracted service. When the vendor provides this information, an administrator may demand something entirely different to make the disbursal. Then, once the vendor furnishes this new information, another administrator requests an entirely different piece of information. This process continues such that, by the end, the vendor has had to provide A, B, C, D, E, and F pieces

[9] The emerging demands are often different from those prescribed in terms of payments, and stem from corruption, poor knowledge of the financial system, or even a desire to prevent payment to escape scrutiny in case something goes wrong.

of information even though they were only mandated to provide the first three.

These types of burdens are not unique to private vendors supplying the government. Programme managers that rely on approval from other administrators for fund releases are embroiled in the same cumbersome process. One of the authors of this book regularly faced these situations as the Principal Secretary of Rural Development in Bihar while trying to get fund disbursals from the Union government, and later as Joint Secretary in the Government of India in the Ministry of Rural Development when he was trying to make payments to contracted agencies for skilling rural youth. We provide examples of Santhosh Mathew's personal experiences in Appendix 5.

Though little research has been conducted to study the problem and consequences of administrative burdens resulting from shifting demands, Accountability Initiative studied how four Bihar Cluster Resource Centre Coordinators (CRCCs) spent time as part of their jobs (Aiyar et al. 2015). Researchers found that the CRCCs spent 17 to 31% of their time dealing with paperwork rather than working on education which was their mandate. Instead of being programme implementers, the Coordinators became another cog in the bureaucratic wheel of filing and paperwork. Similarly, in their study of time use of Block Education Officers, Aiyar and Bhattacharya (2016, 64–65) note that officers spent a majority of their time chasing data for higher-level district officers. Dasgupta and Kapur (2020, 1323) again find similar problems. In a time-use survey of Block Development Officers, bureaucrats reportedly allocated 22% of their time on planning and budgeting for programmes. If funds are to reach beneficiaries and have expected outcomes, we need to ease the administrative burden stemming from the financial architecture so that programme managers can actually get to focus on what they were hired for and perhaps what they are motivated to do in the first place—programme implementation (Aiyar et al. 2015).

Other research by Rajaraman and Gupta (2016) further highlights systematic burdens. From 2010 to 2015, the Thirteenth Finance Commission reformed the system of statutory grants from the Government of India to local governments. The reforms sought to bypass the procedural impediments to the fund flow for grants, which limited the ability of local governments to access grants even when eligible. The reforms made two-thirds of the grant unconditional and one-third conditional upon the performance of the state government. While these reforms were expected to reduce the shortfall of receipts against allocations, FY 2011–12 and FY 2012–13 shortfalls peaked at 13.75%. During the period of the reforms, 10 states failed to receive rural grants in full. Rajaraman and Gupta (2016, 80) explain as follows:

> Even though the basic grant was wholly unconditional, the administrative guidelines issued imposed an additional condition—that local body elections should have been held on time (a reasonable insertion, since local grants are disbursable only to local governments within their legitimate term of office). Another requirement was that the previous tranche should have been disbursed by the state to local governments within five days of receipt from the centre with interest added on for delays beyond that. Although not a conditionality, it would have added complexity to certification of utilization and added on to the transactional delay.

The ways in which even unconditional grants were put forth with procedural controls by the Union government made it harder for administrators to meet programmatic goals. In this case, unconditional grants were made available to ease the restrictions imposed by conditional grants. Instead, the financial rules in place meant that the administrators faced administrative burdens regardless.

In an alternate system, observability between fiscal and programmatic events would allow programme managers to recognize the unit-level conditions that were met for payments, and stakeholders (like fund managers) to have near real-time access to this information without expending additional effort; by design, such a system would make all the available information observable to those with authorization. In such a world, programme managers can focus on their core jobs of programme implementation rather than on providing proof for higher-ups to secure payments. Such a system may better motivate their work, and allow bureaucrats to exercise discretion and autonomy in places that matter (Honig 2024).

5.4 Delays and Uncertainty

Administrative burden for bureaucrats has a downstream effect on government functions. This has long been the case, as noted by observations by Paul H. Appleby, who was invited by the Government of India to review the Indian administrative system in 1956. Commenting on the need for local agencies to obtain constant approval of central agencies—even when budget allocations were made—Appleby remarked: '. . . it [the budget] is not determined in fact when enacted, but is actually being made day after day throughout the year. Detailed control after the fact of general determination has the effect of delaying, frustrating and even nullifying decisions made earlier at the highest levels'.[10]

These delays and frustration continue today due to unsynchronized and unconnected PFM systems. The architectural problems necessitate that officials request information up and down the command chain prior to approving funds, snowballing the delays in funds reaching the end recipients. There are several areas where uncertainty and delays are common.

For one, it takes an inordinate amount of time for a financial manager or the treasury to ensure that bills are approved for payments. Treasury approvals typically slow down if there are many bills that require attention. Treasury officials must check if rules are being followed prior to approving a payment. Similarly, financial managers are part of a programmatic department and ask programmatic questions to ensure the payments are for legitimate or plausible reasons.

Sectoral studies in India have documented these delays. Studies on the National Health Mission find that a file requesting the release of funds at the state level goes through a minimum of 22 desks in Uttar Pradesh, 32 in Bihar, and 25 in Maharashtra, resulting in timelines as long as 3–5 months within the state itself (Choudhury and Mohanty 2018, 21; Kapur 2019). While this data is older and may have improved, delays have not completely gone away. Such delays are exacerbated by even more delays in releases from the Union government to the states. In the health sector, Kapur and Vaishnav (2018, 5) find that 59% of the Union

[10] Paul H. Appleby, *Re-examination of India's Administrative System* (New Delhi: Manager of Publications, Government of India, 1956), p. 50, as quoted in Handa (1980, 650).

government's releases to states for the National Health Mission were in the last quarter of 2015–16.

The picture is not very different in the education sector. When tracking the flow of the funds allocated for elementary education between 2009 and 2015, Accountability Initiative found that none of the eight districts across six states in their study had received their total allocated budget within the financial year (Aiyar and Raghunandan 2015):

> ... much of it [the money] makes its way to the district halfway through the financial year ... once money reaches the district, it can take two to six months to travel from the district bank account to the schools, where expenditure actually takes place. As we discovered through an annual national PAISA survey, for over 50% schools in India, even small grants that schools are expected to receive annually for essential purchases, are credited to their bank accounts somewhere between November and December, well over halfway into the school year.

Let us further take the example of Samagra Shiksha Abhiyan (SSA), an education scheme which uses the Society mode of transfers and receives money directly from the central Ministry of Human Resource Development to the state society. Recall that according to the rules described earlier, at least 25% of the funds are disbursed in the first instalment when the financial year begins in April. The percentage, however, can vary. In 2015, for instance, the central government released 57% of the allocated budget to states by September (albeit with notable variance across states). Yet for SSA, only 27% were released from the Consolidated Fund of states to state societies, on average (Kapur and Srinivas 2018, 2). In fact, a December 2015 survey by Accountability Initiative revealed that 31% of schools reported having received no money (Kapur and Srinivas 2018, 2), an issue we delve into in the next subsection. Similarly, by the end of FY 2014–15, state governments had only released 62% of the funds they received (Kapur and Srinivas 2018, 3). Uncertainty in releases to state societies consequently causes the implementing agencies to use up much of the available funds quickly, leading to funding hurdles in later parts of the year until the remainder of the allocations are received. This problem occurs even though funds have been released to the State Consolidated Fund.

The number of layers in the fund flow system that we describe in Chapter 4, and the lack of synchroneity between them, make the simple task of moving money from one level of government to the next a long and tedious process, based on a cascade of Utilization Certificates (UCs) that are aggregated at each level of government from the panchayats right up to the Union government. As Mathew (2018) notes in an op-ed:

> Each level of government must first produce utilization certificates verifying that a majority of its previously disbursed funds have been used. When this is done, the level above the one requesting the funds releases a portion of the annual allocation down the chain. The effects of delayed fund transfers in one part of the chain therefore cascade through the entire system and in the process, amplify failure. For instance, if a district fails to receive its second instalment in time, all panchayats under it will be refused further instalments even though individually many of them may have fulfilled all eligibility conditions.

Compounding these problems is the fact that multiple actors and institutions exercise administrative control over fund releases at each government level. This results in procedural frictions when multiple individuals need to approve UCs, sanction orders, or respond to a myriad of information requests for fund releases within a narrow time frame (Nilachala and Bose 2020). Ultimately, last-mile programme managers and programme recipients feel the effects of strained fiscal circumstances which prevent the timely functioning of government programmes.

Delays have several consequences. For one, they reduce credibility of upstream budgeting processes as well as the effectiveness of downstream spending. As Aiyar and Raghunandan (2015) explain in an op-ed: 'for the average frontline administrator, annual budget allocations are a meaningless figure for she has no guarantee of when and how much of this allocation will in fact reach her bank account'. They also describe how the combination of excessive paperwork to obtain funds and delays in these funds leads to wasteful expenditures: 'Over the years, we've met enterprising headmasters who have used their money to buy fire extinguishers for schools that are yet to construct their buildings; for desks and chairs when they already have them; and to whitewash walls even when the walls look perfectly clean.'

These kinds of activities are typically undertaken in the last quarter, when implementing agencies receive the delayed funds and are under pressure to use them before the end of the financial year. Due to delays, state governments, which now manage more than 60% of the total government expenditure, have historically underspent their budgets. For instance, states ended up spending only Rs. 27.71 trillion in 2017–18 (USD 430 billion at the time) (Reserve Bank of India 2018, 10) against the Revised Estimate of Rs. 30.285 trillion (USD 469.9 billion at the time) (Reserve Bank of India 2018, 6).

Were there observability between PFM and programmatic goals, financial managers need not ask a series of questions about why and where funds are being spent. Instead, they could have a direct view of resource needs. Similarly, were there observability within PFM, treasury managers could understand how much funding is allocated and has been already disbursed, and how much is needed for ongoing tasks. As we describe in the following chapters, such synchroneity and observability can coincide with tools that enable automated checks to verify whether rules are followed prior to fund disbursal, thereby minimizing delays.

5.5 Corruption

There are a wide number of possibilities for corruption in public financial management, much of which happens during the budget execution stage (Dorotinsky and Pradhan 2007). We highlight the ways in which corruption stems from the current PFM architecture itself, which relies on advance, bulk payments through instalments. Because there is a lack of observability and synchroneity within various stages of PFM—as well as between fiscal and programmatic events—officials at each administrative level can seek rents in exchange for file approvals to disburse funds. Alternatively, they can withdraw money in bulk using Abstract Contingent (AC) bills into untracked bank accounts below the state level—and with varying levels of accountability. These features of PFM create situations in which corruption can easily fester, with little accountability until years down the line, if at all, since obtaining data can involve multiple stakeholders and require a long time.

Indeed, the above characteristics of the funding architecture have enabled corruption in schemes throughout India. Two episodes in Bihar lend further insight. In one of the most well-known cases of corruption in the state, government employees from the Animal Husbandry Department submitted false bills to several district-level treasuries for payments to non-existent companies. At times, officers in the department submitted inflated bills that were much higher than the authorized allocation amounts. These bills were for the procurement of cattle feed, medicine, and artificial insemination instruments, leading to the case being known as the 'fodder scam' (Bhatt 2013).

Officials were able to overdraw funds with forged allocation letters. In the 1980s and 1990s when the scam took place, the treasury relied on letters from state government departments allocating funds to different Drawing and Disbursing Officers (DDOs)[11] in various regions of the state. Departmental officials used forged allocation letters to syphon off Rs. 6.44 billion, an amount more than the actual budgetary allocations of over 15 years prior to 1996 (Bhushan 1996). Over the course of two decades, the scam involved over 600 officials and is estimated to have cost the public Rs. 9.5 billion (Bhatt 2013).

During the time of the fodder scam, withdrawals that totalled more than the allocations were feasible because paper-based trails in the form of signed letters were not backed by any easy-to-source information that authenticated budgetary actual allocations by state governments. When Finance Department officials looked at the total expenditures in the Animal Husbandry Department, these expenses could still pass scrutiny if state totals were within the allocated limit, but regional totals were not. Auditors who were doing district-level checks had no way of ensuring that allocations were authentic if they were given forged letters to begin with. To add to the problems, the Accountant General's office at the time destroyed paper vouchers every five years, making it impossible for auditors to trace anything prior to that time frame (Bhushan 1996).

Today, an observable and digitized database allows the state-level and district-level treasuries to verify actual allocation amounts, making it difficult to withdraw past these allocations. However, other issues persist.

[11] A Drawing and Disbursing Officer can withdraw money from the treasury or deposit money into the treasury.

We illustrate another case in Bhagalpur, Bihar that highlights these issues, which result from money stored in various levels of sub-state agencies, without near to real-time synchroneity with state or auditing systems.

Between 2004 and 2017, Manorama Devi, the founder of a women's cooperative society, worked with a network of bureaucrats and politicians to use government funds for investments, profits from which were distributed back to those who colluded with her (Karamakar 2020). This was made possible because in some cases (such as with land acquisition), DDOs in Bihar are allowed to have bank accounts which allow them to withdraw funds from the treasury using AC bills (bulk advances for cash disbursements and expenditures for which details are not known at the time of withdrawal), place them in a bank account, and disburse them later. Such DDOs include officials such as District Animal Husbandry Officers, Block Development Officers, and District Magistrates, amongst others.

In the Bhagalpur case, district officials, including the District Magistrate, withdrew large amounts of money using AC bills from the district treasury under instructions from the state government. These funds were meant for payments such as compensation for land acquisition, social welfare, and health departments (Times News Network 2017). District officials kept these funds in the District Magistrate's official bank account. Through collusion between bank and district administrators, these funds were then taken out of the official bank account with forged signatures and deposited in the bank account of Devi's SRIJAN, a cooperative society engaged in vocational training and promotion of self-help groups for women that claimed to be running a bank for women. The cooperative society, in fact, was not running a bank—it was not connected to any official banking systems. Rather, it was a cooperative that allowed members to take out loans and deposit funds. Using fraudulently acquired government funds, SRIJAN operated as a local loan shark and 'investor', allowing it to accrue interest from others (Kumar 2017).

On paper, government officials in Bihar must maintain their bank accounts with public sector or cooperative banks. After government funds were deposited in the official account, staff would use the District Magistrate's forged signature to write a cheque to the NGO's account. To ensure that the cheques did not bounce, officials and Manorama Devi worked with the bank staff to send the right amount of funds from the

SRIJAN account to the government account before a cheque for payment was to be deposited. At the same time, district officials doctored fraudulent e-statements indicating the money was in the government account. The scam was discovered after Manorama Devi passed away and officials failed to collude fast enough and replenish the government account in time by letting a cheque bounce (Kumar 2017).

The Bhagalpur case reveals a big challenge with fund flow which enables corruption. The lack of observable real-time data on expenditures at each level in tandem with the lack of synchroneity between such data and monitoring agencies means that auditing becomes a lengthy process. It requires a series of checks across files and bank statements, some of which can be easily forged or made to disappear. During the Bhagalpur incident, money was withdrawn and replaced without ledger entries. In today's world, a simple query to fetch the statement of a government-owned bank account can suffice in preventing such cases if access to information in the core banking system is made available to those who need to see it. While reforming fund flow architecture may not eliminate all forms of corruption, it can certainly reduce the incentives that exist in the way PFM is currently architected.

5.6 Poor Auditing

Synchroneity and observability within PFM and between PFM and programmatic events is necessary for effective internal or external audits. Very often, today's auditing systems require auditors to physically check files throughout the state or country or rely on programme managers to obtain copies of UCs—the very managers who used these Certificates to secure funding. UCs and their underlying vouchers are distributed throughout offices of the state and Union government and their implementation agencies, often in poorly labelled bags or boxes, making them difficult to find and track.

Even where digital systems exist, observability isn't always as easy as one would imagine for the government's statutory auditor. Take, for instance, the contentious 2017 debate over whether the CAG could access the Goods and Services Tax Network's digital systems (Joseph 2017; Kumawat 2017), and more recently, the CAG's submission to the

Calcutta High Court in 2021 that audit teams had not been provided any information by officers in West Bengal pertaining to Amphan Cyclone in 2020 (Thakur 2021). These are merely the more publicly known and extreme cases. Auditors routinely face problems accessing data to conduct the job. This reduces the effectiveness of audits in three ways: what can be audited, the representativeness of the audit, and the usefulness of the audit for future action.

For one, auditors cannot always investigate all facts of a case since the information needed to audit is not obtainable without the cooperation of the audited. There are two issues at hand. Though the CAG and state Accountant General offices are tasked with auditing expenditures based on payment vouchers, they often lack the appropriate documents for proper scrutiny. A World Bank study indicated that 52% of the expected statements of expenditures in their study were never provided by implementing agencies (World Bank 2006, 5). This is especially problematic given the PFM system already creates long delays between fund disbursal and expenditures. Missing vouchers further debilitate the ability of auditors to conduct timely and appropriate checks.

There are also instances where programme managers do not give auditors the UCs which are used to check that the managers secured funds only after accurately reporting expenditures. For instance, as of March 2024, the Union government's 22 departments or ministries contained 37,986 missing UCs aggregating up to Rs. 727.6 billion (approximately USD 8 billion), with the earliest missing UCs dating as far back as 1978 (CAG 2025a, 131). State-level audits have also revealed similar irregularities, with some states faring better than others. As of 31 March 2024, Jharkhand had 41,755 UCs totalling Rs. 1,331.6 billion (approximately USD 14 billion) over the last ten years (CAG 2025b, 85); Haryana had 2,037 pending UCs amounting to Rs. 170.2 billion (USD 1.8 billion) (CAG 2025c, 115); and Bihar had 9,649 pending UCs of Rs. 708.8 billion (USD 7 billion) (CAG 2025d, 119). These numbers are smaller in some other states, totalling 22 missing UCs worth Rs. 110.7 million in Kerala (USD 1.2 million) (CAG 2025e, 117), and 111 missing UCs worth Rs. 28.1 billion in Tamil Nadu (USD 312 million) (CAG 2025f, 117) over the same time period.

Such high numbers of missing UCs prevent the CAG from fulfilling its auditing function, especially since the pressure to ensure that UCs are submitted largely lies with each ministry or department rather than with the CAG. The CAG, therefore, cannot push for implementation agencies to provide these documents unless the Public Accounts Committee (PAC) of the legislature acts on it. Given that missing UCs are so prevalent and widespread, the PAC is unable to effectively require ministries and departments to submit UCs and adjoining vouchers. Such disjointed monitoring functions, combined with poor systems for accounting and auditing, create an issue of accountability wherein no agency can ensure that the spent funds meet intended goals. The government has introduced a digitized file management system which may result in some improvements.

Even when UCs are provided, auditors face an additional hurdle: gathering the Certificates and verifying that they were based on actual expenditures is an arduous task. As explained earlier, UCs are merely a statement by a fund manager stating that a certain amount of money has been spent, at peril to their job. The assumption is that the manager is doing their job honestly, so the amount that they claim was used was indeed spent. However, to verify that the expenditures were spent based on the UC, auditors have to ask for documents from that very individual. As one can imagine, many perverse incentives can be at play here, and UCs may pass scrutiny of auditors even with forged or inaccurate expenditure receipts.

In fact, Rajiv Mehrishi, the 13th Comptroller and Auditor General of India from 2017 to 2020, addressed this very problem in a June 2022 speech,[12] calling out the difficulty in obtaining the data needed for audit:

> . . . they [CAG staff] simply cannot get the data because the office just simply says that the file is lost or the file is misplaced. The manually-entered computer data is unreliable. And thus the entire process is not auditable. You get what the department wants to give, not what you want to see. So, naturally you are not going to know whether you actually saw the malfeasance or the wrong spending or not. Honestly speaking, the shape that our accounts and accounting system are in, indicates that Government accounts are not properly auditable.

[12] The speech was given on 10 June 2022 at the India Habitat Centre, New Delhi at a conference on 'Mainstreaming Digitization of Public Expenditure in India'. An excerpt of the speech is available in the report of the conference (Iyer and Chowdhury 2022).

If there was granular-level observability for auditors, and synchroneity between programmatic and fiscal events, then auditors could do their jobs far more efficiently and accurately.

Because of a lack of synchroneity and observability between fiscal and programmatic events, auditors also lack granular details on the 'population' requiring auditing. This prevents auditors from randomly selecting the sample of vouchers to be audited. Instead, they rely on a selective sample that has the required data and frequently change who is audited to finish their goals. This creates two risks. One, the sample chosen for auditing may not represent the population being audited. Since the audit findings do not represent the population or the 'system' being audited, this also limits the ability of audits to comment on systemic challenges. Two, it creates opportunities for collusion between auditors and those being audited. If auditors or agencies select who to audit, the possibility of checks being done with advance warning is not nil.

CAG began Voucher-Level Computerization (VLC), which seeks to digitize vouchers, beginning from July 1987. The programme was at first unsuccessful, and was only later reformed in the late 1990s. Uptake of the VLC, however, has been slow (Kumar 2009). Even with optimum uptake, the system still grapples with a core functionality issue—the auditing system is not integrated with the expenditure management system. Voucher-level computerization can check to see if vouchers are filled out properly and the total amount of spending matches the records in the books of accounts. However, it still relies on individuals to report line-item expenditure rather than being part of a larger system that automates and, by design, generates data on how much money is allocated, how much is spent, and whom or what it is spent on. These functions are key to auditing.

Rai (2020) notes that, recognizing the 'requirement of reliable, accessible, and searchable data related to government revenue and expenditure', the CAG in 2020 recommended to the President of India that India should enact a Digital Accountability and Transparency Act (DATA):

> This advice is about an enabling legislation called DATA that aims to bring transformative change in financial governance. It is based on the principle of 'accountability by design' where all spending and receipts information is required to be made available using an 'open data' and

'data standards based' approach. It advises end-to-end digital transactions in the government resulting in complete and reliable data, which, in turn, will support the establishment of budget baseline, data driven cost estimates, performance comparison of departments and agencies. It aims to ensure completeness of data capture by prescribing 'data standards' for recording and reporting of expenditure and receipts by the government and all agencies or entities including corporations, autonomous bodies, etc., which perform functions on behalf of the government. It is proposed as an enabling legislation that aims to make spending and receipt information of the government available on a centralised portal data which is machine readable, granular, comprehensive, non-repudiable and purpose linked.

As Sabharwal and Phatak (2020) further explain, passing the DATA legislation will not only enable better audits but also allow legislatures to have assurance that the money is spent for the intended purpose. Creating an effectively auditable system is indeed possible, and we spend a considerable portion of Chapters 6–8 expanding on how precisely it can be done.

5.7 Data and Citizen-based Accountability

For citizens and civil society more broadly, knowing how fiscal and unit-level programmatic events are linked is important on two levels. One, these links allow those who experience public services directly to hold the government accountable for their delivery. Two, civil society needs observability to hold the government accountable to its policy goals.[13]

[13] For instance, the Bhilwara Principles offer a useful framework that 'outlines the five essential elements of a Social Accountability Framework, as conceptualized and articulated by people facing lack of accountability' (Swamy 2019, 36). These elements are:

1. *Jaankari* (Information): Access to relevant information to know, act, self-govern, make informed choices, and hold those who govern accountable to their mandate, such as, as an example, data on status of cash transfers.
2. *Sunwai* (Hearing): Right to be heard. For a system to facilitate accountability, there must be adequate, inclusive, and multiple modes for citizens to articulate grievances. An example would be information on processes in place for citizens to file grievances across multiple channels.
3. *Karyawahi* (Proceeding): Time-bound grievance redress when complaints are registered. An example of information required for this would be receiving status updates on grievances.

The lack of observability and synchroneity between fiscal and programmatic events impacts the predictability of public services for its recipients in multiple ways. Foremost, if local governments have to navigate multiple systems or records to inform individuals of their entitlements, beneficiaries can have difficulty assessing what they owe or are owed. This impacts the first stage of benefiting from public services—being able to enrol in the services one is entitled to. Furthermore, if there are no near real-time updates on which beneficiary has completed the necessary requirements to receive their next subsidy or good, citizens must rely on administrators who go through multiple files to update them. Not only can this cause confusion for beneficiaries over their right to entitlements, but it can also lead them to rely on the very administrators who may be mismanaging programmes.

For instance, in Rajasthan, government officials were systematically violating the rights to minimum wage of rural workers during the 1990s. In response to this trend, a social movement emerged that led to the adoption of India's Right to Information Act, allowing citizens to confront government officials about the gap in what they received from the government versus what was reported as being received. As the names of relevant officials and details of expenditures were exposed during public hearings organized by the Mazdoor Kisan Shakti Sangathan (MKSS), more and more cases of corruption started coming out in the open. In a similar vein in Maharashtra, the right to information became an integral part of the citizen's campaign against corruption (MKSS 2014). Ultimately, information is crucial in filing appropriate grievances, and especially important in providing corruption-free access to funds.

Civil society organizations, as well as global organizations such as the WHO, continue to rely on resource-intensive and hard-to-scale tools for tracking resources such as Public Expenditure Tracking Surveys (PETS) and Public Expenditure Reviews (PERs) to obtain information on financial outlays. Think-tanks working with government's financial

4. *Suraksha* (Protection): Protection from harassment or intimidation for complaining and disturbing an established status quo, for instance, information on processes in place to offer protection.
5. *Bhaagidari* (Participation): Institutionalized platforms of participation that enable the voice of communities to reach the state while accessing services, planning for use of public funds, monitoring programme delivery and registering grievances. An example of information required on these platforms and how to engage with them would be important.

data such as the National Institute of Public Finance and Policy and the Centre for Budget and Governance Accountability have to spend an enormous amount of time first obtaining and then digitizing that data from PDFs so it can be used for analysis. Though the CGA manages the Public Financial Management System (PFMS) to process payments, the system only captures data for Union-to-state transfers. Some states have also adopted similar systems such as Khajane in Karnataka, but the data is limited to the few schemes that use these systems.

Ultimately, if financial decisions are to drive accountability in programme delivery, reduce corruption, and enable citizens to obtain payments on time, data observability and, subsequently, data accuracy are key. Similarly, if researchers and civil society actors are to hold sectors of the government accountable or understand their functions, an interconnected set of information systems to collect and share data at the level of use (rather than in the aggregate) is essential. Such data can both improve the amount of information that citizens and civil organizations have and decentralize the access to audit the government and hold it responsible (Mukhopadhyay 2017).

Beyond such data access, citizen-based accountability must also go beyond the reforms we highlight. Scholars have noted, most prominently with the case of MGNREGS reforms, that expenditure management issues have not resolved beyond-the-last-mile challenges that citizens face in accessing money after it reaches their bank accounts. As these challenges fall beyond the scope of our inquiry, we highlight them in Appendix 3.

5.8 Conclusion

India's PFM architecture that allocates, disburses, and processes expenditure has made significant improvements, but more remains to be done. Rather than enabling implementation agencies to produce outcomes through public spending, the architecture—which uses unconnected and unsynchronized data on fiscal and programmatic events—contributes to public service delivery failures. Given that few outside or even inside the government are privy to the entire architecture, the country's PFM is rarely cited as a contributing factor to India's public service delivery or state capacity failures.

As we highlight in this chapter, however, such an architecture has cascading impacts on public service delivery. Today, it contributes to over Rs. 1 trillion (USD 12.35 billion) in float; delayed decisions by bureaucrats to approve payments; administrative burdens for bureaucrats; significant delays in funds reaching implementation agencies; institutional mechanisms for corruption; auditing failures; and an ecosystem that fails to enable civil society or citizens to effectively hold the government accountable. All these elements combine to reduce the efficacy of programme implementation.

PFM can be re-architected to enable rather than stymie public service delivery. In the last section of the book, we highlight the changes that can make this possible.

6

Public Financial Management: From Good to Great

Based on existing public financial management models, the next three chapters of this book propose a vision for an on-demand architecture for public financial management in India. In this chapter, we conceptualize the overall architectural features of a PFM system if they are to be capable of resolving the issues of synchroneity within PFM, and between fiscal and programmatic events discussed earlier. Next, Chapter 7 reviews specific functions that can be built from an improved architecture that would directly address many of the symptoms of the PFM system (as it exists now) that we have reviewed in the book thus far. Chapter 8 discusses how the Finance Departments can foster co-ordination capacity across other government departments to make these reforms happen.

In this chapter, we first turn to discussing how the government can reconfigure the architecture of PFM in section 6.1. We argue that a new architecture can enable further improvements within the PFM system outlined later in the book. Section 6.1 discusses three key architectural shifts of improved PFM: (a) coupling fiscal and programmatic data so they are connected to each other using data exchange and data standards; (b) co-ordinating fiscal events and programmatic events such that fiscal events are triggered when programmatic events occur through Rules as Code; and (c) making fiscal and programmatic data, at the unit level, accessible to authorized personnel to improve policy implementation. We focus section 6.2 on the properties of the overall PFM system to support these architectural shifts. The first fundamental property must be data protections, without which any PFM system risks providing tools to those who wish to further state incapacity. Additional properties include data liquidity, or the ability to transform and reuse data for multiple

Public Financial Management, State Capacity, and Public Services in India. Santhosh Mathew, Bhumi Purohit, and Devesh Sharma, Oxford University Press.
 DOI: 10.1093/9780198973195.003.0006

purposes; the once-only principle for capturing and sharing data; and ensuring data credibility. We summarize this overall vision in Figure 3. The final layer in the figure, outcomes, is detailed in Chapter 7.

The ideas in this chapter are inspired by and aligned with the principles of Digital Public Infrastructure (DPI), an approach to digital transformation that India's G20 presidency brought to the global stage. While the current label for this approach is DPI, it has undergone evolution and relabelling over the years, including Societal Platforms, shared digital infrastructure, and public digital platforms. Mathew (2021) alludes to some of the ideas in this book with reference to shared digital infrastructure, which was the label this approach had in 2021.

6.1 Three Architectural Shifts

In order to create an on-demand architecture, we first discuss the key architectural shifts that serve as the foundational block for further changes: coupling programmatic and fiscal data, co-ordinating programmatic and fiscal events, and making unit-level programmatic and fiscal data available as needed for those authorized to access it.

6.1.1 Principle 1: Coupling Programmatic and Fiscal Events

The first key principle necessary for PFM improvements is the coupling of programmatic and fiscal events to enable two-way visibility. In other words, an expenditure should be linked to the precise item for which it is used, such that finance managers can see the outputs—without necessarily seeing an individual's privately identifiable information, as we detail in section 6.2.1. Similarly, programme managers should be able to see the financial inputs and outflows to guide implementation.

In the budget execution stage, the primary unit of expenditure—the payment—should be linked to an associated activity or output for each programme. During the budget reporting and monitoring stage, every payment should be linked to an associated expenditure report. Such a report would allow stakeholders to examine the aggregate expenditure

01	ARCHITECTURAL SHIFTS	• ***Coupled*** **fiscal and programmatic data so there are two-way linkages between the two; done through data exchange and standards** • ***Co-ordination*** **such that completion of a programmatic event triggers a fiscal event** • ***On-demand*** **data availability at the unit level to those who have data authorization**
02	SYSTEM PROPERTIES	• **Data protection is instituted by design** • **Data liquidity enables data to be recombined and reused** • **Data is collected at a single point of entry, only once** • **Data is credible**
03	OUTCOMES	• **Just-in-time funding** • **Smart payments** • **User-level experience platforms, such as programme-specific IT platforms, auditing platform, and public transparency platform**

Figure 3 Reforms for On-Demand PFM Architecture. This chart summarizes our key recommendations to enact an on-demand architecture for PFM reforms. The first stage outlines the three main architectural shifts for such a system. The second stage outlines the key properties of the system. The third stage outlines some of the outcomes of the overall architecture, including just-in-time funding, smart payments, and an experience layer of platforms.

for a programme as well as the unit-level expenditure for each programmatic output. It would further include information on when, how, and where the money is spent, in addition to whom it is given to and for what purpose. All these pieces of data should not reside in one place, but can be pulled from registries and datasets that are already maintained by the government—say, the beneficiary registry, the master data of programme outputs, or the income information of beneficiaries. However, the component data should be available to those who need it to perform their work, in a protected manner. The ability to fetch data from different places, registries, or servers on a need-to-know basis can allow stakeholders to see individual items for data analysis or auditing as needed. In essence, such linkages dispel the ambiguity and lack of insight that stem from existing expenditure reports, which tend to aggregate data upwards. As we highlight later, though, the assurance for programme managers that outlays, outputs, and outcomes are connected does not need the underlying data to be made available in a manner that violates individual-level data privacy or fundamental data protections.

To expand on this principle, we look at an example of a programmatic event: a beneficiary of the rural housing scheme Pradhan Mantri Awas Yojana (PMAY) becoming eligible for the next tranche of funds. They qualify for the next tranche when they complete the construction of the foundation of the house and a programme manager uploads a geotagged photo of the work in an existing government database. Ideally, a PFM system, at the time of release of the tranche, should be able to connect the payment amount with the beneficiary's name, alongside assurance that a government functionary has verified the location of the work. However, this does not require giving unrestricted access to the geotagged photo itself to everyone in the government machinery.

This issue of data access is essential. As we note in section 6.2.1, zero-knowledge proofs should be a default mechanism available for obtaining assurance. For an auditor to ensure that the expenditure was indeed made for the purpose of PMAY and was released to a low-income household, they do not necessarily need to know the bank details and income levels of the person receiving the funds. Rather, the auditor could gain assurance that the beneficiary was indeed low-income without accessing data on the beneficiaries' actual income. Similarly, the auditor does not need to know the beneficiaries' bank account details—the auditor just needs to know that the money went to an account into which the beneficiary has authorized the government to deposit funds. In this world, once the bank account information has been entered during registration in a government scheme, only the citizen or someone they authorize to access that information should be able to see and modify that bank record. If the auditor or supervisor needs to conduct additional checks and obtain assurance that the geotagged photograph was indeed real, they should request specific access to the data related to the payment they are auditing. This access should be logged in a separate tamper-proof registry.

Keeping data protections in mind, the primary objective that data coupling of programmatic and fiscal events serves is connecting the inputs (i.e. public expenditure) with outputs (i.e. what was done with that expenditure) and outcomes (i.e. policy goals that were met). Granular data would allow policymaker and other stakeholders to observe nuances in the data, such as unexpected inputs and outputs and how to resolve such issues. At the aggregated data level, auditors or programme managers can then easily avoid the currently arduous procedures under the push-and-wait architecture explained earlier to obtain information on where such gaps exist.

To this end, we advocate the idea of 'PFM-driven observability', which ensures that fiscal events—anything to do with money, such as budget estimations, budget approvals, raising an invoice, approval of a payment, fund transfers, and payments—originate from and are coupled with programmatic events at the granular level, such as vaccines delivered, houses built, institutional deliveries co-ordinated, school toilets repaired, consulting services received, conditions for payments to vendors, employees, beneficiaries met, etc. This coupling should be observable to those implementing the programmes—for they would have this information anyway—but the assurance that this coupling exists should be available throughout the chain of accountability in government. Ensuring that stakeholders have observability of these programmatic events as well as their relationship with fiscal events is an essential yet missing part of producing credible insights and enabling accountability for public expenditure in India.

Structurally, this idea fits with the existing power structures and transactions within government, where Finance Departments impose conditions and rules on administrative departments and, in return, allocate budgets and release funds. The change we are advocating is therefore limited; the assurance and evidence expected for the approval and processing of these funds should be clear to anyone in charge of funds. In advocating for this approach of PFM-driven accountability of public service delivery, we take the view that transversal functions of the state such as public financial management can improve observability of inputs and outputs across the government ecosystem.

To envision the benefits of coupling, take the Mahatma Gandhi National Rural Employment Guarantee Scheme (MGNREGS), which provides work for low-income households across India. The scheme's underlying database, which is publicly available at nrega.dord.gov.in, couples the programmatic data of the work done (wells or canals dug, trees planted, etc.) with muster rolls of who did the work and payments they received. This information has allowed beneficiaries, civil society actors, and researchers to verify the authenticity of these programmatic and fiscal claims, contest inclusion or exclusion errors in the scheme, demand action when funds have been delayed, and obtain a nationwide understanding of the scheme and its impacts.

6.1.1.1 Enabling Coupling: Data Exchanges through Data Standards and Platforms

Coupling of programmatic and fiscal data, which may sit in databases housed in varying registries, platforms, and servers, requires data from varying sources to 'speak' to each other through data exchange capability. Data exchanges refer to the myriad ways in which data—both digital and manual—are shared between systems. The most ubiquitous forms of data exchanges exist in modern-day offices: sharing information on paper, through Excel files, over emails, or over e-Office platforms. In more complex renditions, the government uses IT systems to facilitate bilateral data exchange. For instance, the PFMS platform, a technological system used across government, allows data exchange from the Union to state governments and vice versa. Another approach is mandating actors in a system to use a single monolithic software imposed from the top down, onto which data must be entered directly. For instance, MGNREGS requires officials to use its website to enter programmatic events to trigger fiscal events.

Existing methods for data exchanges in India are often limited by the context, programmatic needs, and number of actors involved. In addition, they may not always be easy to scale or allow for asynchronous onboarding (where programmes are able to use platforms at their own time rather than being compelled to use them by an institution). When the number of events within a programme or the typology of expenditures expands, it may be difficult to extend information systems to these newer contexts—have 'extensibility'[1]—particularly in centralized or monolithic

[1] For information systems, Johansson and Löfgren (2009) define extensibility as 'the ability of a system to be extended with new functionality with minimal or no effects on its internal structure and data flow', and posit three design principles that make a system extensible: (a) modifiability, i.e. being able to change and offer newer capabilities with least number of changes to least number of elements, (b) maintainability, i.e. being able to change without risk of adding new errors, and (c) scalability, i.e. the ability of a system to expand in a chosen direction without major modifications to its architecture.

systems. Still, existing approaches encourage centralized or monolithic systems because decentralization can make data unexchangeable; if several actors capture and enter data into their own system in numerous ways, stakeholders may not be able to use these data. Public financial management systems, in their bid to co-ordinate and obtain information, therefore tend to move more towards centralization. Take, for instance, software systems such as Integrated Financial Management Information Systems (IFMISs) that are widely used in public financial management. These systems impose a uniform and contextually agnostic way for actors to manage and report expenditures, making it difficult for specific departments or individuals to visualize or analyse data as per their needs.

We believe that fiscal information management with the use of data exchanges can be (a) decoupled from, (b) orthogonal to, and (c) applicable to a variety of fiscal management practices—and be done without centralization.[2] With the right standards and protocols in place, multiple actors can share fiscal information amongst themselves without needing to work within a uniform system.[3]

This idea is based on a wider notion of DPI and 'platform thinking'. In an array of fields—from payments, to marketing, to trade and commerce—platform-oriented designs reduce friction in interactions between multiple actors (Choudary 2014; Girtu 2018; Kolossovski 2019; O'Reilly 2011; World Bank 2022). While platforms are defined in multiple ways, we think of them as foundations over which a broader number of complementary systems, software, or products can be built. For instance, the Android system as a platform provides a stable foundation above which others can add information and build applications. Other examples include Airbnb, Uber, GPS satellites, and most notably the internet. Closer to home, Aadhaar, UPI, and ONDC were all built with a similar design philosophy.

More recently, advocates have suggested that data exchanges be built as digital public infrastructure (DPI) (The Rockefeller Foundation 2021). The UN Development Programme (UNDP), in conjunction with the

[2] A more decentralized architecture, such as the ONDC architecture based on the Beckn protocol, is possible. Further research on this is needed.

[3] Information is required on government programmes—both within and outside the executive—to understand the state of any given programme, manage, collaborate, course-correct, and hold the actors responsible. It is vital for co-ordination in any of these forms.

Government of India (UNDP 2023), define DPI as having four characteristics, where: (a) the underlying infrastructure is interoperable; (b) it can be built on open technology standards; (c) it works without geographic or demographic restrictions; and (d) it has frameworks, rules, and regulations to prevent misuse. In fact, the TAGUP Report submitted to the Ministry of Finance in 2011 suggested that India build an Expenditure Information Network along similar lines (TAGUP 2011). Physical infrastructure, such as railroads, built through public investments, laid the foundation for private innovation and societal development. Similar systems for the twenty-first century are now built in the digital domain to simplify the flow of people, money, and information, collectively referred to as DPI.

Data exchanges for financial information through platforms can significantly improve the flow of information between, within, and outside of governments. Unlike websites designed to serve actors with specific and narrow needs for information, platforms can use data standards and protocols to exchange data and are by design able to produce a standardized data layer for consumption by multiple actors. A data exchange- and data standards-led approach, in contrast to a monolithic software-led approach or a website, is far more scalable, contextually adaptable, and open to multiple use cases for the same data.

Data standards are especially important to build coherent platforms for data exchange. We explain the value of data standards further below. To illustrate, consider the following lyrics of a Hindi film song.

> *Is Baat Ko Agar Tum, Zara Aur Saja Ke Kahte*
> *Zara Ghuma Phira Ke Kahte To Achcha Hota*
> (Had you told me this in an embellished way,
> In a more roundabout way, it would've been nice)
>
> - Romila Dutta (Preity Zinta) to Karan Gill (Hrithik Roshan). 'Agar Main Kahoon', written by Javed Akhtar in the film Lakshya (2004), directed by Farhan Akhtar.

In the song **'Agar Main Kahoon'**, Karan is trying to express his love to Romila, but the words miss the mark. Romila, through the song, tells Karan that it would have been better had Karan expressed his love in a particular way, in a different standard. The International Organization for Standardization (ISO) might very well be referring to Romila and

Karan's different love languages when it declares that 'great things happen when the world agrees' (ISO 2022).

Standards are all around us, shaping the way different needs are met in a variety of domains. For instance, when shipping container sizes became standardized, it significantly reduced loading and unloading cost at the ports, and an entire ecosystem of ships, trucks, and ports developed around it to make global shipping significantly cheaper than it was in the 1950s (Klose 2015; Levinson 2016; Postrel 2006).[4] Similarly, light bulbs connect to their sockets using the E27 or B22 standard. USB-C is an increasingly popular standard for power cables to connect to smartphones and laptops, allowing phones of different brands to be compatible with chargers of different brands. Standards are the reason it is possible to share a photograph taken on an Android phone to an iPhone. There are standards for food safety, energy transmission, health management, and information technology security. When standards are not duly followed—often and almost always—products become incompatible.

Standards serve a variety of purposes. Because they tend to be established through a consensus based on a collective process, they often meet stakeholder needs. They also protect the users of a system, particularly when those users have little to no choice in the matter; standards can ensure that the product or service delivers as claimed, performs as specified, and is reliable, durable, and safe (Standards Council of Canada 2025).

Open standards go a step further in their value-add to stakeholders. The International Telecommunications Union defines open standards as 'standards made available to the general public [that] are developed (or approved) and maintained via a collaborative and consensus driven process. Open Standards facilitate interoperability and data exchange amongst different products or services and are intended for widespread

[4] We argue that instead of looking at PFM as one of many vertical functions, public finance needs to be embedded throughout the life cycle of a government programme's management, from planning to execution to monitoring. This re-architecture is fundamentally different from point solutions that solve a part of the vast set of PFM problems, such as creating pooled bank accounts for certain agencies at subnational or local level to spend from, transferring money to smart card or mobile money for agencies to incur expenditure more quickly, or drafting a chart of accounts such that comparisons can be made across jurisdictions on the outlay on specific items of expenditure. The challenge with these point solutions is that these interventions may or may not be synergistic with the rest of the ecosystem. Any architecture that seeks to improve upon the current architecture must then attempt to reduce the cost, burden, and time it takes to produce as well as exchange information.

adoption' (ITU 2022). Open standards also support and foster innovation in their environment, leading to what designers call the hourglass architecture. The internet has an hourglass architecture, and uses TCP/IP and HTTP as core standards at the bridge of the hourglass (Akhshabi and Dovrolis 2011; Beck 2019; Georgia Institute of Technology 2011). In our context, an hourglass architecture through open standards means that there is a layer of data standards and exchange protocols in the middle. Ministries, departments, agencies, or even other stakeholders can build layers of services, data, and components below these data standards and exchange protocols in the middle, which is often called the neck of the hourglass. Similarly, institutions can build a number of user-facing services at the top—say, an analytics interface for programme managers to visualize expenditure or programmatic data, or a citizen-facing portal where they can determine the benefits they are eligible for.

The Government of India's policy on Open Standards for e-Governance recognizes the value and characteristics of these open standards. It elaborates on how open standards can be institutionalized within government, creates guidelines for selecting an open standard when multiple options are available, and provides a rationale and framework for the policy (Ministry of Communications and Information Technology 2010).[5]

Government systems are characterized by islands of legacy systems using heterogeneous platforms and technologies that are spread across diverse geographical locations, in varying states of automation. This makes cooperation, co-ordination, collaboration, and integration of information across different departments a hefty task. Open standards, on the other hand, support consistent, standardized, and reliable implementation of e-Governance solutions which meet the objectives of the policy.

In the pre-digital age, in fact, this is how governments have always functioned. There has constantly been a standard format for reporting expenditure or programmatic progress to the Union government, whatever

[5] Open standards serve a variety of benefits. As the National Urban Innovation Stack Digital Blueprint (Institute of Urban Affairs and Ministry of Housing and Urban Affairs 2021) notes, 'a standard is a collection of minimal requirements and definitions which helps solutions to co-exist and interoperate correctly with each other, speed up time to market, meet the necessary regulations/policies, help compare applicability of competing solutions, and allow solution providers the choice of technology'.

the process and general financial rules of incurring the expenditure in the state government might be. However, the curse of top-down mandates of big information technology projects is often that in ill-informed bids to enforce standards, these projects overreach into defining processes and designs beyond their mandate, or requiring the use of tools that do not prioritize data protection.

Take, for instance, the Ministry of Finance's Public Financial Management System (PFMS), which was created to give end-to-end visibility on fund use for Centrally Sponsored Schemes. PFMS is extremely suitable for use cases important to the Union-level Ministry of Finance—such as fund releases to state treasuries and tracking balances in societies that implement Centrally Sponsored Schemes. However, for the most part, it is a closed system that is not designed and incentivized to integrate payments with programmatic events, account for these events, or audit these payments. These tools are maintained by each spending agency separately, so integration with PFMS would allow for data observability for authorized stakeholders.

For instance, PFMS has offered the capabilities to compile detailed bank balances and bank statements of various government agencies across multiple banks into a single view for over a decade and a half. Unfortunately, it is not possible for state governments to use PFMS 'as a service' for just this one function without sharing other data on citizens in a state with PFMS. When data is shared with PFMS, it is inherently visible to the Union government, even for State Sponsored Schemes. Without a specific mandate to support states, teams in PFMS also do not have the incentive to develop, service, and maintain products and services for states. For states, then, using PFMS as a service for particular functions comes at a cost.[6] But knowing that state governments are eager to track government agency expenses in one place, the managers of PFMS have an apt opportunity to create the functionality in its own system, run it as a service with its own product team, and incentivize governments to use it with agreed standards and protocols—without the need for data to be stored with the Union government.

[6] PFMS could be offered as what technologists call Software as a Service (SaaS) with data protection and service standards in mind.

As noted earlier, most expenditure under Centrally Sponsored Schemes currently requires mandatory use of PFMS. Instead, had the CGA defined expenditure standards for reporting and allowed states to share data from their own PFMS-like systems, they could have obtained uniform expenditure data all the while lending states autonomy. Recognizing this challenge, in his letter to the President of India that we referred to earlier, the then CAG Rajiv Mehrishi advocated for data standards for fiscal reporting (Sabharwal and Phatak 2020).

What the world of PFM requires is a set of commonly agreed data standards for fiscal reporting and data exchanges to co-ordinate expenditure across tiers of government, functions of government, and various geographies. This would enable various tiers of government to implement digital solutions and co-ordinate with each other in ways that top-down software solutions do not allow, thereby empowering different government departments to use technologies and processes that meet their contextual and evolving needs. We detail how governments can implement such open standards in Chapter 8.

Ultimately, these data exchanges and data standards are both needed to co-ordinate fiscal and programmatic events. In order for data to speak to each other, two things must be true. One, data should be organized or reported in a uniform manner based on standards and, two, stakeholders should be allowed to capture data in a way that meets their needs. By creating standardized ways of reporting or managing data, various levels of governments or different policy managers can use decentralized data platforms that facilitate exchange through agreed-upon and open data standards. Coupling of fiscal and programmatic data thus becomes feasible across government regardless of the programme, geography, or context.

6.1.2 Principle 2: Co-ordinating Programmatic and Fiscal Events

A second important principle of PFM architecture is the codification of the relationships between programmatic and fiscal events where programmatic events can trigger fiscal events by design. In other words, fiscal events (transfer of funds and recording and accounting of expenditure) must take place where appropriate once the underlying programmatic events occur, and fiscal events are not undertaken without the programmatic events they are associated with taking place. This principle requires that decisions to undertake fiscal events flow directly from programmatic events as per standard procedures, such that there are no errors of omission.

One way to ensure that fiscal and programmatic events are inextricably linked is to use digital technologies to minimize discretion in making payments once a programmatic event occurs. A digital code, for instance, could ensure that as soon as proof of a programmatic event is uploaded and verified, it triggers the fiscal event—including payment, accounting, and data availability for auditing. This process could further include any manual checks that need to happen by managers. Of course, such digital technologies are not necessary for this architectural goal. In today's world, individuals use this same principle, but manually. They check that certain conditions are met and, if they are, they implement the rule that the payment should be met. However, they do so manually, which introduces a greater likelihood of delays and mistakes.

Take the routine task of road measurements as an example. Typically, junior engineers record the length of the work done in the measurement book. Assistant engineers verify this work (the programmatic event), and the payment is made to the contractor (the fiscal event). Ideally, the assistant engineer or their principals should not need to sign off on the payment. Their only concern should be ensuring that the junior engineer's measurement of the completed roadwork is accurate and acceptable for facilitating payment. Once this is established, the payment should be released in one of two ways. One, in a non-digital world, a third party within the government, like a manager or supervisor, can look at anonymized data to determine if the conditions to make a payment are met and, subsequently, makes the payment. This is akin to an escrow, but the payment is drawn from departmental budget funds held within the treasury. In a digital world, the payment is released as soon as the uploaded data is algorithmically verified, thereby triggering a payment—we call it a smart payment. It is comparable to what would happen in a digital escrow; we expand on smart payments in the next chapter.

6.1.2.1 Enabling Co-ordinating: Rules as Code

How can fiscal and programmatic events be co-ordinated such that payments go out automatically once proof of programmatic events is produced? Such co-ordination may seem intricate. As some have argued, policy compliance very much depends on the complexity of a given policy. Easy problems, or what Peters and Tarpey (2019) call 'tame', have clearly defined information which allows stakeholders to make quick policy decisions. Complex or 'wicked' problems, on the other hand, require stakeholders to make connections using information from various sources and, even then, may not always yield clear solutions (ibid.). However, it is important to recognize that all payment decisions are 'tame' problems, in the sense that the conditions under which payments are to be made are always clearly defined. This is true for both complex and simple policies and contexts.

We therefore propose implementing the co-ordination of fiscal and programmatic events through what is known as Rules as Code, or RaC. RaC is an approach for publishing policies, regulations, and legislation wherein a machine-consumable version of a rule complements and coexists with the natural language (or human-made) counterpart (Basu 2020; Observatory of Public Sector Innovation 2020).

To understand RaC, take a very simple example of a policy that provides community healthcare workers (ASHA) financial incentives for a variety of tasks. Some of these tasks are clear-cut and easy to implement, but even more complicated tasks fall within an RaC framework. For instance, ASHA workers receive performance-based incentives for ensuring that a child receives all its immunizations in her first year—a task which can be documented and is verifiable. A claim that an ASHA worker has helped 10 children receive their immunizations can either be self-reported, which is often the case, or could be data that a health facility provides. In either case, the ASHA facilitator—the manager who oversees, on average, 20 ASHA workers—would approve or reject the claim based on her supervision and verification. In instances where the credibility of data is suspect or the immunization is difficult to verify, the task of the ASHA facilitator is complex. For instance, a self-reported claim by an ASHA worker that she has given two drops of an oral polio vaccine to a child at an Anganwadi centre is not easy to verify. The facilitator thus needs to cross-check this information with parents, or use other forms of discretion to verify the payment eligibility—something that RaC cannot fix.

However, a PFM system can have RaC such that once a facilitator verifies that an ASHA worker has done the work, the payment amount reported in the state-level policy for the incentive for a specific task is automatically released to the ASHA worker. In having this RaC, the system deviates from the current one, wherein the process between an ASHA submitting her claim and the facilitator approving it is unobservable to anyone monitoring this interaction—for corruption, for instance. Even after the facilitator approves that the work is done, the block accountant or chief medical officer of the district still needs to approve the payment, adding additional delays and administrative burdens. The RaC can therefore streamline this process.

RaC have multiple benefits. For instance, RaC could clarify exactly what rules agencies or frontline workers can follow to obtain government funds, or which conditions can allow citizens to be eligible for payment through social sector programmes. Using RaC, Ministries of Finance and Integrated Finance Divisions can surpass their focus on transactional functions—such as processing of budgetary payments or exercising internal control—to their policy and regulatory functions such as setting fiscal policy rules or targets, ensuring that the legal framework for budgeting and public finance is respected, and setting the medium-term budget framework.[7]

Other governments have adopted similar reforms. The Better Rules programme (Fraser 2021) in New Zealand and France's early efforts that led to OpenFisca[8] are some of the early examples of this idea. For example, the New Zealand Government's SmartStart programme calculates eligibility criteria for a benefit and, subsequently, the benefit itself, using a digital rules engine (New Zealand Government 2022). Such RaC aid both programme implementers and supervisors. Government supervisors can ensure that rules for a policy are clear and, subsequently, ensure better policy enforcement. At the same time, RaC can unburden implementers from having to continuously interpret rules from different

[7] See Allen et al. (2016) for the classification of functions of Ministries of Finance.

[8] OpenFisca is a 'free, reusable, modular open source project . . . [which] allows modelling of legislation in code to improve the transparency of, and access to, the law' (Mohun and Roberts 2020, 40).

supervisors, who often prescribe the same rules but in different ways. RaC also offer non-government consumers of rules—citizens and vendors—consistency and credibility when engaging with the government.

While rules can be non-digital or machine-readable, machine-readable rules have distinct advantages. They can minimize errors and ensure that all rules are implemented in a uniform way with speed. Furthermore, they can assure beneficiaries and suppliers of government programmes that when agreed-upon actions are completed, they will be guaranteed payments. These two key benefits can reduce the gap between policy goals and their implementation, while expediting the state's execution capacity.

We recognize that proposing the use of information technology can generate strong responses ranging from tech-utopian to tech-phobic. We find the simplified binary of the two extremes unhelpful for our proposed work. These positions do not always demark which parts of a system lend themselves to the use of information technology, and for which use cases. Here, we argue that RaC are apt when policies or rules themselves are simple and clear-cut. However, more complex rules or policies ought to be left to the discretion of relevant stakeholders. In these cases, capturing exact data on policy implementation might serve to be more fruitful for lending policy insights.

Implementing RaC is not without its share of risks. There are three major types of risks with RaC. One, there are policy risks in which the policy that is coded is itself incorrect, leading to inexact payments. Two, there are translation risks, where the policy on which the code is based is correct, but the translation of the policy into code is not, leading to incorrect payments. Three, there are verification risks, meaning that the underlying policy and the translation of the policy into code are both accurate, but the data being used to determine the payment is not. Because the data capture is incorrect, it leads to wrongful payments.

While these risks are significant, they can also be mitigated. Policy risks can be addressed by reforming the policy itself; translation risks by reforming the code. Verification risks can be reduced by improving the data systems and data capture mechanics, so that the underlying data is accurate to enable RaC for payments. Lastly, a major concern may be that, without human discretion, codes may enable wrongful payments because of the absence of manual checks. However, even this risk is not independent of policymakers' desire to wrest control. RaC only acts as a force multiplier for existing rules in any state of the world. Often, policymakers will need to make judgement calls to encode rules and determine what should be left to the realm of discretion and discovery. Our argument is not about eliminating discretion from this equation; it is about calibrating discretion through an iterative process in a way that it aids programmatic objectives.

Overall, RaC can enable the coupling of programmatic and fiscal events for payments by ensuring that they are automatically made once the necessary proof of a rule's completion is available in a system. RaC ensure that rules can be understood and interpreted with consistency and efficiency. They can also allow government officers to finalize tasks that are repeatable and predictable with less administrative burden.

6.1.3 Principle 3: Making Programmatic and Fiscal Data Available on Demand

The third key principle is that information on programmatic and fiscal events at the primary unit of activity should be 'posted' in data systems and be made available on demand for those entitled to have access. Simply put, unit-level data should be observable, with data protection in mind. This principle requires that when programmatic and fiscal events occur, the information regarding them should first be reported or digitally captured in a database. This database should be available to those authorized to access this information. This approach differs from the current structure whereby, often, officials must request data from those with access. Instead of government officers furnishing the information when requested, the data should be made available regardless of requests.

This is a shift from the push-and-wait architecture (defined earlier) to an on-demand architecture. In fact, section 4 of the Right to Information Act requires *suo moto* declarations of particular pieces of fiscal and programmatic information.[9] However, most parts of government do not

[9] According to section 4.2 of the Right to Information Act, 'it shall be a constant endeavour of every public authority to take steps in accordance with the requirements of clause (b) of subsection (1) to provide as much information *suo motu* to the public at regular intervals through

proactively disclose information as required under the law. For one, making such information available requires a huge amount of administrative burden in already understaffed offices. Second, incentives to withhold information, which we discuss in Chapter 5, prevent people from rendering such data available—whether it is an executive engineer maintaining an information monopoly around how much sand was used to repair a road, or whether an ASHA supervisor is taking an executive decision on how many cases of institutional deliveries of a worker are approved for payment.

At first glance, it may seem that making unit-level data available on demand may not substantially shift administrative burden or change incentives. However, examining the root cause of these problems can offer a different perspective. The problem of administrative burden with providing data required by law stems from how information management systems function within implementing agencies. Various parts of the Union and state governments implement various information systems that do not connect to each other. Therefore, programme officers often have to search for data before they can provide it, leading to administrative burden. If unit-level data is available on demand—and through connected open systems, which we discuss in Chapter 8—this burden need not exist.

Similarly, the architectural shift can also reduce the incentives to withhold information. A prime example is the way MGNREGS is implemented, wherein the entire workflow is recorded and made digitally available—who each beneficiary is, the amount of work they do, the materials used—in order for payments to go through. These preconditions for capturing unit-level records prevent programme staff from withholding data, and by the function of the scheme's design, it is made available to all stakeholders, including data to the public. MGNREGS' workflow provides a prime example of how recording unit-level data and making it available on demand can function for other schemes. Indeed, a study on PayDash, a platform that aggregates data from the MGNREGS

various means of communications, including internet, so that the public have minimum resort to the use of this Act to obtain information'. Some of this information includes (a) the budget allocated to each of its agency, indicating the particulars of all plans, proposed expenditures, and reports on disbursements made; (b) the manner of execution of subsidy programmes, including the amounts allocated and the details of beneficiaries of such programmes.

workflow to track and process payments, reduced payment processing time for recipients by 24% and increased the likelihood that households participated in the programme (Dodge et al. 2025). The randomized control trial further found that PayDash improved payment processing by reducing the search cost for scheme managers to track frontline workers' performance; over 60% of scheme bureaucrats reported that they shared performance data with their subordinates—whether to reduce payment delays or to improve their performance (Dodge et al. 2025).

There are further advantages to making a similar function available across schemes. For instance, adopting the principle of making unit-level data available to authorized personnel on demand would enable the Union government or state governments to allocate and report on budgets based on actual expenditure, rather than on estimates from budgetary releases. When expenditure information is available in near to real-time, and not subject to friction of aggregating information across all arms of government, governments can better decide public finance needs and more efficiently allocate resources. In 2011, one of us, Santhosh Mathew, had written to Sumit Bose, the then Secretary of Expenditure in the Ministry of Finance, to request that expenditure reported in the Union Government portal for expenditure management (CPSMS, now PFMS) be made a part of the Union accounts tabled in the Parliament. With PFMS' increased capabilities and India's digital transformation, this is even more possible today than in 2011. A version of this has already begun in the form of Statement 4AA,[10] which is now a part of the Union Budget 2025–26 and shows data on fund balances under Single Nodal Agency (SNA) Accounts for Centrally Sponsored Schemes.

To summarize, PFM in India needs to enable three core architectural features: coupling programmatic and fiscal data through data exchange standards, co-ordinating programmatic and fiscal events through RaC, and making unit-level data available on demand to those with authorizations. Through these shifts, other layers of reforms can be built atop, some of which we detail in Chapter 7.

[10] Form 4AA is titled 'Statement showing Fund balances under Single Nodal Agency (SNA) Accounts with State/eligible UT Governments in respect of select Centrally Sponsored Schemes'.

6.2 System Properties of the Architecture

The chapter thus far first builds on architectural shifts that are important for an on-demand architecture: coupling programmatic and fiscal data, co-ordinating programmatic and fiscal events, and making on-demand unit-level programmatic and fiscal data available. Using these principles, section 6.2 explains the fundamental system properties that can enable these changes: data protections, data liquidity, the once-only principle for data, and data credibility.

6.2.1 Data Protection by Design

A necessary property of PFM systems that capture and share fiscal information is data protection built by design. In fact, we contend that without data protections, our ideas should not be implemented. Data protections are distinct from data privacy; data privacy refers to the legal and regulatory arrangements that govern who has access to data, whereas data protections focus on confirming that systems apply these arrangements to ensure compliance with data privacy. Currently, there is ongoing debate about the ways in which data privacy should be protected—not only with fiscal data but other forms of data more broadly. Given that data privacy is a legal matter, we instead turn to the importance of data protections, which are an institutional feature.

In any on-demand architecture or government programmes more broadly, users should have credible assurances of data protection. To understand why, we can turn to James Scott's argument in *Seeing Like a State* in 1998, which contends that governments create legibility of citizens through means such as assigning surnames or creating standardized measurements in order to exercise control of populations. State controls enable governments to both surveil and manage citizens or their outputs. For instance, revenue collection through taxes requires governments to know who citizens are, their residence, and their occupation. Today, such visibility may be considered acceptable in societies where states have predictable intentions, and where citizens have trust or assurance that governments having visibility to such information will not yield unexpected consequences. However, given no government can guarantee such outcomes in the future, as parties or government powers change, legibility or visibility can also become an authoritarian tool for surveillance or for creating intentional state incapacity (Levitsky and Ziblatt 2019; Suryanarayan and White 2021). For citizens to buy into systems of information management which enable governments to create individual-level databases and for governments to ensure that systems are not used in unintended ways, systems should have data protections as a core and necessary property to ensure that data is used in an authorized and protected manner. Such data protections can take two routes.

One, those whose data is recorded should know how their data is being used. Take, for instance, one of the many investments that the Union and state governments are making in digital tools for PFM—the ability of a core banking server to ping a commercial bank with an account number to obtain the name of the bank account holder. This feature is typically used by governments as an additional check both for intergovernmental transfers and for transfers to recipients such as vendors and beneficiaries. However, recipients are not alerted when government parties use this facility to procure their details.

However, creation of a system that overcomes this limitation is possible, as has been undertaken in Estonia. The country's citizens enrol in nearly 99% of services online. With Estonia's Personal Data Protection Act in place, each citizen can consent to the use of their personal data in the digital space using a digital signature and an encrypted message via phone or email (Priisalu and Ottis 2017). When anyone seeks out a person's record on the government system, the system alerts that the data has been accessed. If a person does not consent to data sharing with the individual seeing it, they can take legal action. Such data protections are not only ideal but necessary because, ultimately, an individual's data ought to belong to that individual.

Two, any given government feature that exchanges data should undergo a third-party certification process that ensures that only authorized individuals can see particular pieces of data, and that it is stored in a secure manner. While we leave the details of these functions to cryptographers who have developed tools to ensure data protection, we do note the importance of this tool. It is useful to prevent hacking; furthermore, even within the government, only individuals who need to see data to do their work can access it—and that too, only government personnel for whom the data is necessary.

Overall, these two types of data protections—an individual's right to consent to the use of their data and measures to ensure that data storage is secure with adequate access protocols—are even more important in a country whose digital literacy landscape is still evolving. We recognize that this requires governments and civil society actors to undertake additional layers of security. For instance, individuals need to be made aware of what data points are being collected and how they are used, especially in areas where citizens only interact with technology in simple forms such as phones or televisions. Data protection structures therefore need immense work to ensure technological reforms are enacted well.

Data protections, and the two aforementioned ideas of limited access to data and citizen consent for sharing data digitally, are especially important for the architecture and principles we discuss, which argue for unit-level records of why funds are utilized and by whom. Unit-level data often includes personally identifiable data—such as names, contact information, demographic data, or government-issued identification numbers—as well as sensitive data. Take, for example, the case of a citizen who goes to a public hospital to be treated for HIV using their entitlement for free healthcare through the Ayushman Bharat Pradhan Mantri Jan Arogya Yojana (PM-JAY). The digital system of a public hospital, in our version of the architecture, may record their PM-JAY number (ABHA ID), their name and contact details, their underlying health history, and the reason for their visit. While this unit-level data may be available, the health history and a citizen's reason for obtaining healthcare are both sensitive and private, and ought not to be shared beyond the administrator and healthcare provider making the entry and the immediate care team of the patient. Similarly, though the system may have a record of the PM-JAY number to obtain the requisite funds for treatment, there would be no need for anyone within the hospital or the administration of the PM-JAY scheme to see the underlying socio-economic caste data to process payments, through which the scheme ensures eligibility. While we detail why unit-level data is useful elsewhere in this chapter, we explain here how data protections could be created to ensure both of these conditions are true in four points.

One, each team in charge of creating a digital platform akin to the ones we propose would need to categorize precisely what data points are personally identifiable and sensitive, and who in the bureaucratic hierarchy needs access to the data. In our example, both types of data will be useful for the immediate care team for the patient, and for the immediate individual in the State Health Agency or a pre-approved third-party provider who is in charge of approving the pre-authorization request submitted by the hospital before they start the treatment. At most, this approving individual's supervisor may need access to the same data. However, if anyone else beyond these two authorities require either personally identifiable data or sensitive information, they would need to obtain authorizations to access the data.

Two, because both personally identifiable information and sensitive information are available at the unit level, information can be selectively given to authorized individuals without revealing the identity of a person. This can especially be possible through zero-knowledge proofs, which is a way of revealing that information is true without showing the underlying information itself. This concept is already used in government. Public officials often need to verify that eligible beneficiaries of certain schemes have incomes below the poverty line but can do so if individuals have a Below Poverty Line (BPL) Ration Card. This card authenticates that the beneficiary meets the criteria without revealing their exact income.[11] Similarly, in our example, the PM-JAY card, along with the biometric authentication of the patient when they enter (if they have an Aadhaar card), would verify that the individual is both the person they claim to be and that they meet socio-economic criteria, without giving anyone information about their biometric data, their income level, their

[11] Murtagh (2023) explains how zero-knowledge proofs work using the 'Where's Waldo' analogy: 'But I'm a *Where's Waldo?* purist, and I don't want to spoil your chance to find him for yourself. Is there a way that I could prove to you that I found Waldo without you learning where he is? … Here's how a zero-knowledge proof can help with my Waldo dilemma: I take an opaque cloth that's much larger than the book, and I cover the page with it. I shift the book around under the cloth to a random location so you lose track of where it is. Then, I cut a small silhouette of Waldo's face in the cloth so that you can see him (and only him) peeking through. You're now convinced that I had indeed found Waldo, but that's all you've learned. You already knew Waldo was somewhere on the page, and you have no clue what part of the book you're looking at, so you've merely learned the truth of my claim—I know where Waldo is hiding.' We are grateful to Kanwaljit Singh for introducing us to this idea.

caste, or any other information. The idea of zero-knowledge proofs can therefore be a central way for selectively and safely providing data in digital systems.

Three, and as we state earlier, unit-level data is risky because unauthorized persons can utilize this information to use private and sensitive data for electoral benefits. This is especially true with digital systems, which reduce the search cost for obtaining data. An incumbent party, for instance, wouldn't need to ask party workers in villages to find out who obtained what schemes if they are in charge of the state apparatus like their competitors would, and thus have an information monopoly. Two types of legal safeguards are thus important: determining for how long personally identifiable and/or sensitive data is stored in digital systems, and citizens' right to privacy that allows them to consent to who can obtain their information, similar to Estonia's Personal Data Protection Act. In the Indian case, Digital Personal Data Protection Rules 2025 under the 2023 Digital Personal Data Protection Act (DPDPA) govern digital personal data.

Four, personally identifiable and sensitive data should automatically be anonymized beyond a certain level of government, beyond which authorized bureaucrats may not need to know such information. For instance, in our example, beyond the State Health Agency official and their supervisor (or similar-level individuals in the authorized third-party administrators), data should only be available in an anonymized and aggregated manner. In this case, authorized state-level PM-JAY officials may be able to see how many beneficiaries participated in each gram panchayat and how much funding was utilized, but not necessarily for what conditions or by whom. Such automated aggregation and anonymization, along with the decentralized nature of how data sits in our reform model, can ensure that data protections are robust enough to prevent state capture of information for non-necessary means.

As this section lays out, both data protections and data privacy measures are important to ensure that any data that is collected and/or stored is used ethically and with consent of the citizen. There are multiple ways to make data protection baked into the design of digital systems. While we leave the details to experts in the field, notable scholarship has argued that measures such as zero-knowledge proofs, differential privacy, and secure enclaves can be used to share just enough data with stakeholders without compromising data protection.[12]

Additionally, it is important to note that while we discuss data protections in relation to individual rights, the same concepts apply to governments. For instance, a state government may be hesitant to use the Union-level PFMS to manage its state finances because, in doing so, it risks having to share its data with other levels of government. In principle, the same data protections afforded to individuals should be expanded to organizational units, so long as they do not infringe on individual rights to data protections.

When a digital system is enacted without data protections in place, citizens face two risks. One, when they interact with a government that enforces a digital system, they risk privacy violations that may lead to unauthorized parties obtaining their information. Two, if they choose to remove themselves from such a system in the absence of data protections, they risk exclusion from public services altogether. Data protections are therefore necessary and may need to be gradually improved with changing needs to avert both risks.

6.2.2 Data Liquidity

The second property that is essential for an architectural change is data liquidity. We define data liquidity by borrowing from Wixom and Piccoli (2021) as the ability of an organization to 'transform data such that it can be reused and recombined for new value creation'. They further note that while data is inherently reusable or recombinable, not all organizations capture or store data so it is 'accurate, complete, current, standardized, searchable, and understandable'. The challenge with public finance data in India is that it stays in siloes in highly illiquid forms that are very context-specific, rendering them unusable in other contexts or in real-time.

[12] The issue of data privacy has particularly been an important one at the time of our writing this book, when the Department of Government Efficiency, an independent initiative of the Trump administration in the U.S., has obtained access to data that was previously only observable to authorized personnel within government.

To be sure, data liquidity need not imply that data must be captured, stored, or analysed digitally. There are indeed ways to ensure that it is captured in a standardized, manual way, and analysed manually as well (or have a situation where either the data capture or analysis are digital while the other is manual). In today's world, though, manual data capture requires incredible co-ordination across offices to ensure accuracy and replicability. Any changes must similarly be communicated across a vast number of actors. As anyone who has ever undertaken manual data collection (such as paper-based surveys) can testify, the process is laborious and time-consuming, with added compliance costs. We suggest digital data capture systems because they help reduce these time, compliance, and co-ordination costs.

We argue that India needs liquid and granular fiscal data at the level of each transaction that captures a variety of information—how much money has been allocated and spent, to whom it has been sent, for what purpose, and when. This data ideally ought to be captured at the individual or unit level, at the site of the spending. As we detail later in section 6.2.4, data liquidity can indeed be undertaken without creating a large database that may violate data privacy, or even with principles that fundamentally protect citizens' data securely. Yet, with data liquidity, when a government officer requires a specific piece of information that they are authorized to access for a specific purpose, temporarily and securely providing that data may drastically increase accountability and improve programme management.

In the last two decades, the Government of India has made advancements in the digitization of revenue collection and, subsequently, revenue data. Tax collection, which used to be a manual process, has been seamlessly integrated into a digitized system that allows the government to collect, monitor, and account for direct and indirect tax collection in real-time. Three key parts of the reform are notable: (a) the Tax Information Network (TIN) for nationwide income tax collection, monitoring, and accounting; (b) the Indian Customs EDI System (ICES) for customs management[13]; and (c) the Goods and Services Tax Network

[13] We do not discuss the ICES at length here. However, along with the TIN and the GSTN, it has been a successful measure of digitizing revenues through customs. Refer to CAG, 2023, *Report No. 14 of 2023 - Union Government (Department of Revenue-Customs).* Available at

(GSTN). While the last of these, the GSTN, has received much attention, the TIN and ICES are important predecessors in their own right, having vastly improved over the last decade.

Let us take the TIN and its use of data liquidity. The government uses the TIN platform to collect, monitor, and audit income tax collection. It has also digitized a variety of services which are available at physical centres. Today, individuals can pay income tax by applying for a PAN (Permanent Account Number), which provides taxpayers with a unique ID number. Once the individual receives their PAN, employers can automatically deduct taxable income and make electronic payments to the government. PAN holders can also use the TIN to pay taxes online and verify the status of their payment or returns. Moreover, the platform allows entities that deduct taxes at the source to apply for TAN (Tax Deduction and Collection Account Number) and obtain quarterly statements of the payments initialized through their banks. Lastly, the platform allows authorized personnel to serve as intermediaries to taxpayers and make payments or review that a PAN holder's details are accurate.[14]

Through the TIN, the Income Tax Department of India has a full view of revenues accrued through direct taxes. Authorized personnel can view itemized records of who paid direct taxes and when; they can confirm whether this information is accurate. While the current process of auditing taxes is manual, the Department is in the process of setting up an automated auditing system that can be manually back-checked for errors. Through machine learning, software can flag the various cases of fraud or mistakes in payments, show possible recourse methods, and

https://cag.gov.in/webroot/uploads/download_audit_report/2023/AR-14-OF-23-IT-Audit-of-ICES-1.5-Eng-dig-sign-064d228f58dbb76.03655566.pdf.

[14] This is a significant change from the way citizens experienced the income tax payment process in the past. Taxpayers frequented physical offices to file paperwork and there was a lack of transparency on the issuance of tax refunds. Digitization has not yet solved these issues completely. In 2019, tax refunds equalling up to 10% of the 2017–18 tax collections had been held back, some for up to four years (Sharma 2019). The government recognized that these problems ultimately hurt revenue collection. According to the Income Tax Act, the Income Tax Department has to pay a 6% interest on delayed tax refund after 90 days of an appeal being filed. To overcome these lags, the Union Cabinet approved reforms which will ensure that refund processing times decrease from an average of 63 days to just 24 hours (Waghmare 2019). This commitment seems increasingly feasible and would signal a transformation in how citizens experience a government service through digital centres. What used to be a complex and uncertain system has transformed into one that citizens, businesses, tax officials, and auditors can use for their own needs and that too through a single service point.

alert the right authority to resolve a case. Such efficiency will speed up an otherwise laborious task with similar or even greater accuracy and improve the revenue of the Government of India.

Following the digitization of income (direct) taxes, the Government of India took on the consolidation of various indirect taxes into a single tax called the Goods and Services Tax (GST).[15] The GSTN (Goods and Services Tax Network)—a collaboration between the central government, state governments, and financial institutions—commenced with the digitization of the payment and monitoring process for the GST.[16] As of 2024, the GSTN has not led to an increase in revenue collection that was expected by some analysts, but here we discuss the implications of the reform for data liquidity (Agarwal et al. 2024). The creation of such a network signalled a significant shift in the government's ability to take on complex IT projects. The GSTN not only integrated a variety of indirect taxes into one uniform taxing system but did so in collaboration with states. Rather than maintaining separate protocols for the Central GST, State GST, and Integrated GST, the GSTN integrates all payments into one system. As with the TIN system, taxpayers have the option of making payments online or through physical counters. Regardless of how the payment is made, taxpayers can view their payment summaries on the GSTN portal. Similarly, authorized tax officials and auditors can view line-item as well as aggregated data to conduct checks.

The TIN and GSTN have transformed how data from revenue platforms is used. One of the biggest accomplishments of these reforms has been that revenue data is now captured at the unit level and all aggregation upwards is automated. This feature has reduced administrative burden and improved internal transparency. At any point, authorized officials from the revenue department can see real-time revenue; however, data does not directly synchronize with the budget or expenditure data. As a result, the government officials creating the annual budget—the key

[15] For a history of the GST and GSTN, please refer to the GST Council's website: http://www.gstcouncil.gov.in/brief-history-gst.

[16] In September 2018, the government proposed a change in the ownership structure such that the GSTN would become a government-owned entity. At the time of writing, this change had yet to take place. For more information, see: https://economictimes.indiatimes.com/news/economy/policy/cabinet-clears-proposal-to-convert-gstn-into-government-entity/articleshow/65965553.cms?from = mdr.

document that guides the following year's policies—in the Budget Office at the Ministry of Finance have revenue data at the time of budget creation or in periodic intervals when they call for it; however, they do not have any data on a real-time basis on expenditure over the financial year.

We argue that transparency gains from the revenue side of fiscal events need to be replicated on the expenditure side. In order to ensure that revenue and expenditure data available to relevant stakeholders (in the revenue, budget, expenditure, and auditing offices) are both accurate and up-to-date, it is vital that the revenue data from various sources such as the TIN, ICES, and GSTN as well as expenditure data from all departments are integrated and streamlined into a single budget platform. The goal is to integrate real-time revenue information with real-time expenditure information, set budget allocations for departments, and have the budget reconcile against real-time revenues and expenditures.

We propose that data liquidity be instituted by design, allowing integration and sharing of real-time revenue and expenditure data across all stakeholders, including those in the revenue and expenditure departments in the Ministry of Finance. With these advancements, governments in India can have far more observability and predictability in managing their often-scarce resources. With these two functions, they can also reduce float, ease administrative burden, improve accountability, and increase internal and external transparency.

To understand what data liquidity can look like, take the instance of a fiscal data exchange platform which the government can create. Such a platform can act as a hub for fiscal information exchange across multiple systems in government or with stakeholders, specifically citizens and auditors, such as exchange of fiscal information between PFMS and programme-specific IT platforms. This fiscal data exchange platform can perform three goals: (a) link real-time revenue data with budget allocation and expenditure data for authorized government officials and auditors to monitor, as well as with citizens; (b) allow the Budget Office at the Ministry of Finance to digitally set budget allocation and authorization limits for all programmatic funds; and (c) upon expenditure from a specific allocation, update remaining authorizations for each programme in real-time, informing frontline officials of how much money is left and how it can be used. Such data sharing can also enable the platform to undertake predictive budgeting analysis to foresee how the money may be spent, assisting in future planning against intended goals.

With such uses, data liquidity provides various advantages for budget and expenditure management. For one, it offers a holistic overview of the country's finances with regards to revenue and expenditures, ensuring that at any given time authorized officials have an overview of when, where, and how much money the government is bringing in or spending. This is an essential function for the annual budget-making process. Much like how businesses can adequately strategize for the future only when they have a clear overview of their finances, the government can only plan well when it knows where it stands.

Data liquidity, in many circumstances, can ensure that all revenue and expenditure data coming in is accurate. This functionality would reduce both the redundancy of data, i.e. duplication across various locations in different ways (digital, manual, etc.), and ensure that errors are minimized. Additionally, the budget allocation function can establish that departments spend time leading up to the budget deadline by strategizing for programmatic success in a financially sound manner rather than calculating numbers and routing information upwards. Data liquidity would also bring changes to expenditure management. This system will dramatically reduce the administrative burdens on public officials to constantly validate and update budgets and fund authorizations. Instead, they can focus on creating evidence-based implementation plans.

A major concern with data liquidity is that it forces agencies—especially at state and local levels—to share data in standardized ways. Required reporting already exists for Central Schemes and Centrally Sponsored Schemes funded wholly or in part by the Union government. However, to guarantee autonomy for states over data sharing for State Sponsored Schemes and state-level non-scheme payments, reporting can be limited to basic expenditure and authorization data—without budget heads—to a central repository. This kind of data sharing will ensure that there is a national picture of the number of scheme and non-scheme programmes, their corresponding authorizations, and their real-time expenditure. While ideally these numbers should be at the granular level, we recognize that this may infringe on states' autonomy. In many cases, then, states may opt to report such data in an aggregate manner that does not promote data liquidity.

While we propose these data collection exercises with both central union and state schemes in mind, it should be noted that they have the potential to benefit local policy implementers, auditors, and citizens as well. Data liquidity inherently makes data usable for different contexts at large.

For policy implementers, data can be used in a variety of ways. One example of the use of data liquidity is predictive budgeting to show policymakers the expected expenditures for upcoming financial years based on past expenditure and goals. If a programme officer today wants to allocate funds for the entire financial year, they must look through separate files, often non-digital, to calculate what was spent for each budget code and the amount remaining. In addition, they must determine future expenditure needs and plan for them. Given this complexity, overburdened administrators rarely take on such data analysis. Rather, agencies often request a rush of expenditure in the last quarter or the last month of the financial year. Data liquidity can enable programmatic systems to use granular-level data for past programmatic and fiscal events to understand future needs.

Data liquidity, if implemented well, can provide a holistic view of policy programming for implementers. It can improve the transparency of the data for both programmatic and fiscal needs. Further, it can be recombined or reused to enable synchroneity across both PFM and programmatic platforms. Without liquid data, synchroneity is possible but limited; data within and across platforms that have different functions may not be captured in a way that can be matched. When data is granular and liquid—and meets open data standards—synchroneity becomes possible, and it can be combined in new ways to assist programme implementers.

Data liquidity also has benefits for auditors. If expenditure and revenue data is liquid, it can be integrated with data from the Comptroller and Auditor General (CAG) office. This is akin to a reformed version of the existing Voucher Level Computerization (VLC) database of State Accountant Generals, which also allows for remote auditing. The VLC database currently sits as a stand-alone system in many states, yet to be integrated with state-level payment systems. Once reformed, data liquidity can facilitate departments to integrate relevant revenue and expenditure data with central and state-level data to provide to auditors. It can also allow policymakers to link revenue or expenditure data with scheme-specific social audit data to examine the effect of outputs on outcomes.

To envision what this looks like, imagine an auditing platform that compiles and sends voucher-level expenditure data to an auditor. This functionality can allow auditors to digitally determine the accuracy of receipts and expenditure against eligibility rules and payment codes soon after they are made. With such visibility, they can take swift action as needed to curb corruption or verify inconsistencies. Moreover, based on big data analysis and machine learning, the auditing platform can automatically identify, isolate, recognize, and flag potentially fraudulent or risky payments. This will ease the task of auditing.

Once payments are verified after auditing, the data can also be used by officers to automate accounting reports. One such core report can be the generation of an electronic Utilization Certificate (eUC)—an electronic or non-electronic version of which is currently required to legally account for funds received from the government. Rather than requiring officers to manually compile UCs, data liquidity can allow officers to automate the process. Once a compilation of vouchers reaches a particular proportion of the authorized budget, a software can generate an eUC in a prescribed format, containing the complete details of each voucher. Such generation, amongst the creation of other automated reports, can create error-free accounts with comparative ease.

While we provide such examples of data use for expenditure management, programme managers, and auditors and accountants, these functions are only possible with data liquidity. They require liquid and granular data that can be compiled and used in a variety of ways for different functions. It also, of course, requires that the data be accurate, a problem we discuss in section 6.2.4.[17]

6.2.3 The Once-only Principle

Earlier, we noted the importance of data liquidity, or the idea that data can be reused and recombined for new value creation. In India, to ensure this for fiscal event data—such as budget estimates, approved allocations,

[17] A simple canon to make this plan operational is to adopt a portfolio view of digital transformation that puts interoperability at the center, so it can meet multiple digital needs. For an example of how this is done, see Hatcher-Mbu et al. (2024).

received invoices, authorized payments, or released payments—we advocate the use of the once-only principle in government (OOP). Indeed, it is possible to extend OOP to programmatic events.

The idea of once-only principle has been gaining ground for a few years now. The goal of the principle is to reduce the administrative burden for both users and governments (Cave et al. 2017; Rashid 2020). We believe that the reduced burden for citizens has been far more of a focus in the conversations around OOP and rightly so. However, OOP also offers the opportunity to significantly reduce administrative burdens for government officials and improve service delivery.

To understand this better, take an incentive payment to a frontline health worker—for ensuring the delivery of a newborn in a health facility as opposed to delivery at home—as an example. The uniquely identified action of securing the institutional delivery would need to be entered in an IT platform. It would refer to the mother who delivered the newborn, the facility where the birth took place, the health worker who helped with the birth, and if relevant for any programmatic purpose, even the medical worker who helped deliver the newborn. Once this entry is made, a digital workflow system would calculate the amount of payment to be made to the health worker based on a predefined formula. In the same digital workflow, the health worker's supervisor would record her agreement or lack thereof to the data reported by the health worker. Based on a supervisor's approval, an authorized officer can decide to send a payment request from the digital workflow system; or else, ideally, the system would trigger an automated payment release without this additional check as soon as the direct supervisor approves the work. As more entries are made for this particular healthcare worker, the total number of institutional deliveries she made would be aggregated and an output report created that can be used for multiple ends—be it to appraise the worker's performance or create reports for state and national governments. The same digital workflow would feed this data to oversight bodies such as auditors. Since the payment is being made through digital budget and payment engines, these transactions would form the basis for financial reports.

In the alternate system that exists today, the health facility records an institutional delivery in its own system. The health (ASHA) worker reports the institutional delivery to the ASHA supervisor separately, independent of the ASHA facility records. The payment to the worker goes through a process that does not necessarily depend on the record that the ASHA supervisor maintains for the work done by an ASHA worker. When the payment is processed, there is no feedback of this payment data to the ASHA supervisor.

As this example shows, OOP and the data generated from the single point of entry provide clear benefits. Such data need only be entered once, but it can be machine aggregated to fit the various reporting and monitoring needs of stakeholders. This ensures that frontline workers and government administrators can focus on programme implementation instead of spending significant time on producing multiple reports, since all the reporting is done based on a single source of data point entered in the primary unit of the transaction.

It is important to note a nuance here. We argue that the data for every programmatic or fiscal event should only be entered once. If a payment is made for, say, *n* number of institutional deliveries facilitated by an ASHA worker, ideally, *n* programmatic events should be made in the system, and only once per event. The corresponding payment, a fiscal event, should be coupled to the output in that it should be clear how much payment is corresponding to each of the *n* number of institutional deliveries. When the state health minister asks for the total payment made to facilitate institutional deliveries, the aggregate numbers should be automatically calculated from the granular data that was entered. Otherwise, if aggregate data are calculated separately, then they risk creating an alternate 'truth'. Essentially, the once-only principle needs to be implemented in a way that granular data entered only once generate the aggregate forms of reporting.

Implementing the once-only principle would benefit from agreements on open data and open standards in government systems,[18] the usage of which allows for something far more fundamental to also be possible.

[18] Examples of using once-only include the following: (a) State Integrated Financial Management System (IFMS) publishes application programming interfaces (APIs) so that digital workflow management systems of various programmes can connect to the Centralized Funds Management System (CFMS) for fund transfer orders and reverse information on status of payments; (b) State IFMS develops capabilities like account validation from bank CBS (core banking solutions) systems, and Aadhaar-based payments, or it uses the facility provided by PFMS if offered as SaaS.

With this endoskeleton of interoperable systems and open data standards in place—in contrast to hard-coded exoskeletons of centralized software—it becomes possible to create far more software applications and solutions to couple programmatic and fiscal events, from which valuable insights can be derived. Going back to hourglass architecture, OOP enables backward and forward integrations. For instance, district hospitals can better appraise healthcare workers and how well they facilitate institutional delivery, allowing them to identify which workers need more support. At the same time, it allows policymakers to understand outcomes which, when combined with qualitative data, can help highlight the models of behaviour and support to healthcare workers that have higher impact on reducing maternal mortality in a particular geography.

6.2.4 Improving Data Credibility at Source

Our focus on data liquidity and OOP prompts a more fundamental question: There is so much administrative data already that governments don't use. Why focus on generating more data? A key concern with existing data is that there is little trust in its credibility, making it difficult for governments or others to use it functionally. To expand, we draw on literature which looks at trust as a key mediating variable between information quality and information usage (Corritore et al. 2003; Denize and Young 2007; Fisman and Khanna 1999; Kelton et al. 2008; Lucassen and Schraagen 2011; Marsh and Dibben 2003; Tomkins 2001).[19] Scholars have recognized the strong need for theoretical and empirical research on trust within the field of information science, and the subject continues to develop.

Building on this literature, we focus our attention on the semantics of information; data is credible when information is accurate, complete, and neutral (Lucassen and Schraagen 2011).[20] In order to understand

[19] Denize and Young (2007) find that information exchange norms that emerge as part of the long-term co-production of the relationship have the greatest positive association with the level of trust. They posit that trust can be manufactured by managing communication.

[20] Lucassen and Schraagen (2011) offer a 3S model to judge trust in data—semantics, surface, and source features of information. Semantic features relate to content, e.g. whether the

how to improve data credibility, let us first distinguish between two types of data: transactional data and master data. Transactional data in government is data that varies by events and relates primarily to 'verbs' of public service delivery—programmatic and fiscal events such as a benefit paid, a registration request received, or an inoculation given. Conversely, master data is slower moving, and relates primarily to the nouns involved in government service delivery—people, organizations, properties, health facilities, equipment, etc.

In a system where master data is not standardly recorded, different individuals entering the data can end up referring to the same 'noun' in varying ways. In a ubiquitous example within government, names of villages are frequently spelled differently across databases. Alternatively, an individual may refer to a village distinctly from another, leaving out a location altogether from an event.

One way to increase the credibility of master data is by creating recognized and accredited registries with unique identification for these nouns.[21] Such registries can take forms such as (a) a database or list of beneficiaries usable by multiple government programmes, (b) a list of government employees, teachers, ASHA and Anganwadi workers, or other users of digital workflow management systems, (c) a list of institutions like primary healthcare centres, hospitals, schools, etc., (d) a list of organizations like District Rural Development Agencies and district health societies, (e) a list of properties or public infrastructure like roads, bridges, buildings, etc. To resolve the issue of village names differing or missing across databases, for instance, the Local Government Directory standardizes the names of these areas across the country, mapped to standardized sub-districts, districts, and state names.

information is accurate, complete, and neutral. For instance, a data table with half the fields missing and others with spelling errors might have less credibility than a table without these features. Surface features deal with presentation, e.g. length of the information, corresponding references and pictures, the writing style, etc. For instance, references in an academic paper would increase its credibility, so long as these references are sound. Source features are about properties like authority or credibility of the source from which information is received. For instance, one may trust a video forwarded by a known contact to their messaging app over one from an unknown number.

[21] This is also referred to as differentiation data, since these data concisely differentiate the particular item being processed from all other items (W3 Computing 2014).

While registries improve credibility of master data, improving the accuracy of transactional or fast-moving data requires different solutions. One approach is to capture data when events occur. As opposed to someone reporting an event occurred, data is captured as part of the process. Take, for instance, the example of a newborn's institutional delivery in a hospital, which is facilitated by an ASHA worker. It is possible to have a system where the woman going to a hospital for the delivery is registered in the hospital's system. Alongside this data, the registry also captures who brought her there, namely the ASHA worker. Rather than the worker needing to report that she took the mother to the hospital, the data is captured in the process.

Where data cannot be captured during an event, another established approach amongst data scientists and researchers to improve data accuracy and completeness is to reduce the distance between an event and its reporting.[22] The credibility of transactional data depends on the distance between the event and the data capture of that event, particularly for fiscal and programmatic events. This distance can take many forms. One is the time-lapse between the event and the data capture. The longer an event remains unrecorded, the higher the probability of the inaccuracy of it being logged due to recall bias. This is well-understood for field surveys, where recall period mediates the accuracy of responses. A second form is the physical distance itself. Data captured remotely by someone—mediated by one or more agents who are closer to the programmatic or fiscal event—is prone to errors of omission and commission, either bona fide or mala fide. The ideal way to improve data credibility in these situations where transactional data cannot be recorded in the process of the event is to ensure that data is reported directly in the shortest time frame possible, soon after an event occurs.

A third way to address data credibility for transactional data is to ensure that self-reported data can be triangulated and checked. For instance, rather than asking an ASHA worker to report whether she helped an individual with an institutional delivery, supervisors can ascertain information by also requesting the data on when, where, and why the event

[22] For a synopsis on creating credible measurement systems, see chapter 4 in Karthik Muralidharan's 2024 book, *Accelerating India's Development: A State-Led Roadmap for Effective Governance.*

occurred. It also allows auditors to conduct accurate back-checks. This can be done using a variety of methods, including quick phone or SMS surveys, or through physical verification.

These ideas are not new in government. NITI Aayog, for example, has been using some of these concepts for some time. The Development Monitoring and Evaluation Office,[23] an attached office of NITI Aayog, has been ranking government departments using a Data Governance Quality Index[24] that incorporates many of the ideas above. Today, these rankings are based on self-declared information by departments; they lack a verification process led by the Development Monitoring and Evaluation Office. Another example is the intent expressed by the then CEO of NITI Aayog, Mr. Amitabh Kant, in his opinion piece in MINT of 17 August 2021[25] in which he discussed plans that, in the Aspirational Districts Programme (ADP), data reported by districts will be subject to randomized back-checks. These nested back-checks use the prescribed guidelines of each scheme to verify data by different levels of officials and are supervised by NITI using its randomization engine. They are also periodically cross-verified by independent agencies. Such efforts, adopted at scale across schemes, can improve data credibility significantly.

To summarize, ensuring data credibility will depend on the type of data. For master data, standardized registries can ensure that data is collected and used in similar ways across contexts. For transactional data, information should ideally be captured 'automatically' or as soon as possible when an event occurs, without the need for an individual to report it. When this is logistically not possible, alternatives include data triangulation and randomized, nested back-checks.

6.3 Conclusion

Overall, this chapter focuses on three architectural shifts that can enable PFM reforms. Governments must first couple data from programmatic

[23] https://dmeo.gov.in/.

[24] https://dmeo.gov.in/content/dgqi-overview.

[25] Available at https://www.livemint.com/opinion/online-views/our-adp-model-will-unlock-the-power-of-data-to-improve-lives-11629133457057.html.

and fiscal events using data standards-based exchanges, so that it is clear exactly which inputs relate to what outputs and outcomes. Two, governments must co-ordinate fiscal and programmatic events using Rules as Code, such that when programmatic events occur, they trigger fiscal events by design. Three, governments must create data systems that ensure programmatic and fiscal data is available at the unit level, and on demand.

To promote the adoption of these principles, we suggest a few system properties of the architecture. One and most importantly, data should be protected by design such that data coupling only takes place in limited ways, using principles such as zero-knowledge proofs, to ensure that authorized personnel have access to data to implement policies without violating data protection or data privacy of citizens. Indeed, adopting on-demand architectural reforms without data protection is risky. Two, data should be liquid such that singular data collected at the unit level can be recombined and reused for multiple purposes. Three, the system must abide by the once-only principle whenever possible such that data is recorded once, when the event happens, and by the very individual 'creating' the event. This allows stakeholders to aggregate data upwards while having enough granularity to make key decisions, like whom to make payments to and when. Four, the system should enact safeguards to ensure that the data is credible at its source.

Ultimately, we argue that three shifts are key to on-demand architecture: coupling programmatic and fiscal data, co-ordinating this data, and making unit-level data available on demand. These principles must be supported through system properties of data protections, data liquidity, data capture using the once-only principle, and data credibility.

7

Outcomes of Proposed Changes in the Architecture

So far, we have explained how poor expenditure capacity of the state affects public service delivery and programme implementation across the public financial management (PFM) cycle. As we outline in Chapter 5, there are significant issues which stem from a weak PFM system: (a) float, (b) incentives for deciding when to pay recipients, (c) administrative burdens, (d) uncertainty and delays in flow of funds and poor budget execution, (e) corruption, (f) poor auditing, and (g) poor data and citizen-based accountability. These problems limit the ability of the state to undertake its tasks, whether it be in the realm of regulation, production, or financing goods and services.

In this chapter, we introduce three of the many possible outcomes that can emerge from the architectural reforms we highlight in Chapter 6. In our experience, these outcomes are essential for a public financial management system if they are to resolve public service delivery problems. To be sure, the architectural shifts we discuss can enable many outcomes beyond these three. However, the shifts we discuss seek to structure PFM systems to increase optionality, so that many reforms can emerge over time in different geographies and in different programmatic circumstances based on local needs.

The first outcome is just-in-time funding, which de-layers the number of government accounts that hold funds. It can ensure that money reaches the end recipient in a timely manner. With just-in-time funding, money comes directly from the Consolidated Fund of India, or the Consolidated Funds of the States based on the structure of the scheme, and is directly delivered to the recipient. The second outcome is smart payments, which digitally and algorithmically execute the decision to pay a recipient based

Public Financial Management, State Capacity, and Public Services in India. Santhosh Mathew, Bhumi Purohit, and Devesh Sharma, Oxford University Press.
 DOI: 10.1093/9780198973195.003.0007

on predefined conditions. The third is a holistic experience layer of PFM which consists of platforms that users directly interact with, namely programme-specific platforms, an auditing platform, and a public transparency platform.

Ultimately, these three outcomes represent a true shift from a push-and-wait architecture—that drives requests for funds, information, and decisions up and down a chain of command from the Union to implementing agencies—to an on-demand architecture. On-demand architecture is designed to ensure that funds, information related to funds,[1] and the ability to take decisions related to those funds are available to authorized stakeholders in near real-time while accounting for public service delivery hurdles.

7.1 Just-in-Time Funding

Despite the progress in PFM reforms, India's current PFM architecture prevents stakeholders in government from making fund allocation, disbursal, and expenditure management decisions in real-time. As we explain in Chapter 4, the system largely continues to function in the way it was envisioned decades ago, well before the advent of modern banking, financial technologies, and internet-enabled workflow management. With the exception of SNA-SPARSH launched as a policy initiative in July 2023, funds are sent through the co-ordination of files and cheques, through various bank accounts, and with the approval of multiple middle managers before being disbursed to the implementation agency. These complications lead to a variety of problems for public service delivery.

As we have noted earlier, nascent changes in the system have emerged that are similar to just-in-time funding. Just-in-time funding refers to making funds available to implementing agencies instantly, as and when they need the funds, for incurring public expenditure. This means that intergovernmental transfers happen without delay, and without the need for advance fund transfers. We enhance our definition of just-in-time funding with an additional variation: when these transfers happen,

[1] By information related to the funds, we mean information around budgets, allocations, transfers, and expenditure.

they should record not just how much is transferred but also why it is transferred.

The Union government has made strides in getting closer to a just-in-time system of intergovernmental fund transfers. Through the Treasury Single Account (TSA) system (created in 2016) and SNA-SPARSH (created in 2023), autonomous agencies and government departments, respectively, can now directly draw funds from the Reserve Bank of India's e-Kuber system. This allows them to digitally draw funds as needed against financial ceilings, a marked improvement from receiving and parking money in advance. The TSA system in the Indian context is for Central Sector Schemes, whereas SNA-SPARSH is for Centrally Sponsored Schemes. These systems of intergovernmental fund transfers have a long history, and have been incorporated in both India and other countries, which we detail in Appendices 1 and 2. In the global context, this idea is generally called Treasury Single Account (TSA) for management of funds in treasuries. Once the instructions from the Prime Minister's Office on just-in-time funding are fully implemented, the full benefits of SNA-SPARSH will become available. Some details of these are contained in our publications, correspondences,[2] and research collaboration while in government (Banerjee et al. 2020; Mathew 2016, 2018; Mathew and Goswami 2016; Mathew and Purohit 2017). While TSA and SNA-SPARSH are much needed and welcomed, we note the progressive scope of different interventions for cash management summarized in Table 1. For one, just-in-time funding needs to extend to State Sector Schemes so that any payment can be drawn directly from the Consolidated Fund of India and/or the Consolidated Funds of the States—regardless of the original funding agency. Two, while using SNA-SPARSH and TSA, governments currently use the treasuries as their banks. In conceptualization of just-in-time funding, departments or autonomous bodies should not have to rely on the treasuries to use a

[2] Santhosh Mathew, then Secretary Rural Development, and Anup Mukherjee, then Chief Secretary, Bihar, advocated for Bihar to be included as one of the first pilot states for CPSMS, which was approved by the then Controller General of Accounts (CGA), C.R. Sundaramurti, and the Planning Commission. Mathew also advocated for the idea of just-in-time funding with NITI Aayog and its then Vice-Chairman, Arvind Panagariya. The minutes of that meeting provide NITI Aayog's assent to the idea of just-in-time funding (NITI Aayog 2016). This was preceded by a note for consultation with state governments on using just-in-time for MGNREGS (Ministry of Rural Development 2015).

Table 1 Progressive scope of different interventions for cash management

	SNA-SPARSH (2023)	Treasury Single Account (2016)	Just-in-Time Funding
Applicable to	Centrally Sponsored Schemes	Central Sector Schemes	Centrally Sponsored Schemes, Central Sector Schemes, State Sector Schemes
Route	Consolidated Fund of India and/or Consolidated Funds of the States → government departments → beneficiary, vendor, or employee	Consolidated Fund of India → autonomous bodies --→ beneficiary, vendor, or employee	Consolidated Fund of India and/or Consolidated Funds of the States → government departments and autonomous bodies → beneficiary, vendor, or employee
Banking System	Treasuries		Treasuries or banking systems
Payment Gateway	Through RBI's e-Kuber system using financial ceilings rather than cash (no parked funds)		Through RBI's e-Kuber system or banking system using financial ceilings rather than cash (no parked funds)
Use of Programmatic Data	None; only records the payment amount to beneficiary, vendor, or employee		Links each payment to unit-level programmatic use

just-in-time funding system; they could use their own banking systems as needed to draw funds against ceilings, with the feature of synchronized records appearing in the corresponding treasuries.

The last component of our reforms is perhaps the biggest departure from SNA-SPARSH and TSA. These two existing systems currently record the final amounts of payments made, but not necessarily with the unit-level programmatic information. Using the architectural shifts described earlier, we argue that just-in-time funding should couple the purpose of payments by linking fiscal data to the programmatic data at the unit level when possible. This means that funds are not pushed lump sum, in the aggregate, to spending agencies but as and when the need arises, with corresponding records being made available for what that need for funds is in the system.

Just-in-time funding has various benefits for public service delivery. For one, if funds go directly from a consolidated fund to the end recipient, they do not need to be held anywhere and therefore can lead to reduced, and ultimately eliminated, float. Furthermore, if the data is adequately recorded in systems that utilize such a funding model, they can enable better fiduciary and performance auditing. As we show in the next subsection, however, for truly easing performance audits, smart payments are the need of the hour to connect programmatic events and fiscal events. Lastly, just-in-time funding can also ameliorate corruption through two means. One, it can reduce the number of individuals involved in transactions, each with varying motivations. Two, it can lessen the need for speedy access to money if funds can directly and instantly be sent from consolidated funds to end beneficiaries. That being said, the literature identifies many forms of corruption. One of these is grand corruption, which involves large sums of money used to distort policies (Bussell 2012; Rose-Ackerman 1996). Just-in-time funding cannot directly reduce some forms of corruption such as grand corruption, but is more useful for addressing cases where inputs are devoured without any tangible outputs.

There are several limitations to the impacts on public service delivery with just-in-time reforms alone. Direct, just-in-time transfers can go a long way in reducing delays in fund transfers, friction in reporting expenditure, and administrative burden—though these reductions can only go so far without smart payments (which we describe in the next subsection). This is because just-in-time funds can eliminate the friction in transfers of funds within various parts of government once the decision to make a payment has already been made, but smart payments are needed to reduce frictions in the decision-making process of making a payment itself.

Additionally, they can partly reduce payment delays to beneficiaries once payment decisions are made, but cannot adequately ease all burdens. As we highlight in an example from Brazil's Bolsa Familia in Appendix 2, direct and timely payments are inadequate if the end beneficiaries have difficulty obtaining information about payments. Bolsa Familia improves transparency to end beneficiaries by sending them an SMS once funds are transferred to their account, ensuring they are aware of when to withdraw their benefits. Transparency, therefore, is not only about improving accountability within the government but doing so at the last mile to prevent leakages at all stages. Achieving transparency should therefore go beyond just-in-time funding.

These benefits are partly visible in a study of the Mahatma Gandhi National Rural Employment Guarantee Scheme (MGNREGS). In 2015, the Union Cabinet approved changes to the funding architecture of the scheme to ensure that beneficiary wages are directed from a central account directly to the recipients. In doing so, they have cut down the need for accountants to maintain thousands of physical files to keep track of each payment and have thus freed up their time for more productive work. These reforms were partly based on an academic study in Bihar, where researchers affiliated with the Abdul Latif Jameel Poverty Action Lab (J-PAL) at MIT conducted an evaluation analysing the impact of financial reforms on corruption and scheme expenditure between 2012 and 2013 (Banerjee et al. 2020). Recognizing that an efficient fund flow mechanism is essential for effective implementation of decentralized social programmes, the study sought to enact e-governance reforms for MGNREGS. Through a tighter fund flow process, researchers expected a larger amount of money would go to local governments and beneficiaries.

MGNREGS, considered one of the largest social protection programmes in the world, worked with 50 million households during the time of the study. Participants of the programme were guaranteed 100 days of work per year, mostly in unskilled manual labour on community projects. The authors of the Bihar study estimated the scheme's unmet demand for work using the 2009–10 National Sample Survey (NSS) data. They found that 35% of households in Bihar requested work but could not find it. Moreover, Imbert and Papp (2015) compared NSS data with NREGA reports and estimated that 20% of the reported work did not match actual work, suggesting leakages through ghost workers. Similarly, Niehaus and Sukhtankar (2013) studied 1,499 households listed as MGNREGS workers in the state of Odisha and found that only 821 existed and had actually worked. Even amongst those who had worked, most had received lower payments than the official reports suggested.

These inefficiencies stemmed from the previous funding architecture, in which local government bodies—namely gram panchayats—were able to request funds in bulk prior to incurring expenditure. Fund requests for anticipated expenditure were made to the block and, upon verification, passed to the district for approval. Thereafter, the state would make the final approval and send the funds to the local government account by releasing them to the district. This created a few substantial problems:

1. Block and district officials could leverage their approval power in order to extract rent from local government officials who sought the scheme funds.
2. Since funds were disbursed as per planned expenditure rather than actual expenditure, funding did not match the actual needs. While certain panchayats accrued float, others lacked funds to operate the scheme.
3. Blocks and district officials could delay funds or make arbitrary requests for information and procedures for fund approval, creating hurdles in scheme implementation.

To reform the system, the Government of Bihar implemented an e-governance mechanism in approximately one-third of the 3,000 villages in the researchers' study. The reforms in these villages were twofold: they linked fund flows to actual expenditures for beneficiary wages while also reducing the number of administrators in the approval process. Village officials could directly request wages for each beneficiary using an electronic platform without requiring intermediary approvals. So long as the information was entered into the IT system (CPSMS, now known as PFMS), the funds were automatically released from the state account to the local government account, bypassing the district.

These two simple changes, however, did not increase the funds available to local governments as was expected. Surprisingly, they reduced the scheme expenditure by 17% in villages. However, community surveys showed that a reduction in expenditure did not impact the number of beneficiaries who participated in the programme nor the wages they received. The disjointed impacts suggest that the reduction in scheme expenditure came from reduced corruption. To confirm the hypothesis, researchers found that MGNREGS functionaries in the villages where the reforms were evaluated experienced a decline in their personal wealth, presumably from a reduced ability to extract rent. The reduced corruption, at least in the form of personal wealth accumulation, was accompanied by a reduction in unused float; the bank accounts of villages that undertook the reforms held, on average, 50% lower balances than those without the reforms, marking increased efficiency in use of public funds.

The reform had an impressive impact on reducing leakages. Yet, initially, implementing these changes was marred by adoption difficulties. As is the case with most interventions that involve deployment of information technology solutions, the initial phase of the reforms additionally experienced technological problems, for example with server capacity, internet bandwidth, and reconciliations between PFMS and banks, causing fund delays to beneficiaries. Delays, however, reduced with time as technological issues were resolved.[3] As MGNREGS officials can attest to, delays in flow of funds can still occur because of a number of issues, such as inadequate budget allocations for the scheme—something an on-demand architecture cannot address—despite the funding guarantees made to citizens.[4]

However, the broader ecosystem has significantly improved since the Bihar study, and some of the problems that occurred are now avoidable. For instance, reconciliation of bank statements with core banking servers used to take up to 60 days, but now takes only two days. Reach of broadband internet in many parts of India has also significantly increased, as has the bureaucracy's familiarity with technological platforms. Other problems that may still occur, like delays to beneficiaries, may emerge from the expenditure architecture but can be further reduced by smart payments, which we describe in the next subsection.

In 2015, the Government of India enacted an improved version of reforms similar to those in the study at the national level. National-level

[3] Building on this experience, we outline how such problems can be addressed in the future, particularly by enacting asynchronous onboarding, in Chapter 8.

[4] A true appreciation and contextualization of the ideas in this book requires a recognition of the enormous number of actors, incentives, and systems involved in making even a single payment to a beneficiary in India. These systems run across Union government, state governments, district administrations, local governments, the Reserve Bank of India, National Payments Corporation of India, banks, payment service providers, payment gateways, National Informatics Centre (NIC) at the Union level, NIC at the state level, corporate banking correspondents, individual banking correspondents, and beneficiaries who actually access their money at any given cash-out point. We therefore acknowledge that the ideas we propose solve specific types of challenges in expenditure management, rather than being a panacea for a range of problems that emerge from the many other interactions between multiple parts of this ecosystem.

MGNREGS reforms, however, were not hastily actualized. In order to ensure proper implementation, the government undertook a phased schedule:

1. **Phase 1:** Rather than funds going from the Union to the beneficiary, funds were sent to each state's consolidated fund, which held the money until it was ready to be released. The consolidated fund has two restrictions: any withdrawal requires legislative approval, and the funds lapse during the end of the financial year in March. While this created float, districts and blocks were done away with as intermediaries for managing the funds. These levels, instead, were able to better focus on monitoring the programme.
2. **Phase 2:** To bypass the restrictions placed by the consolidated fund as well as to reduce float, each state created a State Employment Guarantee Fund (SEGF). Funds were sent to the SEGF and then to beneficiaries. The SEGF was a separate bank account and was non-lapsable regardless of the financial year. Through the SEGF, the Union and states maintained a float for a maximum of 14 days. To do so, the Union would send only two weeks' worth of funding for workers' wages to the SEGF, and refill it based on the balance at the end of those two weeks. Limiting float allowed the Union to retain some flexibility in how funds were spent across states and for varying needs within the scheme. Two weeks' float also allowed for more regular monitoring of funds, allowing officials to see when the SEGF balance for any given state was abnormal.
3. **Phase 3:** The third phase was enacted eight months after the second phase. Here, the float was brought down to zero by sending payment from the Consolidated Fund of India within the Union government to the beneficiary. As funds leave the Consolidated Fund, a contra-entry is made in the SEGF to enable the states to maintain records of funds going in and out. In order to send these funds, an officer at the Pay and Accounts Office of the Ministry of Rural Development at the Union must make a daily approval ensuring that the fund transfers meet codal requirements and are within the authorization limit.

This phase-based approach to implementing just-in-time funding in MGNREGS was a function of appetite for change and productive negotiations between Union and state governments. Whether or not comparable future reforms for all schemes and programmes require similar phases depends on similar variables—whether Union and state governments are willing to integrate PFMS, state-level financial management information systems (FMISs), and implementation agency-level bank arrangements. Equally important would be the necessary political and administrative inclination for such implementation, which we discuss in Chapter 9.

To conclude, direct transfers from the main government account to the end recipient in a timely manner (within 48 hours during business days) is the key idea behind just-in-time funding. In order to make this effective, we also highlight how just-in-time funding can supplement transparency so that end recipients are aware of what they have received, when, and where. Such reforms, while ambitious, are far from a moonshot idea in today's world, where private bank transfers occur within a matter of seconds. We highlight how the architectural shifts we propose in Chapter 6 can support just-in-time funding in section 7.4.

7.2 Smart Payments

Continuing with the spirit of improving basic functions of PFM, we suggest a second key outcome that is possible with the architectural shifts, in addition to just-in-time funding,[5] to truly automate fund transfers: smart payments. By smart payments, we refer to the use of digital technology to automate payments as soon as pre-approved conditions for those payments are fulfilled.

Despite sounding futuristic, smart payments are not new. Indeed, payments have consistently grown smarter over time: from paper currency to instruments like cheques and drafts, electronic bank transfers (NEFT/RTGS/IMPS), and, most recently, UPI-based mobile payments.[6] Ultimately,

[5] While just-in-time funding seeks to make government-to-government transfer of funds and authorization to spend those funds smoother, smart payment seeks to make government-to-person and government-to-business payments smoother.

[6] UPI is the United Payments Interface developed by the National Payments Corporation of India. Launched in 2016, UPI allows for instant inter-bank, peer-to-peer, and person-to-business fund transfers. The technology has enabled mobile-based bank transfers to happen seamlessly.

payments have evolved to be smoother, without disruptions, once the decision to pay is taken. However, despite these technologies, frictions have persisted in government payments because frictions prior to the payment itself—in the decision to authorize payments—have not been eliminated. While just-in-time funding allows funds to be available to implementing agencies so that payments *can* happen in time, smart payments can go further and ensure that payments *do* happen in a pre-defined and immediate manner, once conditions to make payments are met.

Government payment delays, long the norm in the Indian state, are a result of identifiable conditions that characterize India's archaic public finance architecture. The basic problem in government payments arises from a simple fact found in any sequential contract: when payment and the delivery of a good or service occur asynchronously, there is the potential that the payment is not made after receiving the product or vice versa. In any such contract, the parties first rely on the trust that the other party will uphold their terms and, failing so, count on the state and regulatory regimes for the contract to be enforced. This accountability mechanism becomes trickier in government when one of the parties to the contract is also responsible for holding itself accountable and enforcing the agreement.

In addition, various bureaucrats in charge of approving the funds are personally held responsible in case of payment issues, as we describe at length in Chapter 5 (see section 5.2). This may delay their decision-making to approve payments. Delays in government payments affect the economic health of beneficiaries and make programme outcomes weak and indeterminate.[7] However, given that the process of most government

[7] As of July 2016, Business Today and the Construction Industry Development Council respectively estimated that Rs. 1–3 trillion and Rs 2.5–3.5 trillion monies of private contractors were stuck due to government payment delays—a sizable ~1.5% of India's GDP (Mondal 2016). Four years later, as of March 2020, the unpaid dues by the Government of India to various suppliers and private contractors would have likely climbed to around Rs 9.5 trillion or 4.2% of the GDP estimated for the 2021 fiscal year (Sahu 2020). Estimates in other industries cite similar problems. Manivannan and Zaveri (2021), for instance, in studying the financials of centrally owned public sector enterprises, estimate an annual balance of Rs. 1.3 trillion against an annual procurement budget of just Rs. 0.3 trillion, implying that state enterprises owed five times as much as they procured. In the information technology sector, Nasscom calculated that government dues to IT vendors were to the tune of Rs. 50 billion as of July 2016. As of February 2020, the National Highways Authority of India owed businesses Rs. 259 billion. Similarly, the government owed Rs. 918.6 billion to electricity-generating companies by the end of the 2021 financial year (Dey 2019; Sahu 2020). These problems are not specific to businesses. In 2020 and 2021, ASHAs (Accredited Social Health Activist, or community health workers) protested in Bihar,

payments resembles a list of specifications that are checked to receive payment, we advocate for the use of smart payments, which deploy straight-through processing using the principle of 'if this, then that', and eliminate manual intermediation. An algorithm, equipped with well-defined preconditions, can automatically release funds when manual certificates or database-driven reports verify that conditions for payment have been met. Such smart payments would allow for efficient contract management, attract the best quality contractors, preserve liquidity, facilitate better planning, and sharpen accountability.

Smart payments are frequently undertaken in the private sector, albeit sometimes in a manual way. The traditional escrow—a centuries-old financial setup—involves pre-funding a contract, but through a neutral third party who is paid for their services to ensure that obligations to the contract are met. When the vendor delivers and this third party verifies the service or product, the latter makes the payment. The etymology of the word escrow finds its roots in the old French word *escroue*, meaning scroll or a piece of parchment—that is, a checklist. In other words, an escrow is nothing but a checklist and a decision that emerges from the checklist: once checked, it automatically fulfils the contract. Smart payments can be thought of as a digital escrow, which executes the automatic release of payment from the Union or State Consolidated Fund when defined preconditions are verified as met. An algorithm can verify, through reports and policy-based preconditions, that the rules for payment have been met and a recipient's payment can automatically be released.

As we note earlier, smart payments work on the basic principle of 'if this, then that'. For instance, if the officer directly responsible has certified that the supplier has delivered the product and if the person required to verify the quality of the supplied product ascertains that the quality meets standards, then the payment is released by a computer algorithm. This will be true as long as it was clearly articulated beforehand that these were the only two preconditions required to be fulfilled for releasing the

Karnataka, Maharashtra, and Andhra Pradesh over non-payment of dues (The Hindu 2021; Venugopal 2020). Other beneficiary groups as diverse as crop procurement commission agents, farmers, state government employees, railways employees, and doctors have participated in public protest over payment delays as well (Cris 2021; HT Correspondent 2020; Misra 2021; PTI 2020).

payment. The need for discretion and manual decision-making is in setting a well-defined checklist of terms which specify the quality or quantity of the good or service, and terms for how these terms are verified. If there is a set of well-defined preconditions, fund transfer occurs without the need for an approval decision at every layer of government. In such a system, the algorithm may also flag that the preconditions cannot be approved if the evidence provided for approval is unclear, and bureaucrats can exercise their judgment to manually approve or deny these flagged payments.

Smart payments are not new in government. For instance, on a specified day every month in India, millions of government pensioners across the country receive their pensions. These pensions do not require monthly authorizations from any individual. Rather, they are automatically triggered if pensioners have a valid life certificate, renewed for the year. Similarly, when someone cancels their railway ticket, the refund of their ticket fare does not need approval from the Minister of Railways—it happens within seconds of someone cancelling their railway ticket.

Smart payments also have an additional benefit; they would remove a major roadblock in the current system, wherein a senior official must approve payments based on certification by others, without always knowing if the underlying data is accurate. With smart payments, no one would have to aggregate reports, add file notes, and actively choose to make a manual payment approval—thereby significantly reducing administrative burden. Instead, smart payments would allow governments to use a system for non-arbitrary application of rules for releasing payments.

Another important consequence of smart payments is a sharper locus of accountability on those who provide certification that preconditions for payment are met. The digital system would record who provides the data for such certification. The person providing the data should ideally be the person at the implementation level directly engaged with the work undertaken. If this is the case, there is no information asymmetry between the programme officer 'generating' information or observing the event and the officer recording the data in the system. Because the officer's name is attached to the data that is required for certification, and observability of that data is not hampered by any other officer, there is also no information monopoly about either the data or the person providing that data. Lastly, in this system, officials are only responsible for the validity of the information contained in their report and their own certifications, not for the payments based on other people's certifications. There is no armour or shade that can be provided by a senior officer who may be implicated in wrongful contract management. Nor is there a challenge of collective action where officials involved in a contract fail to report others out of fear of being accused of conspiracy themselves because they had approved some of the preconditions for a payment.

In addition, smart payments can also incur benefits for both auditors and beneficiaries. For auditors, clearly seeing why authorizations are made can ease performance audits, especially if smart payments' data is combined with transaction data from just-in-time funding. These datasets can help auditors conduct both fiduciary and performance audits seamlessly. For beneficiaries, smart payments—in addition to just-in-time funding—could both add clarity to what is needed to receive funds and ensure there are limited delays.

Smart payments, of course, need not be absent of scrutiny. Transactions can simultaneously be monitored at an aggregate level and any transaction can still be flagged by reviewing officials for further verification. But what smart payments eliminate is the need to manually check every single instance of payment. Still, they do bring with them their own share of risks and may not be applicable where a decision to pay cannot be reduced to a checklist (though we believe such a case to be infrequent in government). Data protection is a classic concern that is increasingly more relevant in this era of digital governance, even with improved cybersecurity (Abelson et al. 2015). Bugs, performance issues, privacy, legal issues, and mishandling of exceptional cases are some problems that almost seem unavoidable but represent a learning curve in the use of smart payments (Wang et al. 2019). We already advocate that on-demand architecture must be accompanied with strong data protections, which serve to maintain data privacy, in mind. The other risks are technical, and may be manageable with the right execution teams. Additionally, ensuring that bureaucrats can still maintain discretion to verify conditions for payments can reduce some of these risks.

7.3 The Experience Layer of Digital Systems for Various Use Cases

In this section, we propose digital interfaces or systems for a variety of stakeholders who can leverage the data and services that are possible to offer in an on-demand architecture. Specifically, the goal here is to expand on how platforms can be designed to ensure they are legible and clear to the end user and serve the needs of various actors.

With an on-demand architecture, implementing agencies can request and receive funds and information on their budgetary allocations as needs arise, instead of waiting for the Union or state governments to push down funds. Fund releases can occur once the annual sanction order is issued and the conditions for fund release are met. Digital systems, automated bank transfers, and processes can provide end-to-end solutions for creating a smarter, better financial management system. The need to use such digital technologies is especially relevant in a country like India, whose software industry has developed world-class IT systems globally.

Keeping in mind the architecture that we propose in Chapter 6, here we highlight some of the user-facing systems that can connect to this architecture. While the on-demand architecture opens up optionality and many possibilities that we haven't imagined in this book, we specify three particular systems that can use this architecture: programme-specific digital workflow management systems, an auditing system, and a public transparency system.

With programme-specific digital workflow management systems, each programme or scheme should have its own workflow-driven system that will sync with the budget management, expenditure management, and auditing systems such as PFMS, state-level FMISs, and Accountant Generals' systems. Programme managers should be able to use digital workflow management systems to execute all aspects of the programme or scheme that they administer, including payments. Programme-specific digital workflow management systems can also be linked to the Socio Economic and Caste Census (SECC) or other demographic databases to automatically identify which schemes or programmes citizens are eligible for and enrolled in—so long as data protection is an integral part of the design. For more details on the eligibility function of programme-specific digital workflow management systems and our recommended suggestions, see Appendix 4.

All programme-specific digital workflow management systems should sync with an open API-enabled interface of the current PFMS or state FMISs.[8] The PFMS will need to be updated to facilitate direct payments from the Consolidated Funds of the States and/or the Consolidated Fund of India to the end recipients. When payment instructions from programme-specific digital workflow management systems come to PFMS, the latter will trigger payment releases and account for it. At the time of writing, this has already started for some Centrally Sponsored Schemes, even if not through open APIs.

Auditing systems are another type of significant user-facing systems that connect to this architecture. By pulling data from programme-specific IT platforms, PFMS, and FMISs, auditing systems can digitally compile vouchers for various cross-checks, like matching sum amounts with Utilization Certificates, AC bills, or expenditure reports. Such readily available payment information would allow auditors to virtually and digitally scrutinize expenditures. The platform should sit with the office of the Comptroller and Auditor General (CAG) and state-level offices of the Accountant General.

A third important system that should be ensured is for public transparency platforms that leverage data liquidity. This system should connect with programme-specific digital workflow management systems, provide data to the public, and enable them to access the state and its services. The public transparency platform should allow policymakers, bureaucrats, researchers, and individuals to obtain relevant real-time data. Existing platforms such as https://data.gov.in/ or https://ndap.niti.gov.in/ could be upgraded to serve this function. Below, we outline four moonshot ideas for ways in which public transparency platforms can work for citizens and civil society.

One, citizens, civil society organizations, and researchers ought to have access to real-time expenditure and authorization data for all programmes at the macro level, through both raw data files and data

[8] Santhosh Mathew, one of the authors of this book, wrote numerous internal government letters between 2010 and 2016 which speak to the need for open APIs. Further, as noted on the CGA website (accessed 2 January 2024), standard open APIs are now a part of their strategy for PFMS 2.0 (CGA 2024).

visualizations for easy comprehension. Data should also include performance metrics that allow for outputs and outcomes to be measured against expenditures.

Two, individuals participating in public programmes, with their consent, and through a safe, encrypted system, should securely be able to obtain real-time data on the date and quantity of funds deposited to their account for any given programme. Such data should include the amount they have received or are due to receive, and reasons for why a payment is delayed or has failed. While such a complicated database for a national level may seem radical, it is not very different from what MGNREGS already does.

Three, the platform should clearly lay out the requirements for any individual to enrol in a scheme in a simple and easy to understand manner. Information on eligibility is currently available, but often within long texts written with policymakers and bureaucrats in mind, rather than beneficiaries. Making this information available, especially in local languages, can allow citizens to be informed, and better contest inclusion or exclusion errors. A similar platform ought to be made available to vendors and businesses dealing with the government. Such a system shifts the onus of information from the beneficiary or vendor on rights they are legally entitled to, as it is in the current system, to the government who ought to be giving this information as part of its social and commercial contract. Additionally, potential beneficiaries should be able to anonymously answer questions pertaining to their eligibility criteria to see if they ought to be enrolled in any schemes that they are currently excluded from. Wherever they are not enrolled in a scheme they are eligible for, they should be able to initiate a trackable request for enrolment directly on a platform. This has two advantages. One, it ensures that information on all available schemes and programmes is made available through a platform accessible through multiple channels (such as websites, mobile applications, and physical centres) so that citizens do not have to seek out multiple pieces of information from a variety of actors. Two, it prevents local middlemen or officials from fashioning their own criteria, and limits corruption.

Four, data must allow citizens to see which expenditure (outlay), output, or outcome data or decisions have been checked, found correct or incorrect, and approved or not approved by which manager or supervisor of a government scheme—down to the lowest level of scheme administration. Doing so will enable governments to fulfil their obligations under the public disclosure clauses of India's Right to Information Act and reduce the load on officials in responding to individual requests for information. A prime example of use of such disclosures is the Mazdoor Kisan Shakti Sanghathan (MKSS), an organization in Rajasthan that has developed a powerful approach to social auditing. The organization first obtains and collates all relevant documents for a scheme like MGNREGS using tools authorized under the Right to Information Act (which MKSS notably helped to legislate). It then holds a public hearing in each village where it works, where MKSS reads out public records, including expenditures for the scheme, for citizens to contest if need be. After contestations are made, often in front of officers who have had to hand over cash to citizens during the hearing itself, MKSS releases reports of the results to institutions, including the press (Ramkumar and Krafchik 2006).[9]

The very same methods used by MKSS further pioneered the use of social audits within a government programme, MGNREGS, in Andhra Pradesh. With the help of MKSS, the state government created an independent body for audits that worked separately from the rural development bureaucracy, essentially acting as a top-down and independent regulator. In a qualitative study of the social audit programme, Mukherji and Jha (2017) find that these forms of social audits helped curb corruption. The architectural shifts and system properties in the previous section can enable a similar function for bodies within states, organizations, and citizens alike to obtain and collate information using public transparency systems without needing to seek out information.[10]

[9] While social audits have been powerful for MKSS, the results from scholarship indicate their effectiveness is mixed. Several studies have shown that they need not always work (Afridi and Iversen 2014; Björkman and Svensson 2009; Olken 2007). Afridi and Iversen (2014), for instance, examine three rounds of social audits amongst 300 villages in Andhra Pradesh, a state that has championed social audits. They find no changes in employment generation or detection of irregularities. In fact, they observe an increase in some types of irregularities. Ultimately, Afridi and Iversen (2014) suggest that social audits are rendered ineffective if officers face no punishment for engaging in corruption or misreporting.

[10] However, research to date shows that transparency alone is not sufficient in reducing corruption or improving access to entitlements (Niehaus and Sukhtankar 2013; Peisakhin and Pinto 2010; Sukhtankar and Vaishnav 2015). Transparency does, however, increase an individual citizen's bargaining power or improve the opportunities for collective action (Sukhtankar and Vaishnav 2015). Nevertheless, it has not always worked by itself if, ultimately, transparency does not lead to punishment for the officer at fault. If it is to reveal the detection of fraud and corruption, there ought to be consequences for officials who engage in corruption.

While discussing digital technologies and interfaces and their typically higher functionality than non-digital ones, we would be remiss not to add that the way citizens obtain the information outlined in this subsection ought to be consistently provided in both digital and non-digital ways. In other words, digital layers need to be constructed in a way that they can be extended for non-digital use; we discuss the idea of extensibility in Chapter 6, section 6.1.1.1. In a country where technology adoption is not universal (especially amongst those who rely on government benefits), solely digital solutions will not ensure data transparency. A myriad of solutions to this problem have been tried: painting expenditure data on village-level gram panchayat offices for citizens to scrutinize; reading beneficiary lists at gram panchayat meetings; and having liaisons within villages who can relay digitally available information to those without access to technology. Similar solutions must be explored to ensure that citizen- or society-oriented transparency methods are visible, legible, and usable regardless of one's educational levels, geographic proximity to state offices, technological access, or socio-economic background.

To maximize these platforms, and to consolidate the outcomes we have suggested so far, we propose a more unified digital ecosystem for consideration. Ideally, the following chain of events can enable smooth public service delivery and ensure that funds not only reach individuals but that individuals can easily utilize these funds. To facilitate this, a PFM system should have the following features:

1. The system should enable future budget approximates to be created based on previous expenditure data, enable predictive analytics for projected expenditure, and inform each scheme or department's digital platforms of government priorities for budget authorization for the financial year.
2. It should automatically identify which programmes each citizen is qualified to enrol in based on demographic and socio-economic data when possible and in a secure manner, and the amount they ought to receive in payment for these programmes.
3. Automated payments should go out when programmatic events occur (i.e. smart payments), and proof that they have occurred is recorded on programme-specific digital platforms. The payments must come directly from each state's account—the State

Consolidated Fund (SCF)—and/or the central government's Consolidated Fund of India for payments and be credited directly to the end recipient with no intermediaries in between (i.e. funded using just-in-time principles).

4. Data should be aggregated from the primary unit level upwards across both Union and state programmes to consolidate all expenditure data, in real-time.

7.4 Architectural Shifts and Outcomes

How do the architectural shifts, proposed in Chapter 6, support the development of just-in-time, smart payments, and user-facing systems? In reality, not all the architectural features are required to implement each of these outcomes across India. As we highlight in Table 2, fiscal and programmatic events have to be co-ordinated in order for just-in-time

Table 2 Architectural requirements for PFM outcomes

Architectural Feature	**Just-In-Time Funding**	**Smart Payments**	**Experience-Layer Platforms**
Co-ordinating Fiscal and Programmatic Data	Required for JIT to connect payments to programmatic events, and to improve traceability of where funds are used	Required to couple fiscal and programmatic events for smart payments	Required if platforms are to show why funds are used, instead of aggregate amounts of funds used
Coupling Fiscal and Programmatic Events	Required to combine smart payments with JIT, but not for JIT alone	Required, with Rules as Code, for smart payments to function	Not required for programmatic platforms
Making Fiscal and Programmatic Data Available at the Unit Level	Required for co-ordinating and coupling fiscal and programmatic events for JIT in order to trace where funds are used at the unit level	Required for co-ordinating and coupling fiscal and programmatic events in order to trace where funds are used at the unit level	Not required for programmatic platforms, but data available will be at the aggregate level instead of unit level

funding to function. This allows stakeholders to connect funds to particular types of payments. However, these events need not be coupled unless smart payments are also implemented. For smart payments to happen in the way we discuss, Rules as Code are additionally required. Without their presence, individuals would have to enable manual triggers for fund releases, as occurs across most of the system today.

Similarly, none of the outcomes we discuss require unit-level data to be available. However, this feature allows us to move past the aggregate data issues (and therefore non-verifiability issues) which feature in the current PFM system. Without unit-level data, aggregate and macro-level data on fiscal events would mirror the problems found in today's PFM system.

Why, then, do we insist on co-ordinating, coupling, and making unit-level fiscal and programmatic events readily available to stakeholders? The shifts from push-and-wait architecture to on-demand architecture that we discuss allow for a loosely coupled stack of data exchanges and services that allow users across the PFM ecosystem to work within decentralized yet coupled systems. This point goes beyond any one outcome; it applies to any outcomes that emerge from the on-demand architectural shifts. To understand the importance of this feature, let us take MGNREGS' just-in-time funding reform as an example. The government created a centralized software managed by officers in the Ministry of Rural Development in New Delhi to enable just-in-time funding through its internal software. Authorized implementation managers can access the software and approve funds as needed. They can then download reports of this data, which includes information on both payments and programmatic activities.

However, let us assume that a particular feature of the software—for instance, the way financial and programmatic data are coupled after just-in-time funding is enacted—is not suitable for an implementation agency's needs to understand or visualize the data in a district in Rajasthan. In order for them to change how this data is depicted on their dashboard, they would first need to request this modification from their state-level officers for MGNREGS. These officers, who are not authorized to make these changes, would then determine whether the change is important enough to send to the Union-level officers. Union-level officers can then implement the changes, but not for the district alone; they would have to enable features for the entire country. Given the myriad needs of implementation agencies across the country, Union-level officials may face undue workloads and complexity, and be unable to make modifications to the dashboard requested by one district.

Creating systems where data is coupled, co-ordinated, and recorded based on agreed standards and protocols instead of a monolithic and centralized system allows for outcomes such as just-in-time funding to be enacted without forcing users to use an entire stack of available solutions. This allows users to selectively implement any of the outcomes in this chapter, or others, to suit their own needs. For instance, if PFM is designed with our architectural shifts and system properties in mind, every actor in the ecosystem can make their own choices on how to implement policies while meeting contextual needs. Take the example of a citizen who may not want to receive their social benefits in a bank account, for she does not prefer to share her banking information with the government or does not have a bank account. In a stack of loosely coupled systems, it is possible to replace the building block of 'digital payment' with 'physical payment', in which a citizen would receive a token without providing her banking information. This token would allow her to withdraw her benefits from a bank or post office, while being in control of her data.

The combination of the three architectural shifts we discuss in Chapter 6, coupled with the system properties, makes it possible for PFM systems to realize their potential for service delivery across context, geography, policy, and time. Given that rules for funding and implementation arrangements differ not only by policy but also by implementation level, they need to be supported by layers of decentralization that enable robust and credible data collection by policy implementers and managers while meeting agreed standards. Such features will allow stakeholders to develop contextually relevant solutions at the experience layer of public service delivery. The ideas suggested in this book enable precisely this kind of flexibility and can allow complex interventions to be enacted across various levels of government.

7.5 Conclusion

In this chapter, we outline three key outcomes of an on-demand architecture—just-in-time funding that includes direct and timely payments between government agencies, smart payments that automate the decision to pay based on predefined criteria, and a series of systems that users interact with to fulfil PFM functions.

To be sure, it is indeed possible to pursue specific components of one or more of these outcomes on their own, disconnected from the others. For instance, just-in-time funding data can be created on its own without smart payments. While these piecemeal changes may be easier to pilot or solve for, fragmented and disconnected solutions can limit effectiveness, scale, and sustainability of reforms—a problem we address in Chapter 9. At the end of the day, any true shift from the current push-and-wait architecture to an on-demand architecture will require all three outcomes to be effective holistically.

8

Institutionalizing Improvements in Public Finance Management

In the previous two chapters, we discussed the architectural shifts and system properties required to build an on-demand architecture. We then highlighted three possible outcomes of an on-demand architecture: just-in-time funding, smart payments, and experience-level systems. But given the complexity of the task, how can governments facilitate the creation and adoption of such a PFM system?

We describe a way to approach shifts from a push-and-wait architecture to an on-demand architecture by focusing on four points. In section 8.1, we discuss the need to ensure that any changes to the system align with individuals' incentives. Section 8.2 explores the concept of agility, which ensures that changes in PFM happen in a way that maximizes options available for adoption. In section 8.3, we discuss the idea of asynchronous onboarding, which ensures that changes can be adopted in a piecemeal nature while keeping the larger ecosystem in mind. In section 8.4, we look at how ideas of on-demand architecture can be enacted within existing institutions.

8.1 Recognizing and Addressing Incentives

There is an emerging opportunity[1] for those engaged with PFM—whether within Ministries of Finance or finance divisions within line ministries—to address co-ordination challenges across government. By

[1] This is especially true given the momentum in government from a series of events: from the work of the group of secretaries in 2016 recommending that all government fund flows adopt just-in-time methods, the follow-up actions by the Prime Minister's Office and NITI Aayog in 2016, the request from the Prime Minister of India to the CAG in 2019 to improve data usability, the response of the CAG in the form of the presidential advisory in 2020, and the implementation of SNA-SPARSH by the Department of Expenditure in 2023.

Public Financial Management, State Capacity, and Public Services in India. Santhosh Mathew, Bhumi Purohit, and Devesh Sharma, Oxford University Press.
 DOI: 10.1093/9780198973195.003.0008

co-ordination, we echo Wegrich and Štimac (2014), who define it as the ability of a department to speak internally with itself or externally with other departments, governmental or non-governmental vendors, and relevant stakeholders. Co-ordination capacity is important for governments, not only to address complex governance challenges and ensure policy alignment (Peters 2018; Wegrich and Štimac 2014), but also for the completion of basic tasks, including communicating information on an array of items such as outlays, outputs, and outcomes, or changes to procedural guidelines for scheme implementation.

Public financial management serves a universal and horizontal function across all verticals within government. Akin to human resources units within public or private sectors, all public offices have similar core needs that are addressed by the public finance function. While what is financed differs, the core mechanisms in which finances are budgeted, released, and accounted for are similar. This horizontal salience across departments can allow public finance actors to enact changes which can improve not only PFM but also public service delivery, as we contend in this book.

There are two ways in which PFM can ensure improved co-ordination: one, by aligning incentives (the carrot approach) or two, being a forcing function that requires individuals or systems to change by leveraging the formal authority vested in Departments of Finance (the stick approach). Each approach has its own benefit and can work well in particular cases.

To clarify between the two, consider a common problem that power distribution companies encounter—non-payment of electricity bills by local village bodies or gram panchayats. These non-payments are a small part of the larger budget crisis plaguing the power sector, which still runs on state-run grids with varying efficacy despite decades-long attempts at reforms. Historically, free electricity provided a ripe opportunity to shore up votes, making reforms that required payments from consumers difficult to materialize. As demand for electricity grew, states started increasingly relying on cross-subsidization, or industries paying more for their power supply, than individual citizens, who benefited from subsidies. Yet this pushed industries to build their own power sources, leading state electricity boards into further financial disarray. Consequently, the Union government doled out performance-based grants and loans to state electricity boards to spur innovation, with varying and limited success (Tongia 2007).[2]

Against this backdrop, the state of Punjab has sought to undertake reforms to improve their financial outlay for electricity consumption. Specifically, under the Punjab Rural Water and Sanitation Sector Improvement Project, Gram Panchayat Water Supply and Sanitation Committees (GPWSCs) get budget allocations to use electricity that enables them to supply water to households. Budget allocations, though, form only a part of the revenue for GPWSCs. Households also pay the Committees for the water. Using both budget allocations and consumer payments, the GPWSC pays a public electricity distribution company for the electricity used to supply the water.

However, GPWSCs often do not make payments against these bills for a variety of reasons. For instance, they could have low revenues from user charges because of administrative or financial mismanagement; they may only be charging users for part of the fee, may not be charging all users who should be paying, or may not be collecting payments for all water bills. Given that power is supplied by a public electricity company, there are no repercussions, and the power does not get cut for non-payment.

Over a period of time, though, these bills accumulate and create debt for the electricity distribution company. Given the stalemate between GPWSCs and the distribution company, the latter approaches the state government—specifically, the Finance Department or the Department of Water Supply and Sanitation (DWSS)—to clear these dues. Because these state departments have already allocated the money to GPWSCs, however, they cannot make additional budgetary allocations to the company. Even if a one-time payment order is issued for the back dues, the same problem occurs in the future. Ideally, if the state DWSS or Finance Department had visibility into these dues in real-time, alongside each panchayat's budget allocations and the level of usage, then they would be better able to manage their budgets and cash flows from the state government to the local bodies, and subsequently to the electricity company.

A few possible solutions exist to this problem. As we mentioned earlier, the DWSS and Finance Department can instruct village GPWSCs to pay

[2] For detailed accounts on the political economy of the power sector in India, also see: Chatterjee 2012, Dubash and Rajan 2011, and Mukherji 2014.

these bills immediately and withhold future budgetary tranches if they fail to do so. Such orders can be issued time and again whenever the company requests funds from the state. However, this comes at a debt to the power distribution company for a period, requiring stakeholders to continually monitor which GPWSCs are behind on payments. It also requires the DWSS and Finance Department to constantly release orders, and for GPWSCs to respond to them.

Another approach could be to ensure that the DWSS and the Finance Department have visibility into electricity bills across billing, payments, and pending dues. Let's call this the information approach. Two possible ways exist to undertake it. One is a 'stick' approach. DWSS and the Finance Department can issue orders to local village bodies to immediately report all electricity bills as soon as they are received. Alongside, they could very well issue similar orders to the electricity distribution company and monitor which GPWSCs make the payments. While this method may be effective, it comes at a significant monitoring cost for all involved stakeholders, who have to send the right documents and ensure that bills are paid.

On the other hand, a 'carrot' approach is possible wherein the distribution company voluntarily agrees to share its invoices to a specified data repository as they come in. If this repository is also linked to panchayats budgeting, fund releases, and expenditure management, DWSS and Finance Departments can far more easily co-ordinate the payments of these dues by having visibility into whether payments are made.

In this carrot approach, it is in the distribution company's interests to get on board because it increases the likelihood of payments being made. The DWSS and Finance Department benefit because it reduces their monitoring costs. The interests of local bodies are slightly more complicated to explain.

The GPWSCs may not be concerned with making payments to the electricity company if there are no repercussions for non-payments. But if the Committees are assured that getting on a specified system will help them better manage their billing to citizens—and thus increase their revenue—they may feel motivated to use the system. The DWSS can also bring the GPWSCs onto the system by providing additional staff members for onboarding. Additionally, the Committees may be concerned that they have to report varying expenditures differently for all their programmes to the relevant departments. If the repository can be more broadly used for larger expenditure management and reporting, then that provides GPWSCs with an even greater incentive. Of course, not all of these incentives are necessary, and simply producing the right one can be sufficient, but we make a larger point that integrating systems into wider functions can yield better inducements for departments to use new systems.

There is a broader benefit to using systems not micro-focused on one task, a fact which governments have not used to their full advantage. For instance, state-level agencies often insist that payment orders must be entered and processed in a bespoke software, even when there are bugs in the software that prevent users from entering this data. When combined with limited staff, departments and agencies seeking to adhere to these rules end up dealing with greater administrative burden. Similarly, when departments or agencies insist that funds must be kept in one particular bank, these rules ignore the reality that banking relationships and customer service for any given implementing agency vary based on local contexts. Siloed solutions or systems that solve one problem can often end up creating more snags, especially when designed poorly without the end user or broader linkages in mind.

In summary, governments can implement changes by force to solve one particular problem, or they can create structures that encourage broader reforms for recurring issues. Aligning incentives would entail designing systems that spur government agencies or individuals within those agencies to co-ordinate. The incentives come with a big advantage—they not only create trust that new systems are in the best interest of stakeholders but further ensure that systems also meet the needs of stakeholders.

Incentives, of course, may not always work. This may be particularly true if the incentives of stakeholders to support reforms are not met. While 'sticks' or forcing functions are more straitjacketed, they can be effective when needed. Much like the IMF's conditionalities (International Monetary Fund 2021), for instance, the public finance function can enforce programmatic rules and procedures, which, if not followed, can lead to moratoriums on release of funds.[3] Sticks may work well when

[3] For instance, these conditionalities can ensure that the drone stack (government-specified protocols for using drones) is used for public crop insurance; that training and certification of people processing health insurance claims is as per defined standards; that when frontline

the status quo is entrenched and difficult to change using carrots or motivation alone. Using forcing functions in such a case can propel people out of inertia and create momentum for change. Moreover, in situations where legal or regulatory compliance is required, it makes more sense to use sticks than to incentivize change. For instance, governments seeking to promote clean air solutions may find it preferable to impose pollution limits on factories rather than provide subsidies to promote self-adoption.

Nevertheless, solely imposing centrally developed systems to force compliance underestimates the complexities of the system. Therefore, we posit that incentives or 'carrots' offer the best way to improve state capacity through PFM systems, but forcing functions are more apt for procedural needs. For instance, a forcing function such as ensuring the certification and training of those processing health insurance claims for PM-JAY may work better with the expertise of authorized certifying agencies.

There is another case where incentives may not work: learning new skills can be costly for individuals, and they may not have the necessary inducement to do so. For some frontline workers newly responsible for entering data, they may have to learn how to use digital tools altogether. For others, perhaps, it may be essential to learn how to use these tools in a new way, such as learning a new software. The systems we propose do indeed require significant investment in the training of individuals and convincing them to work with new systems. However, such an investment is not a novel change from the current system; governments are already moving in this direction across the board for both PFM and programmatic workflows. With this kind of change, imposing top-down systems may be required, but with adequate training. Moreover, asynchronous onboarding, an idea we detail in section 8.3, allows governments or departments to onboard into a new architecture in their own time, as and when they are ready. First, we discuss agile development, which is a way of developing systems such that they are designed, implemented, and improved iteratively.

healthcare workers are paid their incentives, the information for calculating their entitlements comes from the same data source that is used to report on health programme outputs.

8.2 Agile Development

The types of changes discussed in Chapters 6 and 7 demand not only a co-ordinated system developed by the Union and state governments, but also adaptations for state-level programmes and schemes. We believe our proposed on-demand architecture is possible politically, technologically, and administratively because of functions that have already been implemented across Union and state governments through PFMS, IFMSs, and the use of digital public infrastructure. States in India have had a history of developing their own systems with use of agile methodologies. On its own, the PFMS system developed by the Controller General of Accounts has also used an agile methodology, even if not always with strategic inputs from states. To highlight the bipartisan support for improving PFM and keeping it agile, we highlight some of the work undertaken by Kerala in Appendix 1. While these examples might appear dated, they show considerable energy exists in India's PFM ecosystem to undertake improvements that are responsive to emerging needs. Any Union-level enhancements must therefore ensure that states can integrate into a larger network and that state structures and learnings can benefit from a broader vision of an on-demand architecture.

Our approach to expenditure management is therefore different to other reforms, such as those enacted for the Goods and Services Tax Network (GSTN). The GSTN provides the digital applications for management of Goods and Services Tax in India. However, the approach to implementing the GSTN was centralized; it offered a singular software to all users.

A centralized system was suitable for GSTN and revenue collection more broadly because revenue collection consists of a very different set of political economy and administrative issues compared to expenditure management. As Allen (2015) notes, broadening of the tax base and improving revenue collection efficiency can garner enough political support to overcome bureaucratic inertia and rent-seeking behaviour. Revenue collection, purely as a network of individuals and organizations, is far sparser and more focused, compared to spending. Expenditure management, on the other hand, requires negotiations across multiple political and administrative jurisdictions, which is why a data standard- and digital public infrastructure-led approach such as data standards and data exchanges is far more suitable for India's public financial management, and for countries with similar complexities.

We examine the opportunities for making improvements in PFM in the next chapter. For now, we recognize that not all changes in PFM may be simultaneously implemented across India, particularly because different levels of government fund varying schemes, and each scheme is managed at varying levels—sometimes through different systems. Indeed, while examining over 100 reforms across six African countries, Williams (2025) finds that top-down approaches that focused on structural changes through one-shot projects largely fell short of their goals. Instead, decentralized processes aimed at continually improving systems were more likely to reach their goals. In the Indian case, in order to implement on-demand architecture with the technical expertise and fast adaptation required at its scale, we propose using agile development, which, along with asynchronous onboarding, can make decentralized changes possible.

Agile development refers to a set of approaches in software development that focus on adaptive planning, fast delivery, constant learning and improvement, and rapid response to change. The principles refer to evolutionary software development through constant collaboration and communication with planners, developers, and end users. Rather than solving problems at the end of the project, development is done in 'sprints' of 2–6 weeks. End users are presented with the software at the end of each sprint; they identify problems, allowing developers to iterate on the fly. Agile development therefore creates an environment of adaptive learning when challenges come up, and fast response to such challenges.

We posit that the on-demand architecture we outline in this book can become a reality in 5–10 years provided all departments undertake agile development. In fact, the Ministry of Electronics and Information Technology (2019) offers the Agile India Enterprise Architecture (IndEA) Framework to promote adoption of agile practices by various parts of government.[4]

[4] 'IndEA provides a generic framework consisting of a set of architecture reference models, which can be converted into a Whole-of-Government Architecture for India, Ministries, States, Govt. Agencies etc. The IndEA framework is based on a federated architecture approach and recognizes the need to accommodate both greenfield (new) and brownfield (existing / legacy) e-Governance initiatives' (Ministry of Electronics and Information Technology 2018).

Despite being a concept originating from computer sciences, the agile approach offers significant potential; it need not be restricted to software development. Rather, agile development can be envisioned as a broader strategy for policy-making, like in many other contexts. Instead of developing tools piloted over long time horizons, we advocate that changes in PFM (whether digital or non-digital) be designed, tested, and revised in smaller bursts with the end goal of testing and researching the intended and unintended consequences. In this light, it is important to imagine agility as a property of the administrative system, and not just how software is procured and developed.[5]

One way to support agility as a broader strategy for policy-making is to use digital Lego blocks that may or may not have been made using an agile approach. These blocks—many of which are digital public goods—can be arranged and rearranged more quickly in response to emerging policy-making needs, thus allowing for optionality as we advocate for earlier. Often, we observe that policy-making slows down because the digital tools in which the policy is embedded are too slow to change, where the tail wags the dog. Take the example of Direct Facility Financing (DFF), a model of funding primary healthcare systems that is intended to reduce administrative delays and inefficiencies in allocation of resources for public health. Binyaruka et al. (2024) note that many low-income and lower-middle-income countries transfer funds directly to publicly funded health facilities to facilitate the flow of resources, expand healthcare provider autonomy to allocate and use public funds, use output-based payments to link resources to services, and improve facility financial management.

Without knowing the context of any particular geography, we do not claim that DFF would be appropriate in every single administrative unit in India. However, if policymakers in any particular state government, urban local body, or panchayat choose to implement DFF, the digital tools must be built in a way that allows the policymakers to make that choice, implement it in a way that works in their context, and if the choice doesn't lead to effective public health services, reverse or modify their intervention. Agility in digital systems becomes a precondition for the

[5] Agility also aligns with Problem Driven Iterative Adaptation advocated by Andrews et al. (2013).

ability to iterate on policy and administrative choices. It should be possible for a state government to accept and implement the learning that transfer of funds is more effective via urban local bodies in some places, and directly from state government in other places.

Agile policy development allows tools to keep up with emerging programmatic needs, and policy creation is independent of restrictions imposed by existing digital systems.

8.3 Asynchronous Onboarding

While agile development improves an approach to how digital systems are developed, another question persists: When do different levels of government come on board across such a vast country? How is one to co-ordinate government programmes across a country by letting a thousand flowers bloom? The answer, in part, lies in the distinction between unified and uniform systems, in that solutions can be part of one unified system—all connected to each other—but they do not have to be uniform and can have different forms or functions based on their contextual requirements. As Pramod Varma explains in 'Societal Platform' (2020), such systems are possible when diverse solutions coexist on shared infrastructure. The data exchange mechanisms and data standards that we note earlier are examples of this shared infrastructure.

Shared infrastructure allows various parts of a network of systems to onboard themselves onto newer capabilities and connections at their own pace, or what is known as asynchronous onboarding. Digital infrastructure such as data exchanges can enable multiple parts of various government tiers to onboard themselves to exchange data as and when their policy objectives and implementation capability are aligned.

Interestingly, this approach can lead to even faster deployment of policy solutions, because the bar to be met is not so much the development of capabilities to adopt a solution or a software developed centrally, but satisfying certain requirements of programme outputs and information exchange by deploying existing capabilities to these requirements. For instance, in the case of fiscal information exchange, administrators can use their own existing software rather than being forced to work with a monolithic and centralized system. If their software does not meet data and fund management protocols, then there is need for negotiation as to how these requirements will be met for integration to work. This negotiation would lead to different plans and timelines for different parts of government.

One example of how centralized information systems can create limitations comes from the authors' experience while in government, when we sought to partner with the Forest Department to provide employment to MGNREGS workers by converting waste land into productive land. The MGNREGS software (called NREGASoft) at that time had no provisions for a state government department to provide money to another state government department. Instead, the state could only further allocate funds to a district, block, or gram panchayat. While changes to the software were technically possible, changes could only be made by an IT team in New Delhi. Given the various needs of MGNREGS managers across the country, it was administratively infeasible for a small team to address these varying needs. A federated architecture using open data standards where decentralized changes by state-level programme managers were possible would go a long way to solving the problem.

As illustrated by this example, one way to help various parts of an ecosystem develop new capabilities and be part of a bigger whole is to offer them digital public goods that meet the requirements and are adaptable to local needs, like digital Lego blocks. An important piece of work required in this approach is to understand what is context-specific, and what is context-agnostic for any given government programme. For instance, a notification is a notification is a notification. What is notified could be different across contexts. The notifier and notified would change with each piece of communication, but a notifier and a notified would always exist. Classification of various kinds of notifications would be different, but the fact that classification is required is universal. In fact, the UK's Government Digital Service developed 'GOV.UK Notify' precisely with this approach in mind (Herlihy 2020).

In this context, Principles for Digital Development[6] are a useful complement; we speak to some of these earlier, such as addressing privacy,

[6] A list of endorsers can be found here: https://digitalprinciples.org/endorse/endorsers/.

security, and reuse of digital assets. One of those principles is 'Use Open Standards, Open Data, Open Source, and Open Innovation' (Digital Impact Alliance 2019). To this end, open application programming interfaces (APIs) facilitate information sharing across platforms or devices.

Open APIs are definitions, protocols, and tools that allow different software to talk or sync across platforms. API-led systems already exist in India. For instance, the Passenger Reservation System allows travel agents and individuals to directly book tickets on a government portal, or through third-party websites and apps that link to the government portal and exchange data. The Ministry of Electronics and Information Technology has also already issued implementation guidelines for the adoption of open APIs (Ministry of Electronics and Information Technology 2015, 2020). PFMS and the Office of the Controller General of Accounts have already noted the need for an open API framework for integration with multiple systems (CGA 2024). It would be wonderful to see governments taking this idea forward.

When data is protected adequately, open APIs can allow stakeholders to onboard asynchronously, and softwares developed across departments and states can talk to each other. With quick, iterative development and open communication across platforms, states and departments will be able to create fast and federated solutions to their PFM reforms that meet their needs and, at the same time, can talk to other state and department platforms.

One key idea that should not be lost with asynchronous onboarding using open APIs is that of holistic reforms. It is indeed possible to pursue specific components of reforms on their own that are disconnected from other pursuits. For instance, creating just-in-time funding can be done without smart payments or data liquidity. Adopting solo systems may resolve some issues—in this case, ensuring that funds reach in time to some parts of the ecosystem—but not others, such as ensuring that programme officers are incentivized to make swift decisions to authorize payments. Of course, piecemeal changes are easier to pilot or solve and may allow individuals to onboard with time. However, if changes are made individually, and without a unified and systemic end goal of larger PFM reforms in mind, they may limit the effectiveness, scale, and sustainability of reforms.

8.4 Institutional Changes

With the right incentive structures and the principles of agile development and asynchronous onboarding in mind, who in government can enact changes in the architecture of PFM? Because we propose that the core work be undertaken around the creation of data standards, protocols, and regulations for individual PFM systems to speak to each other, the question of who enacts the changes relates to who creates these protocols. This pertains not only to legal authority, but also to a more fundamental question of state autonomy. Any reforms will have to ensure that states have a say in how data protocols and standards are created.

There are two pathways to enacting reforms: by creating new structures or using existing structures. The typical way the government does so is through entities called National Information Utilities (NIUs). An NIU is a jointly owned entity between the Government of India, state governments, and private financial institutions that allows capable IT talent to be hired for government projects. While the government can focus on policy creation and its enforcement, the NIU can oversee the development and implementation of IT systems to enable policy-instituted structures (TAGUP 2011).

A primary example of an NIU is the Goods and Services Tax Network (GSTN), a non-government, private limited company[7] that created and maintains the infrastructure for the Goods and Services Tax (GST), a single, comprehensive tax in India. With the guidance of state and central government officials who sit on the Board of Directors, GSTN undertook the development of a front-end GST Portal which serves as the overall IT system for the project. Through open APIs, the system interacts with those created by the Central Board of Excise and Customs (CBEC) and states to enable auditing, approvals, and assessments.

The GSTN, as a jointly owned entity, benefits from both governmental and non-governmental expertise. The composition of its Board and relaxation of rules regarding deputation of government officials allows

[7] Notably, it is subsequently in the process of being converted to a fully owned government company, owned by both the state governments and the Union government. See https://www.gstn.org.in/about-us.

for strategic control and guidance by the government. At the same time, GSTN has independent management structures, freedom to recruit top talent, and a self-generating revenue model. These functions allow the organization to build complex IT systems with speed and reliability, as envisioned by the TAGUP report.

As described above, an NIU is one pathway for enacting an on-demand architecture for PFM. However, there is a simpler solution. The Office of the Controller General of Accounts, despite being a government department in the Ministry of Finance, is already an authorized agency for enacting PFM-related changes, at least for Union government funds. As the institution that has created PFMS, the platform we discuss in this book, it can take on the necessary changes to convert PFMS to work with open APIs or other data exchange mechanisms such that it can integrate with other systems.

The CGA currently lacks the authority to set data standards and protocols for expenditures funded by states. However, in its work on Union government fund transfers, NITI Aayog is already working on data standards to connect PFMS with state FMISs. The CGA can work with institutions like NITI Aayog—which has a representative from each state[8]—to create a working group for setting data standards for nationwide expenditure recording and reporting. These standards can generate the underlying digital infrastructure to co-ordinate and couple programmatic and fiscal data, particularly within PFMS—the primary expenditure management pathway for Union government-funded expenditure. Existing departments and programmes can then create programme-specific digital workflow management systems that connect to PFMS. Similarly, the Comptroller and Auditor General (CAG) can design an auditing platform based on their own needs. Data from these information systems can be used to develop public transparency systems. There are several models for creating public transparency systems, including data.gov in the United States, or NITI Aayog's National Data and Analytics Platform (which currently lacks real-time, unit-level programme and expenditure data).

[8] Each state became a member through Resolution No. 511/2/1/2015-Cab dated 1 January 2015.

In a similar vein to this model, there are institutions that already focus on data quality that can expand their work to improve data credibility standards. Most prominently, NITI Aayog has created a Data Governance Quality Index (DGQI), which assesses Ministries and departments of the Union government for data generation, data quality, and data use practices based on self-reported data. In the absence of back-checks by NITI Aayog of the self-reported data, the developed indicators do not go far enough to ensure data credibility but are a good step forward in reaching this goal. Working with states' officials, NITI Aayog can create best practices for ensuring data credibility at the unit level. Moreover, some states are currently in the process of adopting DGQI. NITI Aayog can take the data credibility work further by creating certifications for data quality by performing random checks that do not rely on self-reports. These steps can help government bodies, departments, and agencies ensure that data meet certain credibility standards.

Of course, departments will have to take on significant work to define processes for the exchange of information more broadly, and information about funds more specifically. What we propose is not a simple task, nor do we expect it to be. As Andrews (2013) points out, any changes in PFM systems call for a network of agents to align. In India, this may involve a wide range of actors such as the Ministry of Finance in the Union government, line ministries, the Controller General of Accounts, the Comptroller and Auditor General, state finance departments, state line departments, the Accountant General of the state, district administration, and local administration. Unsurprisingly, this is no easy feat to achieve. However, by suggesting the use of existing institutions instead of insisting on creating new entities, and building on existing good work, we address the pitfalls of centralized, monolithic, and top-down changes. Instead, this approach allows for agile development and asynchronous onboarding as different levels of government become ready to take on reforms.

Whichever of the two pathways is adopted, a sustainable and useful approach will be to fund technology as teams, not projects. Often, improvements in public financial management, especially ones that involve deployment of digital technologies, are undertaken by funding a project with a limited time horizon with rigid contours. These tend to fail or underperform, for this funding model is not conducive for agile methodologies that can meet user needs as they continue to emerge. When software is procured as if it were a static product, it becomes both worse and more expensive (Pahlka 2024). And that happens because 'software is never done' (McQuade et al. 2019), and projects are engines of legacy creation (Loosemore 2018). Sometimes, this approach is also called 'fund products, not projects' (Narayan 2018; Thomas 2018). Two successful examples of funding teams, not projects are the Government Digital Service in the UK and 18F within the U.S. General Services Administration. To stay agile and meet the needs of stakeholders across the spectrum of public service delivery over time, governments need to fund teams to acquire ongoing, persistent, and flexible capabilities, not static things (Mann et al. 2021; Pahlka 2024).

8.5 Conclusion

In this chapter, we discuss broad and concrete ways to enact an on-demand architecture in India. We first advocate for using PFM systems that are unified across public departments and ministries as a tool to co-ordinate an on-demand architecture. Given that finance functions are similar across the varying activities of government, changes within these functions can either nudge departments to enact PFM changes through incentives or require them to adopt changes. While the former method is productive in most cases, forcing changes through rules and orders may be necessary in others.

In addition to using finance as a forcing function for improvements, we advocate for three core ideas. One, even smaller, piecemeal changes should be implemented with the broader big picture in mind. Two, enhancements ought to be enacted using agile development, which allows policymakers to make, test, and revise new capabilities in short time frames. And three, changes can be enacted in a way that allows stakeholders to onboard asynchronously using open APIs, standards, and protocols. These approaches can make on-demand architecture a reality.

The architectural shifts we discuss—including the system properties—are not a pipe dream and have already been adopted by the Ministry of Health via the Ayushman Bharat Digital Mission, Ministry of Housing and Urban Affairs via the National Urban Innovation Stack, and Ministry of Education via the National Digital Education Architecture. These systems use ideas such as open data standards; registries; extensible systems that can evolve with time; unified systems over uniform systems; interoperability; digital Lego blocks; platform thinking; and observability. Such ideas have yet to be incorporated into the architecture of PFM, which can expand on the foundational work started by these developments.

9

Conclusion: Risks and Opportunities

Scholars and practitioners alike have long sought to improve public service delivery, largely by fixing issues that are visible at the surface—bureaucratic monitoring, incentives, and institutional reforms. However, there exists an underlying plumbing of public financial management (PFM) that fuels public service delivery through financial resources. How do flow of funds and information in PFM affect public service delivery, and how can they help facilitate rather than hinder delivery?

In this book, we highlight the many ways that PFM can be connected to public service delivery failures. At its core, these issues result from two architectural problems in PFM. One, the fiscal data in different information systems is often asynchronous and unconnected, creating uncertainty in the availability of funds. Two, fiscal and programmatic data from public services is not always synchronized, making it unclear how expenditure relates to outputs and outcomes. Moreover, data is not observable to stakeholders across systems, meaning those who are authorized to have fiscal and programmatic data do not have ready access to it, and must rely on intermediaries.

These dual challenges of synchroneity and observability—both within and between fiscal and programmatic events—result in several key issues for public service delivery. They result in float or parked and unused funds; institutional opportunities for corruption; administrative burdens for officials implementing schemes; disincentives for bureaucrats to approve payments; inadequate data sharing to enable adequate auditing; and poor transparency on how much government spends and its outcomes. Alongside, programme administrators and anyone working with the government face uncertainty on when they will receive funds for any particular government programme. Such problems are hindrances to developing state capacity, and more specifically, to the efficient delivery

Public Financial Management, State Capacity, and Public Services in India. Santhosh Mathew, Bhumi Purohit, and Devesh Sharma, Oxford University Press.
 DOI: 10.1093/9780198973195.003.0009

of public services. After summarizing our proposals for changes in this chapter, we end the book by examining the risks and opportunities in pursuing PFM reforms.

We offer three key architectural shifts to resolve the issues of synchroneity and observability in India's PFM. Overall, these shifts, combined with changes to system properties in the way that data is managed, can ensure public finance is in service to public service delivery itself. One, we propose that fiscal and programmatic data should be coordinated so that there is two-way visibility for both types of information. This can be achieved through data exchange standards and protocols, which allow different systems to share data in agreed-upon formats and be linked. Two, fiscal and programmatic events should be coupled in a way that when programmatic events occur, they automatically trigger corresponding payments. This is feasible through an idea known as Rules as Code, which uses uploaded records and algorithms to determine if payment conditions are met. Three, fiscal and programmatic data should be observable such that authorized officials can readily see data at the primary unit of activity.

These three architectural shifts can transform India's PFM and its public service delivery. Given that these shifts are at their core related to data, we offer four system properties that are essential to data management to support these changes. One, and crucially, the system should have built-in data protection by design, meaning that it should comply with robust data privacy and protection standards. Data should also be stored, viewed, and reviewed with repeated consent of individuals whose information is in government systems, and data should only be selectively available to authorized government personnel without personally identifiable information when possible. Two, the system should have data liquidity to the extent possible, such that data can be recombined and reused in new ways. This is feasible when data is captured at the unit level. Three, data capture should follow the once-only principle, meaning that data is only captured once when a unit-level event occurs instead of in duplicative forms. Four, data should be credible, achievable through a range of efforts such as randomly back-checking samples of data, or using foundational registries for data use and analysis when possible.

These architectural shifts and system properties can lead to a variety of new outcomes in PFM. We review three such possibilities in the book. One such outcome is just-in-time funding, which entails direct transfer of funds to recipients from the Consolidated Fund of India and States without the inefficiencies caused by float. In addition to similar recent changes such as SNA-SPARSH which are limited to certain types of schemes, we argue that just-in-time funding should be a feature of every scheme in the country and should entail recording the unit-level programmatic data for activities that funds are supporting. Another outcome is smart payments, which are automated payments that are processed once there is credible proof that all the prerequisite conditions for payment have been met and this has been recorded in a system. A third outcome is creation of user-level platforms for programmes, auditing, and public transparency which can link fiscal and programmatic data and offer a range of data analyses.

Architectural changes like the ones we discuss may seem daunting at first. Not only do they require a government to examine all its existing financial rules and regulations, but they also need it to determine the least disruptive ways to enact the changes. Once enacted, any new features must be constantly monitored to ensure new systems work as intended, without negative externalities. Yet many states and national governments have overseen reforms that required similar high-stakes planning—take the Goods and Services Tax, the National Urban Innovation Stack, the Ayushman Bharat Digital Mission, or the National Digital Education Architecture, for instance.

In order to ensure that changes in the PFM architecture in India are enacted with ease in the future, we offer three principles. One, changes in the architecture should incentivize users to come onboard whenever possible rather than force new systems in place. Along these lines, two, stakeholders should be allowed to onboard a system asynchronously using open application programming interfaces (APIs), meaning they need not be required to use all parts of a new system, nor forced to use parts of the system all at once. At the same time, changes should limit the costs to stakeholders that stem from having highly centralized digital systems. To address this challenge, we thirdly propose that both technology and policy-making incorporate the design philosophy of agile development. Here, agile development is the testing of new programmes, policies, or technological platforms with constant research, extensive trials, and iterative adaptation. This approach, which aligns with research that finds decentralized but continuous changes work better than one-shot, top-down ones (Williams 2025), can allow stakeholders to create policies and digital platforms with adaptive planning, fast delivery, and rapid response to change. With these ideas in mind, changes can come to fruition with reduced risk to public services and government functions.

Though any systemic changes in the PFM system are by nature cumbersome and complicated, we argue that PFM needs to be transformed to incorporate on-demand architecture, with the goal to modernize the pipeline that fuels all government policies and, ultimately, support quality public service delivery. Why, though, would governments enact such complicated changes in the architecture? We address this very question next.

9.1 Addressing Risks

At its core, we recognize that an on-demand architecture seems like a significant risk for governments for two major reasons. One, even after implementation, it will take years for changes to bear fruit, potentially beyond the term of any elected government. Without parties knowing with certainty that they will remain in power for years to come, why would they risk such changes? Two, the proposed architecture could shrink financial sources that support electoral campaigns and the maintenance of patronage networks. While the mix of these financial sources is hard to decipher, the risk of losing streams of revenue could strongly deter actors from undertaking institutional change (Bussell 2018; Kapur and Vaishnav 2018). In this section, we address these two concerns.

There could be adequate incentives for those in government to take on institutional changes because it can signal that the state is serious about 'good governance', which in some contexts has shown success in bringing investments (Fazio and Chiara Talamo 2008; Li and Resnick 2003). Several governments in India have used good governance as a plank to push policies such as Direct Benefit Transfer (DBT) and the Goods and Services Tax Network (GSTN).[1] While the comparative efficacy of these

[1] For instance, a major plank of the BJP government during its 2014 election campaign was that they encourage investment, case in point the Make in India initiative. Similarly, the

campaigns to other strategies is up for debate, what is clear is that a public goods, investment-driven strategy has been part of many political parties' electoral platforms in recent times.

The public finance reforms we propose are a step towards good governance, and have improvements in data connectedness and data sharing at their core. Whereas policies such as the Goods and Services Tax (GST) show that the government is serious about streamlining revenue collection, an on-demand architecture can make certain that expenditures are better managed and streamlined. Smarter, data-driven expenditures can help ensure that revenues are used wisely while limiting excess borrowing. Additionally, they can ensure that investors who fulfil government contracts receive timely and accurate payments, and the outcomes are visible to the polity through transparent systems. Ultimately, an on-demand architecture provides a host of incentives for those who look at good governance as a strategy. This is increasingly clear from recent moves such as SNA-SPARSH for expenditure management.

For states that rely on broad public support, the proposed on-demand architecture can provide incentives to ensure good public service delivery. Some literature suggests that politicians rely on clientelism (Stokes et al. 2013)—the exchange of non-programmatic goods for votes in the form of vote-buying or patronage, for instance—to gain public support during election time. Others suggest that a reliable and productive reputation can be built by informing voters of their policy promises and actually delivering on them. Others rely on credible intermediaries working in local communities and use them to provide goods in exchange for votes (Keefer and Vlaicu 2005).

Research by Keefer and Vlaicu (2005) and Heath and Tillin (2018) shows that voters will engage in clientelism, particularly by accepting low-cost goods, when state institutions are weak. When institutions are stronger, voters forgo short-term clientelistic goods for long-term policy promises. This important finding indicates that there is an opportunity in investing in institutions. Doing so can provide an alternative to costly forms of campaigning that require huge amounts of fundraising to finance vote-buying and monitor the quid pro quo of goods for one's vote

Aam Aadmi Party has regularly campaigned for good public service delivery, while the Bharat Rashtra Samithi has campaigned on improving investments.

(Kitschelt and Wilkinson 2007). Additionally, as Kitschelt and Wilkinson (2007) show in their edited volume, as incomes rise in a country, voters are more easily able to overcome short-horizon goods and wait for larger programmatic goods that benefit more people and deliver long-term benefits.

One instance where politicians invested in bureaucratic institutions to reduce the forces of clientelism occurred in Andhra Pradesh, where the early implementation of the Mahatma Gandhi National Rural Employment Guarantee Scheme (MGNREGS) was an unexpected achievement. The scheme faced threats from various groups, but found early support from the Chief Minister in 2004, who saw an opportunity in strengthening rural development, ameliorating corruption, and reducing the use of patron–client relationships in the scheme's implementation. By bolstering the rural development bureaucracy and equipping it with resources that both shielded it from local interests and built up its capacity,[2] the state saw a rare and successful early implementation of the scheme (Mukherji and Jha 2017). This success followed from a larger reform strategy in the state that included need-based redistribution as well as the evolution of self-help groups, both of which allowed bureaucrats to gradually reduce clientelism and improve the delivery of the scheme (Mukherji and Zarhani 2020).

As the above research indicates, public service delivery can provide an alternative way to campaign, especially in societies where people are increasingly coming out of poverty. However, making public services a credible platform requires the establishment of strong institutions like the one we discuss in the book, and which have been credibly and successfully taken on by other governments that face similar levels of clientelism, most prominently in Brazil (see Appendix 2).

That being said, research also shows that the uptake of reforms is more likely in some states over others. As highlighted by Bussell (2012), uptake of technology reforms depends on how much a party relies on grand versus petty corruption and whether a state has one-party dominance or

[2] Key institutional changes included creating labor-intensive work that did not incentivize contractors to get involved; creating a software to manage funds for the scheme; transacting funds from the post office, then directly through a bank; assigning a local-level functionary to keep track of the scheme; creating thorough records at the village level; and creating a social audit system (Mukherji and Jha 2017).

a multi-party coalition. Therefore, it is plausible that parties that depend on petty corruption may be less likely to enact on-demand architecture. Similarly, it is likely that states with one-party dominance can benefit much more from reforms that allow for improved, broad-based public service delivery. Parties reliant on coalitions may still prefer using clientelistic goods or goods aimed at particular ethnicities, religions, or classes to build group-based support to ultimately form a coalition with other parties (Chhibber and Nouruddin 2004). Despite varying uptake, the reforms do present a case for creating incentives that allow for programmatic delivery of goods.

9.2 Other Risks

Changes in architecture of something as pervasive as PFM also pose administrative and financial risks.

Administratively, an on-demand architecture runs the risk of getting limited support from bureaucrats, especially those situated closer to the frontline in blocks and panchayats who might experience changes to how they do their jobs. The lower the bureaucrats are in the administrative hierarchy, the more likely they are to be affected by increasingly digitized reforms that require the uptake of a new set of skills. Similarly, Treasury and Pay and Account Office employees may get the sense that they will be made redundant given that the technology we propose may operate some of these functions with much less oversight than is currently needed. Such discontent may eventually affect government functions.

Though such administrative pushback is likely, it is fruitful to look at how these risks have been mitigated in Kerala, Brazil, and Mexico (see Appendices 1 and 2). Reforms there were accompanied by a significant amount of effort in obtaining buy-in from government officials at all levels. It is crucial that those who will be working on a new system agree to reforms and understand how to operate computers and any new software. Such buy-in and training is crucial—personnel whose job functions will change should approve of the upcoming changes.

Changing systems also create opportunities for reduced administrative burdens on already overworked government employees. In our interviews in Gujarat with MGNREGS staff members, for instance, all employees we spoke with said they appreciated the reduced administrative burdens that come with digitized payments in the new system. They particularly noted their increasingly more manageable work hours and greater focus on monitoring progress rather than continually checking physical paperwork as they had previously done.

A second administrative risk, however, is that of feasibility. The momentum of adopting digital technologies and an on-demand architecture should be in lockstep with administrative capability. Technological reforms require vast administrative capability that addresses challenges across the entire value chain of affected programmes, without which governments risk shifting the locus of the problem rather than addressing the problem itself (Khera 2017). For instance, direct benefit transfers to bank accounts can address the delays from the point of payment to point of receipt of a benefit, but more is required to resolve rejected payments, diverted payments, or locked payments that occur when bank accounts are frozen. To create more holistic solutions, local administrators would need to be looped into the design architecture, so they can both understand these tools and use them to address localized problems with their jobs.

While we address solutions to these concerns—particularly those concerning delays—in Chapter 7 with just-in-time funding, Khera and other policy thinkers accurately note that technocratic solutions must be aligned with administrative feasibility. If not, beneficiaries risk paying the ultimate cost. However, it is our hope that when new PFM policies are enacted with agile development and asynchronous onboarding, as outlined in Chapter 8, stakeholders, including civil society actors and researchers, can engage with these policies, to ensure that past problems are not repeated. Agile development and asynchronous onboarding also allow those creating the systems to constantly check that the systems work and fix them where needed before fully enacting them, ensuring growing pains are addressed.

Moreover, a crucial part of the phased approach is that the teams in charge of changes to PFM architecture and developing user interfaces can undertake rigorous testing through human-centric design, directly with end users and beneficiaries. Only by understanding key problems and identifying potential solutions prior to implementing them—whether technological or not—can we ensure that changes serve beneficiaries and not merely policymakers or policy implementers.

Lastly, on-demand architecture may seem like an expensive ordeal, dissuading policymakers. Technical changes that take years, ultimately, may be deemed unaffordable in a country focused on economic growth and fiscal sustainability. However, it is important to note that in the long run, the benefits significantly outweigh the costs. In addition to potentially saving Rs. 1 trillion (USD 12.35 billion) in float if float is eliminated, these changes will decrease the need for borrowing, thus impacting GDP growth. On top of such benefits, the government will also save money by reducing corruption and improving auditing. Such reductions have been witnessed in other governments. The government of Mexico, for instance, saved USD 17 billion per year, reducing the risk of high initial costs (Babatz 2013). With proper planning, changes in the architecture of PFM in India can see similar success.

9.3 The Need for Further Research

Political, administrative, and financial risks can be mitigated, but only with adequate research, monitoring, and evaluation. In order to ensure that an on-demand architecture is implemented with rigour, it is crucial to ensure that researchers are welcomed into the process to monitor and evaluate the different phases of implementation. Taking an evidence-based approach can ensure that lessons from one phase are carried into improvements for another. Similarly, rigorous monitoring and evaluation in one part of the system can lead to drastic improvements in an existing system and carry over into a new one.

We also enthusiastically invite scholars to test the claims we posit in this book. As we state in the introduction, this book seeks to describe the current PFM, lay out the consequences we have observed based on the state of current research, and propose a possible set of solutions based on existing knowledge and context. We in no way claim to provide the panacea of solutions for all PFM issues, and we acknowledge that the solutions we posit are based on existing evidence, much of which is not causally identifiable due to the limitations of how PFM currently operates and is studied. Based on our experiences and learnings from others, these ideas have also evolved over time, and we recognize they will continue to in the future as more knowledge and learnings offer new insights. While we posit that the ideas we discuss will yield good public service delivery, we hope policymakers will ensure that scholars have access to the process and can test the costs and benefits of the changes.

We particularly hope that researchers will test how PFM reforms affect field-level bureaucrats and beneficiaries. Ultimately, these groups will be most affected by the way in which funds for public services are delivered. It is crucial that rigorous research accompanies the changes to ensure that a new system works for these groups, and where it does not, adequate policy changes can be made.

Overall, the proposed on-demand architecture outlined in this book is ambitious, but not impossible. The architecture builds not only on reforms taken on in other countries, but on those enacted by various states in India. Recent advancements such as the Goods and Services Tax (GST) infrastructure, the Ayushman Bharat Digital Mission, and the National Urban Innovation Stack have elucidated the government's ability to set up institutions that allow for changes in the architecture.

To end where we started, the tragedy in August 2017 that unfortunately occurred at the Baba Raghav Das Medical College Hospital near Gorakhpur, Uttar Pradesh—to the extent it relates to delayed payments—could have been avoided if an on-demand architecture existed at the time. When the hospital's supply of oxygen started running low, a software could have alerted the hospital to request a payment. The hospital officials could have directly requested the payment with the necessary paperwork attached for oxygen supply through the software, which would have been reviewed by a local government official before being pulled from the Consolidated Fund of the State of Uttar Pradesh. These funds would be sent to the oxygen supplier's bank account within 24 hours of the submission of the request. The hospital officials would be alerted to the payment and could follow up with the suppliers to ensure the delivery before their current supply of oxygen runs out.

Such a system is not too dissimilar from how many private institutions operate, nor should it be a pipe dream for public institutions. In fact, a system such as the one we describe in this book is crucial for governments in India on whom millions of citizens rely for the most basic of services, from food rations and water to emergencies such as children's hospitalizations. If the government is to ensure that some of the world's most progressive social policies are in fact implemented as intended, successfully updating its public financial management systems is a must.

APPENDICES

APPENDIX 1

Learning from Existing Systems in India

India's public financial management architecture is more than due for an overhaul, from the Union down to the grass roots. In order to make public services work, it is crucial to ensure that every rupee allocated for a public service is spent as intended. Though much of the government lacks the infrastructure and systems to make such claims, some states have taken the initiative to build their own integrated financial management information systems (IFMIS). We highlight two such states with advanced PFM systems, Karnataka and Kerala, and focus on key lessons from them for developing IFMIS. These case studies were developed in 2015; while they are not comprehensive of developments that states have undertaken since then, they provide a snapshot of the challenges that Karnataka and Kerala faced, and the reforms they implemented. Thereafter, we offer a proposal on how to use agile development to incorporate existing systems into a new on-demand architecture.

Karnataka's Khajane

Karnataka first began talks of modernizing its treasury in 1997, when it formed a committee to examine the state finance and treasury codes. The committee found major gaps in the manual system, which was failing basic functions, and recommended computerization of the treasury in 1998. According to a case study on the Karnataka treasury, 'validations were discriminatory, the processing time was erratic, accounts were replete with mistakes leading to rampant misclassifications and a system amenable to fraud and corrupt practices' (Singh and Bhattacharya 2013, 4). The committee discovered further problems with fund allocations to Drawing and Disbursing Officers (DDOs), lag in payment delays being reported, over-withdrawal of funds above the allocated limit, and failures in the reconciliation of funds (Singh and Bhattacharya 2013, 3). At the same time, the size of the treasury resulted in redundancy and ineffectiveness. The workforce included 30,000 Drawing Officers using over 27 types of billing forms, with 10,000 heads of accounts (Singh and Bhattacharya 2013, 4). Given the quantity of paperwork created by this architecture, calculating the total budget allocations and expenditures alone required approximately 300 individuals (Haidear 2016). This resulted in duplication of work, creating problematic and delayed monitoring and auditing.

Overall, the manual system created corruption and delays in programme finances, lacked transparency or accountability, and failed to provide the full picture of allocations against expenditures. These issues problematized the task of auditing, which is pivotal for ensuring public financial management is corruption-free. Such problems, as we have noted, are major sources of stagnation or failure of public service delivery.

To resolve these issues, the Government of Karnataka created Khajane, a first of its kind project in India to digitize the functions of the treasury. First established in 2001 as a pilot, Khajane provided a decentralized IT system to each of the 215 treasuries within the state that communicated to the main server at the Finance Department.

Khajane sought to resolve the problems created by the manual system. Each of the 215 treasuries' applications included modules for receipts, payments, deposits, pensions, administrative needs, and maintenance works. Through these modules, financial transactions and monitoring were digitized, allowing payments to take place within a few days. Additionally, all transactions were transparent and available on the application, allowing for easy monitoring, reconciliation, and auditing. The real-time and transparent nature of Khajane meant that DDOs' actions were visible immediately, thereby reducing corruption. Such transparency further allowed all government departments to monitor their financial status, allowing for adaptive budgeting. Lastly, the system allowed accounts to be rendered as early as three days to as late as one month, compared with the six days to three months it took in the manual system (Singh and Bhattacharya 2013).

Khajane's development not only created significant improvements in the architecture of the financial system, but also paved the way for good practices. For one, building Khajane required improving the efficiency of the financial system. In order to facilitate this, the Finance Department was restructured to 21,000 DDOs from 30,000; head of accounts were reduced to 3,432 budget lines from 5,000; and object codes were reduced to 174 from 500 (Singh and Bhattacharya 2013). These reforms involved reviewing and updating legislation and codes to allow for the use of IT. As with the restructuring, the Khajane development process also ensured efficiency. Each module in the project was developed with the supervision of an individual treasury team, allowing for agile and focused progression in programming the software. Upon development, an electronic system required both buy-in and training from treasury staff. Given that Khajane's development was partially motivated by the need to reduce fraud and corruption, the team sought buy-in by focusing on the improvements brought about by computerization. Namely, computerization removed 'the drudgery of manual accounting' and reduced both time and effort necessary to maintain accurate accounts and registers (Singh and Bhattacharya 2013, 11). Upon buy-in, the staff was thoroughly trained on basic computer use as well as Khajane.

Given that Khajane was a first of its kind system in India, it was accompanied by barriers. For one, transparency in the system was high, allowing officials to see allocations against expenditures. When payment processes went into disarray, however, DDOs had to enquire through an Interactive Voice Response System (IVRS) why payments were stalled or rejected, and by whom. Furthermore, many system processes were decentralized, along with vendors who provided them. Given that providers for hardware, operating systems, and uninterruptible power supply (UPS) were all different, along with application coders, implementation was delayed due to lack of coordination. Realizing such challenges, the government undertook a revision of Khajane to create Khajane II.

Khajane II underwent a 10-year development period and incorporated many of the learnings from Khajane.[1] The main advantage of the system was the integration of

[1] All information on Khajane II comes from a meeting with the Khajane II team on 28 June 2016 in Bangalore, Karnataka.

the servers sitting at various treasuries throughout the state. The Finance Department integrated a single e-platform for all internal stakeholders. Doing so allowed for centralized deployment and management of funds. Moreover, Khajane II is integrated with its external stakeholders—banks, the RBI, the Accountant General, and the Government of India—through PFMS. In order to serve its 3,600,000 recipients, Khajane II integrates 400 processes in real-time through its platform, and makes payments within five days, directly to the account of the beneficiary.

Learning from Khajane, Khajane II has automated many of the processes and options for bill generation and approval to minimize human error. When bills are generated by the Drawing Officer, they are automatically sent to each verifier and approved until they reach the agency bank. The agency bank transfers the money to the end recipients and informs RBI. In case a bill is not approved, a DDO can trace the source and reason for the problem and resubmit it. Such automation and consolidation under one server have further reduced delays and improved monitoring. The monitoring dashboards have also improved from Khajane to better fit user needs, amongst which is a citizen platform. Through an e-payment platform, citizens can request and track payments for government schemes, allowing for greater individual-level monitoring.

With better insight, Khajane II was developed by a single vendor, Tata Consultancy Services, to resolve issues of co-ordination. Though this reduced delay caused by co-ordination, it resulted in delays due to problems of institutional learning. Given that the time lag between Khajane and Khajane II was 10 years, many of the staff involved in the former project were not working with the latter. The software developers on board were new and unfamiliar with a public financial management system, hence developing Khajane II required relearning the steps, along with much trial and error.

Khajane II comes closest to parts of our vision, though many of its functions remain manual. As of 2016, for instance, the system was yet to be integrated with the Accountant General's office. Given that the Accountant General's codes and rules do not allow for automated file submission, the office mandates physical copies of documents. Internally, not all modules are integrated or automated. With the exception of e-payments and e-procurement, other efforts require manual entry. Auditing and human resources teams, for instance, must transfer data from hard copies to the application, creating redundancy in the process. However, Khajane II will undertake improvements over the next few years to fully enable automation and digitization.

The experience of Karnataka provides three critical lessons. One, buy-in from district, block, and local-level staff is a difficult task and requires much time and effort. Learnings from Khajane point to the need to incorporate buy-in early on in the programme, with the inclusion of local officials in the process of software development. Doing so is vital to ensuring that manual functions can not only be digitized but also be managed and implemented by staff.

Two, full and fast automation is a pipe dream. Karnataka's experience since 1997 shows that the process of digitization is a multi-stage process that begins from a deep study of existing structures, rules, and business processes. Digitization requires changing existing codes and re-engineering procedures before manual processes can be replaced. Even then, automation is undertaken over incremental stages—by function or department—to facilitate agile development. Only then can lessons from one stage be improved upon in the next. Doing so requires project managers to co-ordinate across various teams, each working on automating their own processes.

Three, institutional learning and adaptation are vital. Tasks must be consolidated to meet programme-specific needs, but oversimplification of a complex process can also create challenges. In order to overcome such barriers, it is crucial that the system runs through open application programming interfaces (APIs) to enable IT systems across programmes to be able to speak to each other. This would allow each programme to create its own digitization processes and application through an external vendor in a way that connects to an open API. This is crucial to ensure that state-level or national-level digitization can be achieved in under a decade. More importantly, this is key to agile development that allows each programme to ensure its application works for its own functions as well as for its own users.

Kerala's Reforms[2]

Kerala's experience with PFM has been unique for two reasons. One, Kerala's local governments are more independent and autonomous than those elsewhere in the country,[3] and two, the state has had a strong participatory planning process (Chathukulam and John 2002). These features of the political process, which are much more focused on grass-roots processes than most states, have translated into the public financial management system. Kerala's PFM consists of three distinct but interconnected pieces—the Integrated Financial Management System (IFMS), administered by the Kerala State Department of Finance; PlanSpace, administered by the Kerala State Planning Board; and a large suite of application packages for local government bodies, administered by the Information Kerala Mission.

Integrated Financial Management System (IFMS)

Computerization commenced in Kerala in the Directorate of Treasuries in the late 1990s, but was limited to the digital preparation of accounts.[4] In 2000–01, the Government of Kerala, facing a severe financial crisis, approached the Asian Development Bank with a proposal for assistance to comprehensively reform financial management in the state (Asian Development Bank 2000; Jeromi 2005). An analysis of the state's fiscal management processes found problems in budget management, expenditure management, and local government finances, amongst other areas. The institutional framework for budgeting and financial management suffered from 'an imbalance between policy objectives and available resources' and 'inadequate

[2] This section, 'Kerala's Reforms', is authored by Vikram Srinivas and reproduced here with his permission.

[3] There are a number of anecdotes to confirm this, though systematic, all-India comparative studies of decentralization are few. Consider, however, the Devolution Report (Ministry of Panchayati Raj and Indian Institute of Public Administration 2024) in which Kerala ranks number 2 in the nation as of 2024.

[4] Based on interviews with stakeholders.

fiscal transparency' (Asian Development Bank Board of Directors 1999). Decision makers who needed credible intelligence to increase efficiency had limited means of obtaining such information. Similarly, local governments lacked the resources and capacity for financial management. To overcome such crucial defects of the system, the Kerala Government committed to an ambitious reform package.

First, the government computerized all the treasuries in Kerala by 2004. Initially functioning as stand-alone systems, these treasuries were linked through dial-up connections to a central server in the Directorate of Treasuries by 2006. For monitoring, the Directorate of Treasuries received summary statistics of transactions on a daily basis (Raghunathan and Nisarudeen 2007). These statistics were available centrally, but the rest of the data was far from fully integrated into a state-wide digital system. The modes of payment remained undigitized, and most sub-treasuries remained manual.[5] Regardless, the reforms improved operational efficiencies. Procedural reforms also accompanied the computerization. For instance, treasury receipt forms were rationalized from 19 to 9.[6] The reforms also led to a new integrated payroll and HR management system—the Service and Payroll Administrative Repository for Kerala, or SPARK. Initially introduced on a pilot basis, it was progressively adopted by a large number of departments by 2007 (Kerala State Government 2000).

This system remained in place, substantially unchanged, till the next phase of reform started in 2014–15. The Government of Kerala declared their intention to introduce an Integrated Financial Management System (IFMS), 'combining various online systems on receipt and expenditure by integrating State Budget with treasury transactions relating to line departments, Accountant General, RBI and others' (Finance [Streamlining] Department 2016). The IFMS was developed mainly by the National Informatics Corporation (NIC), which had executed similar projects in other states and at the national level.

IFMS significantly improved upon the earlier model. Unlike the earlier system of stand-alone servers, the new system has one central server located in the Directorate of Treasuries, allowing all other treasury and sub-treasury offices to access this server on a real-time basis. This system, called the core-TIS (Treasury Information System), serves as the backbone for the financial transactions of the government. All types of receipts and payments into government accounts are processed by the TIS. Thus, stakeholders can access real-time data on the government's expenditure progress on a real-time basis.[7]

Moreover, a number of different modules have been introduced under the new system. Major modules[8] in the IFMS include:

- Budget Management System (BMS), which allows officials to submit budget proposals to the finance department.
- The Budget Allocation Monitoring System (BAMS), which helps in allocating and monitoring budgets. Once budgets are approved by the legislative

[5] From interviews with stakeholders.
[6] From interviews with stakeholders.
[7] From interviews with stakeholders.
[8] From internal presentations of the Kerala Finance Department and interviews with stakeholders.

assembly, the Finance Department posts appropriate allocations to secretaries of line departments. Secretaries may then further allocate funds, as appropriate, to controlling officers, who may further allocate funds to DDOs. As and when expenditure occurs, the available allocations in the BAMS are automatically reduced, so that any officer can monitor the remaining budget for various activities in their jurisdiction.

- The Bill Information Management System (BIMS), which helps automate authorization and payment of contingent bills. When a bill is presented to a treasury officer to be duly authenticated by a DDO, the DDO may only authorize it if there is sufficient unspent allocation for the particular head of account in the DDO's jurisdiction. Once authorized, payment is electronically made to vendors, citizens, government officials, etc.
- A system for online receipt of money due to the government, called e-Treasury, which functions as a government payment gateway. Any department can integrate e-Treasury into their website to enable payments for a wide variety of services, including tax payments, fines, stamp duties, etc.

In addition, existing software such as SPARK was integrated with the IFMS. Modules were developed for a number of functions such as the management of pensions, ways and means, stamps, guarantees given by the state government, and other services.

Another useful feature of the core-TIS is its ability to pay electronically through the RBI-administered e-Kuber system. All bills presented at a treasury are translated into electronic pay order cheques (e-POCs) digitally transmitted to the Directorate. The Directorate sends a consolidated e-POC to the RBI through e-Kuber, which is integrated with the transfer systems administered by the National Payments Corporation of India. Thus, the amount is debited from the state government's account with the RBI and transferred to the respective banks, and then to accounts, whether of beneficiaries, vendors, or officials.[9]

Local Governments

Financial management reforms for local governments in Kerala started in 2002, with the introduction of the Model Accounting System. The system was developed by the Local Self-Government Department with technical guidance from the Comptroller and Auditor General of India, who had published similar model guidelines. This was a single-entry, cash-based accounting system, unaccompanied by any computerization of accounts. After extensive training, it was rolled out to all panchayats in Kerala in 2004.[10]

In 2010, a cross-departmental team was created to draft rules to enable a double-entry accrual process. These rules came into effect as the Kerala Panchayati Raj Accounts Rules in March 2011, supplemented by an accounting manual with

[9] From internal presentations of the Kerala Finance Department and interviews with stakeholders.

[10] From internal presentations of the Kerala Finance Department and interviews with stakeholders.

statutory backing. The introduction of double-entry accounting on an accrual basis, however, necessitated computerization. The Information Kerala Mission (IKM), an autonomous arm of the local self-government department, developed an accounting software called Saankhya based on the new rules. The system was rolled out phase-wise from 2012; it managed to cover all panchayats in Kerala by 2016. Since its development, all panchayats are obligated to discontinue manual financial statements, and instead prepare them through the use of Saankhya. To enable further digitization, the IKM has developed a suite of 16 software applications to meet the various needs of panchayats. After Saankhya, the most commonly used software is Sulekha, which enables digital means of planning. Others include Sevana for registrations and social welfare pensions, and Sanchaya for revenue processes. While Saankhya, the accounting software, is a stand-alone software, all other applications are networked to a central server.

The move to accrual-based accounting systems has been a long-standing goal of Indian public finance, with the 12th, 13th, and 14th Finance Commissions recommending that all levels of government move away from the present cash-based system. In Kerala, local governments have successfully made this shift. While an exhaustive evaluation of the shift and the management information system (MIS) used to manage local finances is yet to be conducted, there is scope for more work to be done.[11] In particular, a stronger training in accounting practices is necessitated to significantly strengthen budgeting practices and fiscal responsibility of local bodies.

PlanSpace

An additional reform in Kerala is the use of PlanSpace, an information system for planning and monitoring progress of plan schemes. PlanSpace was first conceptualized in 2009 and was developed by the Indian Institute of Information Technology and Management, Kerala. At the time, plan monitoring was divided between the State Government's Central Plan Monitoring Unit (CPMU) and the State Planning Board. Today, PlanSpace is administered by the Kerala State Planning Board, an advisory board for the government which prepares an annual comprehensive economic review of the state.

PlanSpace maintains data on physical progress and financial progress of the budgets for various schemes. Heads of departments post information on this system, usually on a monthly basis. Each scheme can be subdivided into an arbitrary number of components, which line departments can define based on the scheme requirements. Departments then assign each component a financial allocation and a physical target. PlanSpace data, entered by heads of departments, is used for state-level progress review meetings headed by the Chief Secretary. To guide local implementation, district officers can also enter data in a parallel system. This data is not collated with the centralized data entered for the whole state by heads of departments. Instead, it is used at the district level for monthly District Development Committee meetings

[11] This is no easy feat; it took over 10 years for UK government to transition from cash accounting to accrual-based budgeting and accounting.

headed by the District Collector and the Zilla Panchayat President. In addition, when annual or five-year plans have to be prepared, the data from PlanSpace is invariably an input into the discussions between the Planning Board and the line department's ministers or secretaries. Overall, the software serves an important function of providing district- and state-level policymakers with up-to-date data on schemes.

PlanSpace's functionality, however, is limited by the fact that it is yet to be integrated with IFMS (or other software) for monitoring and evaluation that are operated by schemes. PlanSpace data, therefore, must be manually entered either at the district or state level and leads to duplication of work. Additionally, the bulk of the data in the system concentrates on financial indicators. While provisions for physical progress and annual highlights of schemes exist, these are much less comprehensive. This is more of a process issue than of the PlanSpace platform. Financial progress for schemes is easily measured in terms of rupees, while physical output indicators need to be defined for each component of a scheme and may not lend themselves to easy quantification or measurement. Without the latter, however, assessing outputs and outcomes of financial resources becomes difficult. Despite these issues, PlanSpace is still a progressive step towards monitoring governance.

Kerala's IFMS is a significant step towards the vision we highlight in this book. A system which integrates budget preparation, allocation, monitoring, and execution goes a long way in reducing the load on officers at all levels—from state to block. Areas of improvement, however, exist. As of 2016, the budget preparation or monitoring process has not integrated systematically with programme data. Indeed, physical indicators (of any sort) are missing from the IFMS altogether. In addition, linkages with the Accountant General's office for accounting and audit were still primarily manual, though progress was being made towards integration.

The existence of three similar systems in their largely independent domains comes with its share of both advantages and disadvantages. The primary benefit: since these systems were being developed specifically for the needs of particular sectors, they were more in tune with the requirements than a single monolithic system could have been. For instance, revenues raised and spent by autonomous bodies such as the state electricity company or universities were outside the ambit of the IFMS, since these revenues did not touch the Consolidated Fund of Kerala. However, PlanSpace captured these revenues since they were part of the State's Plan. Similarly, IKM has developed a vast range of custom applications catering to various needs of local governments, which would have been improbable if they had depended on IFMS alone.

The weaknesses of the multiplicity of systems are a degree of fragmentation. Linkages between the three systems have remained weak—for example, IFMS data measures real-time expenditures, while PlanSpace is dependent on data entry for expenditure data. However, IFMS data is not used for PlanSpace as it is not available in a high degree of disaggregation and omits certain data and includes others. More seriously, planning and resource allocation are deeply integrated sides of the same coin, and separate systems for this go against budgetary best practices.

The success of these systems in Kerala also provides valuable lessons for implementation. First, all software applications were embedded in a functioning system with clear, demarcated roles. Where required, statutory changes, government orders, accounting manuals, and other rule changes were made in advance of the software being rolled out. Second, training was a high priority as evidenced by the various institutions dedicated to this function. The State Finance Department has a Center for Financial Training and Management, the local self-government department has IKM and the Kerala Institute for Local Administration, and PlanSpace has employed permanent project assistants embedded in each district planning office as well as a number of departments. Third, each system used one dedicated vendor. This ensured that a pool of skilled talent understood both the domain issues as well the technological issues involved in software creation and implementation. Lastly, the role of leadership and strategy has ensured the success of Kerala's reforms. Many officers in the government interviewed by the authors discussed the importance of not only adopting an incremental strategy and building upon existing systems where possible but modifying them where necessary as well.

Conclusion

The experiences of Karnataka and Kerala bring forth important lessons for country-wide reforms. In each case, starting with a comprehensive review of financial rules and regulations provides a robust basis for determining which components of PFM require modernization. Upon understanding the rules, moving forward requires a phased approach which can deliver a steady pace for adaptation and change when necessary. Moreover, ensuring that all stakeholders understand reforms and buy into changes is indispensable for smooth implementation.

Adapting these learnings to a national level is not a simple task. An on-demand architecture calls not only for the Union government to create a system that others can integrate into, but also for states to create their own adaptations for state-level programmes and schemes. Given that both states have taken at least a decade to enact reforms, following their exact path can lead to significant delays in implementing any on-demand architecture.

In order to implement the reforms that we suggest in this book (with the technical expertise and fast adaptation required at their scale), we propose using agile development and open application programming interfaces (APIs). Open APIs, which allow data to be shared across platforms or devices, are essential so that software developed across departments and states can talk to each other. With quick, iterative development and open communication across platforms, states and departments will be able to create fast and federated solutions to their PFM reforms that meet their needs and, at the same time, can talk to other state and department platforms.

Kerala and Karnataka have already undertaken the complex task of developing IT-based solutions to PFM problems. Lessons from these states point to three core needs: reviewing finance rules and regulations, implementing a phased approach, and guaranteeing buy-in from all levels. In order to scale up to a country-wide system, however, a careful approach to IT development is necessitated. With asynchronous onboarding (see Chapter 8) and agile development, states with existing systems can simply modify them and integrate into other reforms using open APIs, open standards, and open data protocols rather than needing to redevelop from scratch. Using these core principles, national-level and state-level reforms can occur simultaneously.

APPENDIX 2

Learning from Existing Systems in Latin America

Brazil and Mexico have made various reforms to improve their payments systems, and we highlight these reforms as of 2015 in this section. In Brazil, reforms started in the early 2000s and have centred on two key areas: government-to-person (G2P) payments and government-to-business (G2B) payments. Of these, the most documented reforms have been for Bolsa Familia, a G2P conditional cash transfer programme for families living below the poverty line. The programme started in 2003, when the government consolidated four pre-existing cash transfer programmes across as many ministries to improve efficiency and coherence of programming goals. Under Bolsa Familia, households are entitled to cash grants so long as they allow programme administrators to monitor the health and nutritional outcomes of family members, keep kids in school, and participate in educational activities administered by the programme. The monthly stipends for the programme range from USD 7 to 45 per household, depending on the income and conditional requirements fulfilled by the family.

The programme makes payments through the *Caixa Econômica Federal*, which manages payment processes for all social programmes in the country. Each month, the Ministry of Social Development provides authorizations to *Caixa Econômica*. The authorized funds are then transferred from the central Treasury to a pooled account of *Caixa Econômica*. They are then credited to the beneficiaries' electronic benefit cards (EBCs) on a monthly basis (Lindert et al. 2007, 51). Beneficiaries receive SMS alerts regarding payment dates and any changes in programme rules or payments (CGAP 2011). They can withdraw from this account through multiple points, including 2,000 agents, more than 2,000 banking correspondents, 9,000 lottery points, and *Caixa Econômica* ATMs (Cirasino et al. 2012, 43).

The electronic cash management and consolidation of four programme funds into one has accrued significant benefits. While electronic cash management has decreased corruption by removing intermediaries from controlling funds, fund consolidation has led to reduced float and increased efficiency. Together, these two reforms have decreased administrative costs of the schemes from 14.7% in 2001—when the four programmes ran separately—to 2.6% in 2005 (Lindert et al. 2007, 113). Moreover, beneficiaries have responded positively to electronic benefit cards, which allow them to access funds at ATMs, lottery points, and other commercial establishments. A 2005 survey found that 96.3% of survey respondents found the cards easy or very easy to use, with an average travel time of 22 minutes to the nearest withdrawal point (Lindert et al. 2007, 53). Given improved efficiency, convenience to beneficiaries, and government savings through the consolidation of four programmes, payments for 21 other federal social programmes are under consideration for further consolidation (CGAP 2011, 23).

Similar digital reforms to promote transparency and efficiency for government-to-business (G2B) payments have also been enacted. Most notably, the Brazilian government started a new initiative called *Cartão de Pagamento do Governo Federal* (CPGF) or the Federal Government Payment Card. CPGF replaced an old fund transfer system that was not unlike India's. Prior to G2B reforms, any expenses for low-value procurements that did not require bidding were transferred through multiple layers. First, the central Treasury would transfer funds to current accounts. These accounts were administered by an approved federal employee at various agencies, akin to the Drawing and Disbursing Officer (DDO) in India's system. The federal employee could then spend the funds within 90 days through cash or cheques. The system had low levels of transparency because the central government could not track the float until expenditures were incurred, especially given various accounts sat across multiple agencies. Even if they could track the float, doing so incurred time, costs, and significant administrative burdens (CGAP 2011; Cirasino et al. 2012).

In order to overcome these problems, the government introduced CPGF, which allowed authorized employees to transact directly from the central Treasury through a payment card. To maintain revenues and expenditure at a holistic level, the Ministry of Planning, Budget and Administration agreed upon a time period during which funds would be debited from the central account (Cirasino et al. 2012, 49). Moreover, the CPGF was linked to an online transparency system,[1] allowing agencies and cardholders to view real-time expenditures, along with the date and type of merchant. Following these reforms in 2001, the government launched a new programme, the Civil Defence Payment Card (CDPC), in 2011. The card allows state and municipal governments to make purchases for goods and services during natural disasters and relief operations, with inbuilt features to limit the type of transactions and surcharges.

Akin to reforms in Bihar, these reforms illustrate how technology can simplify system processes to limit redundancy, float, and inefficiency. At the same time, big data from the system can maximize transparency to improve accountability for timely and authorized payments.

Fund Consolidation in Mexico

Outside of G2B reforms, Mexico tried to improve the financial system in two ways. One, the government began reforms by enabling electronic requests for payments across departments. Starting in 1997, *Tesofe* (the Treasury) and the Ministry of Finance worked with departments to create an IT system which enabled *Tesofe* to make payments as requests were filed. Prior to such a system, *Tesofe* would wait for paper requests, creating inefficiency, delays, and poor accounting. These reforms focused on all individual payments for government employees, pensioners, and citizens who received government transfers. Once built, the IT system aimed to make payments the day of the request, increasing the interest earned on cash by the central

[1] Available at http://www.transparencia.gov.br/.

government (Babatz 2013). In 2001, the bank of Mexico—Banxico—launched its real-time gross settlement system, allowing direct transfers between two bank accounts upon instructions through an electronic request. By 2006, almost all suppliers were being paid through electronic transfers. This expanded to some government employees in 2008 and pensioners in 2012 (Babatz 2013).

Lessons from Previous Cases

Programmatic and national-level reforms in India and Latin America offer several solutions to manual fund management. A common solution provided by all the cases is the automation of fund transfers. Rather than sending funds to local administrators or cheques to beneficiaries, all programmes enabled direct bank transfers from the government to the end recipient, leading to drastic reduction in delays. Similarly, for procurement-based reforms, the government provided authorized administrators with electronic cards, enabling real-time payments with real-time monitoring of expenditures. Such automation of transfers and payments drastically improves accountability and transparency, allowing funds to be monitored as they are transferred. In an era where such automation is the norm rather than an exercise in innovation and experimentation, its advent in a country like India is more of a matter of implementation than capability.

Additionally, the case of Bolsa Familia shows beyond-the-last-mile focus which improves transparency to end beneficiaries, by sending them an SMS once funds are transferred to their account, ensuring they are aware of when to withdraw their benefits. Transparency, therefore, is not only about improving accountability within the government but doing so at the final stretch to prevent leakages at all stages. These transparent systems further establish that beneficiaries have the necessary means to hold governments accountable.

APPENDIX 3

Beyond-the-Last-Mile Connectivity

While the budget execution and budget monitoring and evaluation processes create significant hurdles, the buck does not stop there. At the end of the day, any PFM system with an eye towards effective public service delivery must ensure that funds reach the end recipient in a transparent and speedy manner. The vast complexity of the financial and socio-political landscape in India, however, makes this task difficult. Below, we summarize the problems with what we call beyond-the-last-mile problems, or problems once the funds have reached the account of the end recipient.

Funds for public service delivery usually focus on the first mile—routing state and central funds to implementation agencies—and the last mile—effectively sending funds and services to the beneficiary or vendor's bank or post office account. In the current system, there are three ways in which funds can reach the last mile. Prior to 2014, it was common for funds to reach beneficiaries or vendors through cash disbursals from programme workers. While such cash disbursals are rare today, the other two methods are used more frequently. Second, beneficiaries can use public and state-provided bank accounts called the India Post Payments Bank (IPPB) to access banking services at their post offices all across India,[1] which allows individuals to take part in banking. Third, since 2014, the Government of India has transitioned to using Jan Dhan Yojana, a scheme that helps citizens between the ages of 10 and 65 open zero-balance bank accounts and requires schemes to deposit benefits in these or other personal bank accounts. Similarly, vendors today typically receive funds in their bank accounts.

While transitioning from cash to bank accounts is a positive step in reducing corruption, recent studies examine a new set of problems that arise after the money has been credited. These beyond-the-last-mile issues focus on challenges beneficiaries face in withdrawing and utilizing their funds even when they have been deposited to their accounts.

Beneficiaries, especially in poorly banked areas, face significant hurdles in withdrawing funds credited to their bank accounts, which ultimately increases the opportunity cost of obtaining these deposited funds. For one, those living in remote areas often lack banking services, requiring them to travel long distances to withdraw their cash. According to the International Monetary Fund (2021), India has 21.44 ATMs per 100,000 adults, on par with lower middle income countries with an average of 20.26 (International Monetary Fund 2021). This not only increases the cost of obtaining cash, but also mandates high opportunity costs if beneficiaries have to forego daily wages in order to travel to the bank.

[1] This is an upgraded version of post office savings accounts.

The disincentive to access bank accounts is further exacerbated by the fact that beneficiaries often have little knowledge of when the funds, and of what amount, have reached their accounts. It complicates their assessment of their willingness to meet the costs of fund withdrawal. Take, for instance, a beneficiary, Lakshmi, who is a daily wage labourer in rural India. Like most of the rural poor, Lakshmi recently opened a bank account, given that government schemes were transferring payments to the bank rather than through a local post office. She only uses the bank account for public programmes; therefore, she may find it cost-effective to make a visit to the bank only if she has the assurance that money has reached her account. This is because she may need to spend Rs. 50 in travel costs and another Rs. 100 in a foregone daily wage undertaken because of travel. However, given that Lakshmi does not know how much has been deposited in her account until she makes the trip, she may have to end up making three trips to collect Rs. 200 per visit, ultimately costing her Rs. 150 in travel and Rs. 300 in foregone wages, reducing her net benefit to Rs. 150 (USD 1.85). Even if the government deposits the total sum of Rs. 600 in one go and Lakshmi can withdraw the full amount in one visit, the travel costs and foregone labour for one day alone will reduce her net benefit to Rs. 450, a significant cost for any individual relying on daily wages.

In fact, similar issues hindered the implementation of the Government of India's first Direct Benefit Transfer (DBT) pilot in Alwar, Rajasthan. In 2011, the government sought to change the old system that allowed individuals to buy subsidized kerosene at fair-price shops. This old system led to inclusion errors and diversion of kerosene to the black market. The pilot programme therefore required consumers to purchase the kerosene at the full retail price from a fair-price shop (Kumar 2012). Eligible beneficiaries of the scheme obtained a three-month advance for their subsidy in bank accounts. The DBT did lead to fewer kerosene sales, leading observers to initially connect the decline to reduced corruption (Soni and Ramakrishnan 2014). But research by both IISD-TERI (Merrill 2014) and our field interviews in May 2015 revealed that, rather than checking corruption, the pilot in fact led to a reduction in scheme participation. Given the high cost of accessing bank accounts, beneficiaries were dropping out of the kerosene scheme altogether. The opportunity costs of foregone wages and travel costs were higher than the subsidy they were receiving, making it illogical for eligible beneficiaries to continue participating in the programme.

Even in cases where the opportunity costs may not be high, cash flow problems may still hamper the ability of beneficiaries to withdraw their deposits. India's cash-dependent economy creates high demand for cash as individuals opt out of digital transactions in favour of withdrawing cash. A study by the Institute for Business in the Global Context estimates that India's ratio of currency to GDP is 12.04%—four times higher than Brazil, Mexico, or South Africa (Mazzotta et al. 2015, 12). Such a high cash to GDP ratio incurs high printing and distribution costs, estimated to be as high as 1% of the GDP (Zandi et al. 2003). With such costs and distribution needs at play, while the funds may be successfully deposited to a bank account, the bank may not necessarily have the cash to give to a beneficiary, requiring the individual to make multiple trips until the cash can be withdrawn. In fact, reports from Jharkhand and Chhattisgarh have acknowledged that cash delivery mechanisms such as the ones described here increase the opportunity cost for beneficiaries to obtain their subsidies (Gupta 2013).

Transparency issues highlighted earlier in the chapter exacerbate these problems in cash withdrawal. When beneficiaries do not know how much money is deposited in their bank accounts, they are more prone to corruption. During our field visit to the Rajsamand district of Rajasthan, individuals working with the Mazdoor Kisan Shakti Sangathan (MKSS) confirmed that bank agents took bribes or paid beneficiaries fewer funds than they were owed. Without full knowledge of money owed to them for particular schemes, subsidy recipients lacked the information to demand proper withdrawal amounts.

Recent government efforts in financial inclusion are slowly attempting to address this challenge. The Jan Dhan, Aadhaar, and Mobile (JAM) trinity has particularly strived to increase individuals' access to bank accounts. *Jan Dhan*, short for the *Pradhan Mantri Jan Dhan Yojana*, provides zero-balance bank accounts and a corresponding debit card for the poor, with the gradual build-up of facilities for accident insurance, life insurance, micro-insurance, pension, and overdraft. As of January 2021, the government reported that 7.5% of *Jan Dhan* accounts had zero balance, suggesting that the accounts are being used to hold cash (Press Trust of India 2021). A study by Evidence for Policy Design (EPoD) in Madhya Pradesh, though, highlights the gendered nature of bank use. The study found that only 16% of female *Jan Dhan* account holders had attempted to access their funds with a visit to the bank in six months prior to their survey (Field et al. 2016, 16).

These efforts to ensure that each household can use financial services have recently expanded. After demonetization in 2016 and COVID-19, mobile payments have become increasingly commonplace, allowing individuals to exchange money without going to ATMs or banks. Between 2015 and 2020, mobile payment use as a percentage of private consumption rose from virtually 0 to approximately 42% (Nariyanuri 2021, 5). Various policy changes in these years have enabled mobile payments to take off. These include *Jan Dhan* (zero-balance) accounts, UPI (Unified Payments Interface), payments banks, Aadhaar-enabled Payment System (AePS), and Direct Benefit Transfer. Most notably, UPI has enabled expansion of mobile payments.[2] UPI is a digital payment infrastructure that allows users to access money in their bank accounts through a variety of phone applications, irrespective of the banks where the recipient and sender hold money. In this system, mobile numbers are linked to bank accounts and when individuals use applications to send money to a number, the funds are instantly and directly deposited in the recipient's account.

Another promising solution is the use of banking correspondent agents, individuals who typically work for a bank but can undertake tasks such as small deposits and withdrawals directly where the customer resides. This branchless model of banking has had mixed results in Brazil, where it found some success in attracting customers but without scale, incurring high costs as a result (Sanford 2013; Siedek 2008). The proliferation of banking correspondents in India faces similar challenges, along

[2] Payments banks have differentiated banking licences with a lower level of regulatory compliance, given that they cannot give loans using the deposits they receive. Aadhaar-enabled Payment System (AEPS) is a payment service that allows a bank customer to use Aadhaar as her identity to access her bank account and make balance enquiries, withdraw cash, or transfer money. Direct Benefit Transfers are digital transfers from the government into a beneficiary's account for a variety of schemes.

with regulatory problems (Kolluju 2014). To overcome some of these challenges, India's Post Office system has recently launched its own public bank. The India Post Payments Bank facilitates doorstep banking at a nominal fee (India Post 2018). In doing so, the Post Office hopes to use the sizeable scale of the postal service to penetrate remote areas and overcome the challenge of scaling that private banks typically face.

Although the government has started implementing DBT, it must first resolve the internal fund flow mechanisms within government, and at the same time, create strong digital payment ecosystems—with transparency in where, when, and how much funds are sent—to ensure that the poor have access to remote banking and digital payment facilities.

APPENDIX 4

Eligibility Criteria in Programme-Specific Platforms

The eligibility process on programme-specific IT platforms (PSIPs) would serve three functions. One, it would reduce inclusion error by making sure that beneficiaries receiving the payments are authenticated in some form of a registry—be it Aadhaar, the Socio Economic and Caste Census (SECC), Electoral Rolls maintained by the Election Commission of India, or other databases such as Bhamashah Resident Data Hub maintained by the Rajasthan state government—to reduce false or duplicated payments or beneficiaries. Two, using registries as the backbone, the eligibility component of PSIPs would identify potential beneficiaries who are not enrolled in a scheme but are eligible to be included, based on their pre-existing demographic data. This function would allow programme officials to reduce the exclusion error by ensuring anyone who ought to be participating in a programme is included. Third, it would reduce corruption by verifying that enrolled government employees, pensioners, beneficiaries, or vendors meet predefined criteria for the scheme and are registered on the PSIP for a geographic area before receiving payments.

In order to facilitate these functions, each programme's IT platform, the PSIP, would have to be synced with either one database defined above or a centralized and regularly updated social security database that consolidates and cross-verifies data across various databases such as Aadhaar, SECC, and PAN. Upon cross-checks, this data can further be updated through community-driven checks. The primary benefit of this exercise is to reduce inclusion errors. Using algorithms, the database could flag if an individual's information seems to be duplicated. Officials could verify whether the issue stems from minor errors such as spelling mistakes and work with a beneficiary to resolve such duplications. Alternatively, such information could help identify cases of corruption where individuals may be obtaining money through various accounts despite having been eligible for only one payment.

Two, registries, which can include demographic data often used for participant identification, can be synced with systems receiving payments for a given programme. Combining eligibility data with demographic data, the system could alert programme functionaries if someone may be eligible for a programme but is not enrolled. This would allow functionaries to cross-check their records to verify this and seek more information to enrol the individual. Such automated checks would ensure that exclusion errors are limited.

The usage of Aadhaar, SECC, or other means of identification does pose privacy or surveillance threats. It is, therefore, crucial for data protection laws to be put in place prior to use of any personal data, as we highlight in Chapter 6. Such a legal framework is a precursor to the establishment of eligibility databases. Only then can the government ensure that citizens or residents are obtaining services on time, with their consent.

Third, the eligibility component of the PSIP could serve a core function of ensuring that vendor payments and government salaries go to those who are vetted and pre-registered on the platform. When vendors or new employees sign contracts to work with the government for a particular programme, the unit at which they are receiving their payment would enrol them onto the PSIP. Both their eligibility criteria and personal information would be cross-checked. Thereafter, when payments need to be made, they could be automated for regularly salaried employees or triggered when vendors or temporary employees meet pre-specified criteria. For sending payments, this system would allow programme officials to search for a name or select a vendor from a database rather than enter information each time.

This eligibility-focused component of PSIP can enable full transparency, both on the part of programme officers who face a complex system of rules, as well as for beneficiaries and community groups. It can further help reduce corruption in the system, where programme officers may create fake or 'ghost' payment recipients or make duplicate payment requests.

The proposed reforms for eligibility identification on programme-specific IT platforms address a core concern of ensuring the correct beneficiaries are targeted by government programmes and, upon enrolment, receive funds. However, a much larger process of eligibility identification and reform is needed to establish that individuals are not excluded from obtaining a scheme due to corruption, politics, or simply neglect. Currently, there is no single method of identifying all the schemes an individual is eligible for based on their background conditions, nor is there a platform that allows individuals to enrol in any scheme for which they meet eligibility requirements. While eligibility reforms are beyond the scope of our book, we propose two avenues that ought to be considered for safeguarding the full inclusion of individuals in government services.

One, the Government of India can create a digital platform that allows individuals to enter their background information, anonymously if they wish, and learn which schemes they are eligible for but are not enrolled in. This platform can then allow individuals to directly apply for schemes for which they are eligible, remotely, without needing to physically go to various government offices. This proposal may run the risk of violating data privacy if the platform collects demographic information that individuals enter on a website.

An alternative to a government-created platform is a government-facilitated system of societal platforms. Unlike commercial, for-profit platforms, societal platforms are intended for social impact. Through a shared digital infrastructure, societal platforms bring together a diverse community to scale up solutions to problems (Societal Platform 2017). In the instance of eligibility requirements, the government can take eligibility data for all schemes and create open application programming interfaces (APIs)—definitions, protocols, and tools that allow software to talk or sync across platforms. Through open APIs, these criteria can then sync to apps created by non-governmental organizations to screen individuals or households for eligibility and allow them to apply for schemes. An analogous case of such a system is the existing railway ticket booking system in India. The Passenger Reservation System allows travel agents and individuals to directly book tickets via a government portal such as IRCTC or through a service provider such as Cleartrip or Make My Trip. The government system has built-in systems to ensure that the reserved tickets are used as intended.

While the alternatives to a government-run centralized system reduce issues of data privacy or surveillance, they exacerbate privacy violation concerns by non-governmental parties. Any reform, therefore, must come with a robust legal framework for data privacy. Governments must scrutinize any non-governmental organization and prevent them from storing individuals' privately identifiable data that may be misused for other purposes prior to collaboration. It must also ensure that any data entered into a centralized platform—be it governmental or non-governmental—is collected with clear consent and kept strictly confidential.

While these alternatives may provide viable pathways for democratizing the eligibility requirements and process, they require a robust legal and privacy framework for operationalization. Reform of the enrolment process is beyond the scope of our book but remains a fundamental part of the process to make certain schemes reach intended beneficiaries, particularly those who are vulnerable. Maximizing the potential of an on-demand architecture should therefore include reforms to maximize transparency in knowledge and ways in which individuals can claim their entitlements.

APPENDIX 5

Personal Account

To see how these individual issues in budget execution coalesce, we illustrate, as an example, the personal experience of one of us—Santhosh Mathew—during his tenure as the Secretary and, later, the Principal Secretary in the Rural Development Department in the Government of Bihar from 2010 to 2013. Each year during this time, the state department sent several crore rupees (hundreds of thousands of dollars) back to the state Finance Department because it was unable to spend all of it. Most of the money was allocated to the 534 sub-district block offices for workplace expenditure. In order for the allocation to be fully used, district bureaucrats (District Magistrates) had to send expenditure reports for the budget head allocations to the state for its approval for additional funds. District Magistrates, however, never submitted these expenditure reports, preventing them from receiving additional fund allocations.

To explain further, the finance process began at the start of the financial year, during which the Rural Development Department of the state would receive the entire allocation for the office expenses budget head from the state Finance Department following budget approval by the State Legislature. It would then sub-allocate around 75% of the total allocation it had to the office of the District Magistrate. The District Magistrate office would further sub-allocate the budget to different block offices at its discretion. The Rural Department and the District Magistrate did not allocate all the money because it behoved them to keep some reserved in case the most active offices used up their funds by the end of the financial year and they needed to allocate more.

Block offices typically incur office expenditures in spending for a range of items, such as distribution camps, office stationery, supervision of law and order, security arrangements for VIP visits, etc. Under the law, these are all legitimate activities with legislative authorization for expenditure. When the blocks exhausted their allocated funds but needed quick funding at the last minute—say, to plan a field visit to resolve an issue or to arrange a visit from the Minister—they were in a bind. They could produce the paperwork for an itemized breakdown of how each rupee was used and present it to district accountants. The district officers could send this to the state and request further allocations to be made. In the middle of the myriad things that bureaucrats and officials have to do, checking forms and allocating funds took time. This rarely happened due to co-ordination and communication problems across the three branches. Looking for a fix, there was an easier way for block officials to get funds.

At the time (and persisting even today, for many schemes), block officials could pull funding from the district treasury to block bank accounts in advance of using the funds through AC bills. They then used the money from these block bank accounts for day-to-day exigencies. When these expenditures exceeded the allocation for the year, they would be left with a voucher that they could not book against an accounting head. This resulted in what is called an 'unadjusted voucher' in local parlance. An unadjusted voucher, technically speaking, is proof that expenditure has been incurred but not accounted for; in other words, it is an AC bill like we describe in Chapter 4. They are essentially expenditures awaiting to be accounted for using line-item receipts, also known as adjusted vouchers, submitted to the treasury as part of DC bills. As we explained, the allocations for office expenditures were already exhausted, so these unadjusted vouchers did not come from the office expenditure budget head. Instead, they were created by spending money by pulling funds lying in the bank account of the block meant for various schemes. Given that a variety of schemes pulled advance funds into the several bank accounts maintained by the block, they were sitting there with the intention of being used for these schemes. But they also could in the interim be used by block staff for office expenditure in case there was an administrative exigency. The problem was, the voucher could not be adjusted with line-item justification for the expenditure, because the utilized funds were never meant for office expenditure in the first place.

Herein lay the irony—block offices were using funds allocated for various schemes, but for office expenditure. At the same time, the state Rural Development Department was turning over their unallocated funds back to the Finance Departments because they went unspent. To an outsider, this issue seems like a simple one—why did the Rural Development Department not catch this? And why did the block officers not ask for more allocations? To make sense, consider the overburdened administration we mention in Chapter 2. India's bureaucrats are tasked with far more work than they are able to do with the resources they have. Budget heads for office expenditure, as one can imagine, are not something that Rural Development Departments monitored over time to realize the problem. At the same time, in light of the fact that the similarly overburdened block offices did not have visibility into the unallocated funds at the state level, they found an easier route.

This system, of course, led to many problems. For one, scheme budgets were being used for unintended purposes in block offices, which could lead to a potential shortfall of funds. Two, the system also meant that block offices could never produce the receipts for the expenditures they made. Instead, these expenditures remained as unadjusted vouchers, because they did not have the allocation under the office expenditure budget head to absorb all the expenditures. This meant that the Finance Department had zero visibility on how a portion of the fund advances was used. Had appropriate systems existed to show how much each block was allotted and how much they used in real-time—achievable through even a simple live Excel spreadsheet—District Magistrates and the officers at the Rural Development Department could simply allocate more money as needs arose. Instead, money shuffled between the Finance Department and the Rural Development Department, never being used.

References

Chapter 1

Ahluwalia, Isher Judge, Prasanna Kumar Mohanty, Om Mathur, Debarpita Roy, Ayush Khare, and Shreya Mangla. 2019. 'State of Municipal Finances in India: A Study Prepared for the Fifteenth Finance Commission'. New Delhi: Indian Council for Research on International Economic Relations. Available at https://fincomindia.nic.in/asset/doc/commission-reports/15th-FC/reports/studies/State%20of%20Municipal%20Finances%20in%20India.pdf.

Aiyar, Yamini, and T.R. Raghunandan. 2015. 'Pull, Not Push, to Open up Spending'. *Livemint*, 23 February. Available at https://www.livemint.com/Opinion/uuVwhsKE8wKNa13sjI2tHN/Pull-not-push-to-open-up-spending.html.

Banerjee, Abhijit, Esther Duflo, Clement Imbert, Santosh A. Mathew, and Rohini Pande. 2020. 'E-Governance, Accountability, and Leakage in Public Programs: Experimental Evidence from a Financial Management Reform in India'. *American Economic Journal: Applied Economics* 12 (4): pp. 39–72. doi: 10.1257/app.20180302.

Bhatia, Harbans Lal. 2018. *Public Finance*. Noida: Vikas Publishing House.

Cangiano, Marco, Teresa Curristine, and Michel Lazare. 2013. *Public Financial Management and Its Emerging Architecture*. Washington, D.C.: International Monetary Fund. Available at https://www.elibrary.imf.org/display/book/9781475531091/9781475531091.xml.

CGA. 2012. 'Sundaramurti Committee Report on Revision of List of Major and Minor Heads of Accounts'. Ministry of Finance, Government of India. Available at http://finmin.nic.in/reports/sundaramurti_committee_report.asp.

Jena, Pratap Ranjan. 2013. 'Improving Public Financial Management in India: Opportunities to Move Forward'. Working Paper No. 2013-123. New Delhi: National Institute of Public Finance and Policy. Available at https://www.nipfp.org.in/media/medialibrary/2013/04/WP_2013_123.pdf.

Levitsky, Steven, and Daniel Ziblatt. 2019. *How Democracies Die*. New York: Crown.

Lindvall, Johannes, and Jan Teorell. 2016. *State Capacity as Power: A Conceptual Framework*. STANCE Working Paper Series; Vol. 2016, No. 1. Lund: Department of Political Science, Lund University.

Lipsky, Michael. 2010. *Street-Level Bureaucracy: Dilemmas of the Individual in Public Service*. New York: Russell Sage Foundation.

Mangla, Akshay. 2022. *Making Bureaucracy Work: Norms, Education and Public Service Delivery in Rural India*. Delhi: Cambridge University Press.

Mohanty, Ranjan Kumar, and N.R. Bhanumurthy. 2021. 'Assessing Public Expenditure Efficiency at the Subnational Level in India: Does Governance Matter?' *Journal of Public Affairs* 21 (2): e2173. doi: 10.1002/pa.2173.

Nilekani, Nandan, and Viral Shah. 2015. *Rebooting India: Realizing a Billion Aspirations*. Gurugram: Penguin Books Limited.

NITI Aayog. 2023. 'The National Multidimensional Poverty Index: A Progress Review 2023'. New Delhi: NITI Aayog. Available at https://www.undp.org/india/publications/national-multidimensional-poverty-index-progress-review-2023.

NITI Aayog. 2024. 'Multidimensional Poverty in India Since 2005-06'. New Delhi: NITI Aayog. Available at https://www.niti.gov.in/sites/default/files/2024-01/MPI-22_NITI-Aayog20254.pdf

Potter, Barry H., and Jack Diamond. 1999. *Guidelines for Public Expenditure Management*. Washington, D.C.: International Monetary Fund. doi: 10.5089/9781557757876.071.

Press Information Bureau. 2016a. 'Two Groups of Secretaries Present Ideas and Suggestions to PM'. *Press Information Bureau*, 15 January. Available at https://pib.gov.in/newsite/PrintRelease.aspx?relid=134529.

Rao, Menaka. 2017. 'Gorakhpur Hospital Deaths: Doctor Explains Why Oxygen Payments Were Delayed'. *Scroll.in*, 13 August. Available at https://scroll.in/pulse/847105/adityanaths-visit-to-gorakhpur-proved-deadly-as-oxygen-payments-were-delayed-by-a-day-says-doctor.

Rao, M. Govinda. 2022. *Studies in Indian Public Finance*. Noida: Oxford University Press. doi: 10.1093/oso/9780192849601.003.0008.

Rashid, Omar. 2017. 'Gorakhpur Hospital Ignored Warnings on Oxygen Supply'. *The Hindu*, 13 August. Available at https://www.thehindu.com/todays-paper/gorakhpur-hospital-ignored-warnings-on-oxygen-supply/article19484166.ece.

TAGUP. 2011. 'Report of the Technology Advisory Group for Unique Projects'. New Delhi: Ministry of Finance. Available at https://comtax.uk.gov.in/files/pdf/70_TAGUP_Report.pdf.

Watal, Ratan P. 2014. 'Resolution to Constitute an Expenditure Management Commission'. New Delhi: Department of Expenditure, Ministry of Finance, Government of India. Available at https://doe.gov.in/files/circulars_document/Constitute_Expenditure_Management_Commission.pdf.

Chapter 2

Agnihotri, Anustubh. 2022. 'Transfer Preferences of Bureaucrats and Spatial Disparities in Local State Presence'. *World Development* 159: 106023. doi: 10.1016/j.worlddev.2022.106023.

Aiyar, Yamini, and T.R. Raghunandan. 2015. 'Pull, Not Push, to Open up Spending'. *Livemint*, 23 February. Available at https://www.livemint.com/Opinion/uuVwhsKE8wKNa13sjI2tHN/Pull-not-push-to-open-up-spending.html.

Andrews, Matt. 2013. *The Limits of Institutional Reform in Development: Changing Rules for Realistic Solutions*. Cambridge: Cambridge University Press. Available at https://www.cambridge.org/core/books/limits-of-institutional-reform-in-development/C7E4CCE47EE4FC81EBCE0AA78F8A60FE.

Andrews, Matt, Marco Cangiano, Neil Cole, Paolo de Renzio, Philipp Krause, and Renaud Seligmann. 2014. 'This Is PFM'. CID Working Paper No. 285. Cambridge: Center for International Development at Harvard University.

Available at https://www.hks.harvard.edu/sites/default/files/centers/cid/files/publications/faculty-working-papers/285_Andrews_This+is+PFM.pdf.

Auerbach, Adam Michael, Shikhar Singh, and Tariq Thachil. 2024. 'Who Knows How to Govern? Procedural Knowledge in India's Small-Town Councils'. *American Political Science Review* 119 (2): pp. 1–19. doi: 10.1017/S0003055424000297.

Bersch, Katherine, Sérgio Praça, and Matthew M. Taylor. 2017. 'State Capacity, Bureaucratic Politicization, and Corruption in the Brazilian State'. *Governance* 30 (1): pp. 105–24. doi: 10.1111/gove.12196.

Besley, Timothy, and Torsten Persson. 2009. 'The Origins of State Capacity: Property Rights, Taxation and Politics'. *American Economic Review* 99 (4): pp. 1218–44. doi: 10.1257/aer.99.4.1218.

Brierley, Sarah. 2020. 'Unprincipled Principals: Co-opted Bureaucrats and Corruption in Ghana'. *American Journal of Political Science* 64 (2): pp. 209–22. doi: 10.1111/ajps.12495.

Cangiano, Marco, Teresa Curristine, and Michel Lazare. 2013. *Public Financial Management and Its Emerging Architecture*. Washington, D.C.: International Monetary Fund. Available at https://www.elibrary.imf.org/display/book/9781475531091/9781475531091.xml.

Cashin, Cheryl, Danielle Bloom, Susan Sparkes, Hélène Barroy, Joseph Kutzin, and Sheila O'Dougherty. 2017. 'Aligning Public Financial Management and Health Financing: Sustaining Progress toward Universal Health Coverage'. Health Financing Working Paper No. 4. Geneva: World Health Organization. Available at https://iris.who.int/bitstream/handle/10665/254680/9789241512039-eng.pdf?sequence=1.

Chandra, Kanchan. 2007. *Why Ethnic Parties Succeed: Patronage and Ethnic Head Counts in India*. New York: Cambridge University Press.

Chhibber, Pradeep, and Irfan Nouruddin. 2004. 'Do Party Systems Count? The Number of Parties and Government Performance in the Indian States'. *Comparative Political Studies* 37 (2): pp. 152–87. doi: 10.1177/0010414003260981.

Commonwealth Secretariat. 2005. 'Guidelines for Public Financial Management Reform'. London: Commonwealth Secretariat. Available at https://dl.icdst.org/pdfs/files2/06459f7c40725300452a21d9c4d52f48.pdf.

Cruz, Cesi, Julien Labonne, and Pablo Querubin. 2020. 'Social Network Structures and the Politics of Public Goods Provision: Evidence from the Philippines'. *American Political Science Review* 114 (2): pp. 486–501. doi: 10.1017/S0003055419000789.

Cruz, Cesi, Philip Keefer, Julien Labonne, and Francesco Trebbi. 2024. 'Making Policies Matter: Voter Responses to Campaign Promises'. *The Economic Journal* 134 (661): pp. 1875–1913. doi: 10.1093/ej/ueae008.

Das, Raya, Ranjana Roy, and Ashok Gulati. 2024. 'Rationalising Public Distribution System in India'. Policy Brief #28. New Delhi: Indian Council for Research on International Economic Relations. Available at https://icrier.org/pdf/pb-27.pdf.

Dasgupta, Aditya, and Devesh Kapur. 2020. 'The Political Economy of Bureaucratic Overload: Evidence from Rural Development Officials in India'. *American Political Science Review* 114 (4): pp. 1316–34. doi: 10.1017/S0003055420000477.

Dunning, Thad, Guy Grossman, Macartan Humphreys, Susan D. Hyde, Craig McIntosh, and Gareth Nellis. 2019. *Information, Accountability, and Cumulative Learning: Lessons from Metaketa I*. Cambridge: Cambridge University Press.

Fukuyama, Francis. 2004. *State-Building: Governance and World Order in the 21st Century.* Cornell University Press.

Gadenne, Lucie, and Monica Singhal. 2014. 'Decentralization in Developing Economies'. *Annual Review of Economics* 6: pp. 581–604. doi: 10.1146/annurev-economics-080213-040833.

Gibbons, Robert, and Kevin J. Murphy. 1992. 'Optimal Incentive Contracts in the Presence of Career Concerns: Theory and Evidence'. *Journal of Political Economy* 100 (3): pp. 468–505. doi: 10.1086/261826.

Goryakin, Yevgeny, Paul Revill, Andrew J. Mirelman, Rohan Sweeney, Jessica Ochalek, and Marc Suhrcke. 2020. 'Public Financial Management and Health Service Delivery: A Literature Review'. In *Global Health Economics: Shaping Health Policy in Low- and Middle-Income Countries*, edited by Paul Revill, Marc Suhrcke, Rodrigo Moreno-Serra, Mark Sculpher, and Maria Goddard, pp. 191–215. Singapore: World Scientific Publishing. doi: 10.1142/9789813272378_0007.

Grossman, Guy, and Tara Slough. 2022. 'Government Responsiveness in Developing Countries'. *Annual Review of Political Science* 25: pp. 131–53. doi: 10.1146/annurev-polisci-051120-112501.

Gulzar, Saad, and Benjamin J. Pasquale. 2017. 'Politicians, Bureaucrats, and Development: Evidence from India'. *American Political Science Review* 111 (1): pp. 162–83. doi: 10.1017/S0003055416000502.

Hanson, Jonathan K., and Rachel Sigman. 2013. 'Leviathan's Latent Dimensions: Measuring State Capacity for Comparative Political Research'. *The Journal of Politics* 83 (4): pp. 1495–1510. doi: 10.1086/715066.

Healy, Andrew, and Neil Malhotra. 2013. 'Retrospective Voting Reconsidered'. *Annual Review of Political Science* 16: pp. 285–306. doi: 10.1146/annurev-polisci-032211-212920.

Hendrix, Cullen S. 2010. 'Measuring State Capacity: Theoretical and Empirical Implications for the Study of Civil Conflict'. *Journal of Peace Research* 47 (3): pp. 273–85. doi: 10.1177/0022343310361838.

Hicken, Allen, and Noah L. Nathan. 2020. 'Clientelism's Red Herrings: Dead Ends and New Directions in the Study of Nonprogrammatic Politics'. *Annual Review of Political Science* 23 (1): pp. 277–94. doi: 10.1146/annurev-polisci-050718-032657.

International Institute for Population Sciences (IIPS), and ICF. 2021. 'National Family Health Survey (NFHS-5), 2019–21: India: Volume 1'. Mumbai: IIPS.

Jain, Manoj, and Srinivas Gurazada. 2023. 'Eliminating Bottlenecks in Public Finance to Improve the Delivery of Health Services'. *World Bank Blogs*, 29 December. Available at https://blogs.worldbank.org/governance/eliminating-bottlenecks-public-finance-improve-delivery-health-services.

Keefer, Philip, and Stuti Khemani. 2004. 'Democracy, Public Expenditures and the Poor'. World Bank Policy Research Working Paper 3164. Washington, D.C.: Development Research Group, World Bank. Available at https://documents1.worldbank.org/curated/en/393101468779410906/pdf/WPS3164.pdf.

Levi, Margaret. 1989. *Of Rule and Revenue*. Oakland: University of California Press.

Lewis-Faupel, Sean, Yusuf Neggers, Benjamin A. Olken, and Rohini Pande. 2016. 'Can Electronic Procurement Improve Infrastructure Provision? Evidence from Public Works in India and Indonesia'. *American Economic Journal: Economic Policy* 8 (3): pp. 258–83. doi: 10.1257/pol.20140258.

Mann, Michael. 1984. 'The Autonomous Power of the State: Its Origins, Mechanisms and Results'. *European Journal of Sociology/Archives Européennes de Sociologie* 25 (2): pp. 185–213. doi: 10.1017/S0003975600004239.

Miller, Mark, Tom Hart, and Sierd Hadley. 2021. 'Public Finance and Service Delivery: What's New, What's Missing, What's Next?' Public Finance and Service Delivery Series. London: ODI. Available at https://odi.org/en/publications/public-finance-and-service-delivery-whats-new-whats-missing-whats-next/.

Ministry of Finance. 2023. 'Economic Survey 2022–23'. New Delhi: Department of Economic Affairs, Ministry of Finance. Available at https://www.indiabudget.gov.in/economicsurvey/doc/echapter.pdf.

Mookherjee, Dilip. 2015. 'Political Decentralization'. *Annual Review of Economics* 7 (1): pp. 231–49. doi: 10.1146/annurev-economics-080614-115527.

Muralidharan, Karthik, and Venkatesh Sundararaman. 2013. 'Contract Teachers: Experimental Evidence from India'. Working Paper 19440. Cambridge: National Bureau of Economic Research. doi: 10.3386/w19440.

Niehaus, Paul, and Sandip Sukhtankar. 2013. 'The Marginal Rate of Corruption in Public Programs: Evidence from India'. *Journal of Public Economics* 104: pp. 52–64. doi: 10.1016/j.jpubeco.2013.05.001.

NITI Aayog. 2023. 'The National Multidimensional Poverty Index: A Progress Review 2023'. New Delhi: NITI Aayog. Available at https://www.undp.org/india/publications/national-multidimensional-poverty-index-progress-review-2023.

NYU Wagner, and Overseas Development Institute (ODI). 2021. 'An Intergovernmental Perspective on Managing Public Finances for Service Delivery: Assessing Neglected Challenges in the Health Sector and Beyond'. New York: NYU Wagner; ODI. Available at https://wagner.nyu.edu/files/faculty/NYU-ODI-Intergovernmental-Perspective-on-Managing-Public-Finances-for-Service-Delivery-May-2021-Final.pdf.

Press Information Bureau. 2024. '24.82 Crore Indians Escape Multidimensional Poverty in Last 9 Years'. *Press Information Bureau*, 15 January. Available at https://www.pib.gov.in/PressReleaseIframePage.aspx?PRID=1996271&reg=3&lang=2.

Purohit, Bhumi. 2024. 'Bureaucratic Resistance against Female Politicians: Evidence from Telangana, India'. Working Paper, OSF, pp. 1–75. Available at https://osf.io/wervs.

Raffler, Pia J. 2022. 'Does Political Oversight of the Bureaucracy Increase Accountability? Field Experimental Evidence from a Dominant Party Regime'. *American Political Science Review* 116 (4): pp. 1443–59. doi: 10.1017/S0003055422000181.

Raghunandan, T.R. 2017. 'Re-Energizing Democratic Decentralization in India'. In *Rethinking Public Institutions in India*, edited by Devesh Kapur, Pratap Bhanu Mehta, and Milan Vaishnav, pp. 464–507. New Delhi: Oxford University Press. Available at https://carnegieendowment.org/2017/04/21/rethinking-public-institutions-in-india-pub-68221.

Raghunandan, T.R., Swaroop Iyenger, Tanvi Bhatikar, Uthara Narayanan, and Padmapriya Janakiraman. 2016. 'Tracking Fiscal Devolution to Local Governments: A Case Study from Kolar District, Karnataka'. PAISA for Panchayats—Policy Brief 2016. New Delhi: Accountability Initiative & Center for

Policy Research. Available at https://cprindia.org/wp-content/uploads/2021/12/PAISA-for-Panchayat-2016-Brief.pdf.

Ravishankar, Nirmala, Christina Banks, Pura Angela Co, and Boniface Mbuthia. 2023. 'Use of Digital Public Infrastructure to Improve Health Resource Tracking'. ThinkWell. Available at https://thinkwell.global/wp-content/uploads/2023/12/dPFM-for-HRT.pdf.

Shen, Ce, and John B. Williamson. 1997. 'Child Mortality, Women's Status, Economic Dependency, and State Strength: A Cross-National Study of Less Developed Countries'. *Social Forces* 76 (2): pp. 667–700. doi: 10.1093/sf/76.2.667.

Singh, Manjari, Biju Varkkey, Sunil Maheshwari, Promila Agarwal, Shrihari Sohani, Jatin Pandey, Jatinder Jha, Jaydeepsingh Rao, Shivali Rathore, Sunanda Menon, and Avijeet Vijay. 2015. 'A Study for Comparing Salaries/ Emoluments in the Government Sector Vis-à-Vis Central Public Sector Undertakings/ Private Sector in India'. IIM-Ahmedabad.

Singh, Nirvikar. 2017. 'Reforming India's Institutions of Public Expenditure Governance'. In *Rethinking Public Institutions in India*, edited by Devesh Kapur, Pratap Bhanu Mehta, and Milan Vaishnav. New Delhi: Oxford University Press, pp. 180–224. Available at https://carnegieendowment.org/2017/04/21/rethinking-public-institutions-in-india-pub-68221.

Skocpol, Theda. 1985. 'Bringing the State Back In: Strategies of Analysis in Current Research'. In *Bringing the State Back In*, edited by Peter B. Evans, Dietrich Rueschemeyer, and Theda Skocpol, pp. 3–38. Cambridge: Cambridge University Press. Available at https://edisciplinas.usp.br/pluginfile.php/4508121/mod_resource/content/1/Sckopol.pdf.

Soifer, Hillel. 2008. 'State Infrastructural Power: Approaches to Conceptualization and Measurement'. *Studies in Comparative International Development* 43: pp. 231–51. doi: 10.1007/s12116-008-9028-6.

Stokes, Susan C., Thad Dunning, Marcelo Nazareno, and Valeria Brusco. 2013. *Brokers, Voters, and Clientelism: The Puzzle of Distributive Politics*. New York: Cambridge University Press.

Sukhtankar, Sandip, and Milan Vaishnav. 2015. 'Corruption in India: Bridging Research Evidence and Policy Options'. *India Policy Forum* 11: pp. 193–261. Available at https://ssrn.com/abstract=2685819.

Svensson, Arne, and Allan Rosenbaum. 1996. 'Local Governance: Report of the United Nations Global Forum on Innovative Policies and Practices in Local Governance'. Gothenburg: United Nations Global Forum on Innovative Policies and Practices in Local Governance. Available at https://digitallibrary.un.org/record/1488164?ln=en.

Tilly, Charles. 1992. *Coercion, Capital and European States, AD 990–1992*. Cambridge, MA: Blackwell.

Wade, Robert. 1982. 'The System of Administrative and Political Corruption: Canal Irrigation in South India'. *The Journal of Development Studies* 18 (3): pp. 287–328. doi: 10.1080/00220388208421833.

Welham, Bryn, Tom Hart, Shakira Mustapha, and Sierd Hadley. 2017. 'Public Financial Management and Health Service Delivery: Necessary, But Not Sufficient?' London: ODI. Available at https://media.odi.org/documents/Public_

financial_management_and_health_service_delivery_necessary_but_not_sufficient.pdf.

Williams, Martin J. 2021. 'Beyond State Capacity: Bureaucratic Performance, Policy Implementation and Reform'. *Journal of Institutional Economics* 17 (2): pp. 339–57. doi: 10.1017/S1744137420000478.

Chapter 3

Agarwal, O.P., and T.V. Somanathan. 2005. 'Public Policy Making In India: Issues and Remedies'. Available at https://www.igntu.ac.in/eContent/IGNTU-eContent-353062467985-MA-PoliticalScience-4-Dr.GeorgeT.Haokip-Paper401PublicPolicyandDevelopmentinIndia-Unit2.pdf.

Andrews, Matt. 2010. 'Good Government Means Different Things in Different Countries'. *Governance* 23 (1): pp. 7–35. doi: 10.1111/j.1468-0491.2009.01465.x.

Andrews, Matt. 2013. *The Limits of Institutional Reform in Development: Changing Rules for Realistic Solutions*. Cambridge: Cambridge University Press. https://www.cambridge.org/core/books/limits-of-institutional-reform-in-development/C7E4CCE47EE4FC81EBCE0AA78F8A60FE.

Asian Development Bank. 2017. 'Implementing Results-Based Budget Management Frameworks: An Assessment of Progress in Selected Countries'. Manila: Asian Development Bank. doi: 10.22617/TCS179179-2.

Bhunia, Abhirup. 2020. 'PFM Blog: Delivering on India's Public Finance Reform Promises'. *Public Financial Management Blog*, 21 January. Available at https://blog-pfm.imf.org/pfmblog/2020/01/-delivering-on-indias-public-finance-reform-promises-.html.

CABRI. 2013. 'Exploring the Missing Links in Public Financial Management Reforms'. 9th Annual CABRI Seminar, Kenya, 28–30 August. Available at https://www.cabri-sbo.org/uploads/files/Documents/report_2013_cabri cabri_seminar_cabri_9th_annual_seminar_english_cabri_annual_seminar_2013_english.pdf.

Charlesworth, Neil. 1985. 'The Problem of Government Finance in British India: Taxation, Borrowing and the Allocation of Resources in the Inter-War Period'. *Modern Asian Studies* 19 (3): pp. 521–48. doi: 10.1017/S0026749X0000771X.

Committee on Change in Financial Year. 1985. 'Report of the Committee on Change in Financial Year'. New Delhi, India: Department of Economic Affairs, Ministry of Finance. Available at https://dea.gov.in/files/budget_division_documents/Report_LKJhaCommittee_ChangeinFY_1985_1.pdf.

Dener, Cem, Joanna Alexandra Watkins, and William Leslie Dorotinsky. 2011. *Financial Management Information Systems: 25 Years of World Bank Experience on What Works and What Doesn't*. Washington, D.C.: World Bank. Available at http://hdl.handle.net/10986/2297.

Department of Budget and Management. 2023. 'Strengthening Local Public Financial Management through Digital Transformation'. Keynote Message, PHILLBO Annual Convention, Boracay, Aklan, 22 March. Available at https://www.dbm.gov.

ph/index.php/the-secretary-2/speeches/1129-strengthening-local-public-financial-management-through-digital-transformation.

Dutta, K. 1975. 'Programme Budgeting: Evolution, Practice and Trends'. *ASCI Journal of Management* 4 (2): pp. 29–44. Available at https://drive.google.com/drive/folders/1O3J-F6nTSKNl_vYP7FrzSa5YRd-HXR3w.

Dwivedi, Onkar Prasad, Randhir Jain, and Bhagwan Dua. 1989. 'Imperial Legacy, Bureaucracy, and Administrative Changes: India 1947–1987'. *Public Administration and Development* 9 (3): pp. 253–69. doi: 10.1002/pad.4230090303.

Gayithri, Karnam. 2015. 'Performance-Based Budgeting: Subnational Initiatives in India and China'. Working Paper 334. Bangalore: Institute for Social and Economic Change. Available at https://www.isec.ac.in/wp-content/uploads/2023/07/WP-334-K-Gayithri-Final.pdf.

Gurkan, Asli, Kai Kaiser, and Doris Voorbraak. 2009. 'Implementing Public Expenditure Tracking Surveys for Results: Lessons from a Decade of Global Experience'. Washington, D.C.: World Bank. Available at http://hdl.handle.net/10986/11104.

Guthrie, James, Olov Olson, and Christopher Humphrey. 1999. 'Debating Developments in New Public Financial Management: The Limits of Global Theorising and Some New Ways Forward'. *Financial Accountability & Management* 15 (3–4): pp. 209–28. doi: 10.1111/1468-0408.00082.

Handa, K.L. 1985. 'Reforms in Financial Administration'. *Indian Journal of Public Administration* 31 (3): pp. 638–51. doi: 10.1177/0019556119850314.

Handa, K.L. 1990. 'Administrative Culture for Financial Administration in India—Retrospect and Prospect'. *Indian Journal of Public Administration* 36 (3): pp. 507–18. doi: 10.1177/0019556119900315.

Kelkar, Vijay, and Ajay Shah. 2019. *In Service of the Republic: The Art and Science of Economic Policy*. Gurugram: Penguin Random House India.

Khera, Reetika. 2019. 'Aadhaar Failures: A Tragedy of Errors'. *Economic and Political Weekly* 54 (14): pp. 2–10. Available at https://www.epw.in/engage/article/aadhaar-failures-food-services-welfare.

Long, Cathal, Marco Cangiano, Emily Middleton, and James Stewart. 2023. 'Digital Public Financial Management: An Emerging Paradigm'. ODI Working Paper. London: ODI.

MicroSave Consulting. 2022. 'Decoding India's Public Financial Management System: Challenges, Best Practices & Way Forward'. Available at https://www.microsave.net/building-state-capacity/wp-content/uploads/2021/02/Decoding-Indias-Public-Finance-Management-System-Challenges-Best-Practices-Way-Forward_Summary-notes-1.pdf.

Ministry of Finance. 2006. 'Economic Survey 2005–2006'. Chapter 2. New Delhi: Ministry of Finance, Government of India. Available at https://www.indiabudget.gov.in/budget_archive/es2005-06/chapt2006/chapter2.pdf.

Monga, P.K. 2010. 'Mission Mode Project (Treasury Computerization) under the National EGovernance Plan (NeGP)'. Office Memorandum, July 2010. Available at https://www.scribd.com/document/259277547/TreasuryComp-NeGP.

Muralidharan, Karthik, Paul Niehaus, and Sandip Sukhtankar. 2016. 'Building State Capacity: Evidence from Biometric Smartcards in India'. *American Economic Review* 106 (10): pp. 2895–929. doi: 10.1257/aer.20141346.

Natsios, Andrew. 2010. 'The Clash of Counter-Bureaucracy and Development'. Center for Global Development. Available at https://www.cgdev.org/publication/clash-counter-bureaucracy-and-development.

O'Reilly, Tim. 2011. 'Government as a Platform'. *Innovations: Technology, Governance, Globalization* 6 (1): pp. 13–40. doi: 10.1162/INOV_a_00056.

Pandey, Subhash Chandra. 2005. 'Guidelines for Preparation of Outcome Budget 2006–07 and Performance Budget 2005–06'. New Delhi: Controller General of Accounts, Department of Expenditure, Ministry of Finance, Government of India. Available at http://cga.nic.in/writereaddata/Guidelines_Outcome_Budget.pdf.

Paso, Lorena Rivero del, Sailendra Pattanayak, Gerardo Uña, and Hervé Tourpe. 2023. 'Digital Solutions Guidelines for Public Financial Management'. *Technical Notes and Manuals*. Washington, D.C.: International Monetary Fund, pp. 1–69. Available at https://www.imf.org/en/Publications/TNM/Issues/2023/10/06/Digital-Solutions-Guidelines-for-Public-Financial-Management-537781.

Planning Commission. 2011. 'Report of the High Level Expert Committee on Efficient Management of Public Expenditure'. New Delhi, India: Planning Commission, Government of India. Available at http://164.100.161.239/reports/genrep/rep_hle.pdf.

Pope, Richard. 2024. *Platformland. An Anatomy of Next-Generation Public Services*. London: London Publishing Partnership.

Press Information Bureau. 2016b. 'Government today constituted a Committee headed by Dr. Shankar Acharya (former Chief Economic Adviser) to examine the desirability and feasibility of having 'a new financial year'; The Committee to submit its Report by 31st December, 2016'. *Press Information Bureau*, Ministry of Finance, 06 July. Available at https://www.pib.gov.in/newsite/PrintRelease.aspx?relid=146883&reg=3&lang=2.

Sabharwal, Manish, and Deepak Phatak. 2020. 'A Three-Phase Govt Project Uses COVID Policy Window for a Much Needed Reform'. *The Indian Express*, 9 July. Available at https://indianexpress.com/article/opinion/columns/digitising-the-state-6496692/.

Satpathy, Madhubrata. 2018. 'Performance Budgeting'. *International Journal of Humanities and Social Science Invention* 7 (10): pp. 49–57. Available at https://www.ijhssi.org/papers/vol7(10)/Version-3/G0710034957.pdf.

Shah, Ajay. 2006. 'Improving Governance Using Large IT Systems'. In *Documenting Reforms: Case Studies from India*, edited by Seema Narayan, pp. 122–48. New Delhi: Macmillan India. Available at https://www.mayin.org/ajayshah/PDFDOCS/Shah2006_big_it_systems.pdf.

Sitharaman, Nirmala. 2020. 'Explanatory Memorandum as to the Action Taken on the Recommendations Made by the Fifteenth Finance Commission in Its Report for the Financial Year 2020–21 Submitted to the President on December 5, 2019'. 30 January. Explanatory Memorandum. Available at https://fincomindia.nic.in/asset/doc/commission-reports/Explanatory_Memorandum.pdf.

Sitharaman, Nirmala. 2025. *49th Civil Accounts Day Celebrations at New Delhi—1st March 2025*. YouTube. Available at https://www.youtube.com/live/IV7E5s9on3A?t=1055s.

Somanathan, T.V. 2023. 'TV Somanathan: PFMS Reforms Will Lead to Higher Savings in FY23 Compared to Last Year'. *Moneycontrol*, 1 March. Available at https://www.

moneycontrol.com/news/business/economy/tv-somanathan-pfms-reforms-will-lead-to-higher-savings-in-fy23-compared-to-last-year-10183511.html.

Sundaramurti, C.R. 2012. 'Report of Sundaramurti Committee on Revision of List of Major and Minor Heads of Accounts'. New Delhi: Controller General of Accounts, Ministry of Finance, Government of India. Available at https://dea.gov.in/report-sundaramurti-committee-revision-list-major-and-minor-heads-accounts.

Taub, Richard Paul. 1969. *Bureaucrats under Stress*. Berkeley and Los Angeles: University of California Press.

Thakur, Pradeep. 2020. 'CAG Seeks Data Accountability Law'. *The Times of India*, 6 August. Available at https://timesofindia.indiatimes.com/india/cag-seeks-data-accountability-law/articleshow/77398039.cms?from=mdr.

The Comptroller and Auditor General of India. 2015. 'Report for the Year 2014–2015'. Accounts of the Union Government No. 50 of 2015 (Financial Audit). The Comptroller and Auditor General of India. Available at https://cag.gov.in/sites/default/files/audit_report_files/Union_Finacne_Report_50_2015_0.pdf.

Uña, Gerardo, and Carlos Pimenta. 2016. 'Chapter 7. Integrated Financial Management Information Systems in Latin America: Strategic Aspects and Challenges'. In *Public Financial Management in Latin America: The Key to Efficiency and Transparency*, edited by Mario Pessoa and Carlos Pimenta. Washington, D.C.: Inter-American Development Bank, pp. 281–319. doi: 10.5089/9781597822268.071.

Uña, Gerardo, Richard I. Allen, and Nicolas M Botton. 2019. 'How to Design a Financial Management Information System: A Modular Approach'. International Monetary Fund. Available at https://www.imf.org/en/Publications/Fiscal-Affairs-Department-How-To-Notes/Issues/2019/05/15/How-to-Design-a-Financial-Management-Information-System-A-Modular-Approach-46818.

Wencai, Zhang. 2016. 'Global Trends in Public Financial Management'. Text presented at the 40th anniversary celebrations of the Indian Civil Accounts Service, New Delhi, India. Available at https://www.adb.org/news/speeches/global-trends-public-financial-management-wencai-zhang.

XIV Finance Commission. 2014. 'Report of the XIV Finance Commission'. New Delhi: XIV Finance Commission. Available at https://fincomindia.nic.in/asset/doc/commission-reports/14th-FC/14thFCReport.pdf.

XV Finance Commission. 2019. 'Report for the Year 2020–21'. New Delhi: XV Finance Commission. Available at https://fincomindia.nic.in/asset/pdf/commission-reports/XVFC_202021%20Report_English_Web.pdf.

Chapter 4

Ahluwalia, Isher Judge, Prasanna Kumar Mohanty, Om Mathur, Debarpita Roy, Ayush Khare, and Shreya Mangla. 2019. 'State of Municipal Finances in India: A Study Prepared for the Fifteenth Finance Commission'. New Delhi: Indian Council for Research on International Economic Relations. Available at https://fincomindia.nic.in/asset/doc/commission-reports/15th-FC/reports/studies/State%20of%20Municipal%20Finances%20in%20India.pdf.

CAG. 2012. 'Central Plan Schemes Monitoring System (CPSMS)'. Controller General of Accounts, Ministry of Finance. Available at https://pfms.nic.in/Static/Notifications/cpsms-pamphlet%20for%20Competition-Feb%202012.pdf.

CAG. 2015a. 'Report of the CAG on Union Government Accounts 2013–14'. New Delhi: Comptroller and Auditor General of India. Available at https://www.saiindia.gov.in/uploads/download_audit_report/2015/Union_Government_Accounts_1_2015_Chap_5.pdf.

CAG. 2015b. 'Report for the Year 2014–2015'. Accounts of the Union Government No. 50 of 2015 (Financial Audit). New Delhi: Union Government. Available at https://cag.gov.in/webroot/uploads/download_audit_report/2015/Union_Finacne_Report_50_2015_0.pdf.

CAG. 2016. 'Compliance Auditing Guidelines'. New Delhi: Comptroller and Auditor General of India. Available at https://cag.gov.in/uploads/guidelines/Compliance-Guidelines-approved-final-preface-05de4efef9159d0-85033036.pdf.

CGA. 1982. '(Receipts and Payments) Rules'. Controller General of Accounts, Ministry of Finance. Available at http://cga.nic.in/DownloadPDF.aspx?filenameid=2.

Department of Expenditure. 2017. 'General Financial Rules, 2017'. New Delhi: Department of Expenditure, Ministry of Finance, Government of India. Available at https://doe.gov.in/files/circulars_document/FInal_GFR_upto_31_07_2024.pdf.

Handa, K.L. 1980. 'Budget Management Techniques: A Perspective for India'. *Indian Journal of Public Administration* 26 (3): pp. 648–62. doi: 10.1177/0019556119800315.

Lienert, Ian. 2009. 'Modernizing Cash Management'. *Technical Notes and Manuals*. Washington, D.C.: International Monetary Fund. doi: 10.5089/9781462304127.005.

Ministry of Finance. 2020. 'Key to Budget Documents—Budget 2020–2021'. New Delhi. Ministry of Finance, Government of India. Available at https://www.indiabudget.gov.in/budget2020-21/doc/Key_to_Budget_Document_2020.pdf.

Ministry of Finance. 2022. 'India Budget | 2022–23'. February 2022. New Delhi: Ministry of Finance. Available at https://www.indiabudget.gov.in/.

Mishra, Neelkanth, and Prateek Singh. 2017. 'A Deep Dive into State Budgets in India'. In *India Policy Forum Volume 14*, edited by Shekhar Shah, Barry Bosworth, and Karthik Muralidharan. National Council of Applied Economic Research. New Delhi: SAGE Publications, pp. 53–99. Available at https://www.ncaer.org/wp-content/uploads/2022/09/1550144909IPF-Volume-2017-18.pdf.

Moe, Terry M. 1984. 'The New Economics of Organization'. *American Journal of Political Science* 28 (4): pp. 739–77. doi: 10.2307/2110997.

National Academy of Customs, Indirect Taxes & Narcotics (NACIN). 2018. 'GST Flyers'. New Delhi: Central Board of Indirect Taxes and Customs. Available at https://gstcouncil.gov.in/sites/default/files/e-version-gst-flyers/Excise_51%20GST_Flyer_1030-Corr-Binder-230819.pdf.

Panda, Manoj, Purnamita Dasgupta, and William Joe. 2019. 'Resource Sharing between Centre and States and Allocation across States: Some Issues in Balancing Equity and Efficiency'. Draft Report. A Study Prepared for The Fifteenth Finance Commission of India. New Delhi: Institute of Economic Growth. Available at https://fincomindia.nic.in/asset/doc/commission-reports/15th-FC/reports/studies/Resource%20sharing%20between%20Centre%20and%20States%20and%20allocation%20across%20States.pdf.

Patashnik, Eric M. 1996. 'The Contractual Nature of Budgeting: A Transaction Cost Perspective on the Design of Budgeting Institutions'. *Policy Sciences* 29 (3): pp. 189–212. doi: 10.1007/BF00136103.

Planning Commission. 2011. 'Report of the High Level Expert Committee on Efficient Management of Public Expenditure'. New Delhi, India: Planning Commission, Government of India. Available at http://164.100.161.239/reports/genrep/rep_hle.pdf.

Rao, M. Govinda. 2019. 'Redesigning the Fiscal Transfer System in India'. *Economic and Political Weekly* 54 (31): pp. 7–8. Available at https://www.epw.in/journal/2019/31/special-articles/redesigning-fiscal-transfer-system-india.html.

Rao, M. Govinda. 2022. *Studies in Indian Public Finance*. Oxford: Oxford University Press. doi: 10.1093/oso/9780192849601.003.0008.

Rao, M. Govinda, and Richard M. Bird. 2010. 'Urban Governance and Finance in India'. Rotman School of Management Working Paper No. 1568858. Rochester, NY: Social Science Research Network. doi: 10.2139/ssrn.1568858.

Rath, Anita. 2013. 'Growing Centralisation of Social Sector Policies in India'. *Economic and Political Weekly* 48 (4): pp. 62–70. Available at https://www.epw.in/journal/2013/04/special-articles/growing-centralisation-social-sector-policies-india.html.

Raymus, Prashant. 2012. *Mapping Budget Processes in the Public Health Sector in Maharashtra*. Mumbai: Centre for Enquiry into Health & Allied Themes. Available at https://www.academia.edu/5919313/Mapping_Budget_Processes_in_the_Public_Health_Sector_in_Maharashtra.

Sharma, Harikishan. 2020. 'CAG: Centre Broke the Law, Used Funds for GST Compensation Elsewhere'. *The Indian Express*, 25 September. Available at https://indianexpress.com/article/india/cag-centre-broke-the-law-used-funds-for-gst-compensation-elsewhere-6609749/.

Singh, Abhishek. 2017. 'AC-DC Bills: Analysis and Dashboards'. *Journal of Government Audit and Accounts* 7. Available at https://saiindia.gov.in/uploads/journal/journal_journal_August_2017/acdc-bill-article.html.

Singh, Nand Kishore. 2019. 'Speech of Shri N.K. Singh, Chairman, XVFC at the Launch of the Book "Indian Fiscal Federalism"'. IIC, New Delhi, 28 March. New Delhi: Press Information Bureau, Finance Commission, Government of India. Available at https://pib.gov.in/Pressreleaseshare.aspx?PRID=1569807.

Sundaramurti, C.R. 2012. 'Report of Sundaramurti Committee on Revision of List of Major and Minor Heads of Accounts'. New Delhi: Controller General of Accounts, Ministry of Finance, Government of India. Available at https://dea.gov.in/report-sundaramurti-committee-revision-list-major-and-minor-heads-accounts.

Tiwari, Suyash, and Saket Surya. 2019. 'State of State Finances'. New Delhi: PRS Legislative Research. Available at https://prsindia.org/files/policy/policy_analytical_reports/State%20of%20State%20Finances%202019-20.pdf.

Wildavsky, Aaron B. 1964. *The Politics of the Budgetary Process*. Boston: Little, Brown.

Williams, Michael J. 2010. 'Government Cash Management: Its Interaction with Other Financial Policies'. *Technical Notes and Manuals*. Washington,

D.C.: International Monetary Fund. Available at https://www.imf.org/en/Publications/TNM/Issues/2016/12/31/Government-Cash-Management-Its-Interaction-with-Other-Financial-Policies-23936.

Williamson, Oliver E. 1998. 'Transaction Cost Economics: How It Works; Where It Is Headed'. *De Economist* 146: pp. 23–58. doi: 10.1023/A:1003263908567.

XIV Finance Commission. 2014. 'Report of the XIV Finance Commission'. New Delhi: XIV Finance Commission. Available at https://fincomindia.nic.in/asset/doc/commission-reports/14th-FC/14thFCReport.pdf.

XV Finance Commission. 2019. 'Report for the Year 2020–21'. New Delhi: XV Finance Commission. Available at https://fincomindia.nic.in/asset/pdf/commission-reports/XVFC_202021%20Report_English_Web.pdf.

Chapter 5

Agarwal, O.P., and T.V. Somanathan. 2005. 'Public Policy Making In India: Issues and Remedies'. Available at https://www.igntu.ac.in/eContent/IGNTU-eContent-353062467985-MA-PoliticalScience-4-Dr.GeorgeT.Haokip-Paper401PublicPolicyandDevelopmentinIndia-Unit2.pdf.

Agarwal, Surabhi. 2020. 'IT Companies Want Government to Release Past Dues to Tide over Coronavirus Crisis'. *The Economic Times*, 21 April. Available at https://economictimes.indiatimes.com/tech/ites/it-companies-want-government-to-release-past-dues-to-tide-over-coronavirus-crisis/articleshow/75276920.cms?from=mdr.

Aiyar, Yamini, Ambrish A. Dongre, and Vincy Davis. 2015. 'Education Reforms, Bureaucracy and the Puzzles of Implementation: A Case Study from Bihar'. *SSRN Working Paper*. doi: 10.2139/ssrn.2695991.

Aiyar, Yamini, and Shrayana Bhattacharya. 2016. 'The Post Office Paradox: A Case Study of the Block Level Education Bureaucracy'. *Economic and Political Weekly* 51 (11): pp. 61–69. https://www.jstor.org/stable/44004104.

Aiyar, Yamini, and T.R. Raghunandan. 2015. 'Pull, Not Push, to Open up Spending'. *Livemint*, 23 February. Available at https://www.livemint.com/Opinion/uuVwhsKE8wKNa13sjI2tHN/Pull-not-push-to-open-up-spending.html.

Ayub, Jamal. 2019. 'Private Hospitals Seek Payment of Ayushman Bharat Dues'. *The Times of India*, 17 July. Available at https://timesofindia.indiatimes.com/city/bhopal/pvt-hospitals-seek-payment-of-ayushman-bharat-dues/articleshow/70252262.cms.

Bhatt, Sheela. 2013. 'All You Need to Know about the Fodder Scam'. *Rediff*, 1 October. Available at http://www.rediff.com/news/report/all-you-need-to-know-about-the-fodder-scam/20131001.htm.

Bhushan, Ranjit. 1996. 'Crores from Chicken Feed'. *Outlook India*, 28 February. Available at https://www.outlookindia.com/magazine/story/crores-from-chicken-feed/200871.

Bosio, Erica, and Simeon Djankov. 2020. 'How Large Is Public Procurement?' *World Bank Blogs*, 5 February. Available at https://blogs.worldbank.org/developmenttalk/how-large-public-procurement.

Budget Division. 2018. 'Government Debt: Status Paper'. New Delhi, India: Department of Economic Affairs, Ministry of Finance. Available at https://dea.gov.in/sites/default/files/Status%20Paper%20on%20Govt%20Debt%20%20for%202017-18.pdf.

CAG. 2025a. 'Report of the Comptroller and Auditor General of India on Accounts of the Union Government for the year 2023-24'. Union Government, Ministry of Finance No. 16 of 2025 (Financial Audit). The Comptroller and Auditor General of India. Available at https://cag.gov.in/en/audit-report/details/122496.

CAG. 2025b. 'Report of the Comptroller and Auditor General of India on State Finances for the year'. Government of Jharkhand No. 2 of 2025 (State Finances Audit Report). The Comptroller and Auditor General of India. Available at https://cag.gov.in/ag/jharkhand/en/audit-report/details/122657.

CAG. 2025c. 'Report of the Comptroller and Auditor General of India on State Finances for the year 2023-24'. Government of Haryana No. 1 of 2025 (State Finances Audit Report). The Comptroller and Auditor General of India. Available at https://cag.gov.in/ag/haryana/en/audit-report/details/123117.

CAG. 2025d. 'Report of the Comptroller and Auditor General of India on State Finances for the year 2023-24'. Government of Bihar No. 1 of 2025 (State Finances Audit Report). The Comptroller and Auditor General of India. Available at https://cag.gov.in/en/audit-report/details/121585.

CAG. 2025e. 'Report of the Comptroller and Auditor General of India on State Finances for the year 2023-24'. Government of Kerala No. 3 of 2025 (State Finances Audit Report). The Comptroller and Auditor General of India. Available at https://cag.gov.in/en/audit-report/details/123437.

CAG. 2025f. 'Report of the Comptroller and Auditor General of India on State Finances for the year 2023-24'. Government of Tamil Nadu No. 2 of 2025 (State Finances Audit Report). The Comptroller and Auditor General of India. Available at https://cag.gov.in/en/audit-report/details/122036.

Choudhury, Mita, and Ranjan Kumar Mohanty. 2018. 'Utilisation, Fund Flows and Public Financial Management under the National Health Mission'. No. 227. NIPFP Working Paper Series. New Delhi: National Institute of Public Finance and Policy. Available at https://www.nipfp.org.in/media/medialibrary/2018/05/WP_2018_227.pdf.

Choudhury, N. 1982. 'Aspects of Accounting and Internal Control—India 4th Century BC'. *Accounting and Business Research* 12 (46): pp. 105–10. doi: 10.1080/00014788.1982.9728795.

Das, Sourish, and Rabia Khatun. 2020. 'The Problems of Public Procurement and Payment Delays: A Review of the Recent Literature'. *The Leap Blog*, 23 November. Available at https://blog.theleapjournal.org/2020/11/the-problems-of-public-procurement-and.html.

Dasgupta, Aditya, and Devesh Kapur. 2020. 'The Political Economy of Bureaucratic Overload: Evidence from Rural Development Officials in India'. *American Political Science Review* 114 (4): pp. 1316–34. doi: 10.1017/S0003055420000477.

Deka, Kaushik. 2020. 'Why the States Are Broke'. *India Today*, 29 June. Available at https://www.indiatoday.in/magazine/the-big-story/story/20200629-why-the-states-are-broke-1690929-2020-06-21.

Department of Expenditure. 2017. 'General Financial Rules, 2017'. New Delhi: Department of Expenditure, Ministry of Finance, Government of India. Available at https://doe.gov.in/files/circulars_document/FInal_GFR_upto_31_07_2024.pdf.

Dhorajiwala, Sakina. 2018. 'MGNREGA: Honest Labour's Love Lost as Government Makes Payment Process Complex'. *The Economic Times Blog*, 25 April. Available at https://economictimes.indiatimes.com/blogs/et-commentary/mgnrega-honest-labours-love-lost-as-government-makes-payment-process-complex/.

Dodge, Eric, Yusuf Neggers, Rohini Pande, and Charity Troyer Moore. 2025. 'Can Reducing Information Frictions Strengthen State Capacity? Experimental Evidence from a Very Large Public Works Program'. Working Paper. Available at: https://bpb-us-w2.wpmucdn.com/campuspress.yale.edu/dist/7/2986/files/2025/02/PayDash_yls.pdf.

Dorotinsky, William, and Shilpa Pradhan. 2007. 'Exploring Corruption in Public Financial Management'. In *The Many Faces of Corruption: Tracking Vulnerabilities at the Sector Level*, edited by J. Edgardo Campos and Sanjay Pradhan, pp. 267–94. Washington, D.C.: The International Bank for Reconstruction and Development (IBRD) and World Bank.

ET Bureau. 2019. 'Government Meets Fiscal Deficit Target of 3.4% for 2018–19'. *The Economic Times*, 1 June. Available at https://economictimes.indiatimes.com/news/economy/indicators/fy19-fiscal-deficit-is-at-rs-rs-6-45-lakh-crore/articleshow/69596926.cms.

ET Now. 2019. 'Ayushman Bharat: 50% of Empanelled Hospitals Are from Private Sector, Says CEO'. *The Economic Times*, 3 April. Available at https://economictimes.indiatimes.com/markets/expert-view/ayushman-bharat-50-of-empanelled-hospitals-are-from-private-sector-says-ceo/articleshow/68703794.cms.

Fischer, Stanley, and William Easterly. 1990. 'The Economics of the Government Budget Constraint'. *The World Bank Research Observer* 5 (2): pp. 127–42. doi: 10.1093/wbro/5.2.127.

Goliya, Kshitiz. 2016. 'IT Firms Complain of Payment Delays by the Government'. *Livemint*, 19 October. Available at https://www.livemint.com/Industry/urXvP8YFo0vOHsSHgjOWlJ/IT-firms-complain-of-payment-delays-by-govt.html.

Gupta, Moushumi Das. 2020. 'Modi Govt Tells Ministries to Be Careful in Releasing Money to States for Central Schemes'. *ThePrint*, 10 May. Available at https://theprint.in/india/modi-govt-tells-ministries-to-be-careful-in-releasing-money-to-states-for-central-schemes/418314/.

Hazarika, Bhabesh, and Pratap Ranjan Jena. 2017. 'Public Procurement in India: Assessment of Institutional Mechanism, Challenges, and Reforms'. NIPFP Working Paper No. 204. New Delhi: National Institute of Public Finance and Policy. https://www.nipfp.org.in/publications/working-papers/1797/.

Herd, Pamela, and Donald P. Moynihan. 2019. *Administrative Burden: Policymaking by Other Means*. New York: Russell Sage Foundation. doi: 10.7758/9781610448789.

Honig, Dan. 2024. *Mission Driven Bureaucrats: Empowering People to Help Government Do Better*. New York: Oxford University Press.

HT Correspondent. 2020. 'Karnal Commission Agents Protest against Delay in Payment for Paddy'. *Hindustan Times*, 28 October. Available at https://www.hin

dustantimes.com/cities/karnal-commission-agents-protest-against-delay-in-payment-for-paddy/story-gvCgKzWHWE2q3Km4Iuz9RN.html.

Iyer, Indira, and Soumi Roy Chowdhury. 2022. 'Mainstreaming Digitization of Public Expenditure in India to Enhance Accountability and Transparency: Pushing the Frontiers through Real-Time Innovations'. Washington, D.C.: George Washington University; Bill and Melinda Gates Foundation.

Joseph, Josy. 2017. 'Confusion Reigns on CAG Audit of GST Data'. *The Hindu*, 5 April. Available at https://www.thehindu.com/business/Economy/confusion-reigns-on-cag-audit-of-gst-data/article17819889.ece.

Kapur, Avani. 2019. 'Towards "Cooperative" Social Policy Financing in India'. *Accountability Initiative: Responsive Governance*, 15 June. Available at https://accountabilityindia.in/blog/towards-cooperative-social-policy-financing-in-india/.

Kapur, Avani, and Vikram Srinivas. 2018. 'Sarva Shiksha Abhiyan (SSA): GOI, 2017–18'. 9 (2). New Delhi: Centre for Policy Research.

Kapur, Devesh, and Milan Vaishnav. 2018. *Costs of Democracy: Political Finance in India*. New Delhi: Oxford University Press.

Karamakar, Debashish. 2020. 'Ex-Bhagalpur District Magistrate Signed a Cheque for over Rs 12 Crore in Favour of NGO'. *The Times of India*, 11 January. Available at https://timesofindia.indiatimes.com/city/patna/ex-bhagalpur-dm-signed-a-cheque-for-over-rs12-crore-in-favour-of-ngo/articleshow/73194568.cms.

Kaur, Banjot. 2018a. 'Why 275,000 National Health Mission Workers Have Been Protesting for 3 Years'. *Down To Earth*, 24 January. Available at https://www.downtoearth.org.in/news/governance/why-275-000-national-health-mission-workers-have-been-protesting-for-3-years-59565.

Kaur, Banjot. 2018b. 'Women's Day: These 900,000 Health Workers Have No Reason to Celebrate'. *Down To Earth*, 8 March. Available at https://www.downtoearth.org.in/news/governance/no-end-in-sight-as-900-000-ashas-yet-to-receive-any-remuneration-for-months-59846.

Khan, M.H. 2017. 'Public Procurement Issues with Government of India'. SSRN Scholarly Paper. Rochester: Social Science Research Network. doi: 10.2139/ssrn.2925865.

Kumar, Ruchir. 2017. 'What Is the Rs 880 Crore Srijan Scam Roiling Bihar and Why It's Important: 10 Points'. *Hindustan Times*, 24 August. Available at https://www.hindustantimes.com/india-news/what-is-the-rs-880-crore-srijan-scam-roiling-bihar-and-why-it-s-important-10-points/story-VIpjEp5GLnuE9bqBjRq8jP.html.

Kumar, Vijay. 2009. 'Chapter 17: Accounts'. In *The Comptroller & Auditor General of India: A Thematic History 1990–2007 (Vol. I)*. Office of the Comptroller and Auditor General of India. New Delhi: APH Publishing Corporation.

Kumawat, Subodh. 2017. 'CAG amidst Confusion over Auditing Power under GST'. *SAG Infotech Blog*, 5 April. Available at https://blog.saginfotech.com/cag-amidst-confusion-over-auditing-power-under-gst.

Mathew, A. Santhosh. 2018. 'Lessons from GST for Eliminating the Rs1 Trillion Lying around as Float'. *Livemint*, 16 February. Available at https://www.livemint.com/Opinion/8xLS7tbYhNeusRRB0uevuL/Lessons-from-GST-for-eliminating-the-Rs1-trillion-lying-arou.html.

Ministry of Finance. 2016. 'Central Assistance for State/Union Territories and Centrally Sponsored Schemes'. Union Budget. New Delhi: Government of India.

Ministry of Finance. 2019. 'FM Reviews Capital Expenditure & Payments of Maharatnas and Navratnas CPSEs'. New Delhi: Government of India. Available at pib.gov.in/Pressreleaseshare.aspx?PRID=1586546.

Mitra, Pritha. 2006. 'Post-Crisis Recovery: When Does Increased Fiscal Discipline Work?' WP/06/219, IMF Working Paper. International Monetary Fund.

MKSS. 2014. 'The Right to Information and Corruption Movement in Maharashtra'. *Mazdoor Kisan Shakti Sangathan*. Available at http://www.mkssindia.org/mkssandrti/the-rti-and-corruption-movement-in-maharashtra/.

Mondal, Dipak. 2016. 'Payment Blues—Business News'. *Business Today*, 17 July. Available at https://www.businesstoday.in/magazine/corporate/companies-working-for-the-government-under-financial-stress/story/234234.html.

Mukhopadhyay, Amitabh. 2017. 'Foregrounding Financial Accountability in Governance'. In *Rethinking Public Institutions in India*, edited by Devesh Kapur, Pratap Bhanu Mehta, and Milan Vaishnav, pp. 297–338. New Delhi: Oxford University Press. Available at https://carnegieendowment.org/2017/04/21/rethinking-public-institutions-in-india-pub-68221.

Narayanan, Sudha, Upasak Das, Yanyan Liu, and Christopher B. Barrett. 2017. 'The "Discouraged Worker Effect" in Public Works Programs: Evidence from the MGNREGA in India'. *World Development* 100: pp. 31–44. doi: 10.1016/j.worlddev.2017.07.024.

Nilachala, Acharya, and Mampi Bose. 2020. 'Delay in Fund Flow: Consequences, Causes and Remedies'. Policy Brief: 3. Fiscal Governance Reforms at District Level for Improving Fund Flow and Utilisation in Development Schemes. New Delhi: Centre for Budget and Governance Accountability and Tata Trusts. Available at https://www.cbgaindia.org/wp-content/uploads/2020/02/Delay-in-Fund-Flow-in-Social-Sector-Policy-Brief.pdf.

Ponappa, Shyam. 2019. 'Delayed Cash Flows and NPAs'. *Business Standard*, 3 April. Available at https://www.business-standard.com/article/opinion/delayed-cash-flows-and-npas-119040301417_1.html.

Prasad, Akansha. 2013. 'IT Companies like Infosys, HCL Info Shying Away from Government Projects on Payment Delays'. *The Economic Times*, 7 August. Available at https://economictimes.indiatimes.com/tech/ites/it-companies-like-infosys-hcl-info-shying-away-from-government-projects-on-payment-delays/articleshow/21636883.cms?from=mdr.

Press Trust of India. 2015. 'India Ranks 37 in Open Government Index'. *Business Standard India*, 27 March. Available at https://www.business-standard.com/article/pti-stories/india-ranks-37-out-of-102-countries-in-open-government-index-115032700633_1.html.

Rai, Vinod. 2020. 'Digital Accountability and Transparency Act: Towards Digitising the State – NUS Institute of South Asian Studies (ISAS)'. *ISAS Insights*, 18 August. Available at https://www.isas.nus.edu.sg/papers/digital-accountability-and-transparency-act-towards-digitising-the-state/.

Rajaraman, Indira. 2019. 'Payment Delays and the Fiscal-Financial Ricochet'. *Livemint*, 31 January. Available at https://www.livemint.com/opinion/online-views/opinion-payment-delays-and-the-fiscal-financial-ricochet-1548956945084.html.

Rajaraman, Indira, and Manish Gupta. 2016. 'Preserving the Incentive Properties of Statutory Grants'. *Economic and Political Weekly* 51 (9): pp. 79–84.

Rao, M. Govinda. 2022. *Studies in Indian Public Finance*. Oxford: Oxford University Press. doi: 10.1093/oso/9780192849601.003.0008.

Reserve Bank of India. 2018. 'State Finances: A Study of Budgets of 2017–18 and 2018–19'. Reserve Bank of India.

Sabharwal, Manish, and Deepak Phatak. 2020. 'A Three-Phase Govt Project Uses COVID Policy Window for a Much Needed Reform'. *The Indian Express*, 9 July. Available at https://indianexpress.com/article/opinion/columns/digitising-the-state-6496692/.

Saikat, Das. 2019. '3 States Make Instant Gain on Rs 10,000-Cr T-Bills'. *The Economic Times*, 9 February. Available at https://economictimes.indiatimes.com/markets/stocks/news/3-states-make-instant-gain-on-rs-10000-cr-t-bills/articleshow/67910360.cms.

SCRDPR. 2024. 'First Report'. Standing Committee on Rural Development and Panchayati Raj, Ministry of Rural Development, 12 December. Available at https://sansad.in/getFile/lsscommittee/Rural%20Development%20and%20Panchayati%20Raj/18_Rural_Development_and_Panchayati_Raj_1.pdf?source=loksabhadocs.

Sen Gupta, Nandini. 2019. 'Subsidy Schemes Pinch Tractor Companies as State Governments Delay Payments'. *The Times of India*, 19 April. Available at https://timesofindia.indiatimes.com/business/india-business/subsidy-schemes-pinch-tractor-companies-as-state-governments-delay-payments/articleshow/68946744.cms.

Sharma, Harikishan. 2020. 'CAG: Centre Broke the Law, Used Funds for GST Compensation Elsewhere'. *The Indian Express*, 25 September. Available at https://indianexpress.com/article/india/cag-centre-broke-the-law-used-funds-for-gst-compensation-elsewhere-6609749/.

Singh, Abhishek. 2017. 'AC-DC Bills: Analysis and Dashboards'. *Journal of Government Audit and Accounts* 7. Available at https://saiindia.gov.in/uploads/journal/journal_journal_August_2017/acdc-bill-article.html.

Somanathan, T.V., and Gulzar Natarajan. 2022. *State Capability in India*. Oxford: Oxford University Press.

Swamy, Rakshita. 2019. 'The Bhilwara Principles: An Accountability Framework in Action'. *Common Cause* XXXVIII (2): pp. 36–41. Available at https://www.commoncause.in/pdf/The-Bhilwara-Principles.pdf.

Swarup, Anil. 2019. *Not Just a Civil Servant*. New Delhi: Unicorn Books Pvt Ltd.

Thakur, Pradeep. 2021. 'CAG Writes to Bengal, Says Amphan Info Blacked out'. *The Times of India*, 28 February. Available at https://timesofindia.indiatimes.com/india/cag-writes-to-bengal-says-amphan-info-blacked-out/articleshow/81251707.cms.

The Pioneer. 2018. 'Private Hospitals to Get Payments on Ayushman Bharat within 15 Days'. *The Pioneer*, 14 October. Available at https://www.dailypioneer.com/2018/state-editions/private-hospitals-to-get-payments-on-ayushman-bharat-within-15-days.html.

Times News Network. 2017. 'Srijan Scam: Land Acquisition Funds Not to Be Kept in Banks'. *The Times of India*, 1 September. Available at https://timesofindia.ind

iatimes.com/city/patna/land-acquisition-funds-not-to-be-kept-in-banks/articleshow/60313103.cms.

World Bank. 2006. 'India Note on Public Financial Management and Accountability in Centrally Sponsored Schemes'. Washington, D.C.: World Bank. Available at https://openknowledge.worldbank.org/handle/10986/19257.

World Bank. 2020. *Doing Business 2020: Comparing Business Regulation in 190 Economies*. Washington, D.C.: World Bank. Available at http://hdl.handle.net/10986/32436.

Chapter 6

Agarwal, Varun, Devashish Deshpande, Josh Felman, Theerdha Sara Reji, and Arvind Subramanian. 2024. 'GST Revenue Performance'. *Economic and Political Weekly* 59 (37): pp. 35–49. Available at https://www.epw.in/journal/2024/37/special-articles/gst-revenue-performance.html.

Akhshabi, Saamer, and Constantine Dovrolis. 2011. 'The Evolution of Layered Protocol Stacks Leads to an Hourglass-Shaped Architecture'. SIGCOMM'11, 15–19 August, 2011, Toronto, Ontario, Canada. Available at https://www.cc.gatech.edu/~dovrolis/Papers/evoarch.pdf.

Allen, Richard, Yasemin Hurcan, and Maximilien Queyranne. 2016. 'The Evolving Functions and Organization of Finance Ministries'. *Public Budgeting & Finance* 36 (4): pp. 3–25. doi: 10.1111/pbaf.12141.

Basu, Medha. 2020. 'Four Things You Should Know about Rules as Code'. *GovInsider*, 3 March. Available at https://govinsider.asia/inclusive-gov/four-things-you-should-know-about-rules-as-code/.

Beck, Micah. 2019. 'On the Hourglass Model'. *Communications of the ACM* 62 (7): pp. 48–57. doi: 10.1145/3274770.

CAG. 2015. 'Report for the Year 2014–2015'. Accounts of the Union Government No. 50 of 2015 (Financial Audit). The Comptroller and Auditor General of India. https://cag.gov.in/sites/default/files/audit_report_files/Union_Finacne_Report_50_2015_0.pdf.

Cave, Jonathan, Maarten Botterman, Simona Cavallini, and Margherita Volpe. 2017. 'EU-Wide Digital Once-Only Principle for Citizens and Businesses—Policy Options and Their Impacts'. European Commission: Directorate-General for Communications Networks, Content and Technology. doi: 10.2759/393169.

Choudary, Sangeet Paul. 2014. 'A Platform-Thinking Approach to Innovation'. *Wired*, 31 January. Available at https://www.wired.com/insights/2014/01/platform-thinking-approach-innovation/.

Corritore, Cynthia L., Beverly Kracher, and Susan Wiedenbeck. 2003. 'On-Line Trust: Concepts, Evolving Themes, a Model'. *International Journal of Human-Computer Studies, Trust and Technology* 58 (6): pp. 737–58. doi: 10.1016/S1071-5819(03)00041-7.

Denize, Sara, and Louise Young. 2007. 'Concerning Trust and Information'. *Industrial Marketing Management* 36 (7): pp. 968–82. doi: 10.1016/j.indmarman.2007.06.004.

Dodge, Eric, Yusuf Neggers, Rohini Pande, and Charity M. Troyer Moore. 2025. 'From Delay to PayDay: Easing Bureaucrat Access to Implementation Information Strengthens Social Protection Delivery'. NBER Working Paper 33756. doi: 10.3386/w33756.

Fisman, Raymond, and Tarun Khanna. 1999. 'Is Trust a Historical Residue? Information Flows and Trust Levels'. *Journal of Economic Behavior & Organization* 38 (1): pp. 79–92. doi: 10.1016/S0167-2681(98)00123-1.

Fraser, Hamish. 2021. 'What Is Better Rules?' *New Zealand Digital Government Blog*, 7 December. Available at https://www.digital.govt.nz/blog/what-is-better-rules.

Georgia Institute of Technology. 2011. 'How the Internet Architecture Got Its Hourglass Shape and What That Means for the Future'. *Phys.org*, 15 August. Available at https://phys.org/news/2011-08-internet-architecture-hourglass-future.html.

Girtu, Miruna. 2018. 'The Platform Edge'. *The Startup*, 21 September. Available at https://medium.com/swlh/the-platform-edge-dbc541320fa7.

Hatcher-Mbu, Beverley, Fernando Ferreyra, Mihai Postelnicu, Viorel Cihai, Yosef Getachew, Yohannes Fikre, Selam Abraham Kassa, Tom Orrell, Khyatee Tewari, and Octavian Ciubotaru. 2024. 'Demystifying Interoperability: Insights from Our Work in Ethiopia's Agriculture Sector'. Washington, D.C.: Development Gateway. Available at https://developmentgateway.org/publication_landing/demystifying-interoperability-one-building-block-at-a-time-insights-from-our-work-in-ethiopias-agriculture-sector/.

ISO. 2022. 'ISO—International Organization for Standardization'. ISO. 2022. Available at https://www.iso.org/home.html.

ITU. 2022. 'Definition of 'Open Standards'. ITU. 2022. Available at https://www.itu.int:443/en/ITU-T/ipr/Pages/open.aspx.

Johansson, Niklas, and Anton Löfgren. 2009. 'Designing for Extensibility: An Action Research Study of Maximizing Extensibility by Means of Design Principles'. Gothenburg: Department of Applied Information Technology, University of Gothenburg. Available at https://gupea.ub.gu.se/handle/2077/20561.

Kelton, Kari, Kenneth R. Fleischmann, and William A. Wallace. 2008. 'Trust in Digital Information'. *Journal of the American Society for Information Science and Technology* 59 (3): pp. 363–74. doi: 10.1002/asi.20722.

Klose, Alexander. 2015. *The Container Principle: How a Box Changes the Way We Think*. Cambridge: MIT Press.

Kolossovski, Eleanor. 2019. 'A Strategist's Guide to Platform Thinking'. *Medium*, 3 December. Available at https://medium.com/@eleanor.kolossovski/a-strategists-guide-to-platform-thinking-48cc4b666d9a.

Levinson, Marc. 2016. *The Box*. Princeton: Princeton University Press.

Levitsky, Steven, and Daniel Ziblatt. 2019. *How Democracies Die*. New York: Crown.

Lucassen, Teun, and Jan Maarten Schraagen. 2011. 'Factual Accuracy and Trust in Information: The Role of Expertise'. *Journal of the American Society for Information Science and Technology* 62 (7): pp. 1232–42. doi: 10.1002/asi.21545.

Marsh, Stephen, and Mark R. Dibben. 2003. 'The Role of Trust in Information Science and Technology'. *Annual Review of Information Science and Technology (ARIST)* 37 (1): pp. 465–98. doi: 10.1002/aris.1440370111.

Mathew, Santhosh A. 2021. 'The Indian State and Shared Digital Infra'. Available at https://www.linkedin.com/pulse/indian-state-shared-digital-infra-santhosh-mathew/.

Ministry of Communications & Information Technology. 2010. 'Policy on Open Standards for e-Governance'. Available at https://egovstandards.gov.in/sites/default/files/2021-07/Policy%20on%20Open%20Standards%20for%20e-Governance.pdf.

Mohun, James, and Alex Roberts. 2020. 'Cracking the Code: Rulemaking for Humans and Machines'. Paris: OECD. Available at https://www.oecd-ilibrary.org/governance/cracking-the-code_3afe6ba5-en.

Murtagh, Jack. 2023. 'Where's Waldo? How to Mathematically Prove You Found Him without Revealing Where He Is'. Scientific American, 2 June. Available at https://www.scientificamerican.com/article/wheres-waldo-how-to-prove-you-found-him-without-revealing-where-he-is/.

National Institute of Urban Affairs and Ministry of Housing and Urban Affairs. 2021. 'National Urban Innovation Stack (NUIS) Digital Blueprint'. Available at https://web.archive.org/web/20210705171519/https://smartnet.niua.org/sites/default/files/resources/digital_blueprint-digital-4.pdf.

New Zealand Government. 2022. 'Better Rules for Government Discovery Report'. *New Zealand Digital Government*, 8 February. Available at https://www.digital.govt.nz/dmsdocument/95-better-rules-for-government-discovery-report/html.

Observatory of Public Sector Innovation. 2020. 'Cracking the Code: Rulemaking for Humans and Machines – Observatory of Public Sector Innovation'. *Observatory of Public Sector Innovation*, 12 October. Available at https://oecd-opsi.org/publications/cracking-the-code/.

O'Reilly, Tim. 2011. 'Government as a Platform'. *Innovations: Technology, Governance, Globalization* 6 (1): pp. 13–40. doi: 10.1162/INOV_a_00056.

Peters, B. Guy, and Matthew Tarpey. 2019. 'Are Wicked Problems Really So Wicked? Perceptions of Policy Problems'. *Policy and Society* 38 (2): pp. 218–36. doi: 10.1080/14494035.2019.1626595.

Postrel, Virginia. 2006. 'The Container That Changed the World'. *The New York Times*, 23 March. Available at https://www.nytimes.com/2006/03/23/business/the-container-that-changed-the-world.html.

Priisalu, Jaan, and Rain Ottis. 2017. 'Personal Control of Privacy and Data: Estonian Experience'. *Health and Technology* 7 (4): pp. 441–51. doi: 10.1007/s12553-017-0195-1.

Rashid, Naeha. 2020. 'Deploying the Once-Only Policy: A Privacy-Enhancing Guide for Policymakers and Civil Society Actors'. Cambridge, MA: Ash Center for Democratic Governance, Harvard Kennedy School. Available at https://datasmart.hks.harvard.edu/news/article/deploying-once-only-policy-privacy-enhancing-guide-policymakers-and-civil-society.

Sabharwal, Manish, and Deepak Phatak. 2020. 'A Three-Phase Govt Project Uses COVID Policy Window for a Much Needed Reform'. *The Indian Express*, 9 July. Available at https://indianexpress.com/article/opinion/columns/digitising-the-state-6496692/.

Scott, James C. 1998. *Seeing Like a State: How Certain Schemes to Improve the Human Condition Have Failed*. New Haven: Yale University Press.

Sharma, Tarun. 2019. 'Income Tax Department Holding Back Refunds Amounting to Thousands of Crores'. *Moneycontrol*, 10 February. Available at https://www.moneycontrol.com/news/business/corporate-action/income-tax-refunds-amounting-to-thousands-of-crores-delayed-by-up-to-4-years-3498291.html.

Standards Council of Canada. 2025. 'Benefits of Applying Standards'. SCC-CCN. 2025. Available at: https://scc-ccn.ca/standards/benefits-applying-standards.

Suryanarayan, Pavithra, and Steven White. 2021. 'Slavery, Reconstruction, and Bureaucratic Capacity in the American South'. *American Political Science Review* 115 (2): 568–84. doi: 10.1017/S0003055420000933.

TAGUP. 2011. 'Report of the Technology Advisory Group for Unique Projects'. New Delhi: Ministry of Finance. Available at https://comtax.uk.gov.in/files/pdf/70_TAGUP_Report.pdf.

The Rockefeller Foundation. 2021. 'Co-Develop: Digital Public Infrastructure for an Equitable Recovery'. New York: The Rockefeller Foundation. Available at https://www.rockefellerfoundation.org/report/co-develop-digital-public-infrastructure-for-an-equitable-recovery/.

Tomkins, Cyril. 2001. 'Interdependencies, Trust and Information in Relationships, Alliances and Networks'. *Accounting, Organizations and Society* 26 (2): pp. 161–91. doi: 10.1016/S0361-3682(00)00018-0.

United Nations Development Programme (UNDP). 2023. *Accelerating the SDGs through Digital Public Infrastructure: A Compendium of the Potential of DPI*. New York: UNDP. Available at https://www.undp.org/publications/accelerating-sdgs-through-digital-public-infrastructure-compendium-potential-digital-public-infrastructure.

W3 Computing. 2014. 'Effective and Efficient Data Capture'. *W3 Computing*, 18 September. https://www.w3computing.com/systemsanalysis/effective-efficient-data-capture/.

Waghmare, Abhishek. 2019. 'Income Tax Return Processing Time to Reduce from 63 Days to Just 1 Day | Business Standard News'. *Business Standard*, 17 January. Available at https://www.business-standard.com/article/economy-policy/income-tax-return-processing-time-to-reduce-from-63-days-to-just-1-day-119011700086_1.html.

Wixom, Barbara H., and Gabriele Piccoli. 2021. 'Build Data Liquidity to Accelerate Data Monetization'. MIT CISR. No. XXI-5. Available at https://cisr.mit.edu/publication/2021_0501_DataLiquidity_WixomPiccoli.

World Bank. 2022. 'A Digital Stack for Transforming Service Delivery ID-Payments and Data Sharing'. Report Number 170268. Washington, D.C.: World Bank. Available at http://documents.worldbank.org/curated/en/099755004072288910.

Chapter 7

Abelson, Harold, Ross Anderson, Steven M. Bellovin, Josh Benaloh, Matt Blaze, Whitfield Diffie, John Gilmore, Matthew Green, Susan Landau, Peter G. Neumann, Ronald L. Rivest, Jeffrey I. Schiller, Bruce Schneier, Michael A. Specter, Daniel J. Weitzner. 2015. 'Keys under doormats: mandating insecurity by requiring government access to all data and communications'. Journal of Cybersecurity 1(1): pp. 69–79. doi: 10.1093/cybsec/tyv009.

Afridi, Farzana, and Vegard Iversen. 2014. 'Social Audits and MGNREGA Delivery: Lessons from Andhra Pradesh'. IZA Discussion Paper No. 8095. Bonn: Institute for the Study of Labor (IZA). Available at https://www.iza.org/publications/dp/8095/social-audits-and-mgnrega-delivery-lessons-from-andhra-pradesh.

Banerjee, Abhijit, Esther Duflo, Clement Imbert, Santosh A. Mathew, and Rohini Pande. 2020. 'E-Governance, Accountability, and Leakage in Public Programs: Experimental Evidence from a Financial Management Reform in India'. *American Economic Journal: Applied Economics* 12 (4): pp. 39–72. doi: 10.1257/app.20180302.

Björkman, Martina, and Jakob Svensson. 2009. 'Power to the People: Evidence from a Randomized Field Experiment on Community-Based Monitoring in Uganda'. *The Quarterly Journal of Economics* 124 (2): pp. 735–69. doi: 10.1162/qjec.2009.124.2.735.

Bussell, Jennifer. 2012. *Corruption and Reform in India: Public Services in the Digital Age*. New Delhi: Cambridge University Press.

CGA. 2024. 'Public Financial Management System PFMS'. New Delhi: Controller General of Accounts, Ministry of Finance. Available at https://cga.nic.in/Page/Public-Finance-Management-System-PFMS.aspx.

Cris. 2021. 'Kerala Govt Doctors Boycott Duty to Protest Delayed Arrears, Allowances'. *The News Minute*, 29 January. Available at https://www.thenewsminute.com/article/kerala-govt-doctors-boycott-duty-protest-delayed-arrears-allowances-142404.

Dey, Soumik. 2019. 'Finance Ministry to Monitor Payment Dues on Govt Contracts'. *The Week*, 6 September. Available at https://www.theweek.in/news/biz-tech/2019/09/06/Finance-ministry-to-monitor-payment-dues-on-govt-contracts.html.

HT Correspondent. 2020. 'Karnal Commission Agents Protest Against Delay in Payment for Paddy'. *Hindustan Times*, 28 October. Available at https://www.hindustantimes.com/cities/karnal-commission-agents-protest-against-delay-in-payment-for-paddy/story-gvCgKzWHWE2q3Km4Iuz9RN.html.

Imbert, Clement, and John Papp. 2015. 'Labor Market Effects of Social Programs: Evidence from India's Employment Guarantee'. *American Economic Journal: Applied Economics* 7 (2): pp. 233–63. doi: 10.1257/app.20130401.

Manivannan, Pavithra, and Bhargavi Zaveri. 2021. 'How Large Is the Payment Delays Problem in Indian Public Procurement?' *The Leap Blog*, 22 March. Available at https://blog.theleapjournal.org/2021/03/how-large-is-payment-delays-problem-in.html.

Mathew, A. Santhosh. 2016. 'Modernizing Government Payment Systems to Deinstitutionalize Corruption'. *Outlook India*, 11 April, pp. 60–61.

Mathew, A. Santhosh. 2018. 'Lessons from GST for Eliminating the Rs 1 Trillion Lying Around as Float'. *Livemint*, 16 February. Available at https://www.livemint.com/Opinion/8xLS7tbYhNeusRRB0uevuL/Lessons-from-GST-for-eliminating-the-Rs1-trillion-lying-arou.html.

Mathew, A. Santhosh, and Dwijo Goswami. 2016. 'Doing More with Less'. *Economic and Political Weekly* 51 (17): pp. 110–18. Available at https://www.epw.in/journal/2016/17/special-articles/doing-more-less.html.

Mathew, A. Santhosh, and Bhumi Purohit. 2017. 'India's Public Finance Architecture Is the Real Culprit in Gorakhpur and Bhagalpur'. *Livemint*, 25 August. Available at https://www.livemint.com/Opinion/I6UMVabJwiukGs7rB5bmSM/Indias-public-finance-architecture-is-the-real-culprit-in-G.html.

Ministry of Rural Development. 2015. 'Cabinet Note for Consultation with States on MGNREGA-Direct Release of Wages to Workers for Better Implementation and Empowerment of the States'. New Delhi: Ministry of Rural Development, Government of India.

Misra, Shubhangi. 2021. 'Delayed Payment, Less Govt Procurement—How Ghazipur Farm Protest Differs from Singhu, Tikri'. *ThePrint*, 5 January. Available at https://theprint.in/india/delayed-payment-less-govt-procurement-how-ghazipur-farm-protest-differs-from-singhu-tikri/579191/.

Mondal, Dipak. 2016. 'Payment Blues—Business News'. *Business Today*, 17 July. Available at https://www.businesstoday.in/magazine/corporate/companies-working-for-the-government-under-financial-stress/story/234234.html.

Mukherji, Rahul, and Himanshu Jha. 2017. 'Bureaucratic Rationality, Political Will, and State Capacity: MGNREGS in Undivided Andhra Pradesh'. *Economic and Political Weekly* 52 (49): pp. 53–60. Available at https://www.epw.in/journal/2017/49/special-articles/bureaucratic-rationality-political-will-and-state-capacity.html.

Niehaus, Paul, and Sandip Sukhtankar. 2013. 'Corruption Dynamics: The Golden Goose Effect'. *American Economic Journal: Economic Policy* 5 (4): pp. 230–69. doi: 10.1257/pol.5.4.230.

NITI Aayog. 2016. 'Minutes of the Presentation/ Meeting on Just-in-Time (JIT) Fund Management Architecture for Payments in All Government Programs Chaired by Vice Chairman, NITI Aayog on 20th June, 2016, Room No.126, NITI Aayog'. New Delhi: NITI Aayog.

Olken, Benjamin A. 2007. 'Monitoring Corruption: Evidence from a Field Experiment in Indonesia'. *Journal of Political Economy* 115 (2): pp. 200–49. doi: 10.1086/517935.

Peisakhin, Leonid, and Paul Pinto. 2010. 'Is Transparency an Effective Anti-Corruption Strategy? Evidence from a Field Experiment in India'. *Regulation & Governance* 4: pp. 261–80. doi: 10.1111/j.1748-5991.2010.01081.x.

PTI. 2020. 'Railways Employees Protest Delay in Payment of Bonus'. *Deccan Herald*, 20 October. Available at https://www.deccanherald.com/national/north-and-central/railways-employees-protest-delay-in-payment-of-bonus-904758.html.

Ramkumar, Vivek, and Warren Krafchik. 2006. 'The Role of Civil Society Organizations in Auditing and Public Finance Management'. Washington, D.C.: The International Budget Project. Available at https://www.internationalbudget.org/wp-content/uploads/The-Role-of-Civil-Society-Organizations-in-Auditing-and-Public-Finance-Management1.pdf.

Rose-Ackerman, Susan. 1996. 'Democracy and 'Grand' Corruption'. *International Social Science Journal* 149 (48): pp. 365–80. doi: 10.1111/1468-2451.00038.

Sahu, Prasanta. 2020. 'Forget Stimulus, Clear Your Dues: Rs 7 Lakh Crore Unpaid Dues to Industry by Central Govt Depts and PSUs'. *The Financial Express*, 8 September. Available at https://www.financialexpress.com/economy/forget-stimulus-clear-your-dues-rs-7-lakh-crore-unpaid-dues-to-industry-by-central-govt-depts-and-psus/2077404/.

Sukhtankar, Sandip, and Milan Vaishnav. 2015. 'Corruption in India: Bridging Research Evidence and Policy Options'. *India Policy Forum* 11: pp. 193–261. Available at https://ssrn.com/abstract=2685819.

Sundaramurti, C.R. 2012. 'Report of Sundaramurti Committee on Revision of List of Major and Minor Heads of Accounts'. New Delhi: Controller General of Accounts, Ministry of Finance, Government of India. Available at https://dea.gov.in/report-sundaramurti-committee-revision-list-major-and-minor-heads-accounts.

The Hindu. 2021. 'ASHA Workers Demand Release of Dues'. *The Hindu*, 29 January. Available at https://www.thehindu.com/news/cities/Visakhapatnam/asha-workers-demand-release-of-dues/article33690340.ece.

Venugopal, Vasudha. 2020. 'Formalise Conditions, Wages of ASHA and Anganwadi Workers: House Panel on Labour'. *The Economic Times*, 8 August. Available at https://economictimes.indiatimes.com/news/politics-and-nation/formalise-conditions-wages-of-asha-and-anganwadi-workers-house-panel-on-labour/articleshow/77422539.cms?from=mdr.

Wang, Shuai, Liwei Ouyang, Yong Yuan, Xiaochun Ni, Xuan Han, and Fei-Yue Wang. 2019. 'Blockchain-Enabled Smart Contracts: Architecture, Applications, and Future Trends'. *IEEE Transactions on Systems, Man, and Cybernetics: Systems* 49 (11): pp. 2266–77. doi: 10.1109/TSMC.2019.2895123.

Chapter 8

Allen, Richard. 2015. 'Angels and Demons—the Political Economy of PFM Reform'. *PFM Blog*, 30 January. Available at https://blog-pfm.imf.org/en/pfmblog/2015/01/angels-and-demons-the-political-economy-of-pfm-reform.

Andrews, Matt. 2013. *The Limits of Institutional Reform in Development: Changing Rules for Realistic Solutions*. Cambridge: Cambridge University Press. Available at https://www.cambridge.org/core/books/limits-of-institutional-reform-in-development/C7E4CCE47EE4FC81EBCE0AA78F8A60FE.

Binyaruka, Peter, John Maiba, Dastan Mshana, Agnes Gatome-Munyua, and Gemini Mtei. 2024. 'Does Provider Autonomy Work Well in Tanzania? Perspectives of Primary Care Facilities on Budget Execution under Direct Facility Financing and Factors Affecting Provider Autonomy in Singida Region'. *Health Systems & Reform* 10 (3): pp. 1–13. doi: 10.1080/23288604.2024.2432043.

CGA. 2024. 'Public Financial Management System PFMS'. New Delhi: Controller General of Accounts, Ministry of Finance. Available at https://cga.nic.in/Page/Public-Finance-Management-System-PFMS.aspx.

Chatterjee, Elizabeth. 2012. 'Dissipated Energy: Indian Electric Power and the Politics of Blame'. Contemporary South Asia, 20(1): pp. 91–103. doi:10.1080/09584935.2011.646072.

Digital Impact Alliance. 2019. 'Use Open Standards, Open Data, Open Source, and Open Innovation'. Principles for Digital Development. Available at https://archive.digitalprinciples.org/principle/use-open-standards-open-data-open-source-and-open-innovation/.

Dubash, Navroz K. and Sudhir Chella Rajan. 2011. 'Power politics: process of power sector reform in India'. Economic and Political Weekly, 36 (35): pp. 3367–90. Available at https://www.epw.in/journal/2001/35/special-articles/power-politics.html.

Herlihy, Peter. 2020. 'GOV.UK Notify Is Available for the Public Sector to Use for Emergency Staff Communications—Government Digital Service'. *Government Digital Service Blog*, 18 March. Available at https://gds.blog.gov.uk/2020/03/18/gov-uk-notify-is-available-for-the-public-sector-to-use-for-emergency-staff-communications/.

International Monetary Fund. 2021. 'IMF Conditionality'. IMF. 22 February. Available at https://www.imf.org/en/About/Factsheets/Sheets/2016/08/02/21/28/IMF-Conditionality.

Loosemore, Tom. 2018. 'Internet-Era Ways of Working'. Public Digital. Available at https://public.digital/pd-insights/blog/2018/10/internet-era-ways-of-working.

Mann, Robin, Rajive Mathur, Dave Rogers, James Stewart, and Miguel Carrasco. 2021. 'Fixing Digital Funding in Government'. Boston Consulting Group. Available at https://www.bcg.com/publications/2021/fixing-digital-funding-in-government.

McQuade, J. Michael, Richard M. Murray, Gilman Louie, Milo Medin, Jennifer Pahlka, and Trae Stephens. 2019. 'Software Is Never Done: Refactoring the Acquisition Code for Competitive Advantage'. Defense Innovation Board. Available at https://media.defense.gov/2019/Mar/26/2002105909/-1/-1/0/SWAP.REPORT_MAIN.BODY.3.21.19.PDF.

Ministry of Electronics and Information Technology. 2015. 'Policy on Open Application Programming Interfaces (APIs) for Government of India'. Ministry of Communications and Information Technology, Government of India. Available at https://www.meity.gov.in/writereaddata/files/Policy%20for%20API%20for%20GoI.pdf.

Ministry of Electronics and Information Technology. 2018. 'India Enterprise Architecture (IndEA)'. Ministry of Electronics and Information Technology, Government of India. Available at https://www.meity.gov.in/india-enterpise-architecture-indea.

Ministry of Electronics and Information Technology. 2019. 'Agile India Enterprise Architecture (IndEA) Framework'. Ministry of Electronics and Information Technology, Government of India. Available at https://www.meity.gov.in/agile-india-enterprise-architecture-indea-framework.

Ministry of Electronics and Information Technology. 2020. 'Implementation Guidelines for Open API Policy for e-Governance (National Data Highway)'. Ministry of Electronics and Information Technology. Available at https://egovstandards.gov.in/sites/default/files/2021-07/Implementation%20Guidelines%20for%20Open%20API%20Policy%20for%20e-Governance%20%20%28National%20Data%20Highway%29%20V1.0_0.pdf.

Mukherji, Rahul. 2014. *Globalization and Deregulation: Ideas, Interests, and Institutional Change in India*. Oxford: Oxford University Press.

Narayan, Sriram. 2018. 'Products over Projects'. Available at https://martinfowler.com/articles/products-over-projects.html.

Pahlka, Jennifer. 2024. 'Testimony of Jennifer Pahlka, former US Deputy Chief Technology Officer, Senior Fellow, Federation of American Scientists and the Niskanen Center Before the Committee on Homeland Security and Government Affairs, U.S. Senate on Harnessing AI to Improve Government Services and Customer Experience'. Available at https://www.niskanencenter.org/testimony-jennifer-pahlka-on-harnessing-ai-to-improve-government-services-and-customer-experience/.

Peters, B. Guy. 2018. 'The Challenge of Policy Coordination'. *Policy Design and Practice* 1 (1): pp. 1–11. doi: 10.1080/25741292.2018.1437946.

Societal Platform. 2020. *Societal Platform Technology Architecture Principles | Dr Pramod Varma, CTO—EkStep Foundation*. YouTube. Available at https://www.youtube.com/watch?v=1S-599usfjc.

TAGUP. 2011. 'Report of the Technology Advisory Group for Unique Projects'. New Delhi: Ministry of Finance. Available at https://comtax.uk.gov.in/files/pdf/70_TAGUP_Report.pdf.

Thomas, David. 2018. 'Funding Product Teams, Not Projects'. Defra Digital. Available at https://defradigital.blog.gov.uk/2018/10/29/funding-product-teams-not-projects/.

Tongia, Rahul. 2007. 'The Political Economy of Indian Power Sector Reforms'. In *The Political Economy of Power Sector Reform: The Experiences of Five Major Developing Countries*, edited by David G. Victor and Thomas C. Heller, pp. 109–74. Cambridge: Cambridge University Press. doi: 10.1017/CBO9780511493287.

Wegrich, Kai, and Vid Štimac. 2014. 'Coordination Capacity'. In *The Problem-solving Capacity of the Modern State: Governance Challenges and Administrative Capacities*, edited by Martin Lodge and Kai Wegrich, pp. 41–62. Oxford: Oxford University Press. doi: 10.1093/acprof:oso/9780198716365.003.0003.

Williams, Martin J. 2025. *Reform as Process: Implementing Change in Public Bureaucracies*. Oxford University Press.

Chapter 9

Babatz, Guillermo. 2013. 'Sustained Effort, Saving Billions: Lessons from the Mexican Government's Shift to Electronic Payments'. Mexico Case Study. Better Than Cash Alliance. Available at https://www.betterthancash.org/tools-research/case-studies/sustained-effort-saving-billions-lessons-from-the-mexican-governments-shift-to-electronic-payments.

Bussell, Jennifer. 2012. *Corruption and Reform in India: Public Services in the Digital Age*. New Delhi: Cambridge University Press.

Bussell, Jennifer. 2018. 'Whose Money, Whose Influence? Multi-Level Politics and Campaign Finance in India'. In *Costs of Democracy: Political Finance in India*, edited by Devesh Kapur and Milan Vaishnav. New Delhi: Oxford University Press, pp. 232–72.

Chhibber, Pradeep, and Irfan Nouruddin. 2004. 'Do Party Systems Count? The Number of Parties and Government Performance in the Indian States'. *Comparative Political Studies* 37 (2): pp. 152–87. doi: 10.1177/0010414003260981.

Fazio, Giorgio, and Giuseppina Maria Chiara Talamo. 2008. 'How "Attractive" Is Good Governance for FDI?' In *Institutional Approach to Global Corporate Governance: Business Systems and Beyond*, edited by J. Jay Choi and Sandra Dow, pp. 33–54. Bingley: Emerald Group Publishing Limited.

Heath, Oliver, and Louise Tillin. 2018. 'Institutional Performance and Vote Buying in India'. *Studies in Comparative International Development* 53 (1): pp. 90–110. doi: 10.1007/s12116-017-9254-x.

Kapur, Devesh, and Milan Vaishnav. 2018. *Costs of Democracy: Political Finance in India*. New Delhi: Oxford University Press.

Keefer, Philip, and Razvan Vlaicu. 2005. *Democracy, Credibility, and Clientelism*. WPS3472. Washington, D.C.: World Bank Publications. Available at http://documents.worldbank.org/curated/pt/560421468762575480/pdf/wps3472.pdf.

Khera, Reetika. 2016. 'MNREGA: Technology vs. Technocracy'. *Ideas for India*. 17 March. Available at http://www.ideasforindia.in/topics/poverty-inequality/mnrega-technology-vs-technocracy.html.

Khera, Reetika. 2017. 'Impact of Aadhaar in Welfare Programmes'. *Economic and Political Weekly* 52 (50): pp. 61–70. Available at https://www.jstor.org/stable/45132600.

Kitschelt, Herbert, and Steven I. Wilkinson. 2007. *Patrons, Clients and Policies: Patterns of Democratic Accountability and Political Competition*. Cambridge: Cambridge University Press.

Li, Quan, and Adam Resnick. 2003. 'Reversal of Fortunes: Democratic Institutions and Foreign Direct Investment Inflows to Developing Countries'. *International Organization* 57 (1): pp. 175–211. doi: 10.1017/S0020818303571077.

Mukherji, Rahul, and Himanshu Jha. 2017. 'Bureaucratic Rationality, Political Will, and State Capacity: MGNREGS in Undivided Andhra Pradesh'. *Economic and Political Weekly* 52 (49): pp. 53–60. Available at https://www.epw.in/journal/2017/49/special-articles/bureaucratic-rationality-political-will-and-state-capacity.html.

Mukherji, Rahul, and Seyed Hossein Zarhani. 2020. 'Governing India: Evolution of Programmatic Welfare in Andhra Pradesh'. *Studies in Indian Politics* 8 (1): pp. 7–21. doi: 10.1177/2321023020918054.

Stokes, Susan C., Thad Dunning, Marcelo Nazareno, and Valeria Brusco. 2013. *Brokers, Voters, and Clientelism: The Puzzle of Distributive Politics*. New York: Cambridge University Press.

Williams, Martin J. 2025. *Reform as Process: Implementing Change in Public Bureaucracies*. Oxford University Press.

Appendix 2

Asian Development Bank. 2000. 'Technical Assistance to India for Supporting Fiscal Reforms in Kerala'. Asian Development Bank. Available at https://www.adb.org/sites/default/files/project-document/71670/r340-00.pdf.

Asian Development Bank Board of Directors. 1999. 'Report and Recommendation of the President to the Board of Directors on a Proposed Loan and Technical Assistance Grant to India for the Madhya Pradesh Public Resource Management Program'. Vol. 29051. Asian Development Bank.

Chathukulam, Jos, and M.S. John. 2002. 'Five Years of Participatory Planning in Kerala: Rhetoric and Reality'. *Economic and Political Weekly* 37 (49): pp. 4917–26. Available at https://www.epw.in/journal/2002/49/special-articles/five-years-participatory-planning-kerala.html.

Finance (Streamlining) Department. 2016. 'Integrated Financial Management System—Introduction of Budget Allocation Monitoring System—Orders Issued'. Bill Number G.O. (P) No.34/2O16/Fin. 4 April, 2015.

Haidear, Sohan. 2016. 'Case Study: Automation of Treasury Department in Karnataka Government'. Presented at the Institute of Business Administration, University of Dhaka, 16 January. Available at https://www.slideshare.net/sohanhaidear/case-study-automation-of-treasury-department-in-karnataka-government.

Jeromi, P.D. 2005. 'Economic reforms in Kerala'. Economic and Political Weekly 40 (30): pp. 3267–77.

Kerala State Government. 2000. 'G.O. (P) No.222/2000/Fin.'. Kerala State Government.

Ministry of Panchayati Raj, and Indian Institute of Public Administration. 2024. 'Status of Devolution to Panchayats in States: An Indicative Evidence Based Ranking'. Government of India. Available at https://cdnbbsr.s3waas.gov.in/s316026d60ff9b54410b3435b403afd226/uploads/2025/02/202502141128959181.pdf.

Raghunathan, V.S., and A. Nisarudeen. 2007. 'Treasury Information Management (TRIM)'. In *E-Governance: Case Studies*, edited by C.S.R. Prabhu. New Delhi: PHI Learning Private Limited, pp. 433–37. Available at https://www.csi-sigegov.org.in/casestudies/42_trim.pdf.

Singh, Purushotham, and Saurabh Bhattacharya. 2013. 'Khajane—Using Technology for Fiscal Discipline'. Case Studies on E-Governance in India—2012–2013. Available at https://negd.gov.in/hi/case-studies/.

Appendix 3

Babatz, Guillermo. 2013. 'Sustained Effort, Saving Billions: Lessons from the Mexican Government's Shift to Electronic Payments'. Mexico Case Study. Better Than Cash Alliance. Available at https://www.betterthancash.org/tools-research/case-studies/sustained-effort-saving-billions-lessons-from-the-mexican-governments-shift-to-electronic-payments.

CGAP. 2011. 'CGAP G2P Research Project: Brazil Country Report'. Washington, D.C.: Consultative Group to Assist the Poor. Available at http://www.cgap.org/sites/default/files/CGAP-G2P-Research-Project-Brazil-Country-Report_0.pdf.

Cirasino, Massimo, Hemant Baijal, Garcia Garcia Luna, Jose Antonio, and Rahul Kitchlu. 2012. 'General Guidelines for the Development of Government Payment Programs'. Financial Infrastructure Series. Washington, D.C.: World Bank.

Available at https://documents1.worldbank.org/curated/en/883951468000286016/pdf/96463-WP-Box391445B-PUBLIC-ADD-TOPIC-payment-systems-remittances-and-other-related.pdf.

Lindert, Kathy, Anja Linder, Jason Hobbs, and Bénédicte de la Brière. 2007. 'The Nuts and Bolts of Brazil's Bolsa Família Program: Implementing Conditional Cash Transfers in a Decentralized Context'. *World Bank Social Protection Discussion Paper* 709. Washington, D.C.: World Bank. Available at https://documents1.worldbank.org/curated/en/972261468231296002/pdf/398530SP1709.pdf.

Appendix 4

Field, Erica, Rohini Pande, Rigol Natalia, Simone Schaner, and Charity Troyer Moore. 2016. 'On Her Account: Can Strengthening Women's Financial Control Boost Female Labor Supply?' Available at https://economics.yale.edu/sites/default/files/schaner_fba.pdf.

Gupta, Aashish. 2013. 'Old-Age Pension Scheme in Jharkhand and Chhattisgarh'. *Economic and Political Weekly* 48 (34): pp. 54–59. https://www.jstor.org/stable/23528061.

India Post. 2018. 'Doorstep Banking—India Post Payments Bank'. New Delhi: India Post Payments Bank. Available at https://www.ippbonline.com/web/ippb/doorstep-banking.

International Monetary Fund. 2021. 'Financial Access Survey'. Available at https://data.worldbank.org/indicator/FB.ATM.TOTL.P5.

Kolluju, Naveen. 2014. 'Business Correspondent Model Vis-à-Vis Financial Inclusion in India: New Practice of Banking to the Poor'. *International Journal of Scientific and Research Publications* 4 (1): pp. 289–95. Available at http://www.ijsrp.org/research-paper-0114.php?rp=P252174.

Kumar, Ravi N. 2012. 'Pilot Proves Aadhaar Can End Misuse of Subsidy'. *The Hindu*, 8 April. Available at https://www.thehindu.com/business/Pilot-proves-Aadhaar-can-end-misuse-of-subsidy/article12951238.ece.

Mazzotta, Benjamin D., Bhaskar Chakravorti, Rama Bijapurkar, Rajesh Shukla, K. Ramesha, Dhananjay Bapat, and Deepankar Roy. 2015. 'The Cost of Cash in India'. Medford, MA: Institute for Business in the Global Context, Tufts University. Available at https://digitalplanet.tufts.edu/wp-content/uploads/2022/10/Cost-of-Cash-India.pdf.

Merrill, Laura. 2014. 'Power, Gender and Fossil-Fuel Subsidy Reform in India'. Winnipeg: International Institute for Sustainable Development. Available at https://www.iisd.org/gsi/sites/default/files/ffs_india_gender_may_2014.pdf.

Nariyanuri, Sampath Sharma. 2021. '2021 India Mobile Payments Market Report'. S&P Global Market Intelligence. Available at https://pages.marketintelligence.spglobal.com/rs/565-BDO-100/images/2021IndiaMobilePaymentsMarketReport.pdf.

Press Trust of India. 2021. 'Jan Dhan Accounts Cross 41 Crore, Only 7.5% Zero-Balance Accounts'. *Business Today*, 19 January. Available at https://www.businessto

day.in/latest/economy-politics/story/jan-dhan-accounts-cross-41-crore-only-75-zero-balance-accounts-284778-2021-01-19.

Sanford, Caitlin. 2013. 'Do Agents Improve Financial Inclusion? Evidence from a National Survey in Brazil'. Bankable Frontier Associates. Available at https://www.findevgateway.org/sites/default/files/publications/files/mfg-en-paper-do-agents-improve-financial-inclusion-evidence-from-a-national-survey-in-brazil-nov-2013.pdf.

Siedek, Hannah. 2008. 'Extending Financial Services with Banking Agents'. Brief 43846. Washington, D.C.: Consultative Group to Assist the Poor. https://openknowledge.worldbank.org/bitstream/handle/10986/9523/438460BRI0Box31rief1Agents01PUBLIC1.pdf?sequence=1.

Soni, Anmol, and Anjali Ramakrishnan. 2014. 'Evaluation of the Pilot Project on Direct Transfer of Kerosene Subsidies in Kotkasim, Alwar'. International Institute for Sustainable Development and The Energy Resources Institute, Global Subsidies Initiative Report. New Delhi: IISD-TERI. Available at https://www.iisd.org/gsi/sites/default/files/ffs_india_kerosene_alwar_final.pdf.

Zandi, Mark, Sophia Koropeckyj, Virendra Singh, and Paul Matsiras. 2003. 'The Virtuous Circle: Electronic Payments and Economic Growth'. Visa International. San Francisco: Visa International and Global Insight, Inc.

Appendix 5

Societal Platform. 2017. 'A Systemic Method to Resolve Complex Societal Challenges'. 2017. Societal Platform.

Index

For the benefit of digital users, indexed terms that span two pages (e.g., 52–53) may, on occasion, appear on only one of those pages.

Tables and figures are indicated by an italic *t* and *f* following the page numbers